Check Point Firewall Administration R81.10+

A practical guide to Check Point firewall deployment and administration

Vladimir Yakovlev

<packt>

BIRMINGHAM—MUMBAI

Check Point Firewall Administration R81.10+

Group Product Manager: Vijin Boricha
Publishing Product Manager: Preet Ahuja
Senior Editor: Shazeen Iqbal
Content Development Editor: Romy Dias
Technical Editor: Shruthi Shetty
Copy Editor: Safis Editing
Project Coordinator: Ashwin Dinesh Kharwa
Proofreader: Safis Editing
Indexer: Manju Arasan
Production Designer: Nilesh Mohite
Senior Marketing Coordinator: Hemangi Lotlikar

First published: August 2022

Production reference: 1040822

Published by Packt Publishing Ltd.
Livery Place
35 Livery Street
Birmingham
B3 2PB, UK.

ISBN 978-1-80107-271-7

www.packt.com

To my parents. It's all your fault ☺

Foreword

One of my colleagues recently told me that no matter when you get into an industry, you're always getting in on the ground floor of something. For me, that *something* ended up being the early days of Check Point FireWall-1, and what ultimately became the cyber security industry.

I've seen Check Point's various products and services grow and change over the last 26 years. I've helped a lot of people make the best use of Check Point products, both directly and indirectly, including writing my own Check Point books in the early 2000s. While a lot has changed since, including Check Point's corporate logo, the core philosophy behind every Check Point product and service has not.

These days, you need a lot more than just network firewalls to *Secure Your Everything*. Even so, firewalls still play a critical role in most environments by defining boundaries between both private and public networks, enabling controlled access to network resources, blocking malicious content, and preventing both data exfiltration and the unauthorized use of systems.

In the 20 years since Essential Check Point FireWall-1 NG was published, I've been asked numerous times if I was going to write another book on Check Point firewalls. If I were going to do so, I'd probably take the approach that Vladimir has taken in this book. There are concise explanations of the essential features of the Check Point Quantum Security Gateway and Management products, along with step-by-step instructions and annotated screenshots!

If you're just getting started with deploying Check Point Quantum Security Gateways, or you're trying to refresh your knowledge, this book is a great place to start. There's also CheckMates (`https://community.checkpoint.com`), Check Point's official cyber security community, which is full of additional learning resources and discussions to help those who want to continue their learning on Check Point after finishing this book.

Dameon D. Welch (a.k.a. PhoneBoy)
Cyber Security Evangelist
Check Point Software Technologies, Ltd.

Contributors

About the author

Vladimir Yakovlev, CISSP, is an infrastructure and security solutions architect and CTO at Higher Intelligence LLC., with over 20 years of Check Point experience.

He is recognized as a champion in the ISC2 and Check Point CheckMates communities and has been awarded Member of the Year and Contributor of the Year designations by peers, while also speaking at regional and international conferences.

Vladimir has previously held the roles of Sr. V.P. of Technology and CISO, responsible for the design, implementation, and operation of multiple iterations of secure and resilient infrastructures in the financial industry.

He enjoys helping others in the field of cybersecurity and can often be found in the CheckMates, LinkedIn, and ISC2 communities.

This project wouldn't have happened without the encouragement and help from two authors of previous books dedicated to Check Point: Dameon D. Welch (a.k.a. PhoneBoy), my Technical Reviewer, and Timothy Hall, who went above and beyond in engaging with me in deep-dives on a multitude of subjects and sanity checks. Thank you both!

Huge thanks to all members of the Packt editing team and, especially, Romy Dias.

Last but not least, to my family, who tolerated my virtual absence for a year, and, specifically, to my son, Sam Yakovlev. He was (against his will) subjected to the first technical reading of this book, and is mainly responsible for defending the dignity of the English language (and the Oxford comma) from me.

About the reviewer

Dameon D. Welch, widely known as "PhoneBoy," is a Cyber Security Evangelist for Check Point Software Technologies. He is the public face of CheckMates, the Check Point cyber security community.

A recognized industry security veteran, with more than two decades of experience, Welch is best known for his creation of the PhoneBoy FireWall-1 FAQ in the mid-1990s. It was used by Check Point and thousands of its customers worldwide. He is also the author of *Essential Check Point FireWall-1 NG: An Installation, Configuration, and Troubleshooting Guide*.

> *I'd like to thank everyone who has supported and encouraged me over the years.*

Table of Contents

3
Building a Check Point Lab Environment – Part 1

4
Building a Check Point Lab Environment – Part 2

Part 2: Introduction to Gaia, Check Point Management Interfaces, Objects, and NAT

5

Gaia OS, the First Time Configuration Wizard, and an Introduction to the Gaia Portal (WebUI)

6

Check Point Gaia Command-Line Interface; Backup and Recovery Methods; CPUSE

7

SmartConsole – Familiarization and Navigation

8

Introduction to Policies, Layers, and Rules

9

Working with Objects – ICA, SIC, Managed, Static, and Variable Objects

10

Working with Network Address Translation

Part 3: Introduction to Practical Administration for Achieving Common Objectives

11

Building Your First Policy

12

Configuring Site-to-Site and Remote Access VPNs

13

Introduction to Logging and SmartEvent

14

Working with ClusterXL High Availability

15

Performing Basic Troubleshooting

Appendix
Licensing

Index

Other Books You May Enjoy

Preface

Check Point Firewall Administration R81.10+ was written to help security administrators develop the necessary skills for effective deployment and operation of Check Point firewalls or high-availability clusters to improve network segmentation, configure site-to-site or remote access VPNs, and implement airtight access control policies.

Who this book is for

This book is for those new to Check Point firewalls or those who are catching up to the current R81.10++ releases. Although intended for information/cybersecurity professionals with some experience in network or IT infrastructure security, it may also be helpful for IT professionals looking to shift their career focus to cybersecurity. Some familiarity with Linux and bash scripting is a plus.

It may also be useful for technical decision makers as a tool to take Check Point firewalls for a spin before committing resources to proof of concept or in anticipation of purchasing the product. Your security administrators will be better prepared for **Proof of Concept** (**PoC**) or implementation after reading it and building their own lab prior to undertaking formal training and certification.

What this book covers

Chapter 1, Introduction to Check Point Firewalls and Threat Prevention Products, covers the evolution of Check Point security products and capabilities, security management architecture, and the creation of a user account to access relevant software and information.

Chapter 2, Common Deployment Scenarios and Network Segmentation, looks at firewall placement in common network topologies, network segmentation, as well as performance and capacity assessments of existing firewalls.

Chapter 3, Building a Check Point Lab Environment – Part 1, delves into lab topology, components, software, and resources, as well as looking at the installation of Oracle VirtualBox, deployment, and describing a process configuration of a virtual router.

Chapter 4, *Building a Check Point Lab Environment – Part 2*, explains creating Windows Server and Check Point base images and creating and preparing linked clones for the rest of the lab components.

Chapter 5, *Gaia OS, the First Time Configuration Wizard, and an Introduction to the Gaia Portal (WebUI)*, introduces Gaia, the operating system in use by Check Point management servers and gateways. This chapter also covers the First Time Configuration Wizard and Gaia web interface.

Chapter 6, *Check Point Gaia Command-Line Interface; Backup and Recovery Methods; CPUSE*, covers accessing and using the Check Point Gaia command-line interface and expert mode shells. Backup and recovery options and Check Point Update Service Engine are also covered.

Chapter 7, *SmartConsole – Familiarization and Navigation*, provides a detailed examination of SmartConsole features, components, and capabilities and teaches you being comfortable with the Check Point primary management interface.

Chapter 8, *Introduction to Policies, Layers, and Rules*, covers policy packages, blades (features) used in Access Control policies, and their use in layers. The chapter also looks at policy organization methods, rules' structure and capabilities, and their placement based on the packet flows and use of acceleration technology.

Chapter 9, *Working with Objects – ICA, SIC, Managed, Static, and Variable Objects*, looks at the Internal Certificate Authority and Secure Internal Communication and how these factor into the creation of other Check Point managed objects. The chapter also looks at creating your first high-availability cluster and the rest of the objects for lab components, learning about different object types and their properties.

Chapter 10, *Working with Network Address Translation*, introduces network and port address translation using automatic and manual NAT options. The chapter goes on to look at the use of NAT in object properties, policies, and rules, and additional relevant configuration options, as well as NAT logging and interpretation of NAT log data.

Chapter 11, *Building Your First Policy*, defines policy structure while accounting for the most common scenarios likely to be encountered in any infrastructure. The creation of rules and, when necessary, additional objects is also covered, as is expanding a policy's capabilities and granularity by enabling additional features, rules, and objects.

Chapter 12, *Configuring Site-to-Site and Remote Access VPNs*, looks at configuring VPNs for communication with peers, data, or service providers as well as implementing remote access capabilities using Check Point IPSec VPN features. The chapter also looks at utilizing Access Roles for granular remote access.

Chapter 13, *Introduction to Logging and SmartEvent*, explains how logging works in Check Point, and how to use different configuration options to best address your infrastructure's logging requirements. The chapter also introduces SmartEvent, which simplifies the work of Check Point administrators by providing enhanced views, reporting capabilities, and automated reactions.

Chapter 14, *Working with ClusterXL High Availability*, provides an explanation of the ClusterXL HA mechanism, operating a fault-tolerant cluster, and alternative Check Point offerings for high availability and load sharing.

Chapter 15, *Performing Basic Troubleshooting*, looks at troubleshooting constraints and your actions. The chapter introduces typical issue categories, approaches, and tools helpful for solving them. It also looks at initiating and handling service requests interaction with Check Point Technical Assistance Centers. The chapter goes on to detail resources available from, and interaction with, the CheckMates user community.

Appendix, *Licensing*, introduces Check Point licensing terminology, specific information for management servers and gateways, and licensing for a lab environment.

To get the most out of this book

You will need a Windows 10 or 11 PC with 24-32 GB of RAM and approximately 200 GB of free disk space to replicate the VirtualBox lab environment described in the book. If you are experienced in and prefer to use different virtualization platforms, adapt virtual hardware and networking requirements for the lab to a platform of your choice. All software required for the labs is available in free, trial, or evaluation versions. You will be required to register on some of the vendors' portals for access to their products.

Software/hardware covered in the book	Operating system requirements
Check Point R81.10	Windows 10 or 11
Check Point SmartConsole R81.10	Windows 10 or 11
Windows Server 2019 or 2022	Windows 10 or 11
VyOS open source router	Windows 10 or 11

Additional software includes VirtualBox, PuTTY, WinSCP, and Notepad++ and you'll be instructed to install them on relevant physical or virtual hosts throughout the book.

If you are using the digital version of this book, we advise you to type the code yourself or access the code from the book's GitHub repository (a link is available in the next section). Doing so will help you avoid any potential errors related to the copying and pasting of code.

If you are using the digital version of this book, I suggest viewing it in two-page, side-by-side format. This will make it easier to process text referencing screenshots, commands, or code on adjacent pages. Alternatively, download the PDF with figures, referenced later in this document, and use it to look up information referenced in the text.

Download the example code files

You can download the example code files for this book from GitHub at `https://github.com/PacktPublishing/Check-Point-Firewall-Administration-R81.10-`. If there's an update to the code, it will be updated in the GitHub repository.

Download the color images

We also provide a PDF file that has color images of the screenshots and diagrams used in this book. You can download it here: `https://packt.link/ImE2Y`.

Conventions used

There are a number of text conventions used throughout this book.

`Code in text`: Indicates code words in text, database table names, folder names, filenames, file extensions, pathnames, CLI menu choices, commands, and user input. Here is an example: "When logged in to `CPCM1`, execute the command, `set expert-password`."

A block of code or sequential uninterrupted commands is set as follows:

```
add host name host_test1 ip-address 10.0.0.111
add host name host_test2 ip-address 10.0.0.112
add host name host_test3 ip-address 10.0.0.113
```

When commands are shown in the context of a particular shell, are interactive, or are combined with step descriptions, they are shown like this:

```
# Step 1
show installer packages recommended
# Note the Display name of the package you are interested in.
# Step 2
show installer package
# [press spacebar and then press the Tab key]
# Note the Num(ber) corresponding to the Display name of the
package from step 1.
```

Any command-line input or output is written as follows:

```
CPXXX> show date
Date 02/02/2022
CPXXX > show time
Time 18:19:17
$ cd css
```

Bold indicates a new term, an important word, or words that you see onscreen. For instance, words in menus or dialog boxes appear in **bold**. Here is an example: "Once the **Plugins Admin** window is opened, scroll down until you see **Compare** and check the checkbox."

Italics indicates either internal or external references, such as "In *Chapter 7, SmartConsole – Familiarization and Navigation*, we saw how to do that using the management CLI." It is also used to denote a specific keypress, such as "press *Enter*."

Additionally, italics are used to indicate an emphasis on specifics, such as in the following sentence: "Even though the domain objects are defined, created, and modified in SmartConsole, we must use associated CLI tools *on the gateways* where the policies containing these objects are installed, and not on the management server."

[#], [A], and [a] indicate the numerical or letter-based points of interest in figures (screenshots), typically referencing screenshots following the text, unless explicitly noted otherwise, as follows:

"To illustrate how to create additional server objects (also referred to as a Check Point Host object), let's click on the New icon [1] in the Actions menu of the GATEWAYS & SERVERS view, click More [2], and then click Check Point Host… [3]:"

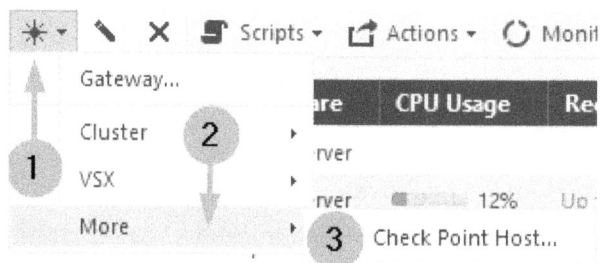

Sample image showing [] instances

Keywords are used whenever a new important term is used in the context of the chapter or a section, such as: "**Access roles** are the ultimate tool for the implementation of the **zero-trust** concept in your environment."

> **Tips or Important Notes**
> Appear like this.

Get in touch

Feedback from our readers is always welcome.

General feedback: If you have questions about any aspect of this book, email us at customercare@packtpub.com and mention the book title in the subject of your message.

Errata: Although we have taken every care to ensure the accuracy of our content, mistakes do happen. If you have found a mistake in this book, we would be grateful if you would report this to us. Please visit www.packtpub.com/support/errata and fill in the form.

Piracy: If you come across any illegal copies of our works in any form on the internet, we would be grateful if you would provide us with the location address or website name. Please contact us at copyright@packt.com with a link to the material.

If you are interested in becoming an author: If there is a topic that you have expertise in and you are interested in either writing or contributing to a book, please visit authors.packtpub.com.

Legal disclaimer

This document was created using the official VMware icon and diagram library.

Copyright© 2010 VMware, Inc. All rights reserved. This product is protected by U.S. and international copyright and intellectual property laws. VMware products are covered by one or more patents listed at `https://www.vmware.com/go/patents`. VMware does not endorse or make any representations about third-party information included in this document, nor does the inclusion of any VMware icon or diagram in this document imply such an endorsement.

All copyrights are property of their respective owners including Check Point®

Share your thoughts

Once you've read *Check Point Firewall Administration R81.10+*, we'd love to hear your thoughts! Scan the QR code below to go straight to the Amazon review page for this book and share your feedback.

`https://packt.link/r/180107271X`

Your review is important to us and the tech community and will help us make sure we're delivering excellent quality content.

Part 1: Introduction to Check Point, Network Topology, and Firewalls in Your Infrastructure and Lab

In this portion of the book, you will be introduced to Check Point products and, specifically, firewalls. We'll look at them in the context of different infrastructure topologies and segments. You will create a realistic lab environment that will be used in subsequent chapters.

The following chapters will be covered in this section:

- *Chapter 1, Introduction to Check Point Firewalls and Threat Prevention Products*
- *Chapter 2, Common Deployment Scenarios and Network Segmentation*
- *Chapter 3, Building a Check Point Lab Environment – Part 1*
- *Chapter 4, Building a Check Point Lab Environment – Part 2*

1

Introduction to Check Point Firewalls and Threat Prevention Products

In this chapter, we will learn about the past and the present of **Check Point Software Technologies** in the context of evolving cybersecurity challenges. We will become familiar with the three main product lines, their components, and their relevance to the threat prevention capabilities of **Check Point firewalls**. We will examine the flexibility and advantages of security management architecture, address the learning process, and go through the user and account creation process in preparation for the following chapters.

In this chapter, we are going to cover the following main topics:

- Learning about Check Point's history and the current state of the technology
- Understanding the Check Point product lineup and coverage
- Introducing the Unified Management concepts and the advantages of security product consolidation

- Familiarization with the **Security Management Architecture** (**SMART**)
- Determining how we learn
- Navigating the Check Point User Center

Technical requirements

For this chapter, we will need a web browser for access to the **Check Point User Center** and a smartphone running either iOS or Android, with an *authentication manager* application of your choice and a time-based, one-time password functionality, such as Google or Microsoft Authenticator, to enable second-factor authentication for access to the User Center.

Learning about Check Point's history and the current state of the technology

To get a sense of the product and the company behind it, it is good to have perspective. When were they founded? How long have they been in business? How consistent is their performance over time? What areas of cybersecurity is the company working in and how well are they rated? To find the answers to these questions, let's look at the past and the present of Check Point Software Technologies.

"In the beginning, there was FireWall-1"

In 1994, **FireWall-1**, released by Check Point Software Technologies, effectively launched the commercial firewall market and, according to Gartner, Check Point has been named a *leader in the Network Firewalls category* 21 times since.

The company received the following mention at the Cybersecurity Excellence Awards for 2016: "*All of the US Fortune 100, and over 90% of the Fortune 500, rely on Check Point solutions to protect their networks and data.*" [1]

Shortly after launching FireWall-1, Check Point released **VPN-1** for remote access and secure connectivity with peers and, over the years, continued to introduce additional components, enhancements, and new products. Since then, the cybersecurity arena has become saturated with many entrants bringing new products to the market. Throughout all of this time, Check Point's expanding product line, and especially their evolving management interface, has been recognized as the gold standard against which all competitors are measured.

Check Point firewalls were originally created to run on multiple operating systems and hardware, hence the name of the company, *Check Point Software Technologies*.

This is an important distinction when compared with the offerings provided by other vendors that were creating their solutions based on specialized **ASICs** (**Application-Specific Integrated Circuits**). When cloud computing ushered in a new era in information technology, Check Point was able to immediately offer the same degree of protection to cloud-based environments as was previously available to traditional infrastructures. Since Check Point enterprise firewalls were running on x86/x64 platforms, they did not require porting or emulation to do that.

Check Point today

Check Point's products are now deployed in 88 countries and more than 100,000 businesses. It has offices in 75 countries, over 3,500 security experts, and a world-acclaimed research and intelligence organization[2]. Its firewall and threat prevention product line has offerings that cover an entire spectrum of clients; from small offices to enterprises, carrier networks, government agencies, and industrial control systems. They are available in the largest number of cloud services, including Amazon AWS, Microsoft Azure, Oracle Cloud, Google Cloud services, Alibaba Cloud, and IBM Cloud.

Check Point Software Technologies was recognized as a *Microsoft Security 20/20 Partner Award Winner* for Most Prolific Integration Partner in 2020, and for Most Transformative Integration Partner in 2021[3].

By protecting networks, hosts, data, workloads on hypervisors, containers, and microservices from advanced threats while using unified management architecture, Check Point remains at the forefront of cybersecurity. It has grown organically and, through judicious acquisitions and integration of complementary products over the years, and is now the largest publicly traded cybersecurity company in Israel, a nation known worldwide for its remarkably strong information security and intelligence capabilities.

With an unparalleled commitment to product evolution and quality, its ever-growing list of partners, dedicated support for automation, and orchestration for organizations adopting DevSecOps practices, it is the best choice for anyone looking to embark on their journey of becoming a member of the cyber defense elite.

Now that we've learned a little about the company's history, let's take a look at Check Point's product line.

Understanding the Check Point product lineup and coverage

The scope of Check Point offerings could be better understood when looking at the following chart depicting the three main branches of products:

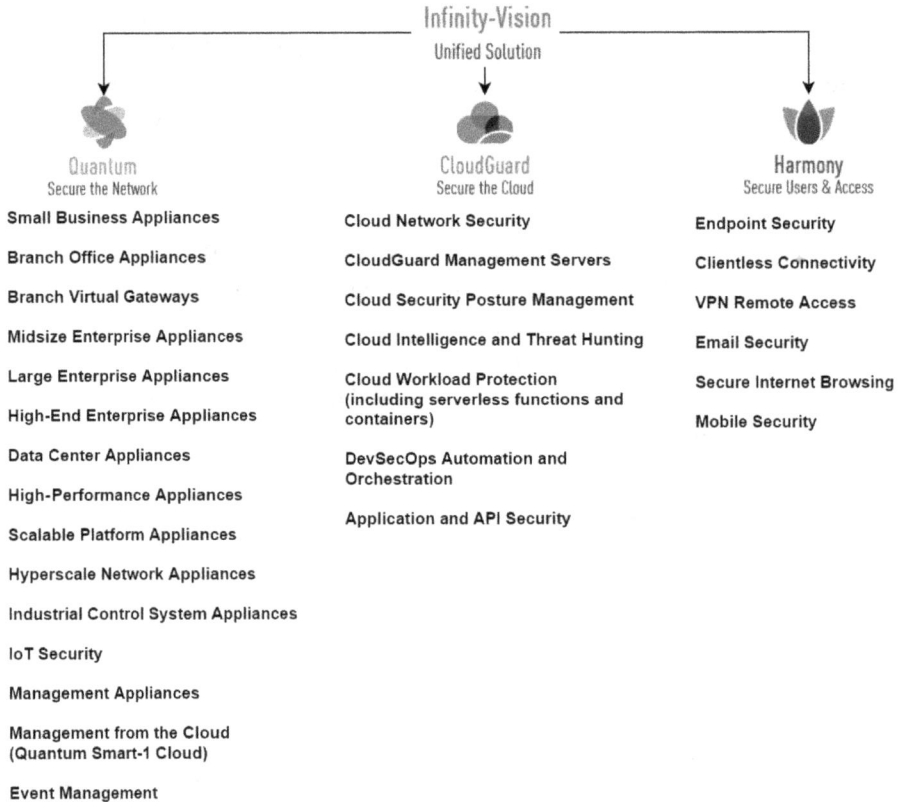

Infinity-Vision
Unified Solution

Quantum	CloudGuard	Harmony
Secure the Network	Secure the Cloud	Secure Users & Access
Small Business Appliances	Cloud Network Security	Endpoint Security
Branch Office Appliances	CloudGuard Management Servers	Clientless Connectivity
Branch Virtual Gateways	Cloud Security Posture Management	VPN Remote Access
Midsize Enterprise Appliances	Cloud Intelligence and Threat Hunting	Email Security
Large Enterprise Appliances	Cloud Workload Protection (including serverless functions and containers)	Secure Internet Browsing
High-End Enterprise Appliances		Mobile Security
Data Center Appliances	DevSecOps Automation and Orchestration	
High-Performance Appliances	Application and API Security	
Scalable Platform Appliances		
Hyperscale Network Appliances		
Industrial Control System Appliances		
IoT Security		
Management Appliances		
Management from the Cloud (Quantum Smart-1 Cloud)		
Event Management		

Figure 1.1 – Check Point unified security architecture components

The **Quantum** branch is primarily concerned with hardware appliances, but it does include Check Point's own cloud-hosted scalable management solution (**Quantum Smart-1 Cloud**).

The **small business appliances** in the **Quantum** branch are running an embedded version of Check Point's firewall. They are different from the rest of the lineup in that category, but they, too, could be managed from the same centralized management solutions as the rest.

The **CloudGuard** branch, while primarily concerned with cloud-based solutions, includes those for the on-premises virtualization environments, such as VMware vSphere, Microsoft Hyper-V, and Nutanix. Additionally, the **management servers** deployed in the cloud as VMs are also considered to be part of the CloudGuard product line.

The **Harmony** branch contains solutions necessary to safeguard endpoints inside, as well as outside, of your organization (including BYOD and mobile devices) and to provide your users with multiple choices for secure remote connectivity.

Now that we have learned about the scope of Check Point products, let's take a look at the benefits of having a single vendor solution protecting your infrastructure and data.

Introducing the Unified Management concepts and the advantages of security product consolidation

Historically, security-conscious enterprises were practicing *defense-in-depth* by layering and combining multiple solutions in the hope of preventing systems and network compromise. While this approach was viable 10 years ago, it is getting progressively more difficult to maintain it.

Let's look at the evolution of the threats over time to get a better idea of why this is so by using the following diagram:

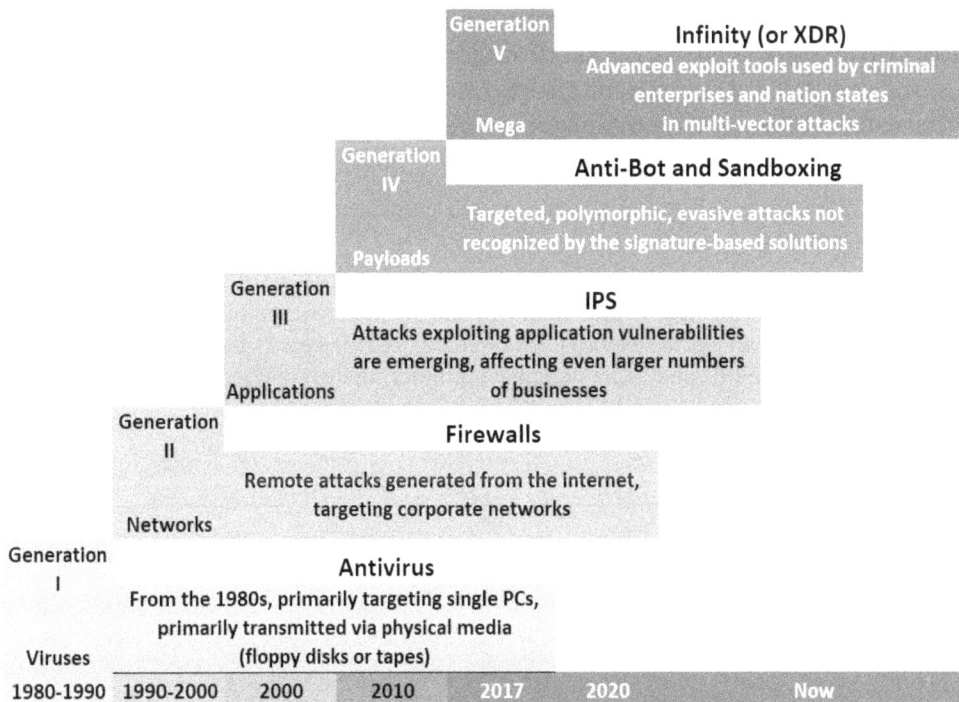

Figure 1.2 – Attack generations and types, escalation, and the response over time

In addition to the complexity and advances of the attacks, the numbers of bad actors, as well as the number of different attacks, are increasing exponentially. The field of offensive cybersecurity is attracting an ever-increasing number of people, not all of them ethical hackers. This contributes to the snowballing effect and the number of compromised systems, networks, and companies. The latest batch of attacks focusing on the supply chain is yet another manifestation of this trend.

The sheer number of cybersecurity vendors and point solutions, each trying to address different problem areas, makes it a virtual impossibility for smaller teams to manage them effectively. It takes years to gain proficiency with a single product, let alone multiple ones. Add to this the rapid development cycles of each vendor trying to keep up with evolving capabilities of cybercriminals and offerings by competition, and you will have to spend most of your time learning about new features and changes in all of these products, while at the same time fighting compatibility issues.

For a while, the combination of **Security Information and Event Management** (**SIEM**) solutions as hubs for the consolidation of logs, their correlation, and **Security Orchestration Automation and Response** (**SOAR**) actions based on pre-defined conditions looked like a possible solution to this problem. However, these options failed to address the multi-vendor cost of human capital, further complicating the operations of smaller security teams. They are now primarily relegated to larger enterprises, carrier networks, and **Managed Security Services Providers** (**MSSPs**) that can afford to keep staffed **Security Operations Centers** (**SOCs**) and dedicated data science and analytics specialists. For most other companies, SIEMs are either becoming *log graveyards* or are mostly used for after-the-fact investigations and audits, but not for proactive threat prevention.

> **Important Note**
>
> For the organizations that do utilize SIEMs, Check Point has *out-of-the-box integration* with ArcSight, LogRhythm, QRadar, RSA, McAfee, Splunk, and Sumologic, and its log exporter can be configured to work with any syslog-, CEF-, LEEF-, and JSON-compliant product. There is also a dedicated Check Point app for Splunk (`https://sc1.checkpoint.com/documents/App_for_Splunk/html_frameset.htm`) for seamless integration.

Serious advances in active prevention or response have also been made by several dedicated **Endpoint Detection and Response (EDR)** vendors. Unfortunately, the EDRs are relying on the installation of their agents on managed endpoints. Components of the infrastructure that do not have the agents remain unprotected.

All the networking gear, printers, copiers, conference room equipment, CCTV, building access and environmental controls, and other innumerable **Internet of Things (IoT)** devices are the shadow army that could be exploited and used for attacks or snooping on your infrastructure. The same applies to all devices on which the OS or firmware is controlled by the vendor or those that are supplied by service providers or business peers.

To compensate, EDR vendors are now actively expanding their integration with partners and going through the rapid acquisitions of complementary businesses to improve the coverage of their products.

Recognizing that the effective prevention of complex modern attacks requires more than just loosely coupled integration between various security tools, in 2017, Check Point developed and introduced its **Infinity architecture**. Tightly integrated products covering all aspects of security infrastructure with common management and enforcement policies dramatically improve detection and prevention rates.

Check Point was perfectly positioned to address these challenges since its **ThreatCloud** is one of the most established and largest commercial worldwide threat detection networks. The likelihood of Check Point encountering new attacks or variants of exploits *closer to home* is pretty good because of its huge global presence. The quality of the data is great since the product coverage extends from networks to endpoints, mobile, cloud, IoT, and industrial systems. Its analytics are supercharged by the ML and AI to identify malware DNA, a set of unique code segments and behavior characteristics that associates each newly encountered malware with a previously known malware family whenever such similarities can be identified. This helps to predict and prevent other, non-immediately apparent attack capabilities and vectors of emerging zero-day threats.

Having all these abilities provided by products from the same security vendor, as well as using common terminology, configuration, logging, analysis, management interfaces, and forensics capabilities, eliminates the complexity and the overhead of multiple point solutions. It also significantly improves your chances of deterrence and the containment of cyber attacks.

In January 2018, the MITRE Corporation released **Adversarial Tactics, Techniques, and Common Knowledge version 1 (ATT&CK v1)**, a framework that validated Check Point's vision for unified security. And in the same month, Check Point announced **Infinity Total Protection**, a simple, all-inclusive, per-user, per-year subscription covering all of its products, including hardware, software, 24x7 premium support, and network security, as well as endpoint, mobile, cloud, and data security with real-time threat prevention.

Competitors realizing the advantages of this approach adopted similar strategies and a new term, **Extended Detection and Response (XDR)**, was coined.

> **Important Note**
> Although it is unlikely that your organization is relying on a single vendor's solutions for all or even most of its cybersecurity needs, strategic consolidation resulting in massive benefits should be considered.

Most likely, Check Point firewalls in your environment are a part of the heterogeneous security infrastructure consisting of multiple point products. In this case, it is imperative to understand their roles, capabilities, and limitations in order to extract maximum value from the product while keeping track of what it is not designed or configured to do, and where complementary security solutions should be applied.

Network segmentation, network access control, threat prevention for individual network segments, categories, and hosts continue to remain some of the key elements of overall sound security posture. Having the benefit of threat intelligence generated by sensors present in all categories of information technology covered by the Infinity architecture makes Check Point firewalls some of the most effective threat prevention and detection tools in your cybersecurity arsenal.

> **Important Note**
> Check Point's mantra is *prevention first*, so it is often the case that engineers must, on purpose, disable prevention in the demo environments to showcase the product's detection capabilities at multiple points in the attack's kill chain.

Now that we know that vendor consolidation may yield better overall results by offering unified visibility of attacks, let's look at what the Security Management architecture can do for the administration of the Check Point infrastructure.

Familiarization with the Security Management Architecture (SMART)

Check Point's **Security Management architecture** is the foundational principle behind the centralized administration of multiple products and devices using common management interface(s).

Smart in the name of Check Point products dates back to when it was used as the acronym for *Security Management Architecture*. It is now present in the name of management servers and services, as well as Smart-1 products and their components: SmartLog and SmartEvent, Check Point's GUI, and the SmartConsole. There is also a migration tool for transition from competing solutions called SmartMove. In a nutshell, SMART could be described as a collection of the administrative stations, management, log, monitoring, and analytics servers that manage a variety of the gateways, endpoints, cloud-based inspection, and threat prevention products designed to seamlessly work together. It is practically infinitely scalable.

The following is a simple diagram depicting a basic implementation of the Check Point gateway and management infrastructure and their components:

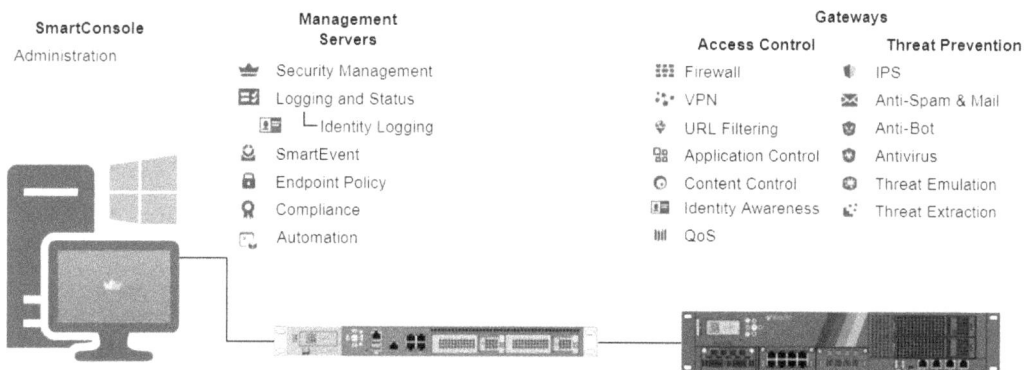

Figure 1.3 – Basic components of Check Point's management architecture

> **Important Note**
> Although we commonly refer to it as a *firewall*, a correct definition would be *gateway*, where a firewall is just one of the components.

SmartConsole is the Windows-based management client application that is connected to all of your management servers, regardless of the components they are running. It will be your primary interface for managing the Check Point infrastructure.

In the context of network architecture, a basic implementation could look as simple as the following:

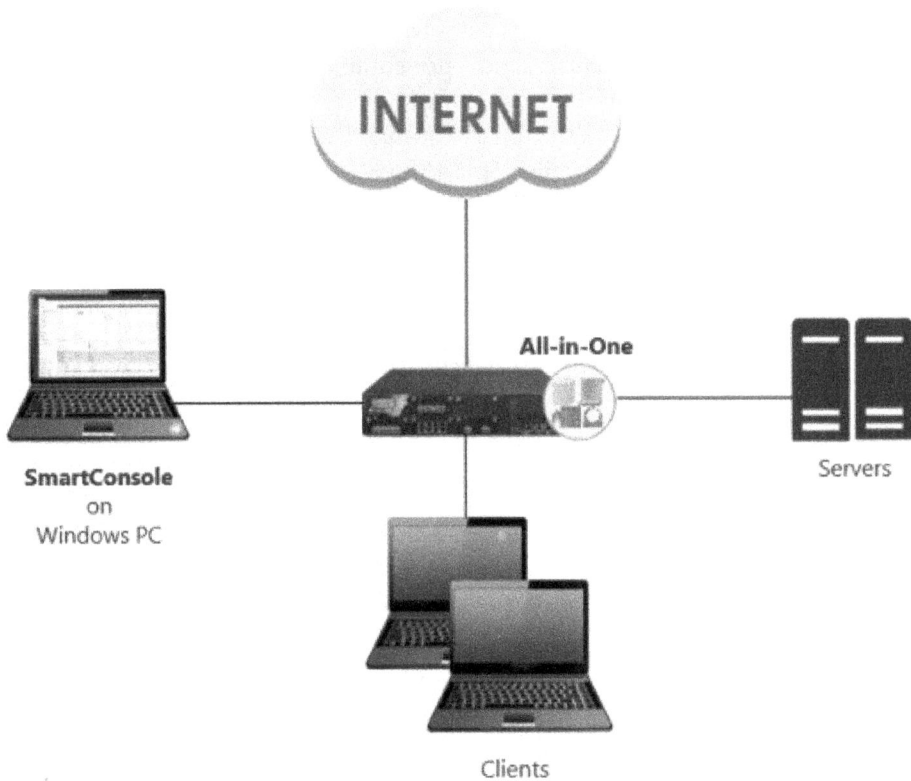

Figure 1.4 – All-in-one implementation. Management and gateway on a single device

In *Figure 1.4*, a single all-in-one device is acting as both a management server and a gateway. This is appropriate for the smallest environments that are aspiring to have world-class protection, but either lack the budget or justification to implement a distributed Check Point environment. It is also appropriate for small-scale labs where you may explore new release features and functionality.

In a slightly more demanding environment, which I would recommend as a good starting point, the management server running all of its constituent components resides on a separate appliance or a virtual machine. In this case, the gateway is running on a dedicated appliance as follows:

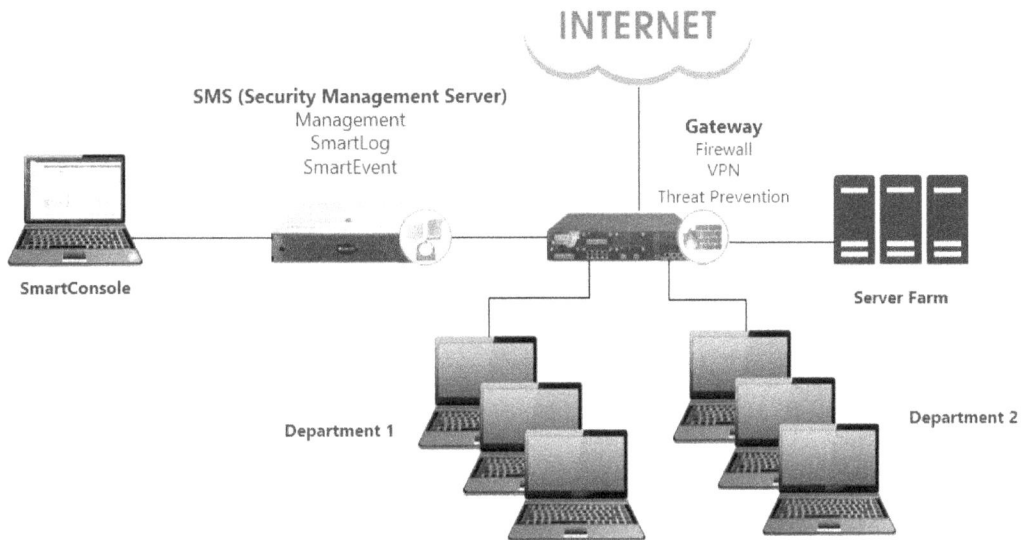

Figure 1.5 – Basic implementation with a dedicated management server

The advantage of this approach is the ability to manage many gateways from a common management server using a **common object database**.

> **Important Note**
> The functionality of the gateway is not dependent on the availability of the management server: even if it is down for maintenance or is being upgraded, the gateways continue to function and are logging locally. Once the management server comes back online, the logs are being automatically ingested by the management server.

Scaled further, **SmartEvent** is split from the management and log server to provide dedicated log correlation and reporting, as follows:

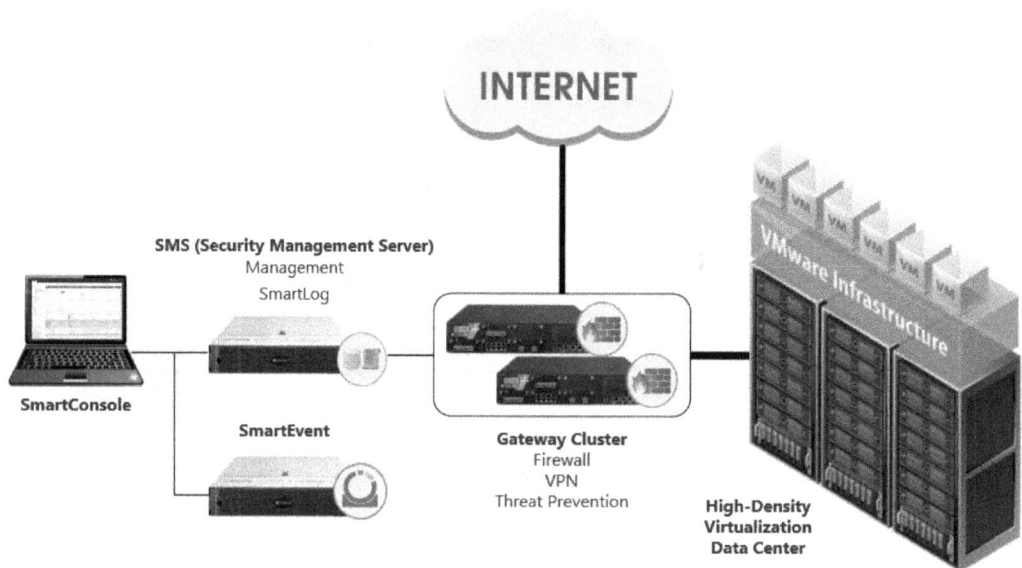

Figure 1.6 – SmartEvent log correlation and reporting on a dedicated server

This is a typical precursor for the expansion to either a *multi-site* or a *hybrid environment*, where a common SmartEvent server is used for log correlation, reporting, and analytics with multiple management and log servers. Since SmartEvent cannot be part of **high-availability management**, it should reside on a separate appliance or virtual machine. This is also one of the ways to offload your existing management server appliance if its utilization is consistently high.

And in a more typical data center environment, high-availability management and site-specific log servers are implemented to manage a larger number of gateways and clusters, as illustrated in the following diagram:

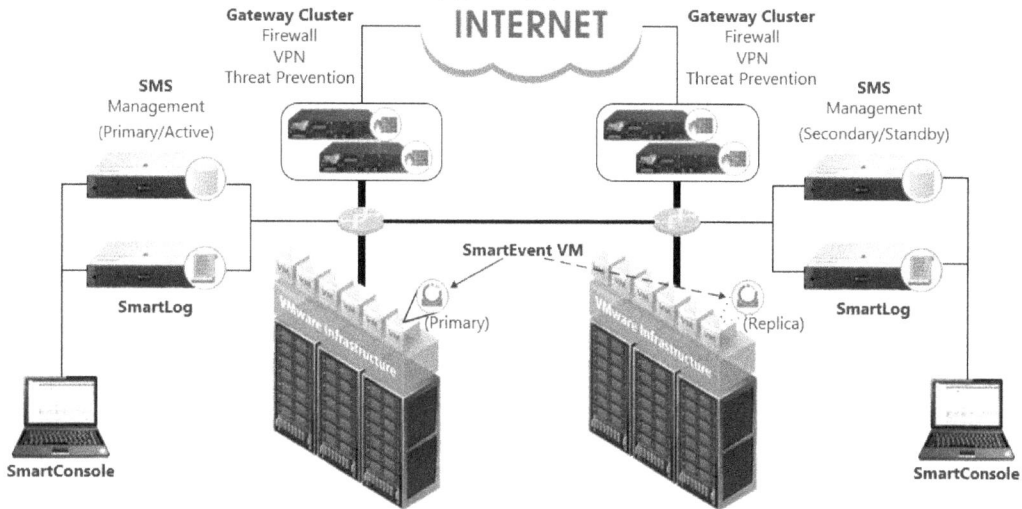

Figure 1.7 – Management high availability with dedicated log servers for multi-site environments

When your gateways and clusters under management generate a massive number of logs, it may be necessary to provide adequately sized log servers for each logically grouped location (typically based on geography or a specific data center).

The environment shown in *Figure 1.7* allows you to ensure that the management servers used to create, manage, and install security policies in your environment will perform consistently, irrespective of the load on log servers.

Scaled even further, it may look like this:

Multiple SmartConsoles and **API clients** → **multi-domain management servers/security management servers** → **multiple enterprise** environments comprised of gateways, clusters, scalable platforms, hyperscale solutions, and/ or endpoints → logging to dedicated log servers with monitoring and analytics provided by SmartEvent servers and forwarding logs to an SIEM of your choice

The more complex and expansive your infrastructure, the more granular implementation of Check Point management you may require to assure the necessary performance and redundancy.

Another option that should be specifically mentioned here is **Quantum Smart-1 Cloud**, a cloud-based management environment that is redundant and scalable on demand to accommodate your enterprise. Check Point takes care of the maintenance updates, upgrades, and high availability in the background, while you are simply administering your infrastructure from it. We will revisit management options in a later chapter to compare the advantages of different choices for specific circumstances.

As you go through this book, you will acquire valuable foundational knowledge equally applicable to all of the implementations described previously.

We now understand that the Security Management architecture allows you to grow your company and maintain the same level of protection, regardless of its scale or model. Before we go all technical, let's look at learning approaches, available options, and the reasoning behind this book's format.

Determining how we learn

There is a widely adopted learning and retention concept representation known as either the **learning pyramid** or the **cone of experience**. This is depicted in the following diagram:

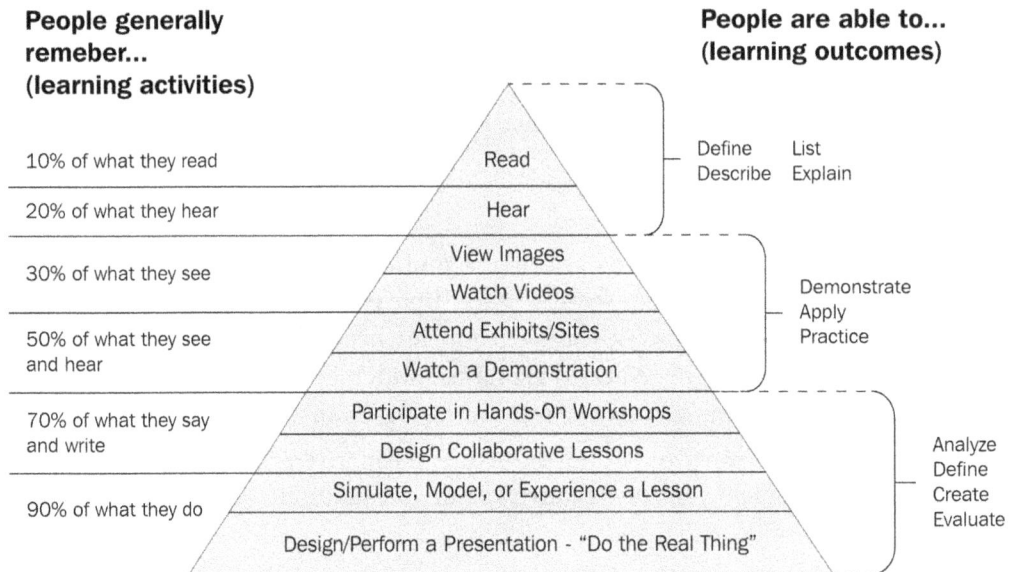

Figure 1.8 – Learning and retention[4]

Not everyone agrees with the exact quantification of the results, but there is no denying that the more of these activities and mediums of learning you are exposed to, the better the outcome.

The comprehension of concepts from reading the material varies depending on how well it is written. If the book or guide is well illustrated, it makes it easier to tie the concepts to real-world applications. Watching someone doing it in a video affirms the validity of the printed or static material. When you are working on a subject in your own lab environment, your confidence in being able to reproduce the results and your level of comfort with expanding your knowledge of the subject grow even further.

Follow the instructions in the book and its complementary online materials to get the most out of it. Choose your own names, emails, and fictional or real company names for registration on portals and access to the products we will be using in the lab.

Some of you may not have the necessary computing capabilities at your disposal to replicate the lab. For them, and those who may be reading this book during your commute, the included screenshots should provide a close approximation of the experience.

We now understand the learning methodologies that impact our comprehension and retention. In the next section, we will familiarize ourselves with the Check Point User Center and configure credentials required for access to resources used in the labs.

Navigating the Check Point User Center

The **Check Point User Center** is the portal for access to a variety of resources, and the place where you will create or manage your Check Point accounts, users, and products. It is also a place where you generate and download licenses and support contracts.

It is accessible at `https://usercenter.checkpoint.com`.

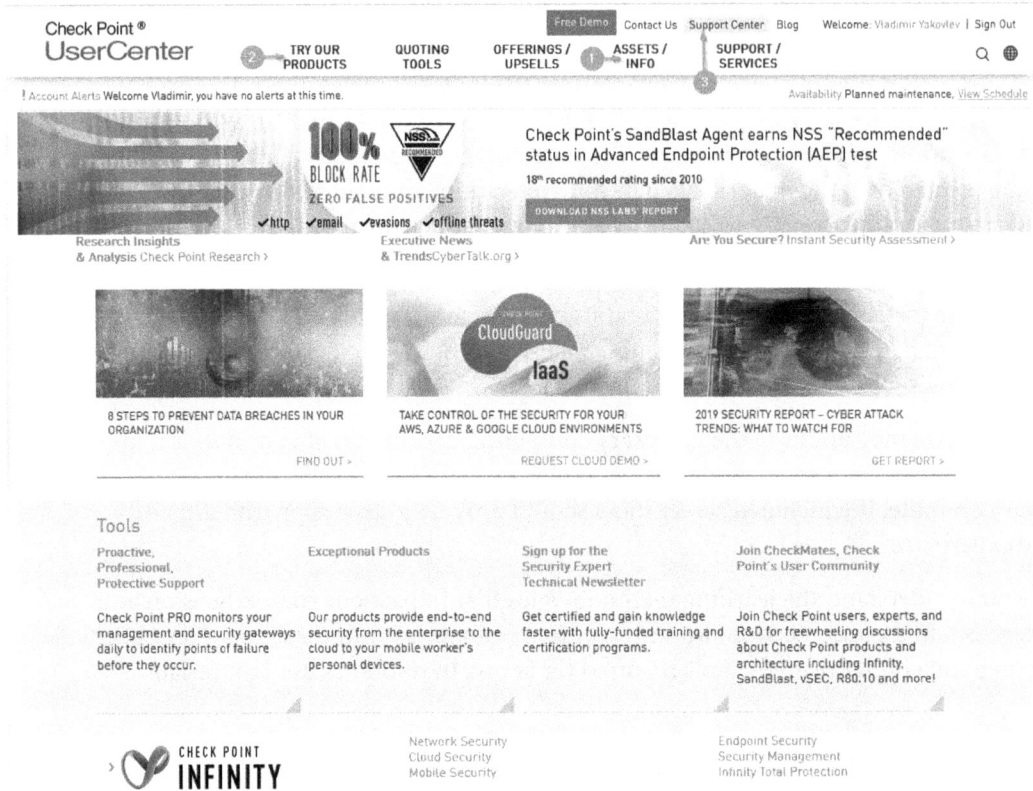

Figure 1.9 – User Center

We will be using the **ASSETS/INFO** and **TRY OUR PRODUCTS** sections to obtain and maintain our lab licenses as we go through the book.

From the User Center, you can get to **Support Center**, a place where you can open and manage **Service Requests** (**SRs**), report security issues, subscribe to or access the **PRO Support** portal (a proactive monitoring and reporting service), and gain access to the technical documentation, alerts, subscriptions, product downloads, and search capabilities across **SecureKnowledge** articles, downloads, documentation, and **CheckMates** community posts and discussion threads.

The following screenshot shows Support Center:

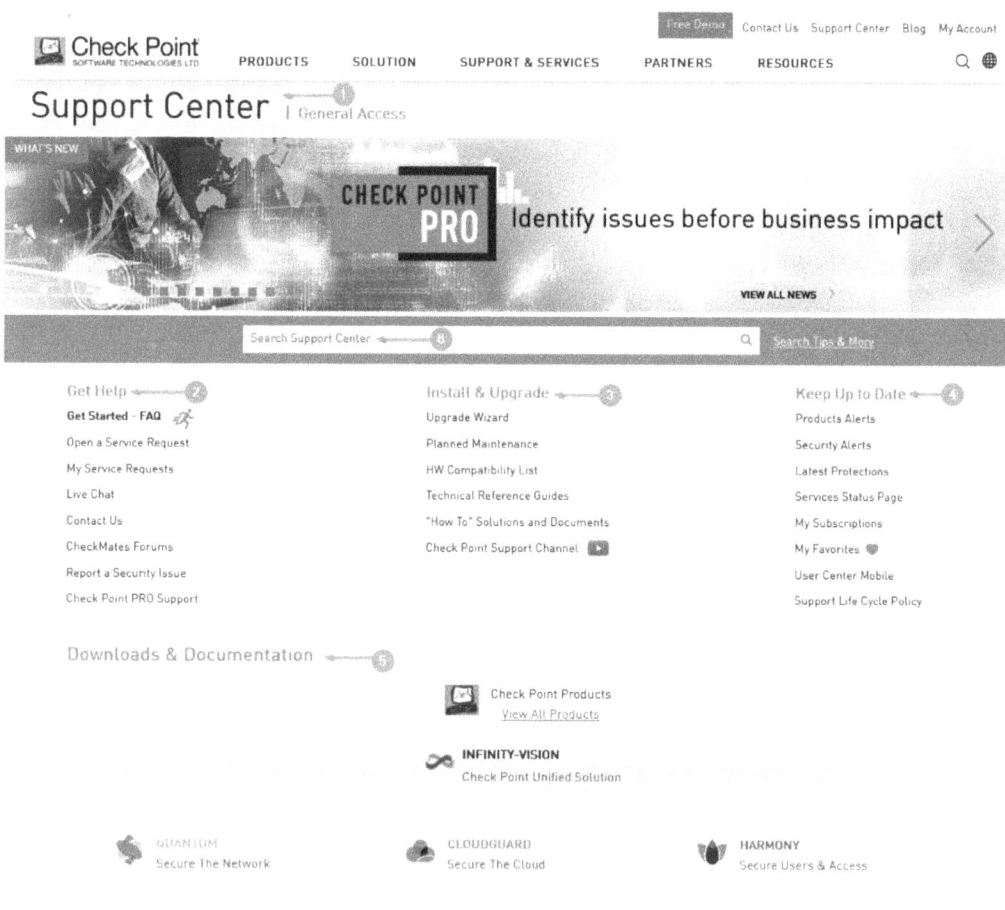

Figure 1.10 – Support portal

Both portals are interlinked, but if you know what you need, it is simpler to get to the right place through a corresponding link.

The Support Center may be accessed at `https://supportcenter.checkpoint.com`.

> **Important Note**
>
> While it is not necessary to register with Check Point in order to download and try their firewall product, this trial will be limited to 15 days. To extend it beyond the initial 15 days, you will have to go through the registration process to request a trial or lab license(s).

Since we must learn how to register and manage users and accounts and how to license the product, we will now start with the registration process.

Follow these steps to register as a portal user and create an account:

1. In your browser, go to `https://usercenter.checkpoint.com`.

2. When prompted with the **Sign In** screen, click on **Sign Up Now**.

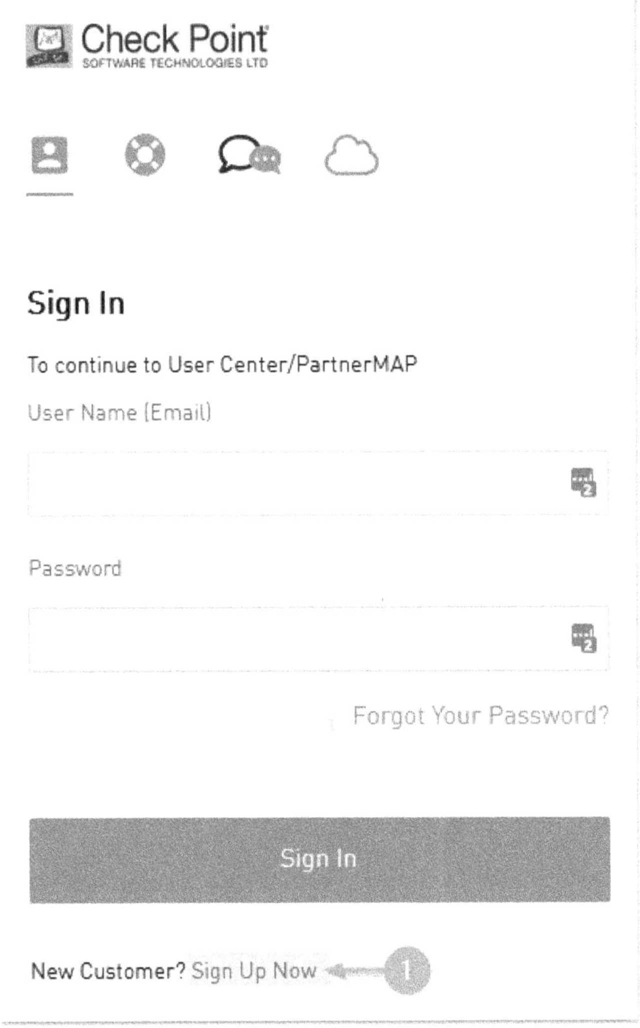

Figure 1.11 – Sign Up Now

Populate the fields with your information and then click **Submit**.

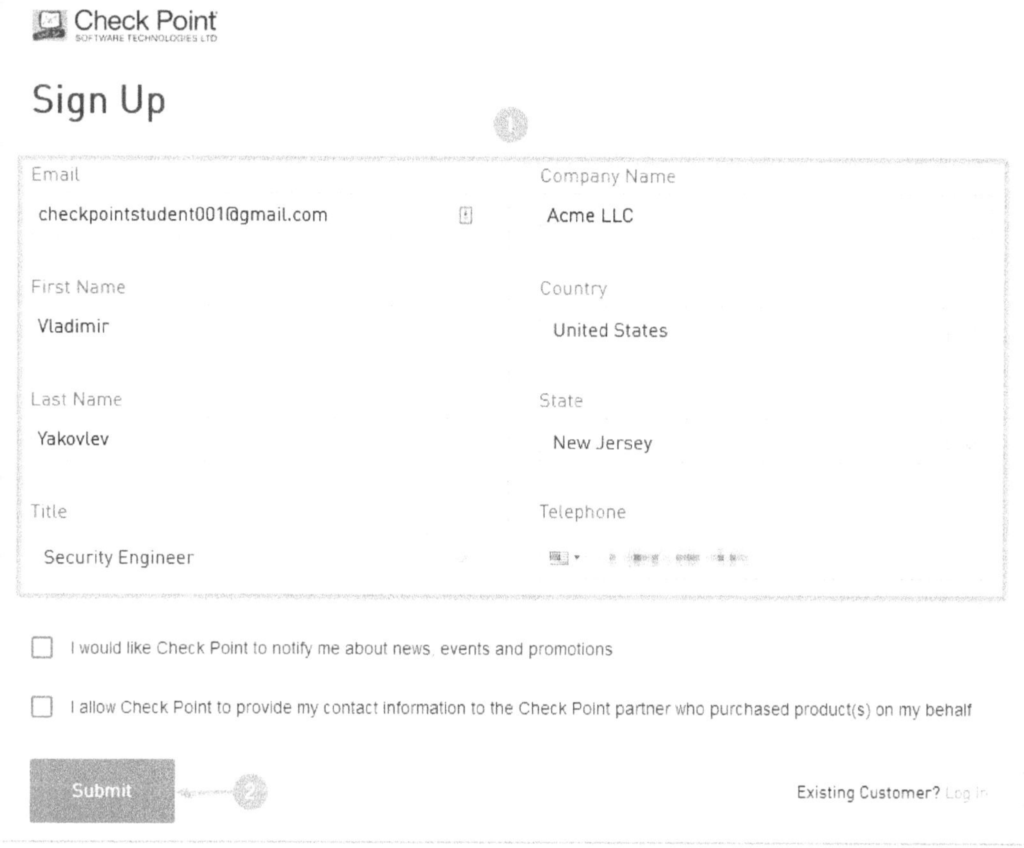

Check Point
SOFTWARE TECHNOLOGIES LTD

Sign Up

Email		Company Name
checkpointstudent001@gmail.com		Acme LLC

First Name	Country
Vladimir	United States

Last Name	State
Yakovlev	New Jersey

Title	Telephone
Security Engineer	

☐ I would like Check Point to notify me about news, events and promotions

☐ I allow Check Point to provide my contact information to the Check Point partner who purchased product(s) on my behalf

Submit

Existing Customer? Log In

Figure 1.12 – Sign Up; user information

3. The **Success!** popup will appear; check your mailbox to continue.

4. Click on **Confirm Email** in the body of the message.

5. Create and confirm a suitably strong password using a combination of uppercase and lowercase letters, numbers, and symbols and then click **Submit**.

6. Click **Sign In**.

7. Enter your username (the same as the email in *step 2* and the password from *step 5*) and then click **Sign In**.

8. Once you are logged in for the first time, click on your username in the top-right portion of the screen and then click on the **Security** shield icon on the left to configure the **Multifactor Authentication (MFA)**.

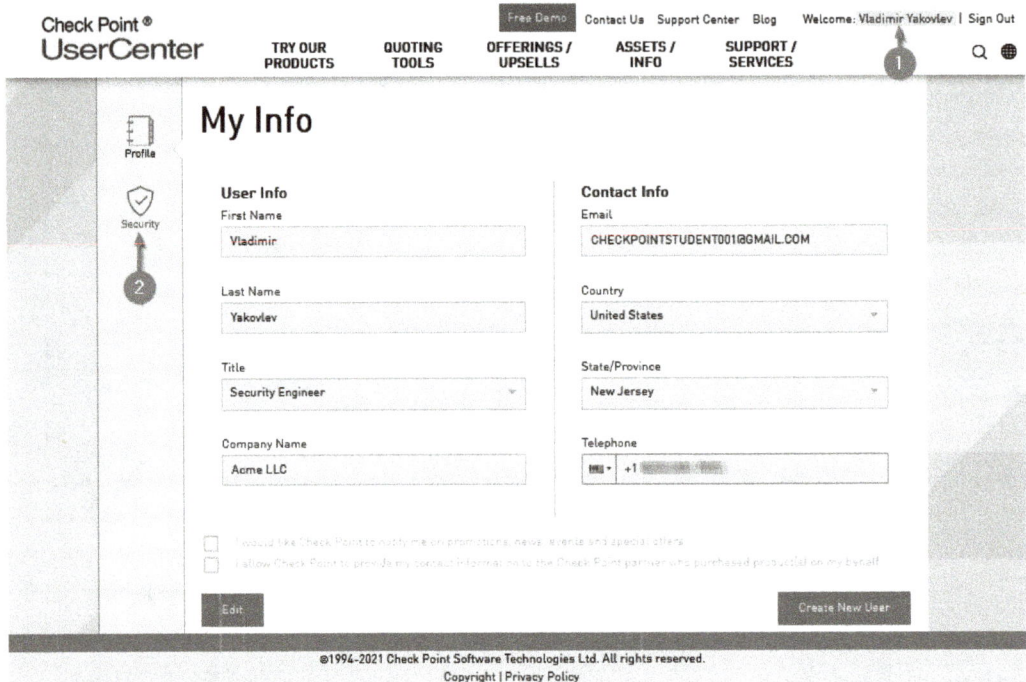

Figure 1.13 – Securing User Center access

9. Toggle the **2-Step Verification** switch to the **On** position:

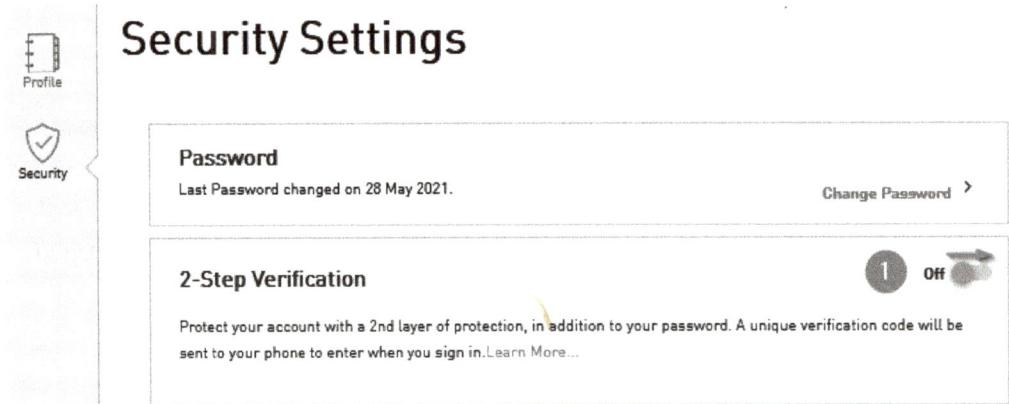

Figure 1.14 – Turning on 2-Step Verification

10. Enter your mobile phone number, verify that the **Text Message** option is selected, and then click **Verify Phone**.

11. Enter the code received via text message and then click **Activate**.

12. Your phone number is now shown as **Verified**. Click on the **Display Backup codes** arrow.

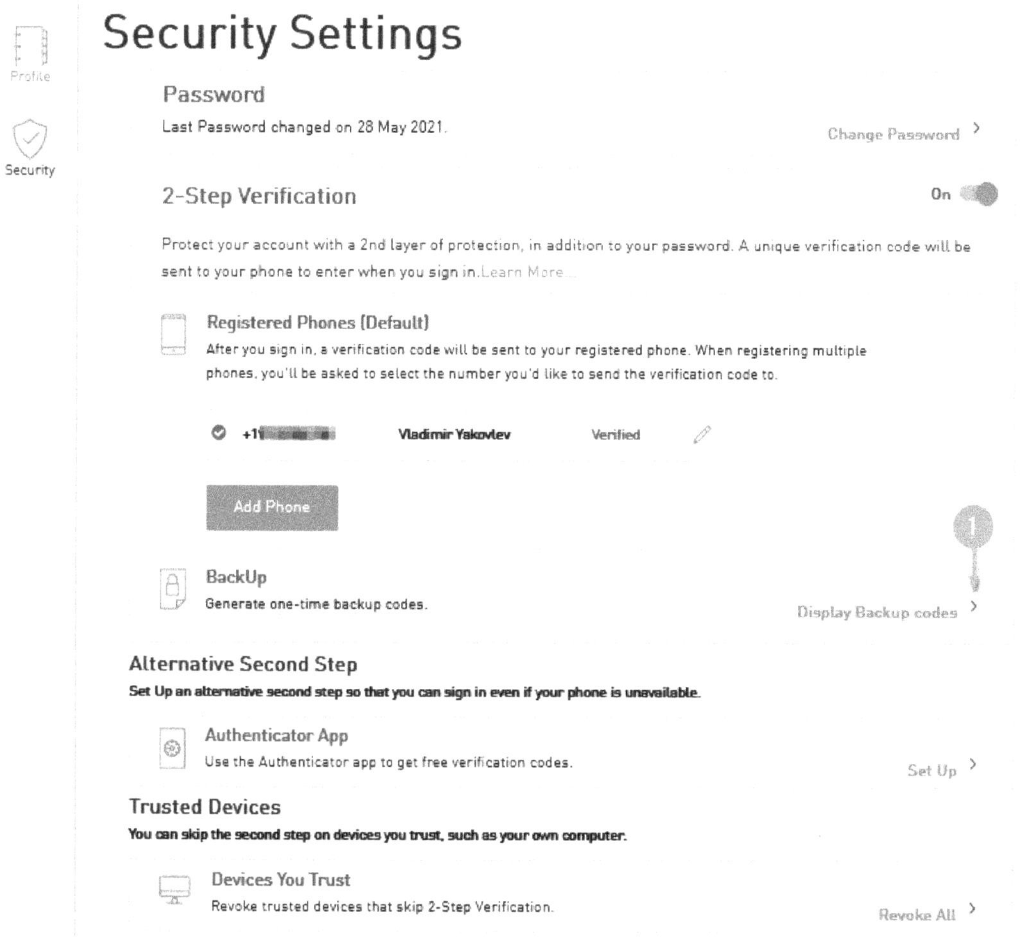

Security Settings

Password

Last Password changed on 28 May 2021.   Change Password ›

2-Step Verification   On ◯

Protect your account with a 2nd layer of protection, in addition to your password. A unique verification code will be sent to your phone to enter when you sign in. Learn More...

Registered Phones (Default)

After you sign in, a verification code will be sent to your registered phone. When registering multiple phones, you'll be asked to select the number you'd like to send the verification code to.

⊘ +1▮▮▮▮▮▮▮   Vladimir Yakovlev   Verified   ✎

[Add Phone]

BackUp

Generate one-time backup codes.   Display Backup codes ›

Alternative Second Step

Set Up an alternative second step so that you can sign in even if your phone is unavailable.

Authenticator App

Use the Authenticator app to get free verification codes.   Set Up ›

Trusted Devices

You can skip the second step on devices you trust, such as your own computer.

Devices You Trust

Revoke trusted devices that skip 2-Step Verification.   Revoke All ›

Figure 1.15 – Backup codes for 2FA

13. Click on **Generate New Backup codes**. When backup codes are displayed, click **Print**. If you do not have a printer connected, print codes to PDF. Click **Close**.

14. In the **Authenticator App** section, click the arrow to the right of **Set Up**.

15. Choose your mobile phone platform and then click **Next**.

16. If you do not have an authentication application on your phone, install either Microsoft Authenticator or Google Authenticator or your preferred **MFA** application. When installed, or if already available, add the new account to it by scanning the QR code and then click **Next**.

17. If the scan fails (observed on very high-resolution monitors with particular brightness and contrast), click on **CAN'T SCAN IT?**, manually enter the key into the authentication manager, and then click **Next**.

18. Enter the dynamically generated one-time code and click **Next**. *Note the time remaining for the action on your phone while doing it.*

19. The authentication app now becomes the default method for the second factor. Let's look at the following screenshot:

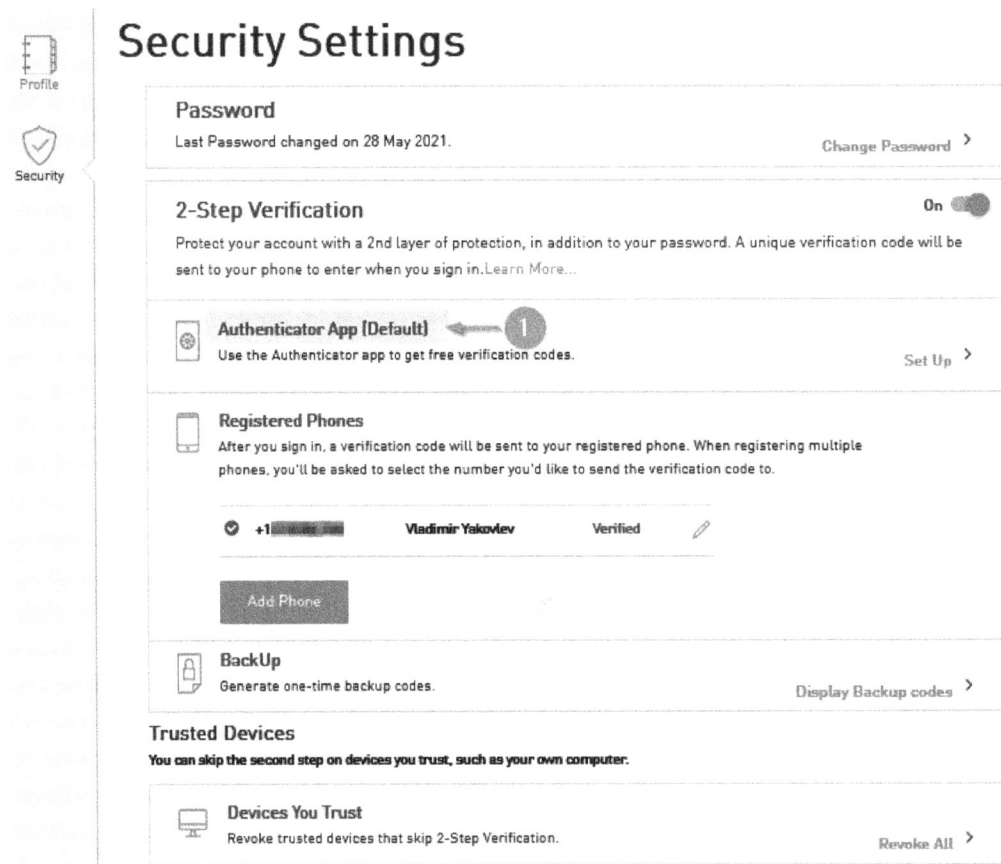

Figure 1.16 – Authentication app as the default 2FA

We can now securely log on to the User Center and access its resources.

Summary

In this chapter, we learned about the history and the present-day state of the technology and services offered by Check Point. We saw the flexibility and scalability of the Secure Management architecture and learned about the advantages of consolidated security solutions, and why they are emerging as the preferred choice for addressing today's complex threat environments. We have also created and secured our User Center account.

Now that we understand the modular nature of Check Point management architecture, we are ready to look at firewall locations within common network topologies and talk about the significance of their placement.

In the following chapter, we will address where and when certain features are better employed for different outcomes. We will learn how to determine the utilization of your currently deployed firewalls and calculate the capacity for new ones.

Further reading

1. `https://cybersecurity-excellence-awards.com/candidates/check-point-software-technologies/`

2. `https://www.checkpoint.com/about-us/company-overview/`

3. `https://www.microsoft.com/security/blog/2021/03/11/finalists-announced-in-second-annual-microsoft-security-20-20-awards/`

4. `https://en.wikipedia.org/wiki/Edgar_Dale`

2
Common Deployment Scenarios and Network Segmentation

In this chapter, we will learn about the importance of knowing your **network architecture** and where the firewalls are deployed. We will look at it from the point of view of administrators inheriting an existing architecture, as well as those who are responsible for making suggestions for the new implementations. We will also briefly discuss the sizing methodology for new devices and the performance of your existing firewalls.

In this chapter, we are going to cover the following main topics:

- Understanding your network topology
- Learning about network segmentation
- Protecting the core
- Protecting the perimeter
- Sizing appliances for new implementations – load on current systems

Understanding your network topology

As a firewall administrator, you must have a thorough understanding of the network in which these firewalls are implemented. The firewall can only control and inspect the traffic that is traversing it.

This brings us to a question about your role as the firewall administrator. Depending on the size of the company you are working for, the maturity of its security practices, its budget, and the size of your security team, your responsibilities may vary greatly.

In a typical large financial organization, there may be dedicated positions for firewall administrators that are limited to the creation and modification of objects, rules, security policies, and troubleshooting. Actual engineering and implementation may be handled by a different team or team member.

In smaller organizations, firewall administration is just one of the functions you are likely to perform, and there are several other security controls you may be responsible for. Networking in these kinds of companies is typically handled by separate teams, but the deployment of firewalls and their integration with the rest of the infrastructure may fall under your purview.

In even smaller companies, or companies with smaller information security and IT teams, you may be wearing multiple hats and, realistically, responsible not only for the firewall administration but also for the complete engineering of network and security controls. In these cases, you may have much greater flexibility in shaping your company's security posture.

Regardless of the circumstances, having access to the network topology data and knowing where the firewalls are located will allow you to perform your job efficiently and minimize the chances of misconfiguration. This will significantly improve your troubleshooting performance.

Let's consider a scenario where you are a firewall administrator unaware of the infrastructure and topology changes being made as depicted in the following figure:

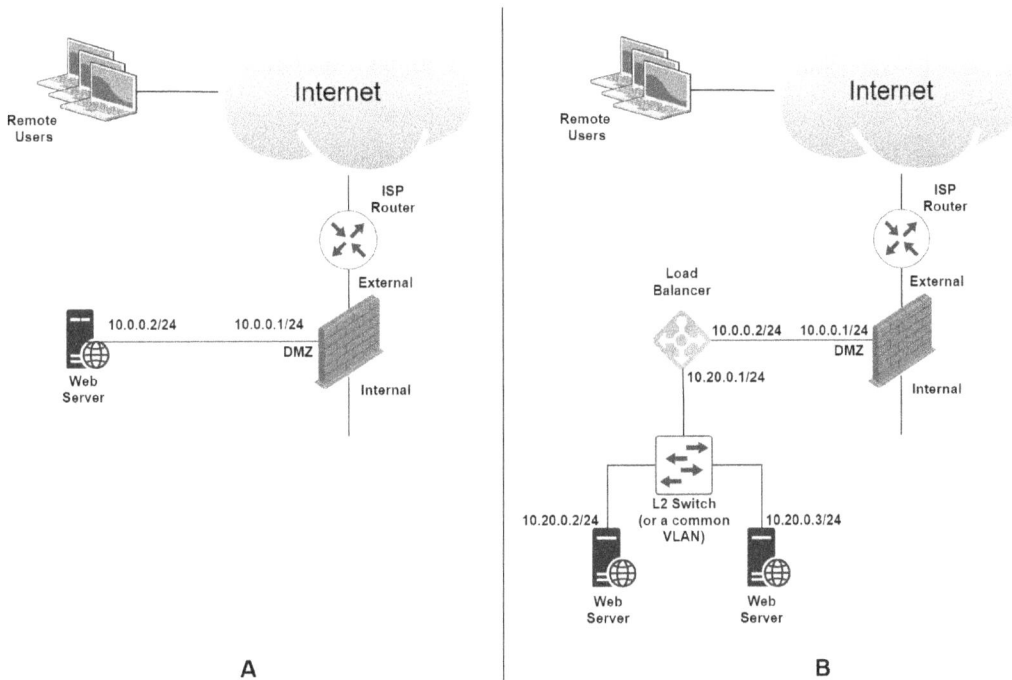

Figure 2.1 – A and B, undocumented topology changes

Scenario A in *Figure 2.1* depicts a firewall/cluster with connectivity to internal networks, the internet, and a **demilitarized zone** (**DMZ**). The DMZ contains a single application server running **Windows Server OS** with **IIS**.

The change is made by the application and networking teams, aiming to improve the performance of the application by deploying a load balancer and adding new instances of the application server.

Scenario B in *Figure 2.1* shows the environment after the change. There is now a new host – a load balancer running a different OS. There is also a new application server. The IP address of the existing application server has changed, and it is now residing on a different network behind a load balancer. The old application server's IP address is now assigned to the load balancer.

A possible outcome here is that if you are to receive a call from application owners claiming issues with access or performance, you cannot identify the IP addresses they are referring to.

Your threat prevention policy, which may have been tailored toward protecting Windows Server OS and IIS, doesn't cover the load balancer.

The example I have just presented illustrates the importance of communication between departments, as well as the necessity of being aware of the changes to the infrastructure you are protecting.

In a larger organization with well-established change controls, you will be in the loop and instructed to make the necessary changes to objects and policies.

In a smaller organization, you are likely to be aware of the pending changes. You will be able to alert the interested parties about the alterations to objects and policies that may be required.

In a mid-sized organization with separate IT and information security departments, poor internal communications, and poorly established change controls, you are likely to be surprised by these kinds of situations. It will then be up to you to create and maintain the network diagrams describing the environment in which your firewalls are operating.

> **Important Note**
>
> For a real-world example of how this can impact your environment, see the summary by **Rapid7** describing recently discovered and disclosed vulnerabilities in the commonly used **F5 load balancers**, as shown in the following link: `https://www.rapid7.com/blog/post/2021/03/18/f5-discloses-eight-vulnerabilities-including-four-critical-ones-in-big-ip-systems/`.

Should you encounter this or a similar situation, write a concise report for your manager that details the issue and suggests improvements in handling topology changes in the future. Establish rapport with your colleagues in IT so that relevant information will be relayed between departments even in the absence of formal procedures.

Your circumstances as a firewall administrator may vary depending on a large number of factors. For instance, you may be designated as an administrator during the initial implementation of the product. Alternatively, you may be hired or appointed long after these firewalls were put in place. Your network infrastructure may be expertly crafted to take maximum advantage of the security controls your firewalls are offering, or it may be lacking in some regard. It may also be that you have inherited an environment that has grown organically over time and its topology and utilization of firewalls are way out of date, and, upon examination, should really be redesigned from scratch.

Regardless of the situation you find yourself in, make the best of it by determining whether security could be improved based on the placement of the firewalls and their capabilities, utilization, and licensed features. If you see an opportunity for improvements, write a detailed proposal and submit it to your manager or to other interested parties, depending on the organization's structure and practices.

When talking about network topology in the context of firewalls, we will inevitably come across the term **DMZ**. DMZ (or demilitarized zone) generally refers to a segment of the network isolated from both internal and external networks. Access to specific services running on hosts located in DMZs, as well as access from DMZs to hosts and services on internal network segments, is controlled by a **security policy**.

Technically, DMZ is a category and not a particular zone – that is, you may have multiple zones serving as individual DMZs, each designated for the type of servers or networks that require partial isolation from the internal network and are accessible either from the internet or one or more external entities.

Having talked about zones, it would be helpful to have a quick look at how Check Point uses them:

- Check Point firewalls can operate as both interface and/or zone-based firewalls at the same time. In Check Point firewalls, a zone could be thought of as a placeholder object that is associated with either single or multiple interfaces of a gateway and the network topology behind those interfaces.

- Each interface can belong to a single zone only.

- Zones can be used as sources or destinations.

- There are four pre-defined zones: **InternalZone**, **ExternalZone**, **DMZZone**, and **WirelessZone**. These zones (and zones in general) *do not* impart any special properties to the hosts behind the interfaces to which they are assigned. Only zones' use in security policies does that.

- You can create additional zones as needed.

- You cannot use a zone in a security policy if there are no interfaces associated with it.

- In rules, you can use either **zones** or their constituent **components** (that is, groups of objects or specific hosts/networks behind the interface associated with the same zone).

- Objects behind the different interfaces associated with the same zone cannot communicate with each other unless permitted by an explicit rule in the policy.

- The same zones can be used across multiple gateways to simplify security policies.

We will discuss zones in more detail later when learning about gateway and cluster objects and when learning about policies and rules.

Now, let's look at several commonly encountered simple network topologies and try to analyze them, figure out the advantages and shortcomings of each, and suggest if and how they could be improved.

Common topology scenarios and exercises

Look at each of the following diagrams and write a summary describing the traffic flows, the content of each **protected enclave** (that is, the topology of the network behind firewall interfaces), possible issues with the topology, and how those issues could be addressed.

Then, read on to compare your conclusions with those given in the chapter.

Scenario 1 – a single gateway with four interfaces

The following figure illustrates *Scenario 1*:

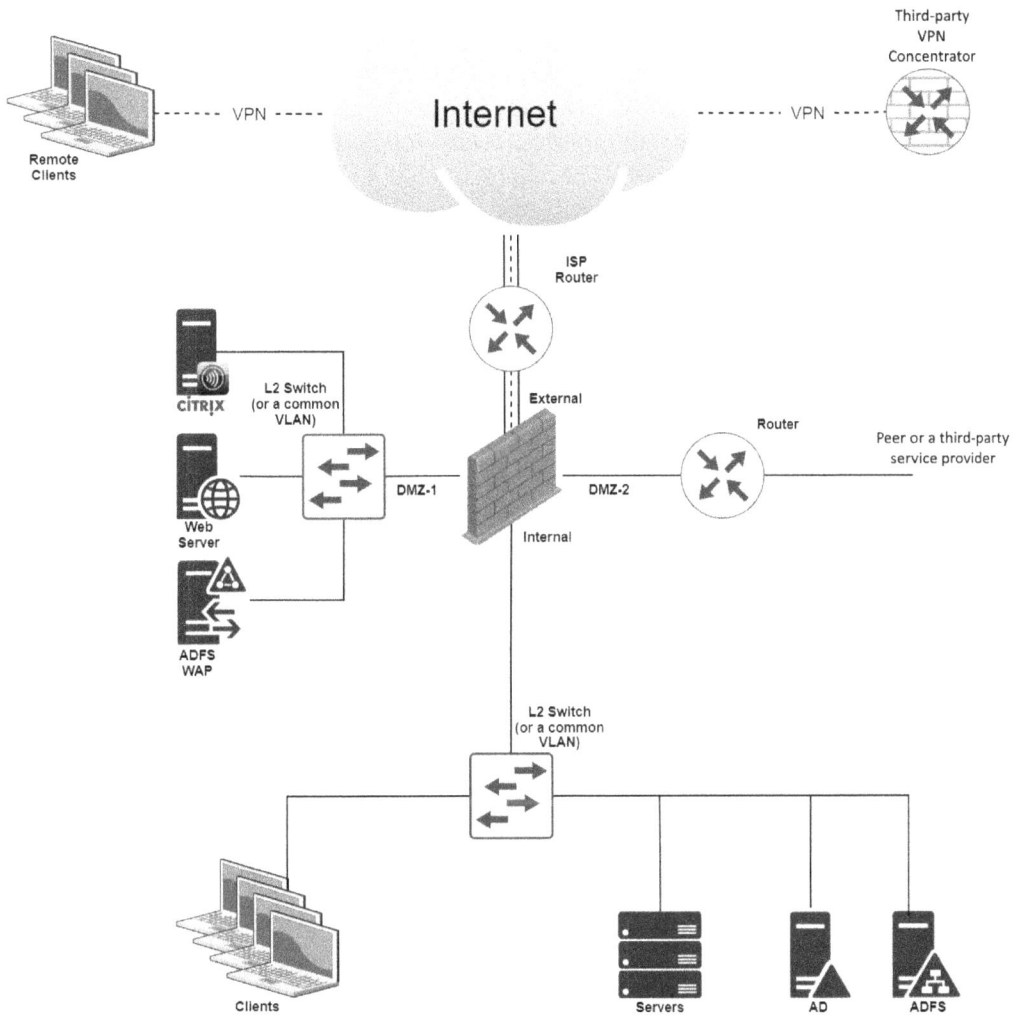

Figure 2.2 – A single gateway with four interfaces

Exercise

Take a minute to study *Figure 2.2*. Identify the protected enclaves.

Use your notepad to describe the possible reasons for this design, as well as its advantages and shortcomings. Suggest improvements to this topology.

The diagram in *Figure 2.2* describes a simple network consisting of the internal, external, and two DMZ segments.

The environment described in *Figure 2.2* is suboptimal for the following reasons:

1. A single gateway is representing a **single point of failure** (**SPOF**) and should be replaced with a **high-availability** (**HA**) cluster. In this case, the use of a single gateway may be acceptable if the **service level agreement** (**SLA**) or IT/information security policy of the company allows for the downtime described in your **Check Point Support Program**. Depending on your support level, the downtime may be as short as four hours or as long as two business days (if **Return Merchandise Authorization** (**RMA**) is initiated after 3:00 PM, local time). A speedy recovery is contingent on having a configuration backup to hand, otherwise, additional time will be required to configure the new appliance.

2. **DMZ-1** contains multiple types of servers, each subject to its own vulnerabilities. Should the attacker compromise one of them, credentials harvested from that one could be used on the others. The attacker's movement between hosts within DMZ-1 will not be traversing the firewall and therefore, the attacker has a much higher chance of evading being noticed.

3. The internal network is not segmented. A single compromised host, where the initial vector may be a USB drive or an infected computer, may allow for a complete network takeover by careful attackers without tripping any alarms on the firewall.

4. DNS traffic between client PCs and name servers is *not inspected by the firewall*. It will always be attributed to the DNS servers performing or forwarding queries. This is important for the identification of the hosts attempting to reach out to the **Command and Control** (**C&C**) of malware, as it predominately relies on the ability to communicate with either compromised or purposely created domains or subdomains. Since all of the internal hosts, including DNS servers, are behind a single firewall interface, you will see *false positives* for the hosts identified as *infected* by threat prevention.

While one of the possible reasons for this implementation is the performance limitation of the firewall, there is no good reason for not segmenting the DMZ into smaller zones. The three hosts depicted in the DMZ are all serving different purposes. Placing them behind separate interfaces will not have a negative impact on performance.

Without altering this topology, you can improve the situation by implementing either **Check Point Harmony Endpoint** or an EDR solution from other reputable vendors. The advantage of **Harmony Endpoint** is the integrated correlated logs and forensic reports from both the endpoints and the firewalls/clusters.

With minimal alterations to the topology, you can either move your domain controllers (if they are running **domain name services (DNSes)**) or introduce additional DNS servers connected to a network behind a different interface of the firewall. Change the DNS configuration of the hosts in your other network segments to use them instead of the servers located on the same segment. This would significantly improve the detection of hosts that are either compromised or are attempting to communicate with malicious resources.

Scenario 2 – a single gateway with DMZ segmentation

The following figure illustrates *Scenario 2*:

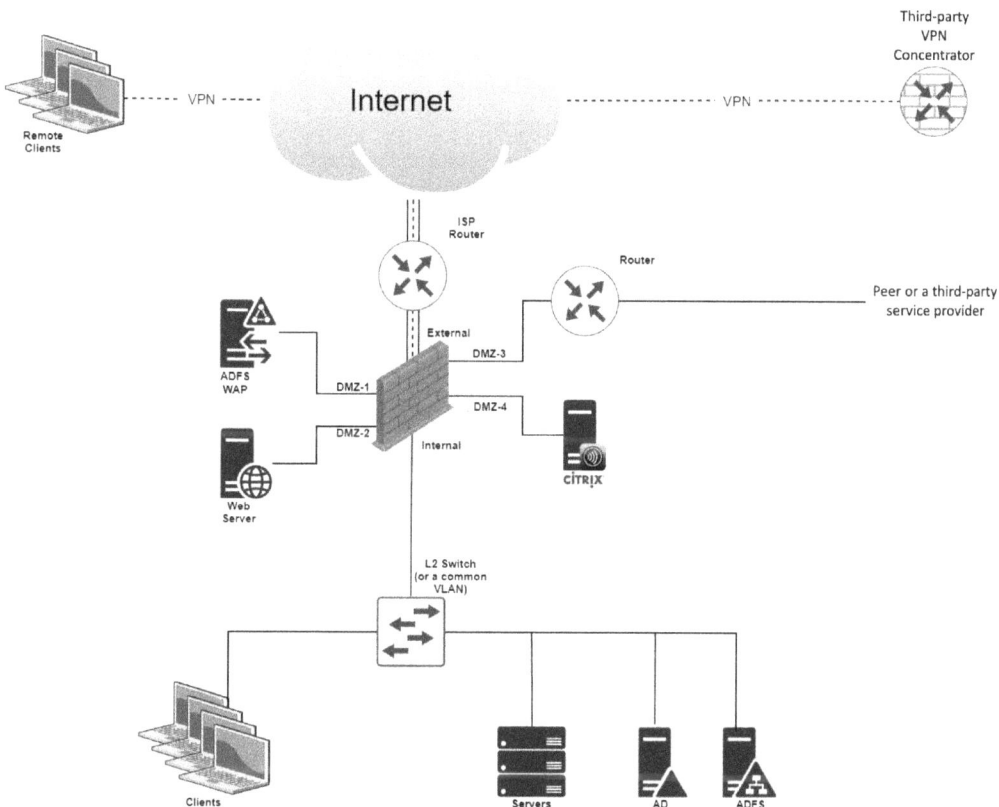

Figure 2.3 – A single gateway with DMZ segmentation

Exercise

Now take a minute to study *Figure 2.3*. Identify the protected enclaves.

Use your notepad to describe the possible reasons for this design, as well as its advantages and shortcomings. Suggest improvements to this topology.

The case described in *Figure 2.3* is similar to the previous one, and most of the same comments are applicable, except those pertaining to DMZ segmentation.

As you can see, all hosts in the DMZs are connected to dedicated interfaces of the firewall, allowing for better access control and a more targeted application of threat prevention.

Scenario 3 – site redundancy via P2P connection and L3 internal segmentation

The following figure illustrates *Scenario 3*:

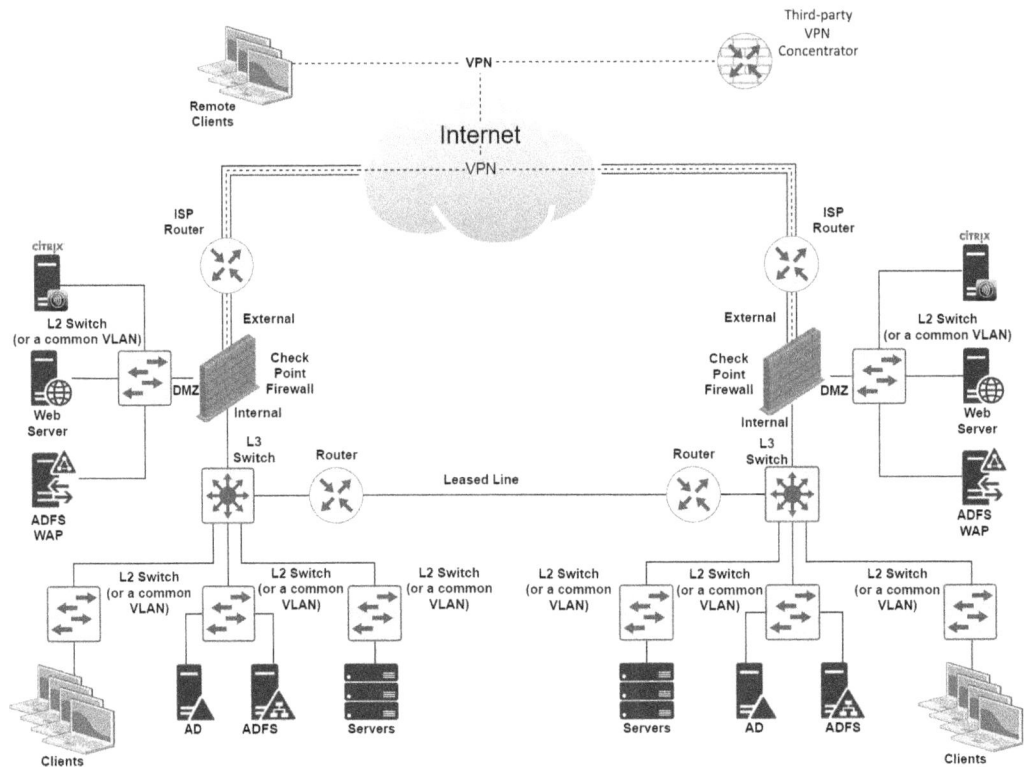

Figure 2.4 – Site redundancy via a P2P connection and L3 internal segmentation

Exercise

Take a minute to study *Figure 2.4*. Identify the protected enclaves.

Use your notepad to describe the possible reasons for this design, as well as its advantages and shortcomings. Suggest improvements to this topology.

The case illustrated in *Figure 2.4* is still lacking in terms of segmentation, but it is marginally better. While the DMZ is still behind a common interface of the firewalls in each site, the internal network segmentation is implemented in the routing (*L3*) switches. We will see how this can help us improve the situation going forward when we cover **Gaia OS** and the **Gateways**, as well as **Cluster** objects.

Additionally, site redundancy is achieved by routing internal traffic to the alternate site via a **site-to-site** or **point-to-point** (**P2P**) link, in case of either a firewall or **Internet Service Provider** (**ISP**) connectivity failure at a single site.

Scenario 4 – a single gateway with DMZs and limited internal network segmentation

The following figure illustrates *Scenario 4*:

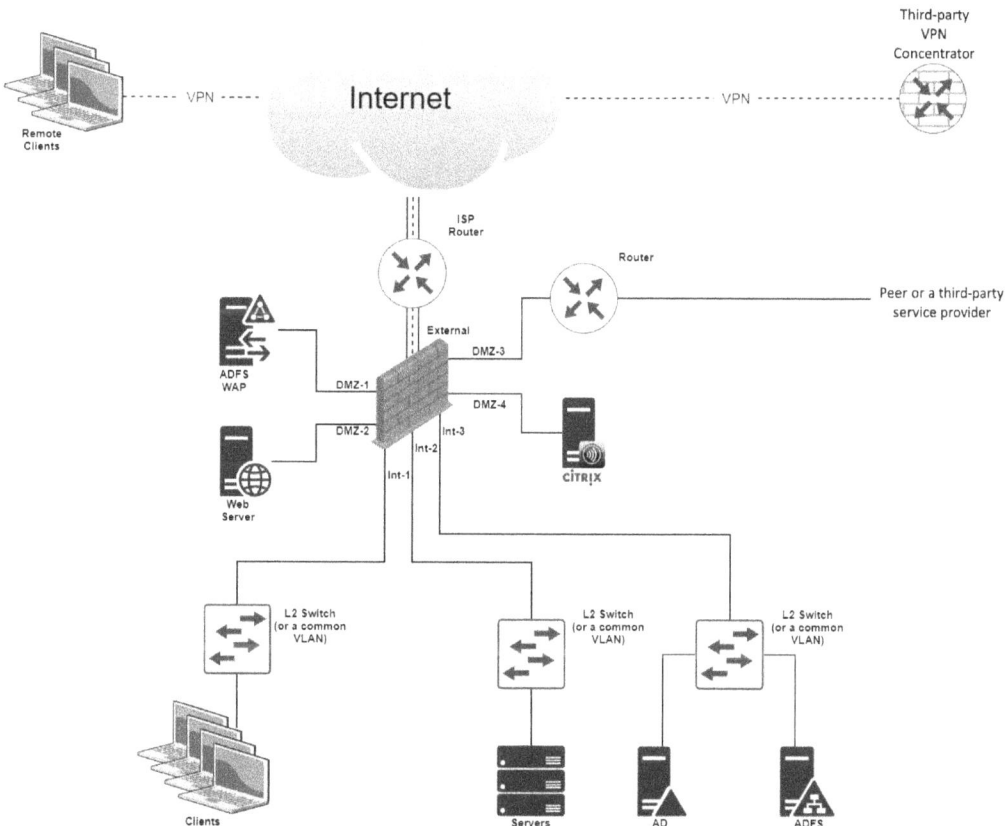

Figure 2.5 – A single gateway with DMZs and limited internal network segmentation

Exercise

Take a minute to study *Figure 2.5*. Circle the protected enclaves.

Use your notepad to describe the possible reasons for this design, as well as its advantages and shortcomings. Suggest improvements to this topology.

The network topology shown in *Figure 2.5*, besides showing a single gateway instead of a better-suited high-availability cluster, achieves a minimum recommended degree of segmentation. Here, you can control traffic between clients and servers. Your AD/DNS is connected to a separate firewall interface, allowing for accurate attribution of the initial DNS queries and the enforcement of the threat prevention policies between all of these segments.

Scenario 5 – site redundancy with MPLS and VPN, L3 internal segmentation

The following figure illustrates *Scenario 5*:

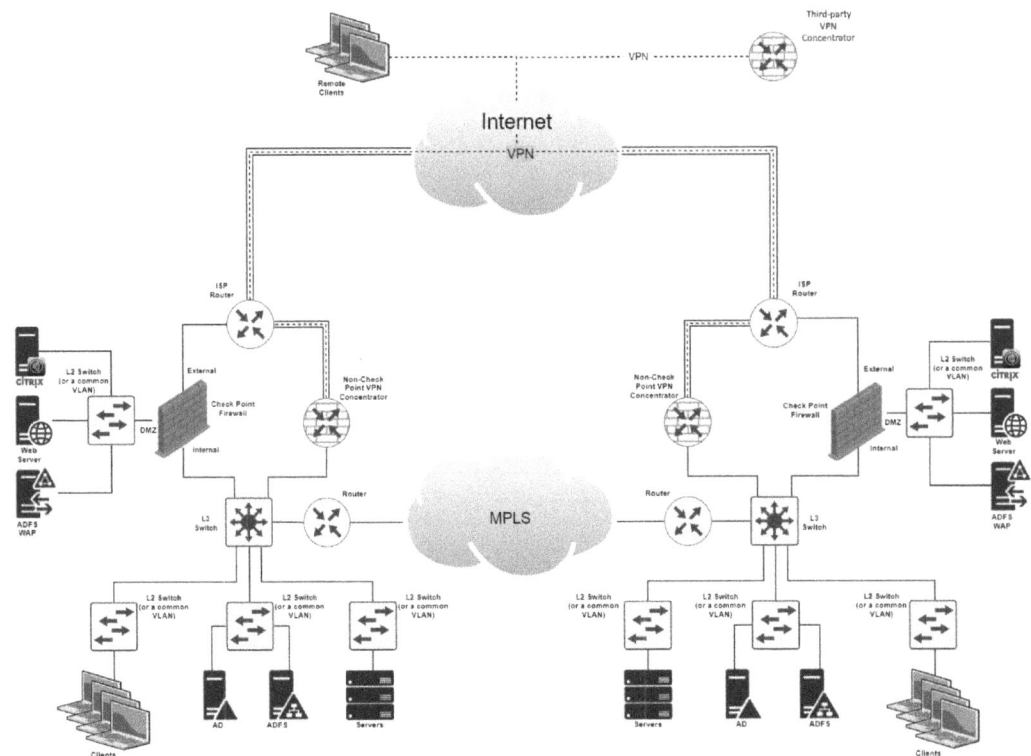

Figure 2.6 – Site redundancy with MPLS and a VPN, L3 internal segmentation

Exercise

Take a minute to study *Figure 2.6*. Identify the protected enclaves.

Use your notepad to describe the possible reasons for this design, as well as its advantages and shortcomings. Suggest improvements to this topology.

The topology illustrated in *Figure 2.6* is not different from that in *Figure 2.4* from the segmentation perspective. The site redundancy is maintained via **Multiprotocol Label Switching** (**MPLS**) and a VPN with a non-Check Point VPN concentrator. This environment is pretty good from a resilience point of view, but its downside is that the decrypted VPN traffic is not inspected or logged by the Check Point firewall unless it is addressed to the hosts behind the DMZs.

This topology (or variants of it) is frequently seen during migration from other vendor solutions (where only a single gateway per site is implemented) to Check Point. It could also be seen in instances where the company is too invested in and/or happy with its existing VPN solution.

Scenario 6 – site redundancy with a stretched cluster, L3 internal segmentation

Finally, let's look at another frequently encountered topology that utilizes stretched **virtual local area networks** (**VLANs**) between two data centers:

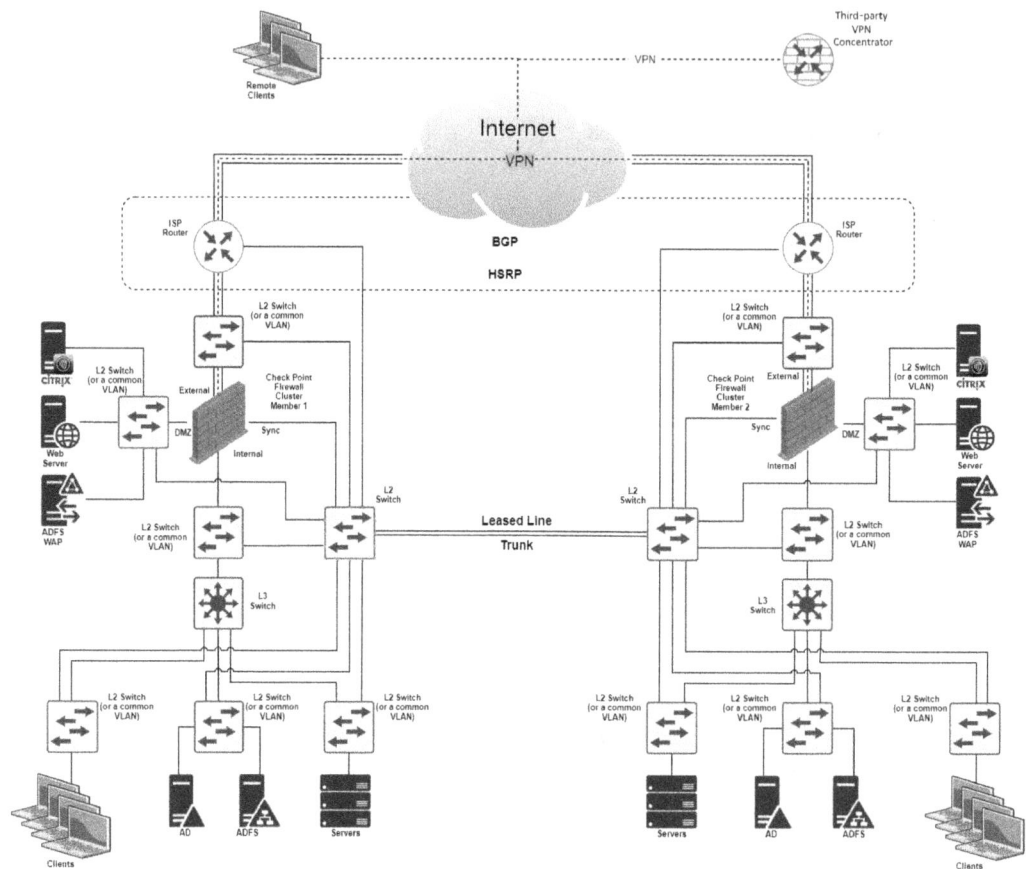

Figure 2.7 – Site redundancy with a stretched cluster, L3 internal segmentation

Exercise

Take a minute to study *Figure 2.7*. Identify the protected enclaves.

Use your notepad to describe the possible reasons for this design, as well as its advantages and shortcomings. Suggest improvements to this topology.

This configuration is often seen when a high-bandwidth and low-latency service is available between data centers. The upside of this design is that it allows for recovery from several infrastructure component failures at each site. Traffic from one site will be making a round-trip via a secondary site before getting to its intended destination, so minor latency increases should be expected. Each site will continue to function even if the connectivity to the ISP is lost, the ISP router is down, the *firewall* (a site-specific *high-availability cluster member*) is down, or (depending on how your L3 switches are implemented), even if there's a loss of the L3 switch at the site.

From the point of view of segmentation, the topology illustrated in *Figure 2.7* is not different from that in *Figure 2.4* and *Figure 2.6*.

> **Note**
>
> Another issue to keep an eye on, in this scenario, is the possibility of a firewall cluster becoming a *split-brain* cluster if the link between sites is lost. In this case, a standby cluster member will determine that there is no active member present and will change its status to active. This will enable both sites to work independently until connectivity is restored. At that point, the active cluster member with the lower priority will enter a standby mode.

Now that we have learned about the importance of having a clear picture of where the firewalls are located in your network topology and how to identify the shortcomings of existing topologies, we should learn about **network segmentation**.

Learning about network segmentation

Today, most of the environments where Check Point firewalls are found are fairly sophisticated and, generally, consist of one or more data centers (or server rooms) that may contain a mix of physical and virtual servers, container hosts, switching and routing equipment, and a dedicated **storage area network** (**SAN**). It may also include a hyperconverged infrastructure where a part of the network, compute, memory, and storage are virtualized and distributed between nodes.

These environments are often a part of hybrid cloud implementations. Access to a public cloud or multiple clouds is achieved either by VPNs or by direct connectivity to cloud service providers (for example, **AWS Direct Access**, **Azure Express Route**, and so on).

Some firms are using public cloud segments of their hybrid infrastructure for elastic computational capabilities while keeping them locked down. Others are using the cloud to host their public-facing applications. Yet, others are extending their entire on-premises infrastructure to the cloud for disaster recovery purposes.

> **Important Note**
>
> Irrespective of the complexity of the infrastructure, firewalls are used to segregate network segments for the enforcement of access control and threat prevention.

Smaller segments allow better granularity in access control and threat prevention. In a hyper-converged infrastructure, technologies such as **Nutanix Flow** and **VMware NSX-T** allow for micro-segmentation of the infrastructure integrated with Check Point's threat prevention capabilities.

Traditionally, networks were segmented based on the criticality of resources. The more critical a resource to the operation of your infrastructure, the more stringent the security controls protecting it were. Additionally, government regulations and industry groups standards require resources with access to certain data to be in a *protected scope*. Examples of this are the **PCI-DSS**[1] and **HIPAA**[2], which define the *in scope* system components.

Later, the role and functionality-based segregation of segments, in addition to criticality and scopes, became a best practice. The US **Cybersecurity and Infrastructure Security Agency (CISA)** describes its recommended network segmentation practices in *Security Tip (ST18-001)*[3], as follows:

> *Security architects must consider the overall infrastructure layout, including segmentation and segregation. Proper network segmentation is an effective security mechanism to prevent an intruder from propagating exploits or laterally moving around an internal network. On a poorly segmented network, intruders are able to extend their impact to control critical devices or gain access to sensitive data and intellectual property. Segregation separates network segments based on role and functionality. A securely segregated network can contain malicious occurrences, reducing the impact from intruders in the event that they have gained a foothold somewhere inside the network.*

The US **National Security Agency (NSA)** document, *Segment Networks and Deploy Application-Aware Defenses*[4], recommends the following network segmentation practices:

> *Segmentation is the practice of dividing a network into sub-networks (segments) of devices that share similar security requirements. Generally, segmentation is done by separating access to the most sensitive and vulnerable services on the network, such as directory services, file-share services, and network management. The network can be further segmented by user groups and determinations made on the level of access each user group requires. Once a network is properly segmented, appropriate application-aware defenses can be utilized to isolate and secure each network segment. Such isolation is essential to blocking an adversary's lateral movement through the network and instilling the principle of least privilege to every network device. Limiting device communication between segments also enables better monitoring and visibility into an adversary's attempts to spread from one segment to another.*

In public cloud environments, you can often see the resources with specific roles residing in their own network segments, where access to and from each segment could be controlled by the **security groups** (AWS) or **network security groups** (Azure).

Now that we have seen the security agencies' recommendations on network segmentation, let's see how we can apply them in the environments we are supposed to secure.

User network segmentation

When considering segmentation for user networks, defining a segment that will reside behind the firewall's interface should generally follow the **principle of least privilege** and the **principle of need-to-know**. If a number of PCs are dedicated to a group of users with identical privileges in your infrastructure, they could be hosted behind a single interface of the firewall (or a cluster of firewalls). If we are talking about a multi-site implementation, we could define a single zone containing all of the interfaces on all clusters, behind which PCs serving users with identical privileges are located.

One possible reason for further splitting these segments is to prevent the lateral threat propagation within common purpose enclaves – that is, if all PCs belonging to your marketing team are in the same segment and their endpoint security solution was insufficient to prevent a compromise of a single PC, then all of the PCs in that segment may be at risk. If you have sufficient resources to split each group into smaller segments, it may be warranted to do so. The access control policy for these segments will remain identical, but the threat prevention system will have a chance to contain the anomaly in a single segment.

While it is possible to achieve a simple segmentation using Layer 3 switches with **Access Control Lists** (**ACLs**), this solution will not include application-aware defenses for individual segments and, therefore, will not adhere to the NSA's recommendations.

The diagram in *Figure 2.8* illustrates how granular segmentation for user networks could be achieved:

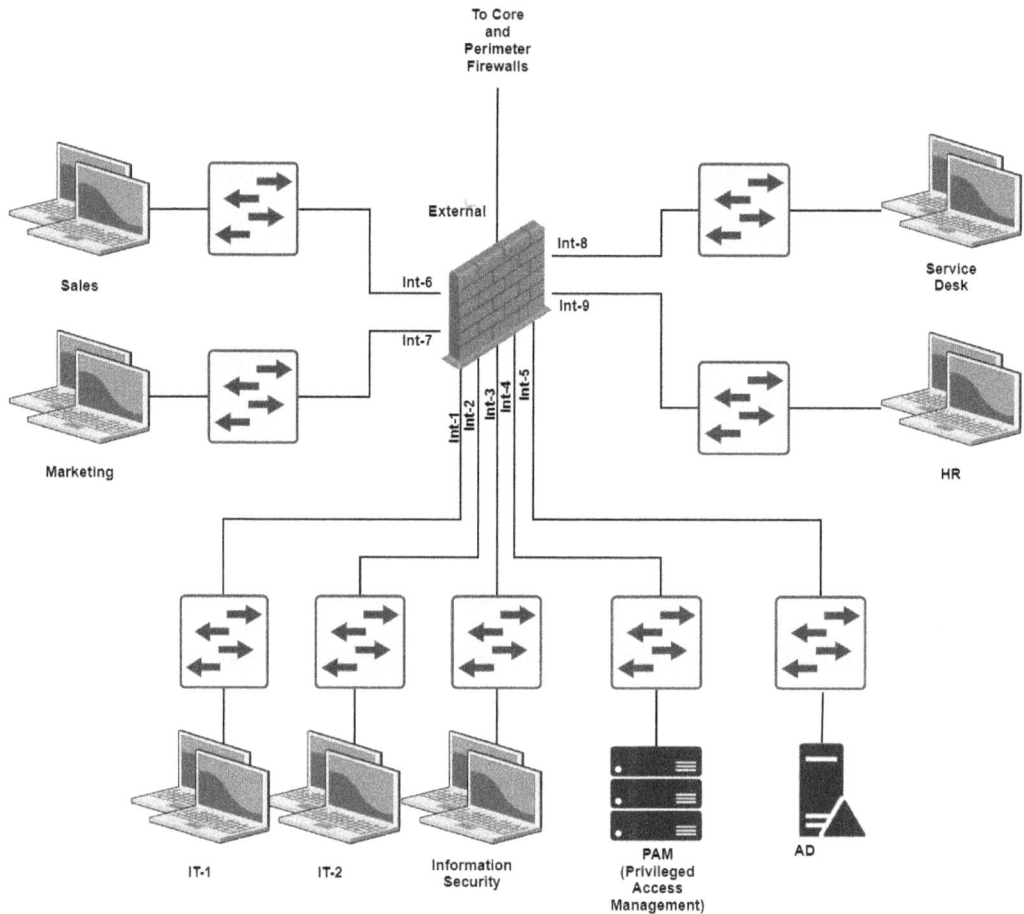

Figure 2.8 – User networks segmentation

> **Important Note**
>
> **Privileged Access Management** (**PAM**) is depicted in *Figure 2.8* for illustration purposes only. PAM typically resides in a data center and is considered a part of the core. The AD server shown in *Figure 2.8* could be a **read-only domain controller** (**RODC**), with or without cached credentials, if the client networks are in remote locations.

Access to a PAM solution is permitted for IT and information security team members for role-based, logged, and recorded access to critical systems. No direct communication between user network segments is permitted by the firewall policy, except between the departments and the **service desk** for possible remote assistance sessions using predefined services.

For example, if a PC in the marketing department is compromised, the attacker would attempt to gain access to computers whose user accounts have higher privileges in the domain or more valuable information, such as those in IT, information security, or HR. Since this traffic is not permitted and not expected under normal conditions, any such attempt will be blocked. In such instances, an alarm could be triggered to alert your security administrators.

Depending on the size and type of the organization, there may be either a physical or virtual separation of resources for different sensitivity levels – that is, you may have dedicated hardware (or a dedicated virtualization environment) for publicly accessible and internal applications and data, or those may reside in the shared virtualization environment.

In the first instance, the implementation of dedicated firewall appliances or clusters may be warranted for the public-facing resources to maintain a hardware-level isolation. In the latter case, **software-defined networking** (**SDN**) is used to isolate the public-facing resources from those intended for internal access, but both can use the same firewall or cluster.

> **Important Note**
> Although it is possible to rely on a single high-availability cluster for the entirety of network segmentation needs, this may not be the optimal solution depending on the complexity, stability, and ownership of the security policy for the segments.

The more complex the configuration of the firewalls you are administering, the higher the probability of unintended complications. It is common enough for firewalls to handle outbound and inbound internet traffic, peer-to-peer VPNs, remote access VPNs, proxies, internal and external dynamic routing, URL filtering and application control, and threat prevention, but generally, these firewalls require more stringent change control with limited maintenance windows. This often leads to much longer implementations of even moderate changes. It also requires much higher proficiency from the administrators, who must consider the specific requirements of each function and possible configuration conflicts or exceptions.

On the other hand, appliances dedicated to limited functions allow for faster modifications, since the likelihood of encountering exceptions is much smaller. These environments are generally more stable than the highly consolidated implementations. They are less demanding in terms of the administrators' proficiency, as the administrators must only deal with a limited number of factors.

In high-security environments, different groups of administrators may be responsible for a subset of the firewalls in the organization, each protecting a different level of security hierarchy using separate security policies. For instance, each group of administrators may be handling either departmental resources, sensitive assets, entire data centers, user access, or public-facing resources. This approach, while adding complexity, will make the unintended or deliberate circumvention of security protocols more difficult.

Now that we have learned about network segmentation best practices and their contemporary definitions by the US security agencies, let's determine how we can secure the core and the perimeter of our infrastructure.

North-South and East-West

North-South and East-West are terms that are commonly used to identify the traffic that enters or leaves the data center (*North-South*) or the traffic within the data center (*east-west*).

Core and perimeter

One of the unanticipated developments of the global pandemic of 2019-2021 (and the ensuing shortages of microprocessors due to the breakdown of supply chains) was the rapid adoption of desktop virtualization technologies. It was easier to provision and secure these environments than to deal with the flood of **bring your own device** (**BYOD**) devices connecting to the corporate infrastructures.

A surprising side effect (from a security standpoint) was that a huge number of enterprises were able to re-impose the standardization and perimeter protection on users while simultaneously reducing the reliance on and expenses associated with the implementation and administration of **mobile device management** (**MDM**).

Unsurprisingly, a lot of the **virtual desktop infrastructure** (**VDI**) and **remote desktop services** (**RDS**) environments ended up running in data centers. For some of these, user network segmentation is a difficult proposition, as they are typically using a single IP address per server for multiple sessions. Even in cases where the IP virtualization features are used, a single DHCP range may be utilized for all sessions. Only a few enterprises (or agencies) can afford separate VDI environments on a per-department basis.

For our purposes, only the non-VDI and non-RDS segments (or any other non-terminal server segments within the data center) are considered part of the **core**. The rest, including publicly accessible resources and extranets, will be identified as a part of the **perimeter**.

The distinction between **core** and **perimeter** does not mean that either requires a lesser degree of protection, but that using different features and security policies for each may result in a better overall security posture and stability of the infrastructure.

Protecting the core

Core firewalls (or clusters) are used to create the enforcement points between assets residing in different data center segments. Here are some examples:

- Internal application servers that belong to a particular department located in one segment, communicating with the database servers for the same department located in another segment, where both are members of Active Directory, with domain controllers in yet another segment. The following diagram illustrates this:

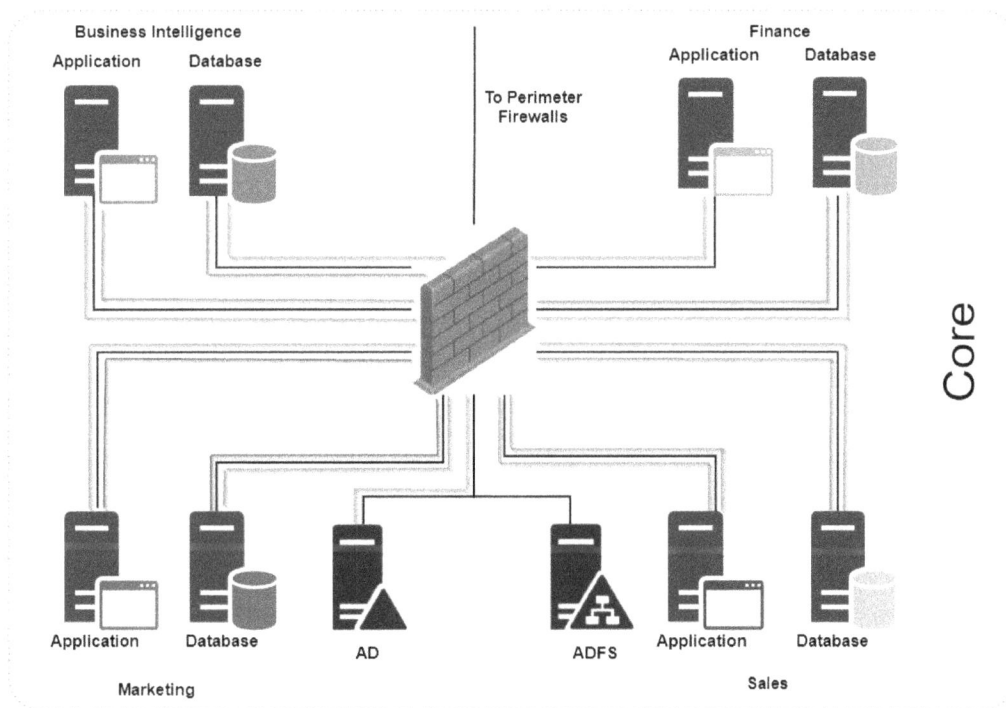

Figure 2.9 – Core networks segmentation

- Different components of a development environment, each located in their own segments, but using the same Active Directory, with domain controllers in another segment.

- Different components of infrastructure management residing in their own segments, such as a **network operations center** (**NOC**) or **security operation center** (**SOC**), forwarding logs to an **SIEM** or **SOAR** server residing in a different segment.

Segments containing VDI instances, RDS instances, and other multi-user terminal servers that may reside in the data centers should be considered part of the **user network** (**perimeter**) segments.

Now that we have seen which components of our infrastructure may constitute the core, let's see what could be defined as the perimeter.

Protecting the perimeter

Perimeter firewalls (or clusters) are used to create the following enforcement points:

- Between internal clients (or hosts) and the core (*Figure 2.8*)

- Between external clients and publicly accessible resources in data centers (as shown in *Figure 2.10*):

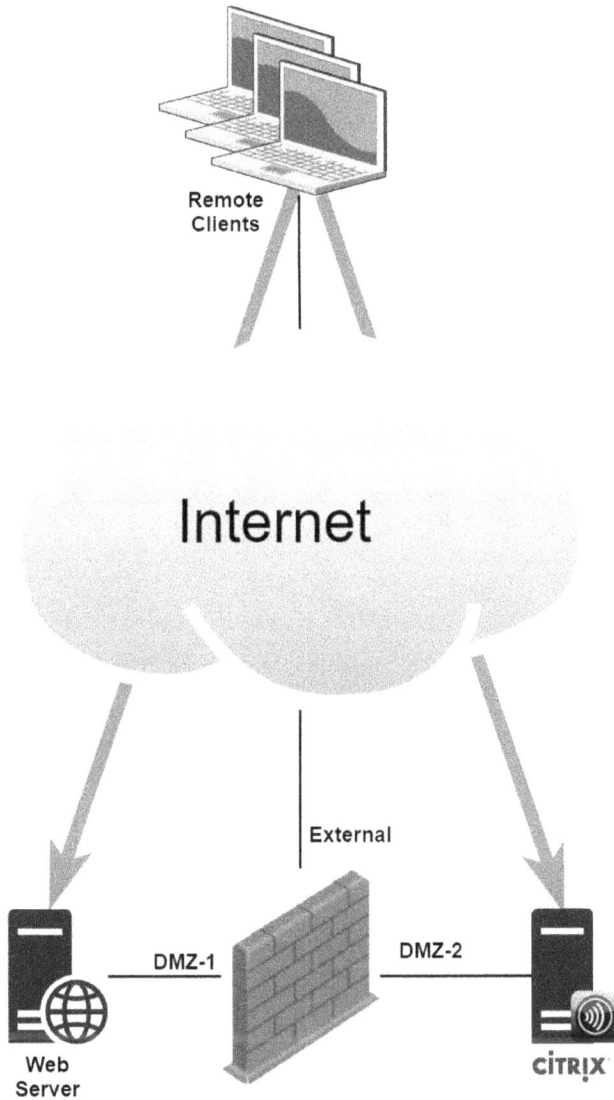

Figure 2.10 – Perimeter networks segmentation

- Between internal clients (or hosts) and resources on the internet and extranets (as shown in *Figure 2.11*):

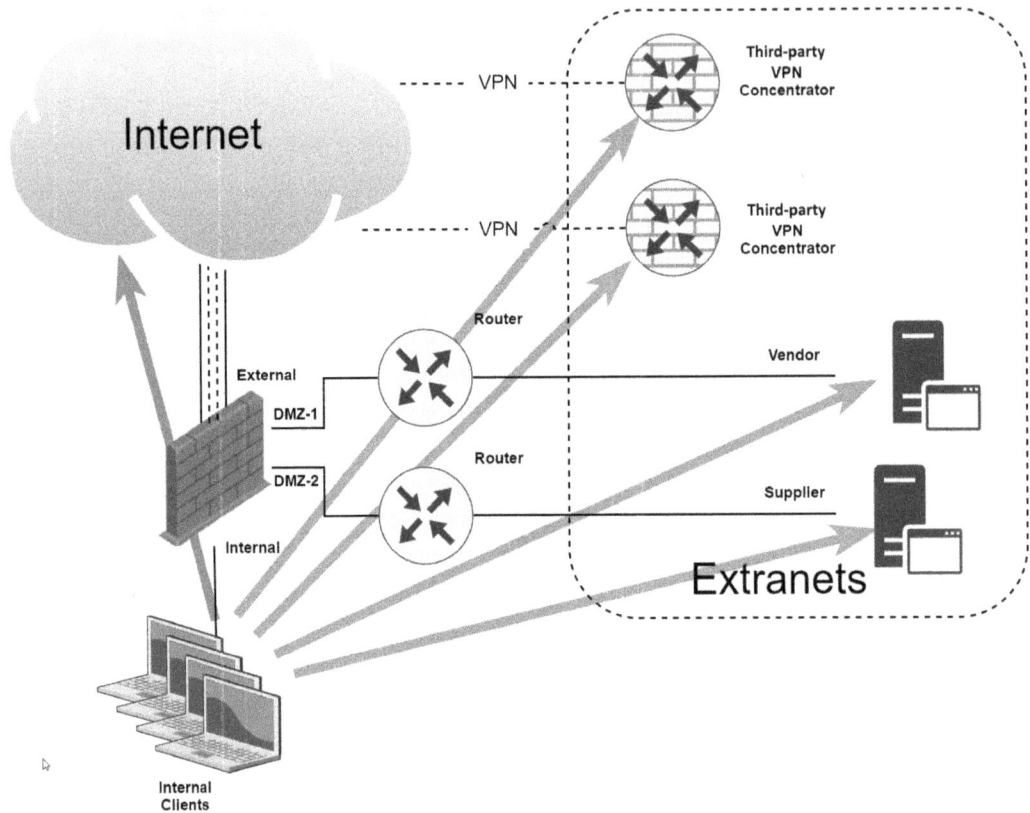

Figure 2.11 – Perimeter networks segmentation

- Between specific segments of data centers and resources on the internet and extranets (as shown in *Figure 2.12*):

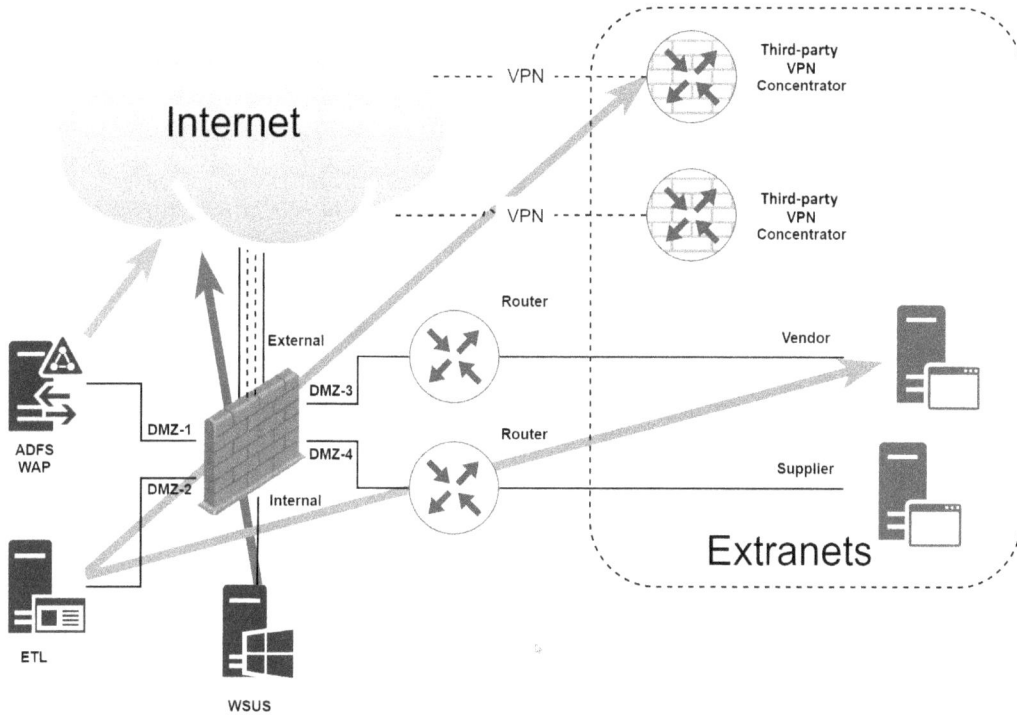

Figure 2.12 – Perimeter networks segmentation

- Between groups of internal clients (or hosts) (as shown in *Figure 2.13*):

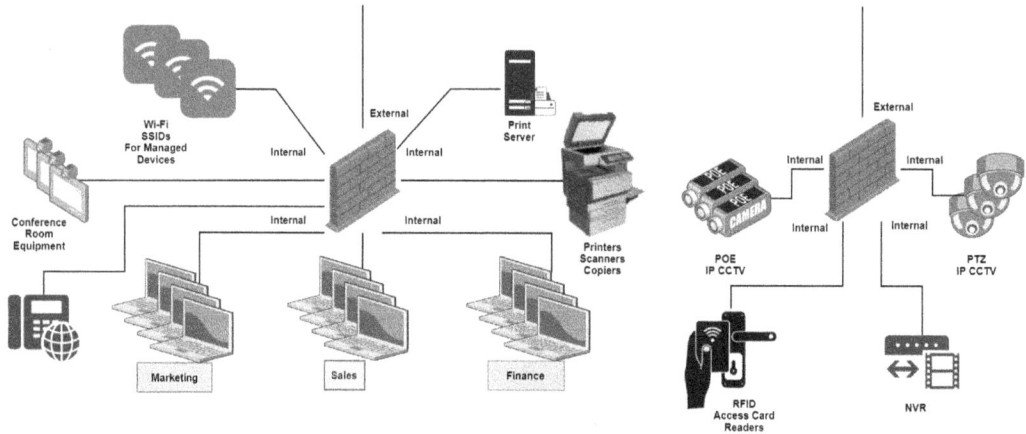

Figure 2.13 – Perimeter networks segmentation

A perimeter is comprised of segments behind firewalls (or clusters) that are one of the following:

- Located outside of a secure physical environment
- Accessible from the outside of your company or agency
- A segment that facilitates connectivity to external resources to perform its functions

Extranets are private networks between your organization and its peers, suppliers, or vendors. Extranets are established either via private (leased) lines or VPNs. In the first instance, each should be terminated on a separate interface of a firewall or cluster to prevent the possibility of a common network segment being used by adversaries that may have compromised one of your peers' networks to exploit possible weaknesses in others.

Hosts could be any IP-enabled device connected to one of your network segments. These include printers, multifunction devices, IP CCTV cameras, physical access control systems, conferencing equipment, environmental controls, wireless access points, and others.

Your overall secure infrastructure may look like the left side of the diagram in *Figure 2.14* and the right side of the diagram in *Figure 2.15* (shown across two figures to accommodate its size):

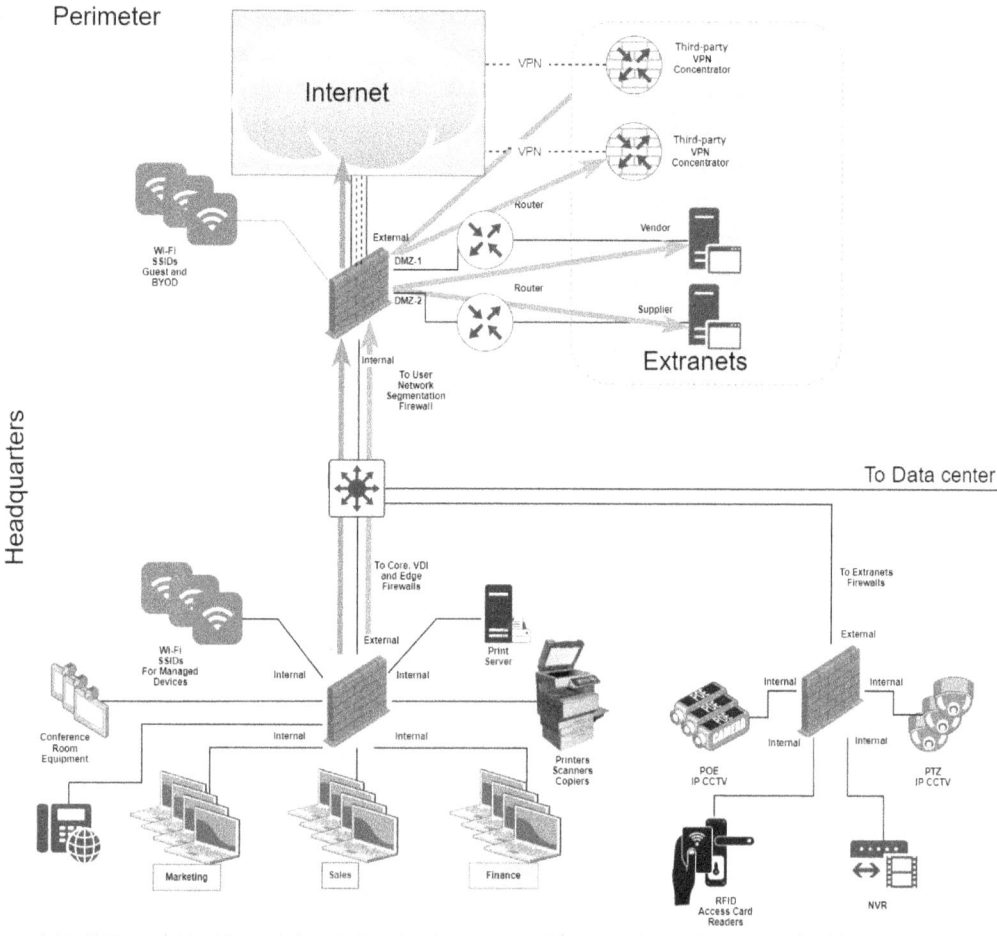

Figure 2.14 – Headquarters networks segmentation

The right side of the diagram (continuation of *Figure 2.14*, shown across two figures to accommodate its size):

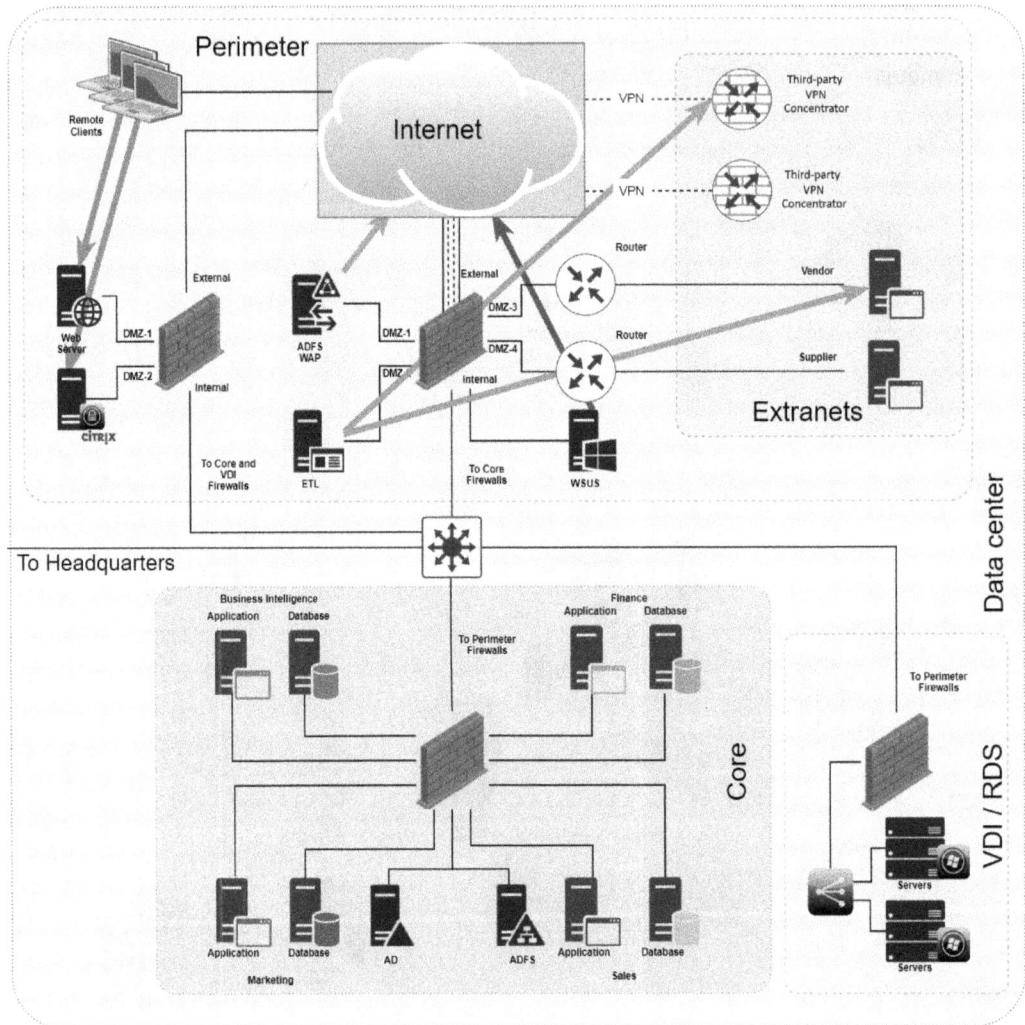

Figure 2.15 – Data center networks segmentation

> **Practical Considerations and Enforcement Consolidation**
>
> For some segments, it is warranted to have an actual dedicated cluster of the firewalls serve as the enforcement point. In other instances, it makes sense to consolidate the enforcement points in a cluster running virtualized instances of the firewalls using a Check Point product called **Check Point VSX**. **Check Point VSX Virtual System Load Sharing** (**VSLS**) offers high availability, load sharing, and scalability. We will learn more about this product in a later chapter.

Now that we have learned about possible perimeter segmentation scenarios, let's take a look at how we can size firewalls or clusters for new implementations, or to determine the utilization of existing Check Point appliances already in our infrastructure.

Sizing appliances for new implementations and determining load on current systems

If you are in a planning phase for the implementation of Check Point firewalls in your infrastructure, you must determine (with some accuracy) the capacity of the appliances that must satisfy your requirements.

A firewall lifespan in a typical infrastructure is between five and seven years and, because of projected growth in data traffic, it is recommended to choose the appliances at around 35% current estimated load utilization.

If you are simply replacing an existing solution from a different vendor, you may have the necessary data available in its performance reports. If they are not locally available, it is possible that this data is collected by the vendor through telemetry, and you can ask their technical support to provide it.

If you are planning to use a Check Point firewall or cluster as a common internet edge gateway, you may be able to obtain historical utilization data from your internet service provider. If there are no existing secure web gateways with HTTPS inspection in use, and no dedicated VPN concentrators or remote access solutions, and if you are planning to use a Check Point firewall or cluster for these roles as well, then the data supplied by the ISP could be used for a rough estimate. For an even rougher estimate, use the current and projected ISP connectivity bandwidths.

In the absence of historical utilization data from an existing solution or, if you are planning a more complex implementation, your best approach would be to look for the network performance monitoring solution logs and graphs over time. It would be helpful to identify, analyze, and document peak bandwidth utilization between sources and destinations that will be moved to separate network segments and segregated by the firewalls. Once you are done, you should be able to quantify the required total throughput of the target appliances. Additional data points that may help are the connection rates, concurrent number of sessions, and total number of users behind each firewall or cluster (if applicable).

Some of the commonly used commercial tools for these purposes are **PRTG**, **SolarWinds Network Performance Monitor** (**NPM**), and **Datadog Network Monitoring**. If you do not have those or similar solutions implemented in your infrastructure, all of them are available as fully functional trials.

If you are trying to determine the utilization of currently implemented appliances for either troubleshooting, scheduling firewall hardware replacement for appliances approaching their **End of Support** (**EoS**) date, planning for growth, or enabling additional features, you can use a Check Point utility called **CPSizeMe**.

CPSizeMe collects performance data over a specified period of time (running it for more than 24 hours is not recommended). It generates a report that can be viewed locally on the appliance or forwarded as a PDF. Details and/or a summary of the results can be sent to Check Point for analysis and to receive possible upgrade suggestions. When you initiate CPSizeMe, you are prompted to answer several questions that help Check Point better interpret the data generated by the utility. These are shown in *Figure 2.16* at [1].

I suggest not skipping over this option, as it may be helpful to you as well. For instance, your estimated gateway throughput [2] may differ from the actual measured data [3], and it is helpful to see both numbers side by side.

That, together with CPU utilization [4], concurrent connections [5], and accelerated packets [6], should help you to establish a baseline for your estimates. I suggest retaining this data for your records.

```
General information
====================
* Email address: cpadmin@mycompany.com
* Name of company / organization:
* Script version: 5.2
* Date & time: 2021-07-11 16:46:24
* Scheduled end: 2021-07-12 16:46:24
* Utility Sampling duration: 1 days
* Appliance: VMware Virtual Platform [1959 MB]
* Active blades: FW MGMT VPN MAB A_URLF AV ASPM APP_CTL IPS DLP IA SSL_INSPECT ANTB MON TE
* Gateway version: Check Point Gaia R80.40
* Gateway name: CPGW
* SecureXL: on
* Clustering:

HA module not started.

* ClusterXL: no


Customer estimation
====================
* Main functions performed by this gateway:
        * Perimeter security: y
        * DMZ security: n
        * Protect the datacenter: y
        * Segment internal networks: y
        * Protect web servers: n
* Estimated number of users: 40
* Estimated gateway throughput [Mbps]: 200
* Size of internet pipe [Mbps]: 130
* Satisfied with gateway performance: y
* Estimated number of remote users: 20
* Estimated number of IPSec VPN remote users: 25
* Additional customer feedback: n


Measured Data
==============
* Maximum gateway throughput: 26.073215 Mbps
* Maximum packet rate: 4029 Packets/sec
* Maximum Total CPU: 76%
        * CPU core 0: 70% (Max core utilization: 100%)
        * CPU core 1: 80% (Max core utilization: 100%)
        * CPU core 2: 47% (Max core utilization: 100%)
        * CPU core 3: 69% (Max core utilization: 91%)
* Maximum kernel CPU: 37%
        * kernel CPU core 0: 29% (Max core kernel Utilization: 34%)
        * kernel CPU core 1: 21% (Max core kernel Utilization: 24%)
        * kernel CPU core 2: 20% (Max core kernel Utilization: 19%)
        * kernel CPU core 3: 78% (Max core kernel Utilization: 91%)
* Estimated number of unique IPs behind gateway: 0
* Maximum concurrent connections: 771
* Average concurrent connections: 220
* Maximum memory utilization: 1314965 KB
* Minimum Free Memory: 1.91208 MB
* Accelerated packets: 0.00%
* VPN traffic: 0.00%
* Detected interface packet drops: no
* Detected install policy: no
* SMT status: Unsupported
* Estimated average of NAT connections: 0% (average concurrent connections:56)
======================================
```

Figure 2.16 – Check Point CPSizeMe, current utilization

Since it is not recommended to run the CPSizeMe utility for longer than 24 hours, we can use another option to determine a current gateway's performance over time.

For older systems (for example, R77.30 or earlier), use the **Check Point SmartView Monitor** utility [1] that is accessible from the **SmartConsole**:

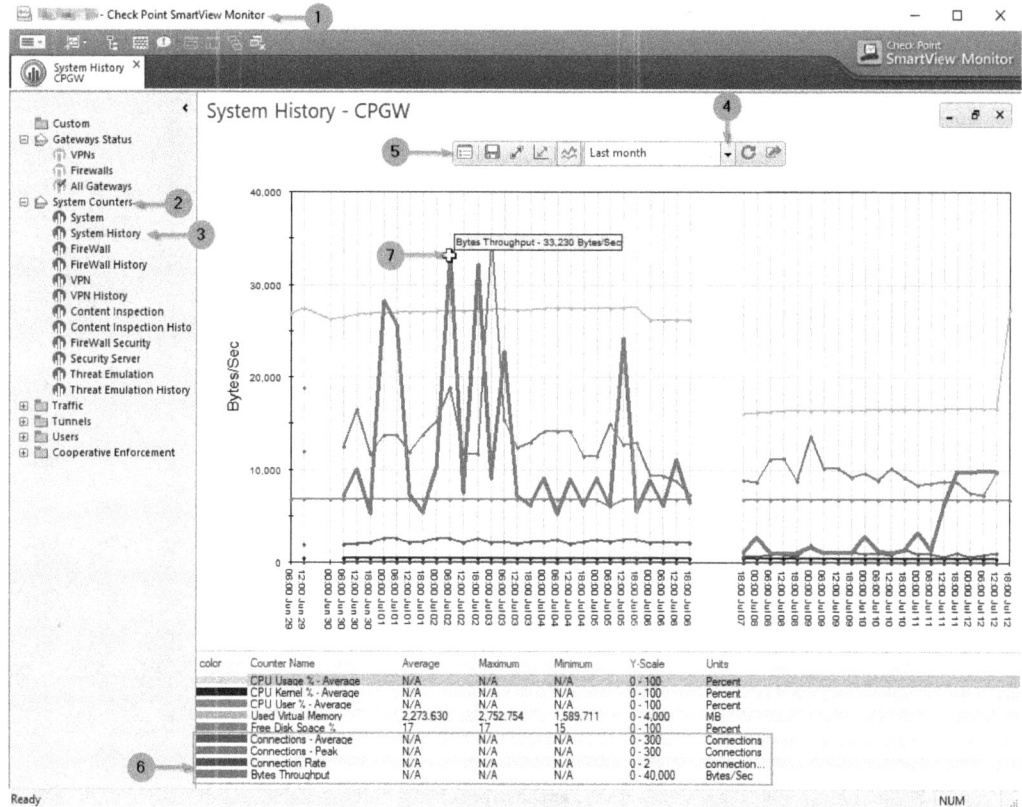

Figure 2.17 – Check Point SmartView, current and historical utilization

Click on **System Counters** [2], and then **System History** [3]. Next, change the duration from the default (Last hour) to Last month [4]. Open the query properties/ counters tab [5] and add these metrics: **Connections – Average**, **Connections – Peak**, **Connections – Rate**, and **Bytes Throughput** [6].

Click **Ok**, and you should see the graph of the historical utilization.

Hover your cursor over the graphs to see the value [7].

For newer systems, we can use a Check Point utility called **Device & License Information**, found in SmartConsole as follows:

1. Open SmartConsole and click on **GATEWAYS & SERVERS** [1]. Click on a line containing a gateway of interest to highlight it [2] and, either click on **Device & License Information…** [3] or on **Monitor** [4]. Either one of options [3] or [4] will open the same utility:

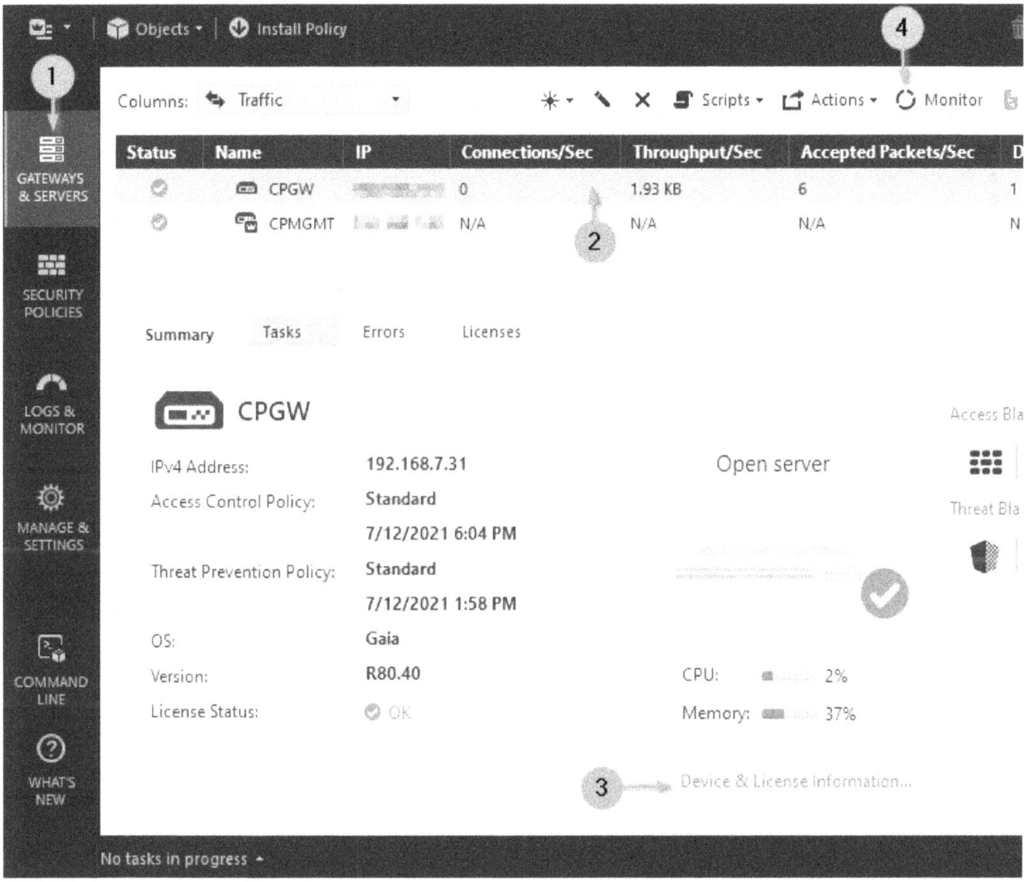

Figure 2.18 – The Device & License Information utility, current and historical utilization

2. The **Device & License Information** utility can be seen in *Figure 2.19*. Note that for these counters to be present, the **Monitoring** blade (feature) should be enabled on the gateways:

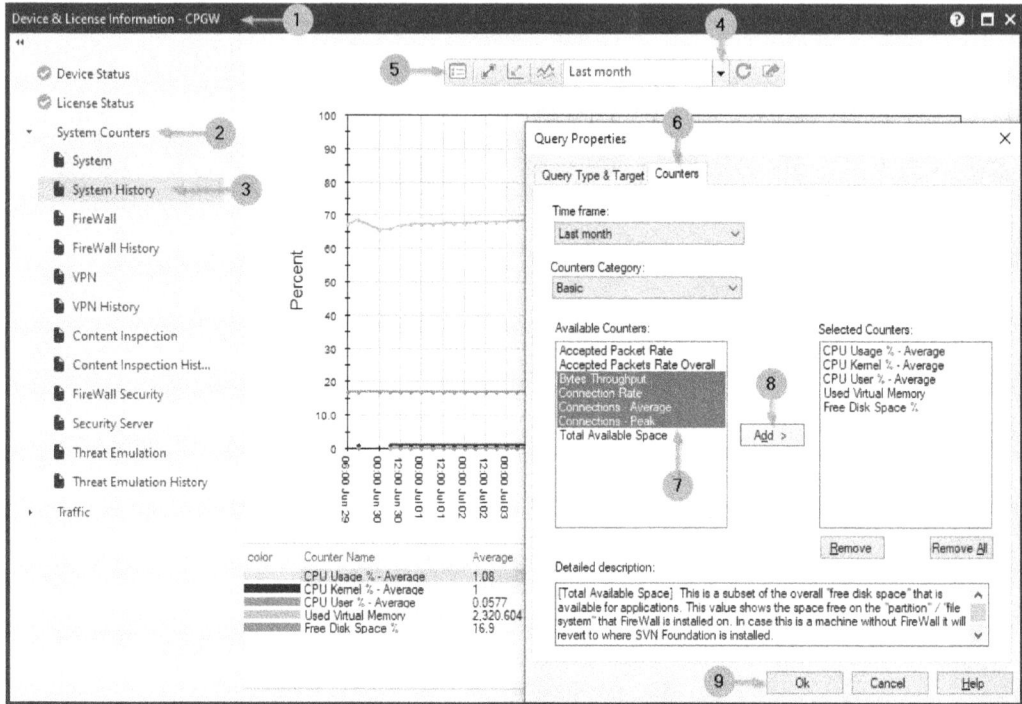

Figure 2.19 – The Device & License Information utility, adding counters

3. Click on **System Counters** [2] and, when it expands, click on **System History** [3]. Change the interval from Last hour to Last month [4] and click the query properties button [5]. When it opens, click on the **Counters** tab [6], select the metrics of interest [7], click **Add >** [8], and click **Ok** [9].

4. In the open graph shown in *Figure 2.20*, the average and maximum values are found in section [1] of the screenshot. Clicking on it will highlight the graph for the relevant counter. Clicking on the graph in the area indicated by section [2] will show the value at the time [3]:

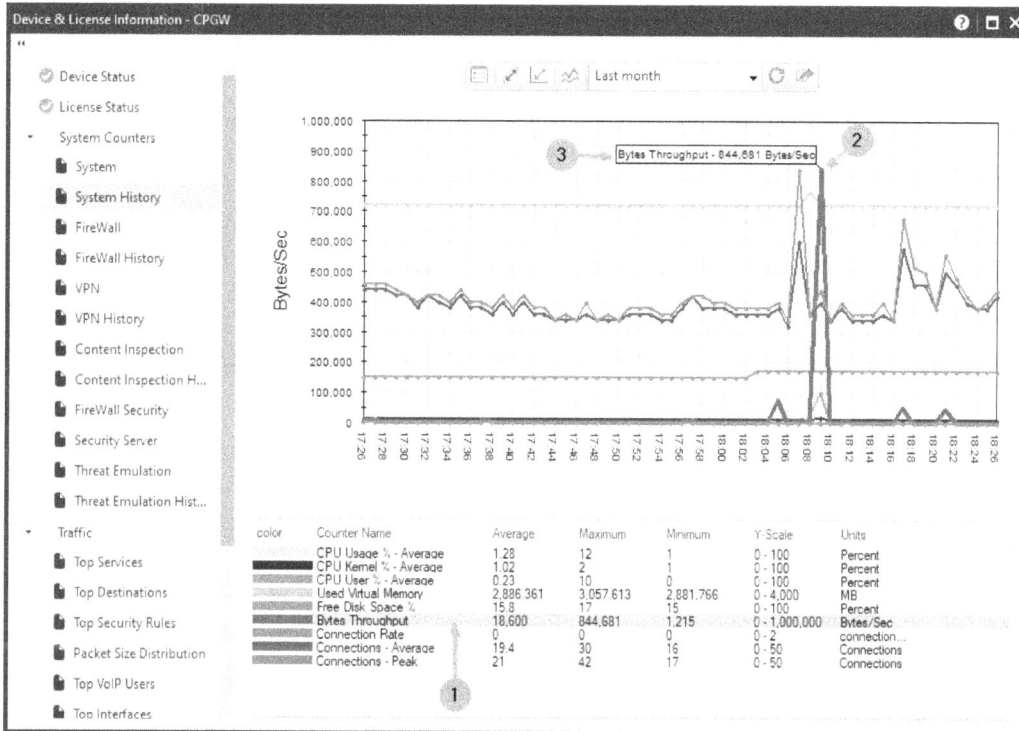

Figure 2.20 – The Device & License Information utility, graphs and values

If you would like to further investigate the utilization of the firewall(s) at a particular date and time (perhaps to take a closer look at the utilization during one of the peaks in the graphs), you can do that using a command-line utility called **cpview**, which retains the last 30 days of performance data in one-minute intervals.

We will review `cpview` later in the troubleshooting section of the book, but *Figure 2.21* demonstrates what the utility's interface looks like:

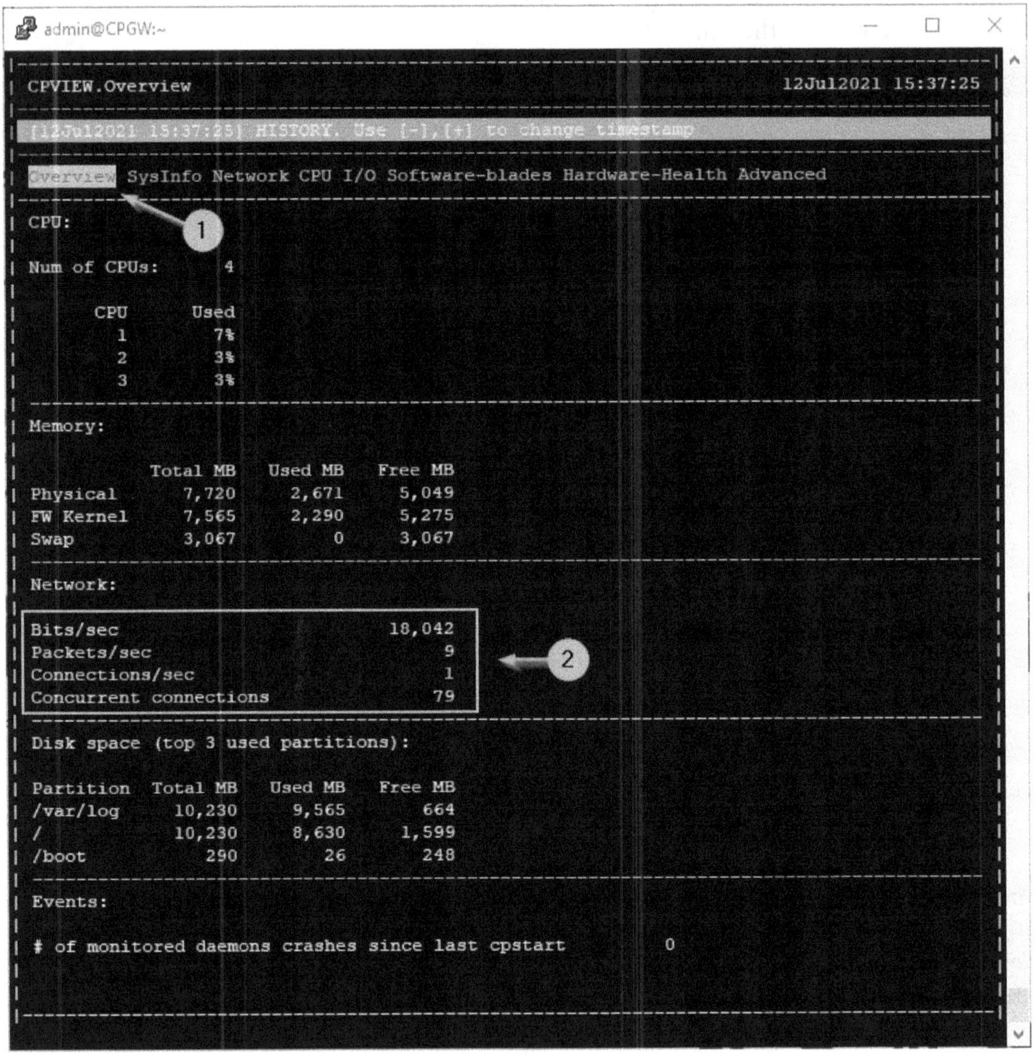

Figure 2.21 – The cpview utility, current and historical utilization

The `Overview` section of the cpview utility [1] contains the data we are interested in [2].

The utility allows us to jump to a particular date and time to start our investigation. From there, we can press and release (or press and hold) the + or - keys to observe the changes in metrics over time.

Once the performance metrics of interest are collected, we must look at the location of the new firewalls in our infrastructure's topology. Depending on the kind of resources they are protecting, we may use a subset of available functions on each firewall or cluster.

Check Point offers three comprehensive software bundles: **NGFW**, **NGTP**, and **NGTP + SandBlast** (also referred to as NGTX). As seen in *Figure 2.22*, each contains a different number of features (also called **Software Blades**. The NGFW and NGTP bundle abbreviations are for Next Generation Firewall and Next Generation Threat Prevention. SandBlast is an aggregate name for proprietary threat emulation and threat extraction technologies):

Security Gateway Feature Sets	NGFW	NGTP	NGTP + SandBlast
Firewall			
Identity Awareness			
IPsec VPN			
Advanced Networking & Clustering			
Mobile Access			
IPS			
Application Control			
Content Awareness			
URL Filtering			
Antivirus			
Anti-Spam			
Anti-Bot			
SandBlast Threat Emulation			
SandBlast Threat Extraction			
DLP			
Security Management Feature Sets			
Network Policy Management			
Logging & Status			

Figure 2.22 – Check Point software bundles and features

Depending on the bundle you have purchased and its current support contract, you will have access to all of its features and may enable some or all of them as needed.

Check Point provides approximate performance data for each hardware appliance model based on the *average enterprise mix* of traffic, and this performance data can be found at the following link: `https://www.checkpoint.com/downloads/products/check-point-appliance-comparison-chart.pdf`.

This performance data is provided for the NGFW and NGTP + SandBlast bundles. For the appliances protecting the core of the infrastructure, NGFW may be sufficient, but for gateways connected to the internet and/or extranets, NGTP or NGTP + Sandblast are a much better choice.

> **Important Note**
>
> For each appliance, the number of virtual systems it is capable of running is provided (should you elect to use them) as VSX hosts or cluster members. VSX requires separate licenses for a fixed number of systems (3, 10, 25, and 50). These licenses are cumulative. Each VSX cluster member must be licensed.

Additional factors that should be taken into consideration are the connectivity requirements and the anticipated growth of utilization (beyond the 60%-65% recommended utilization overhead). For instance, having a requirement to connect the firewall or a cluster to 12 10 GbE links will narrow down the number of available options.

The default and maximum number of *Network I/O Expansion Slots* are shown in the same appliance chart: `https://www.checkpoint.com/downloads/products/check-point-appliance-comparison-chart.pdf`.

Looking at the *Appliance Configurations* tables in that document, you will notice that each is listed with three categories: *Base*, *Plus*, and *Max*. Base is the bare minimum configuration available for orders. The Plus category includes additional I/O and redundant components. Max is equipped to a maximum supported capacity.

If utilization of the infrastructure in which the new appliances are being deployed is expected to grow rapidly and, for whatever reason, the virtualization option is not acceptable, you should consider the **Check Point Maestro** hyperscale security solution. We will touch on this solution later in the book, but for sizing purposes, see the *Maestro Scalability (NGTP Gbps)* table row numbers in the same document.

Summary

In this chapter, we have learned about network topologies and their impact on our ability to implement access control and threat prevention in the infrastructure. We have examined several examples of typical topologies and have learned about the importance of network segmentation. We have also learned about firewalls in core and perimeter segments. Additionally, we learned how to approach the sizing of new appliances for our infrastructure and how to determine the load on the appliances that we may have to manage.

Now that we understand the goals of network segmentation and its impact on the design and security of different network topologies and their segments, in the next chapter, we are going to start building our first Check Point lab environment and learn about the **Gaia** operating system.

Further reading

1. *PCI-DSS Out of Scope definition*: `https://www.pcisecuritystandards.org/documents/Guidance-PCI-DSS-Scoping-and-Segmentation_v1.pdf`

2. HIPAA, *Consider Network Segmentation*: `https://www.himss.org/sites/hde/files/SecurityMetrics%20Guide%20to%20HIPAA%20Compliance_2017.pdf`

3. CISA, *Security Tip (ST18-001)*: `https://us-cert.cisa.gov/ncas/tips/ST18-001`

4. NSA, *Segment Networks and Deploy Application-Aware Defenses*: `https://media.defense.gov/2019/Sep/09/2002180325/-1/-1/0/Segment%20Networks%20and%20Deploy%20Application%20Aware%20Defenses%20-%20Copy.pdf`

3

Building a Check Point Lab Environment – Part 1

In this chapter, we will create our **Check Point** virtual lab environment. If you are studying for your **Check Point Certified Security Administrator (CCSA)** certification using Check Point's **Authorized Training Center**, all the non-Check Point components in the labs are provisioned for you in advance. In our case, we'll have to build the lab from scratch. This should not take long if you follow the provided instructions, and it should be even easier if you follow the accompanying video lessons. The upside of building the lab is that you're likely to gain a deeper understanding of the interaction between the different infrastructure components.

In this chapter, we're going to cover the following main topics:

- Lab topology and components
- Downloading the prerequisites
- Installing Oracle VirtualBox
- Deploying the VyOS router

Technical requirements

For this chapter, you will need a PC with at least 4 CPU cores, 24 to 32 GB of RAM, 200 GB of HDD space (preferably SSD), and a monitor with at least a 1,024 x 768 resolution (1,920 x 1,080 is preferable) running **Windows 10**. References to the software used in the lab are provided in later sections. You will also need **Google Chrome** or **Microsoft Edge** as a browser and an SSH terminal emulator.

Oracle VirtualBox has been chosen because it is free for personal use, can run on either Windows, **Linux**, or **macOS**, and is a capable platform. My personal production labs are built on **VMware** products, but those could either be expensive or require dedicated hardware. The *LabHost* could be running either Windows or any OS supported by Oracle VirtualBox as a host operating system. A full list of compatible host operating systems can be found here:

```
https://docs.oracle.com/en/virtualization/virtualbox/6.0/user/
hostossupport.html
```

> **Important Note**
>
> If you are experienced with (and have access to) different virtualization platforms, such as **VMware Workstation**, **Fusion**, **ESXi**, **Microsoft Hyper-V**, **Linux KVM**, or EVE-NG, you can adapt the lab to any of these with minor modifications and platform-specific scripting.

This chapter (and the following one) takes less than two hours to go through all the steps to have the lab implemented. Screenshots are provided for those who either do not have the necessary hardware to run the lab or are reading the chapter on the go. If you prefer scripted implementation, instructions and jump points are indicated throughout the chapter.

Lab topology and components

Let's start by looking at the topology of the lab, identifying its components, and discussing the sequence in which they are going to be deployed.

Lab topology

Please note that to keep the topology diagram more readable, the IP addresses of the cluster and its members will be depicted as follows:

10.0.0.1(2, 3)/24 =   **10.0.0.1/24 Virtual IP of the Cluster's interface**
10.0.0.2/24 IP of the Cluster Member 1 interface
10.0.0.3/24 IP of the Cluster Member 2 interface

Figure 3.1 – Cluster IP address notation abbreviation

The following diagram (*Figure 3.2*) shows the lab topology, its constituent components, and their IP addresses:

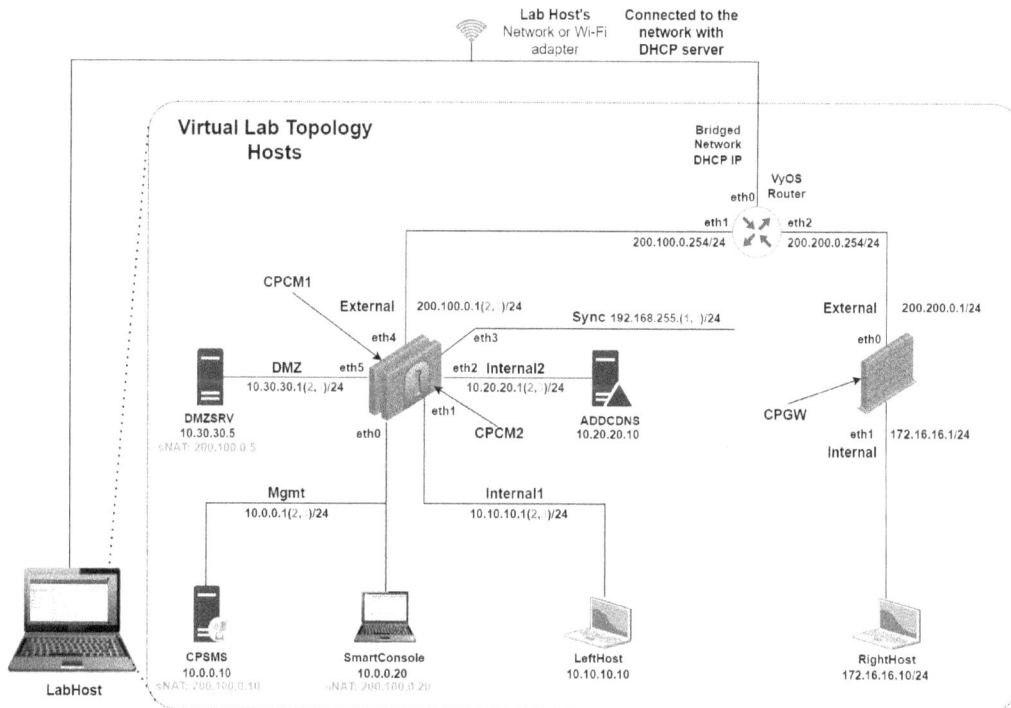

Figure 3.2 – Lab topology, components, and IP addresses

LabHost is the PC on which you are about to create your lab. It does not have to be renamed, as it is used for reference only.

Since we are creating virtual networking, we will be relying on consistently named networks. The network names are shown in the following diagram:

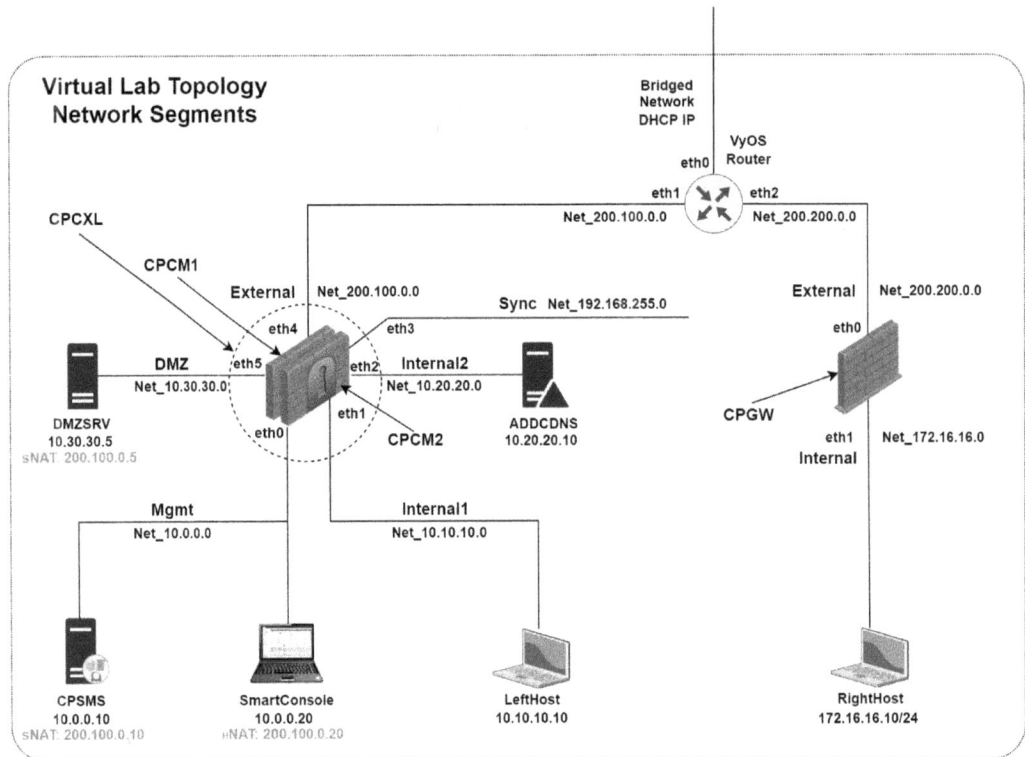

Figure 3.3 – Topology and virtual network segments

Lab components

For easy reference, *Figure 3.4* shows a table of the lab components, the IP addresses of their interfaces, and the network segments they are connected to:

VyOS (Router)			
Interface	**IP**	**Description**	**Network**
eth0	DHCP	Outside	Bridged Adapter
eth1	200.100.0.254	Left	Net_200.100.0.0
eth2	200.200.0.254	Right	Net_200.200.0.0

Cluster (Virtual IPs, CPCM1 and CPCM2)					
Interface	**Virtual IP**	**CPCM1**	**CPCM2**	**Description**	**Network**
eth0	10.0.0.1	10.0.0.2	10.0.0.3	Mgmt	Net_10.0.0.0
eth1	10.10.10.1	10.10.10.2	10.10.10.3	Internal1	Net_10.10.10.0
eth2	10.20.20.1	10.20.20.2	10.20.20.3	Internal2	Net_10.20.20.0
eth3		192.168.255.1	192.168.255.2	Sync	Net_192.168.255.0
eth4	200.100.0.1	200.100.0.2	200.100.0.3	External	Net_200.100.0.0
eth5	10.30.30.1	10.30.30.2	10.30.30.3	DMZ	Net_10.30.30.0

Single Gateway (CPGW)			
Interface	**IP**	**Description**	**Network**
eth0	200.200.0.1	External	Net_200.200.0.0
eth1	172.16.16.1	Internal	Net_172.16.16.0

Security Management Server (CPSMS)		
Interface	**IP**	**Network**
Ethernet	10.0.0.10	Net_10.0.0.0

LeftHost		
Interface	**IP**	**Network**
Ethernet	10.10.10.10	Net_10.10.10.0

ADDCDNS		
Interface	**IP**	**Network**
Ethernet	10.20.20.10	Net_10.20.20.0

DMZSRV		
Interface	**IP**	**Network**
Ethernet	10.30.30.10	Net_10.30.30.0

RightHost		
Interface	**IP**	**Network**
Ethernet	172.16.16.100	Net_172.16.16.0

Figure 3.4 – Hosts, interfaces, IP addresses, and networks

Please note that all subnet masks in the lab are /24 (or 255.255.255.0).

Now that we have seen the topology of the lab, we can move on to obtaining the software.

Downloading the prerequisites

In the `C:\` drive of your Windows host, create a folder, `CPBook`. In it, create a nested folder, `LabShare`. Open `LabShare`, and create the following folders:

- `ISOs_and_OVAs`: Where we will store precompiled appliances and ISO files

- `Scripts`: Where we will download or create the batch files (or scripts) for the VirtualBox configuration and cloning of the **virtual machines** (**VMs**), the configuration parameters for the VyOS router, and the **PowerShell** commands for the Windows VMs

- `Software`: Where the installation files for **SmartConsole**, **PuTTY**, **WinSCP**, **Chrome**, and **Notepad++** will be located

Downloading Oracle VirtualBox and the VirtualBox extension pack

To download Oracle VirtualBox and its extension pack, complete the following steps:

1. Go to `https://www.virtualbox.org/`.
2. Click on the **Download VirtualBox #.#** button in the middle of the screen (where **#.#** represents the currently displayed version).
3. In the **Virtual Box #.#.## platform packages** section, click on distribution for **Windows hosts**. Click on **Save File**.
4. On the same screen, in the **VirtualBox #.#.## Oracle VM VirtualBox Extension Pack** section, click on the **All Supported Platforms** button.
5. Click on **Save File**.

Downloading the Windows Server ISO

To download the Windows Server ISO, follow these steps:

1. Go to `https://www.microsoft.com/en-us/evalcenter/evaluate-windows-server-2019`.
2. Choose **ISO**, and then **Continue**.
3. Fill out the registration form and click **Download**.
4. Choose your language and click **ISO Downloads | 64-bit edition**.

Downloading the Check Point R81.10 ISO and the SmartConsole installation executable

To download the Check Point R81.10 ISO and SmartConsole, complete the following steps:

1. Go to `https://supportcenter.checkpoint.com/supportcenter/portal?eventSubmit_doGoviewsolutiondetails=&solutionid=sk170416`.

2. On the page shown in *Figure 3.5*, scroll down to **Installation** (*Figure 3.5* [1]). Expand each of the three sections ([2], [4], and [6]):

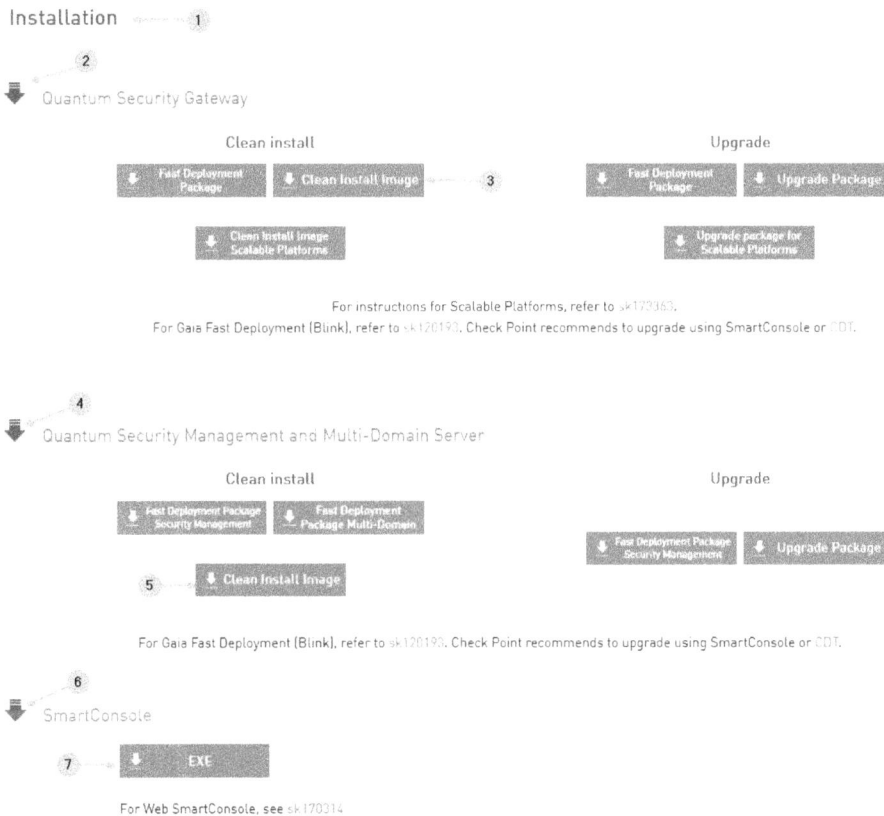

Figure 3.5 – Check Point R81.10 products download landing page

3. Click **Clean Install Image** to download this option. You can use either options [3] or [5] in **Quantum Security Gateway** or **Quantum Security Management and Multi-Domain Server**, respectively, as both result in the download of the same ISO. After clicking the **Clean Install Image** button, a new browser tab will open with a **Download Details** section. Read the **Download Agreement** section, and click the **Download** button.

4. In the **SmartConsole** section, click on **EXE** [7] to download a Windows installation executable for SmartConsole. After clicking the **EXE** button, a new browser tab will open with a **Download Details** section. Read the **Download Agreement** section and click the **Download** button.

Downloading the VyOS virtual router

To download the VyOS virtual appliance, either simply click on `https://s3-us.vyos.io/rolling/current/vyos-rolling-latest.iso` or, if it does not work, follow these steps:

1. Go to `https://vyos.io/`.

2. Click on **Rolling Release**:

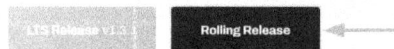

Figure 3.6 – The VyOS virtual appliance download page

3. In the list of the available builds, find and click on **vyos-rolling-latest.iso**.

Downloading the PuTTY terminal emulator

To download PuTTY, complete the following steps:

1. Go to `https://www.chiark.greenend.org.uk/~sgtatham/putty/latest.html`, and under **Package Files** in the **MSI ('Windows Installer')** section, click on the link to the right of the **64-bit x 86** option (*Figure 3.7*):

Download PuTTY: latest release (0.76)

Home | FAQ | Feedback | Licence | Updates | Mirrors | Keys | Links | Team
Download: **Stable** · Snapshot | Docs | Changes | Wishlist

This page contains download links for the latest released version of PuTTY. Currently this is 0.76, released on 2021-07-17.

When new releases come out, this page will update to contain the latest, so this is a good page to bookmark or link to. Alternatively, here is

Release versions of PuTTY are versions we think are reasonably likely to work well. However, they are often not the most up-to-date versi
problem with this release, then it might be worth trying out the development snapshots, to see if the problem has already been fixed in thos

Package files

You probably want one of these. They include versions of all the PuTTY utilities.

(Not sure whether you want the 32-bit or the 64-bit version? Read the FAQ entry.)

MSI ('Windows Installer')

64-bit x86:	putty-64bit-0.76-installer.msi	(or by FTP)	(signature)
64-bit Arm:	putty-arm64-0.76-installer.msi	(or by FTP)	(signature)
32-bit x86:	putty-0.76-installer.msi	(or by FTP)	(signature)

Unix source archive

.tar.gz:	putty-0.76.tar.gz	(or by FTP)	(signature)

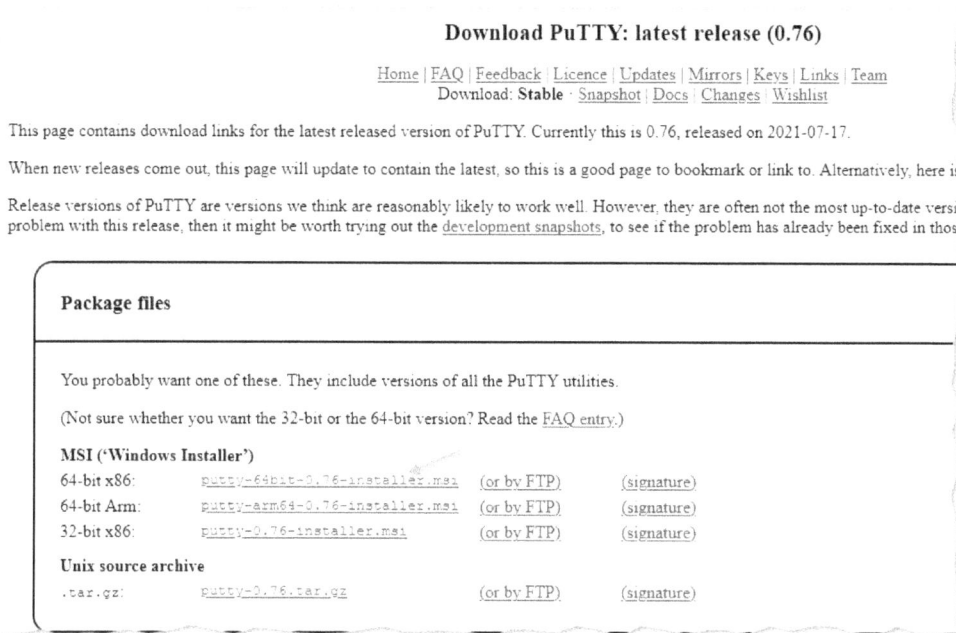

Figure 3.7 – The PuTTY installer download page

Downloading WinSCP

To download WinSCP, complete the following steps:

1. Go to `https://winscp.net/eng/index.php`.
2. Click on **DOWNLOAD NOW**.
3. Click on **DOWNLOAD WINSCP #.##.# (##.## MB)** (the green button on the left).
4. Click on **Save File**.

Downloading the Google Chrome offline installer

To download the Google Chrome offline installer, complete the following steps:

1. Go to `https://www.google.com/intl/en/chrome/?standalone=1`.
2. Uncheck the **Help make Google Chrome better…** checkbox, and click on **Download Chrome**.
3. Click on **Save File**.

Downloading Notepad++

To download Notepad++, complete the following steps:

1. Go to `https://notepad-plus-plus.org/downloads/`.

2. Under the **Downloads** section, click on the top release.

3. On the **Notepad++ #.#.#** release page, in the **Download 64-bit x64** section, click on **Installer**.

4. Click on **Save File**.

Move all ten files you have downloaded for the lab into the corresponding subdirectories in the `C:\CPBook\LabShare\` folder, as shown in *Figure 3.8*:

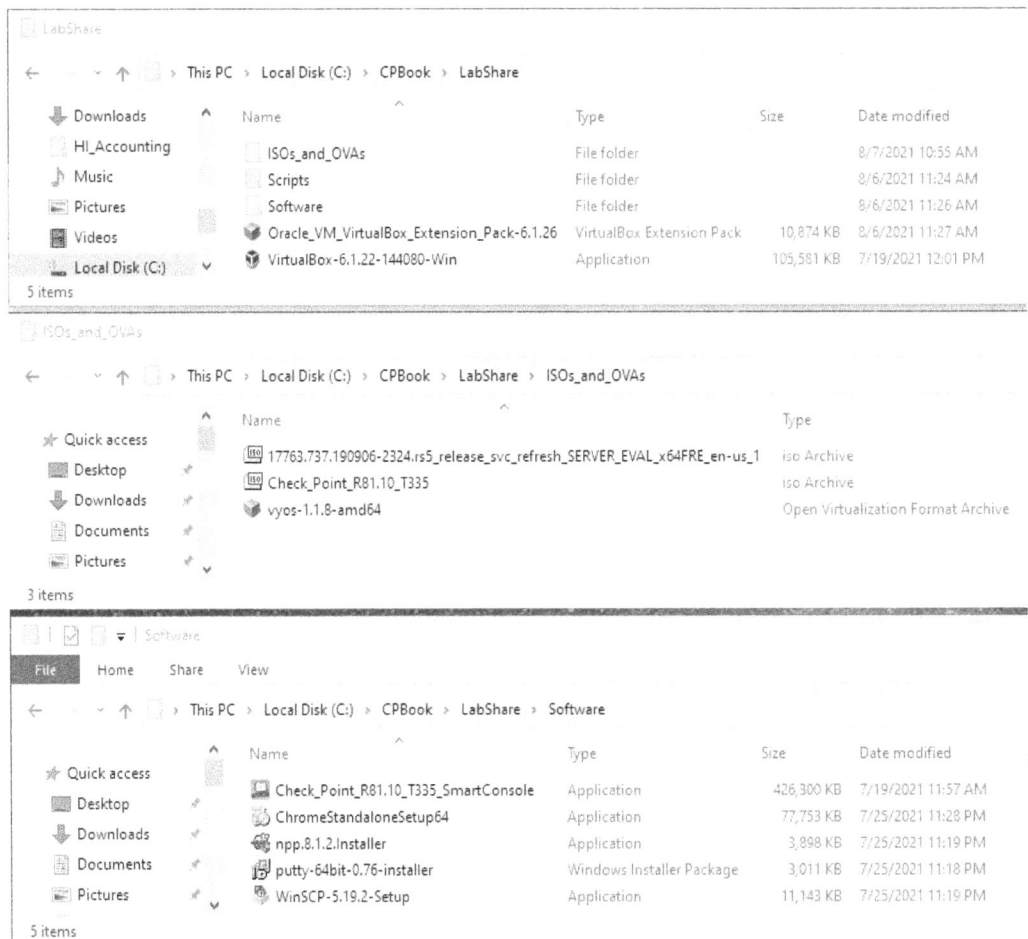

Figure 3.8 – Prerequisite files for the lab

Now that we have downloaded the prerequisite software, we can move on to building our lab environment.

Installing Oracle VirtualBox

In this section, we will install Oracle VirtualBox – a *Type 2 hypervisor* (also known as a *hosted hypervisor,* as it is running on top of the existing operating system). This type of hypervisor is not suitable for production loads but is frequently used for emulation and development.

Follow these steps to install Oracle VirtualBox:

1. Execute the downloaded Oracle VirtualBox installation file (for my Windows host, it is `VirtualBox-6.1.22-144080-Win.exe`).

2. On the **Welcome to the Oracle VirtualBox #.#.## Setup Wizard** screen [1], click **Next** [2]:

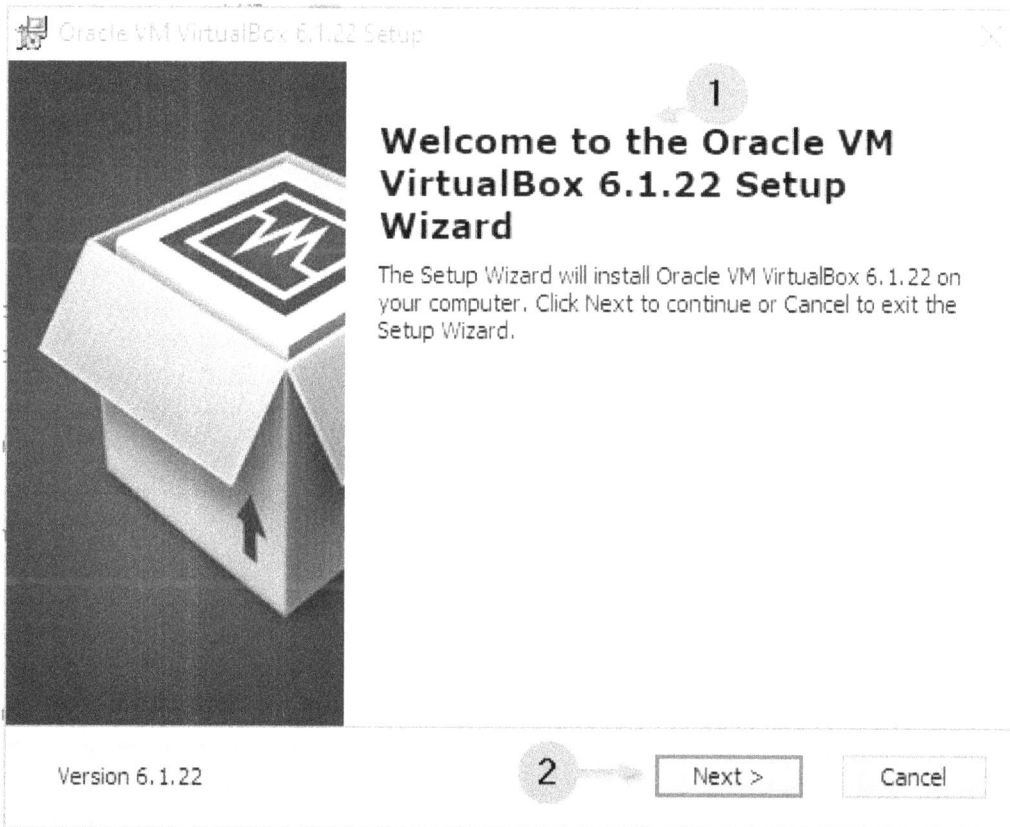

Figure 3.9 – Oracle VirtualBox Setup Wizard

3. In the **Custom Setup** screens shown in *Figure 3.10* and *Figure 3.11* [1], leave the selected options as-is, and click **Next** [2]:

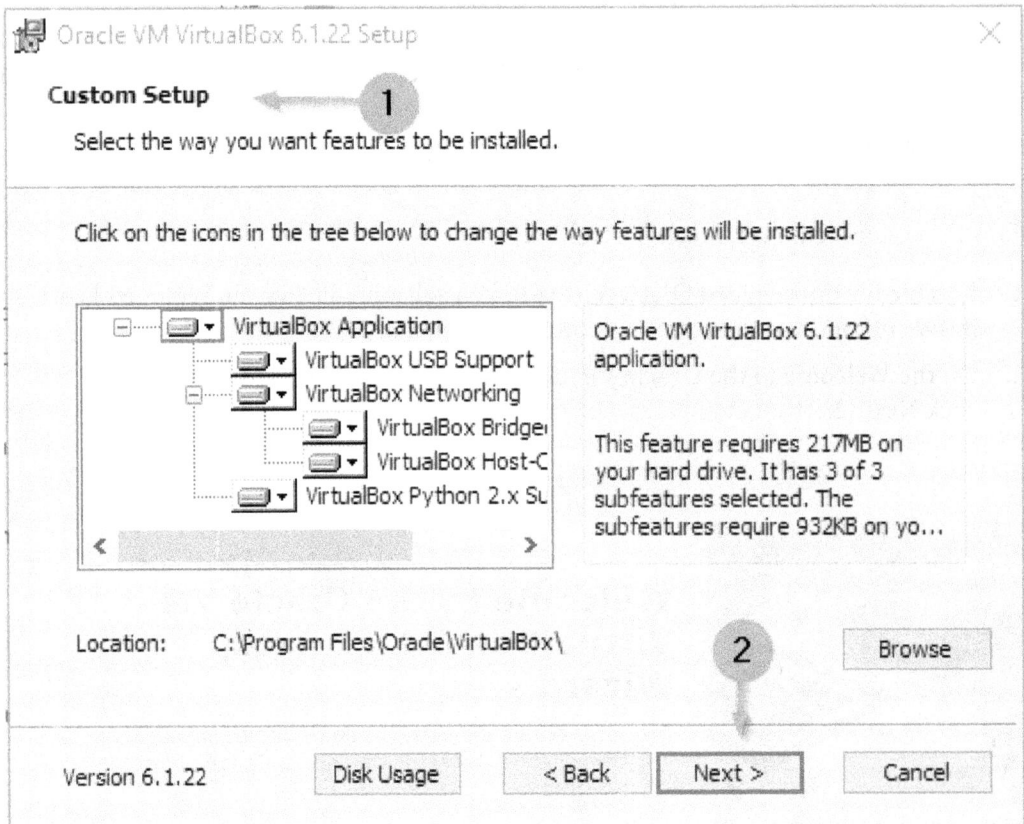

Figure 3.10 – Oracle VirtualBox Setup Wizard, Custom Setup, part 1

Continue this setup operation, as shown on the following screenshot:

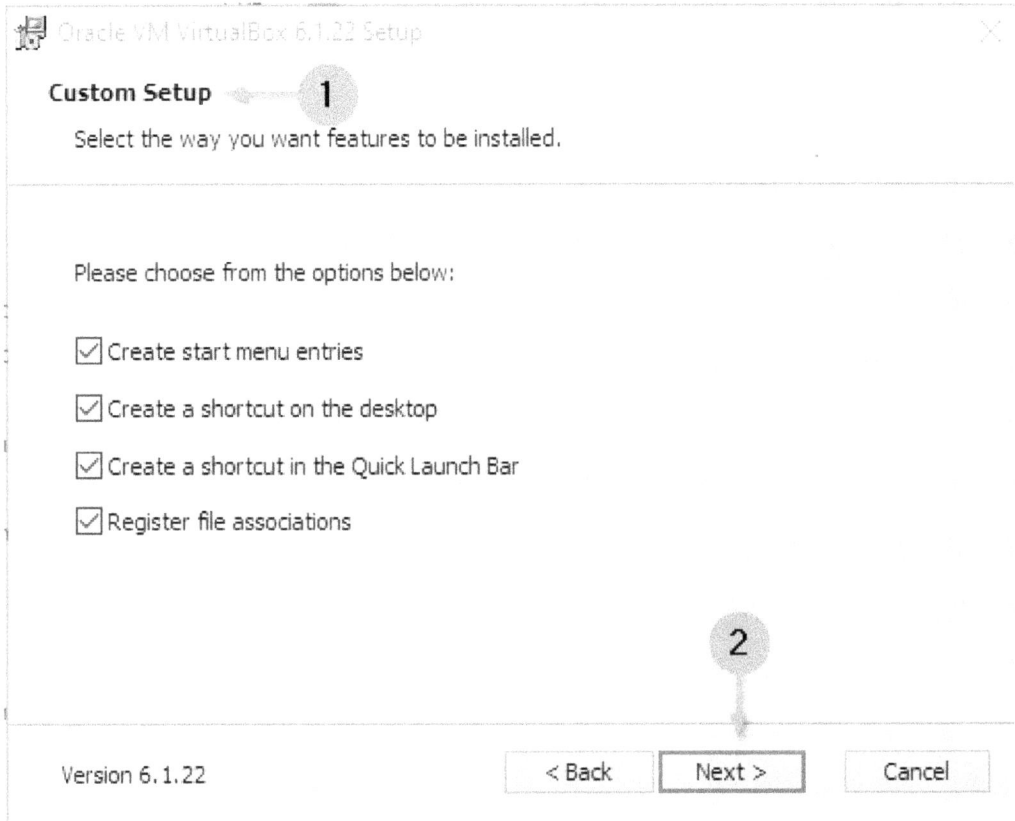

Figure 3.11 – Oracle VirtualBox Setup Wizard, Custom Setup, part 2

4. When prompted with the **Warning: Network Interfaces** message [1], check that you do not have any critical applications or sessions running that, if interrupted, may have a negative impact on unsaved work in progress. If any such sessions or applications are running, save your work and terminate the sessions before continuing:

Figure 3.12 – Oracle VirtualBox Setup Wizard, Warning: Network Interfaces screen

5. When you have saved your work and closed any sessions or applications, click **Yes** on the screen, as shown in *Figure 3.12*.

6. On the **Ready to Install** screen (*Figure 3.13* [1]), click **Install** [2]:

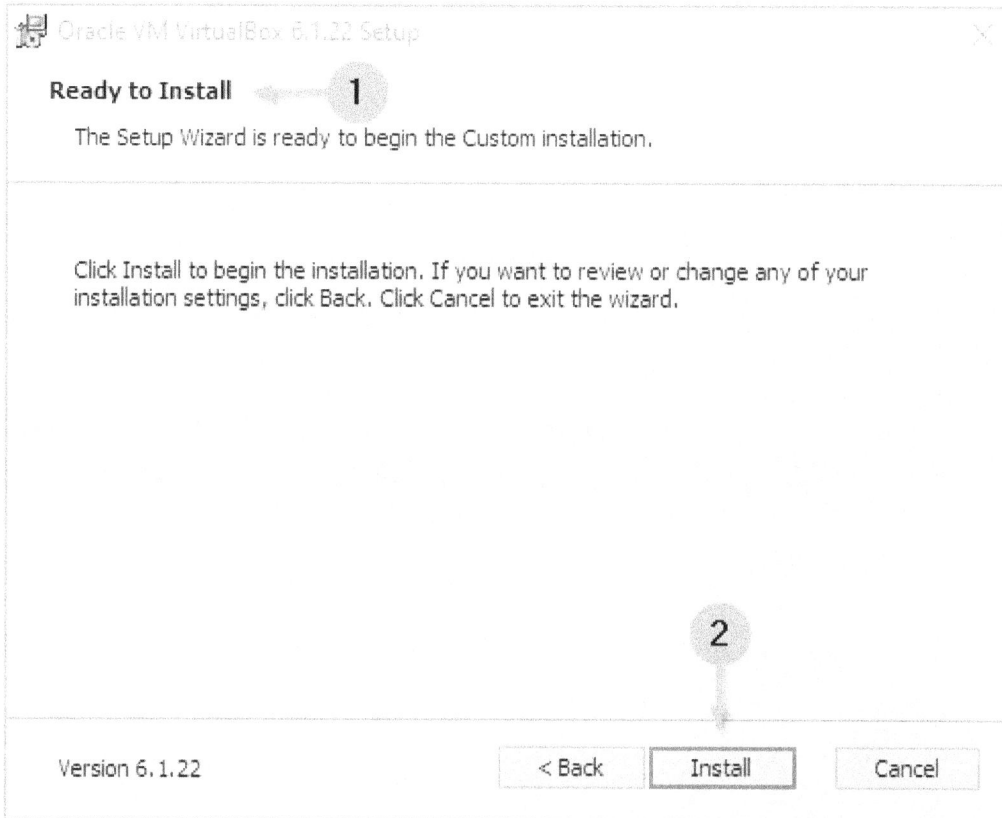

Figure 3.13 – Oracle VirtualBox Setup Wizard, commencing installation

7. When prompted by the **User Account Control** consent prompt to confirm the installation of Oracle VirtualBox, do so.

8. When prompted by the **Would you like to install this device software?** Windows Security popup (*Figure 3.14* [1]), verify that the **Always trust software from "Oracle Corporation"** checkbox is checked [2], and click **Install** [3]:

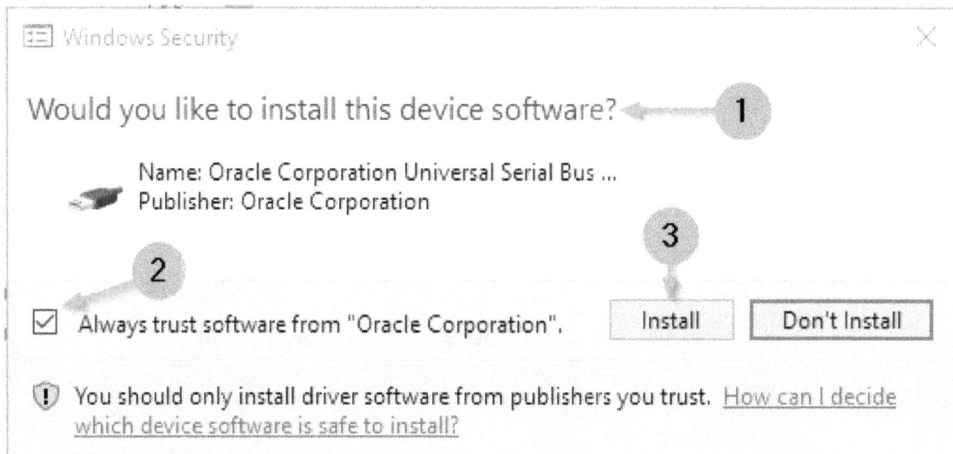

Figure 3.14 – Oracle VirtualBox Setup Wizard, Windows Security prompt

9. When the **Oracle VM VirtualBox #.#.## installation is complete** message is displayed, check the **Start Oracle VM VirtualBox #.#.## after installation** checkbox [2], and then click **Finish** [3]:

Figure 3.15 – Oracle VirtualBox Setup Wizard, finishing installation

6. Select the **Adapter 2** tab [1], and check **Enable Network Adapter**, if unchecked [2]. In the **Attached to** field, choose **Internal Network** [3], and in the **Name** field menu, type **Net_200.100.0.0** [4]. Expand the **Advanced** options [5] and verify that the **Cable Connected** checkbox is checked [6]:

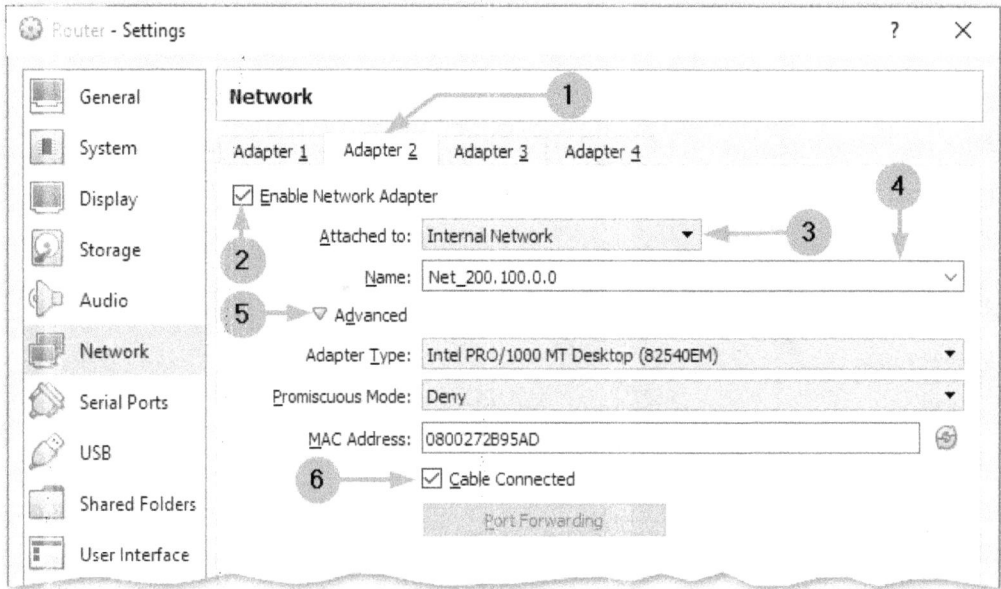

Figure 3.25 – Router VM networking configuration, Adapter 2

7. Select the **Adapter 3** tab [1], and check **Enable Network Adapter**, if unchecked [2]. In the **Attached to** field, choose **Internal Network** [3], and in the **Name** field menu, type **Net_200.200.0.0** [4]. Expand the **Advanced** options [5] and verify that the **Cable Connected** checkbox is checked [6]:

Installing the VirtualBox extension pack

To better interact with some of the VMs and gain the ability to drag and drop files from our LabHost to the VMs via the clipboard and map the LabHost's storage resources to some of the VMs, we need to install the VirtualBox extension pack. This won't only be installed on the LabHost, but will also be installed on the Windows VMs.

To install the VirtualBox extension pack, follow these steps:

1. In your `C:\CPBook\Images_VMs_and_Software\` folder, double-click the `Oracle_VM_VirtualBox_Extension_Pack-6.1.24.vbox-extpack` file (your version may vary, depending on the download date and release availability – you may or may not see the extensions of the files in this folder, depending on the settings of your **Windows Explorer**). When prompted with the choice of opening or saving the file, keep the settings for **Open with** set to **VirtualBox Manager (default)** (*Figure 3.16* [2]), and click **OK** [3]:

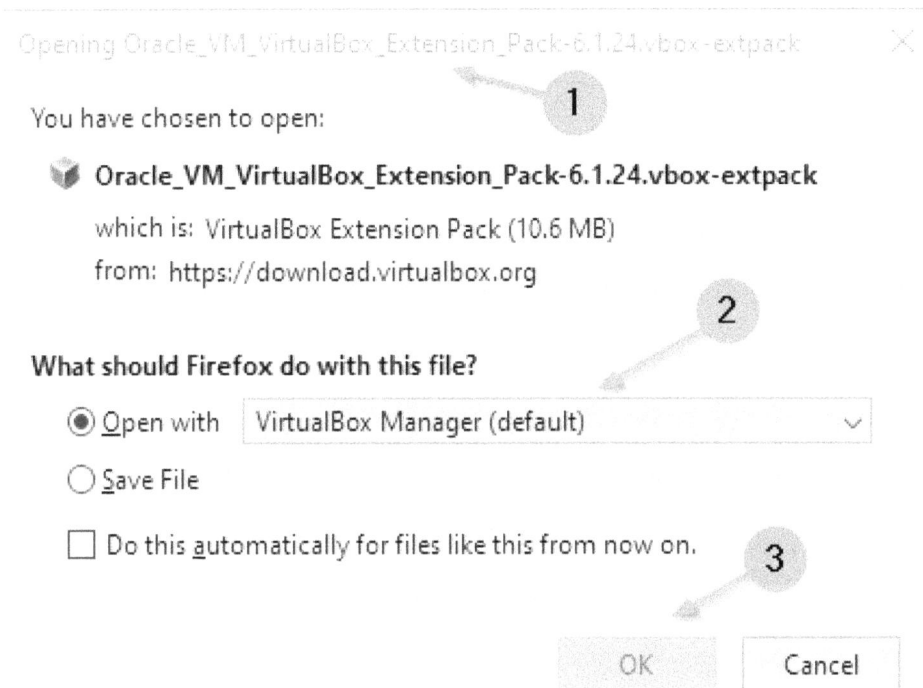

Opening Oracle_VM_VirtualBox_Extension_Pack-6.1.24.vbox-extpack

You have chosen to open:

🗇 **Oracle_VM_VirtualBox_Extension_Pack-6.1.24.vbox-extpack**

which is: VirtualBox Extension Pack (10.6 MB)

from: https://download.virtualbox.org

What should Firefox do with this file?

◉ Open with VirtualBox Manager (default)

○ Save File

☐ Do this automatically for files like this from now on.

OK   Cancel

Figure 3.16 – Oracle VirtualBox extension pack installation, part 1

2. On the **VirtualBox - Question** screen (*Figure 3.17*), click **Install**:

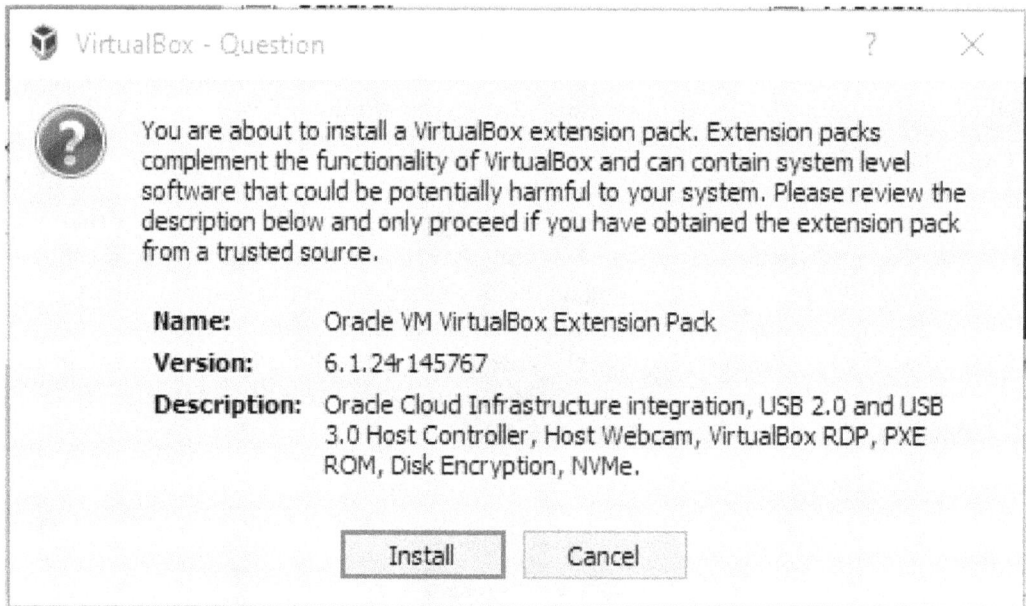

Figure 3.17 – Oracle VirtualBox extension pack installation, part 2

3. Read the terms and conditions in the **VirtualBox License** section by scrolling to the end (*Figure 3.18* [1]), then click **I Agree** [2]:

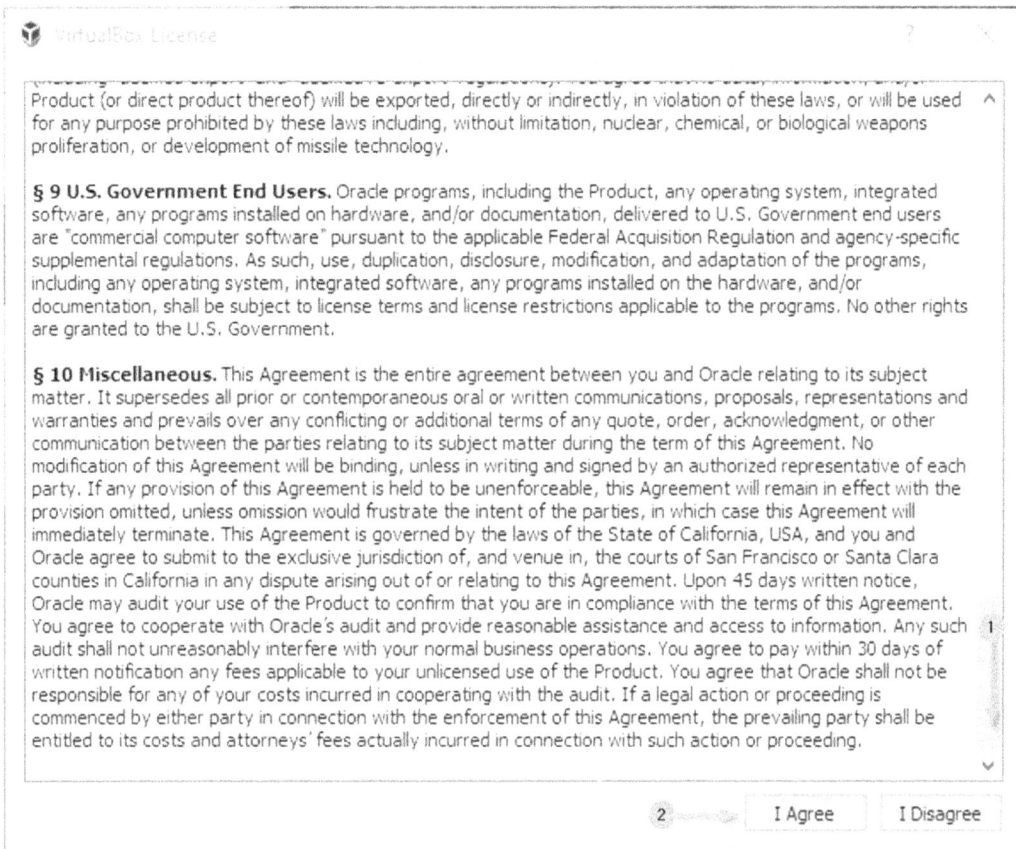

Figure 3.18 – Oracle VirtualBox extension pack installation, part 3

4. Click **OK** to acknowledge the successful installation of the VirtualBox extension pack, as shown in *Figure 3.19*:

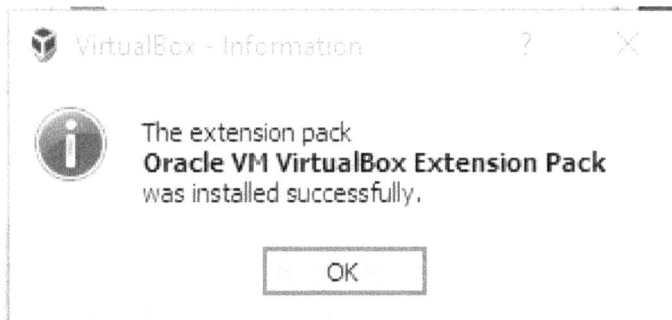

Figure 3.19 – Oracle VirtualBox extension pack installation, completed

With VirtualBox and its extension pack installed on the LabHost, we can move on to the installation of the guest VMs. The first one requires importing a pre-packaged virtual router appliance, and the rest require the installation of products in pre-configured VMs using installation media.

Deploying the VyOS router

In our lab, the VyOS router will act as both the *internet simulator* (by providing routing between the left and the right sides of our setup) and the *internet gateway* (by routing outbound traffic to the destinations outside of our lab environment).

It will be configured with three interfaces. One, connected to the internal bridge on your virtualization LabHost, is configured to obtain the IP address and the default gateway (via DHCP) on your physical network. The second and third interfaces are going to be defined as *internal* interfaces to your LabHost, connected to the external interfaces of the Check Point cluster (on the left) and a single gateway (on the right).

Follow the sequence below to create VyOS router VM and to define its' networking:

1. In **Oracle VM VirtualBox Manager**, click **Machine** [1] and **New** [2]:

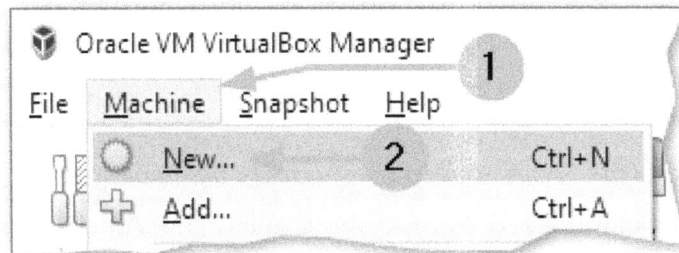

Figure 3.20 – Creating a new VM in VirtualBox

2. In the **Name** field, type Router [1], in the **Type** drop-down menu, select **Linux** [2], and in **Version**, select **Debian (64-bit)** [3]. Click **Create** [4]:

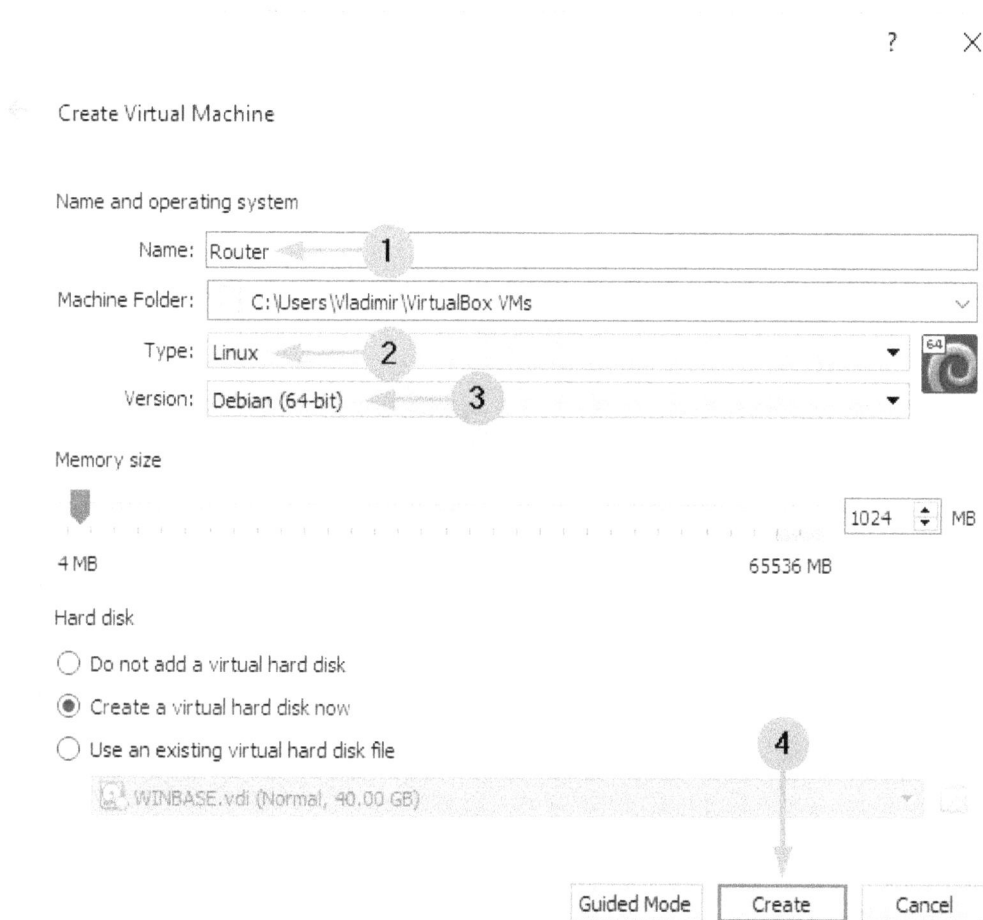

? ✕

Create Virtual Machine

Name and operating system

Name: Router ◀━━━ 1

Machine Folder: C:\Users\Vladimir\VirtualBox VMs ⌄

Type: Linux ◀━━━ 2 ▼

Version: Debian (64-bit) ◀━━━ 3 ▼

Memory size

1024 ⬍ MB

4 MB   65536 MB

Hard disk

○ Do not add a virtual hard disk

◉ Create a virtual hard disk now

○ Use an existing virtual hard disk file

WINBASE.vdi (Normal, 40.00 GB)   4

Guided Mode   Create   Cancel

Figure 3.21 – Defining the new VM's name, OS type, and version

3. In **Create Virtual Hard Disk,** leave all settings on defaults and click **Create** [1]:

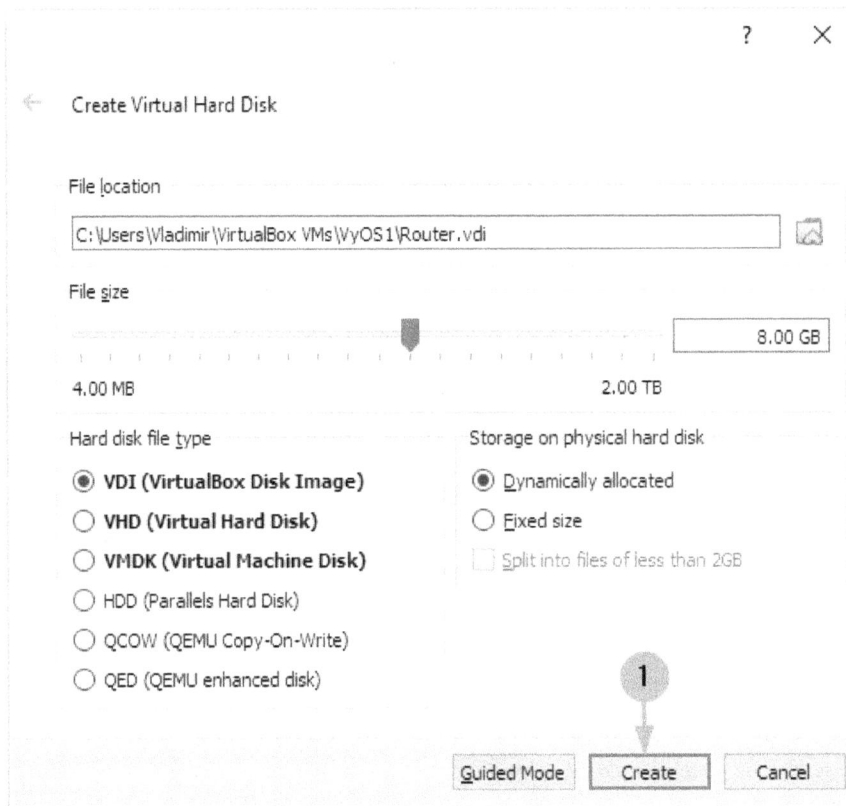

Figure 3.22 – Defining the new VM's storage size and allocation

The VM is now created and is listed on the left side of the Oracle VM VirtualBox Manager.

4. Right-click the newly created **Router** VM icon [1] and click **Settings** [2]:

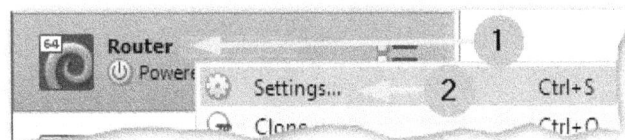

Figure 3.23 – Changing the settings of the VyOS router VM

5. In the **Router - Settings** options listed in the left column, select **Network** [1].
    Adapter 1 is selected by default [2]. Verify that it is enabled [3]. In the **Attached to** field, using the drop-down menu, select the **Bridged Adapter** option [4]. The
    Name field should be populated by your built-in network interface card [5]. Click
    on the **Advanced** expansion arrow [6] to verify that the **Cable Connected** checkbox
    is checked [7]:

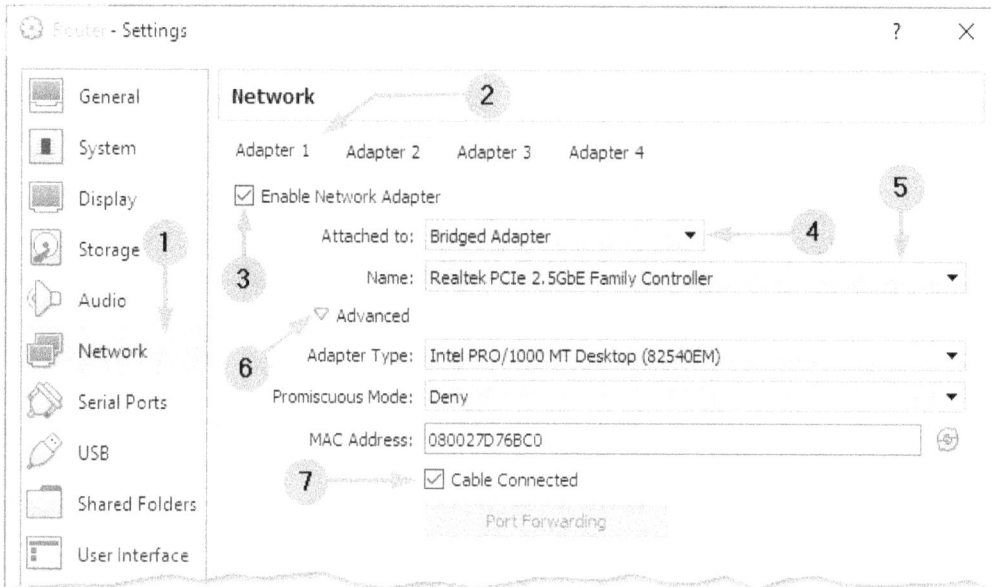

Figure 3.24 – Router VM networking configuration, Adapter 1

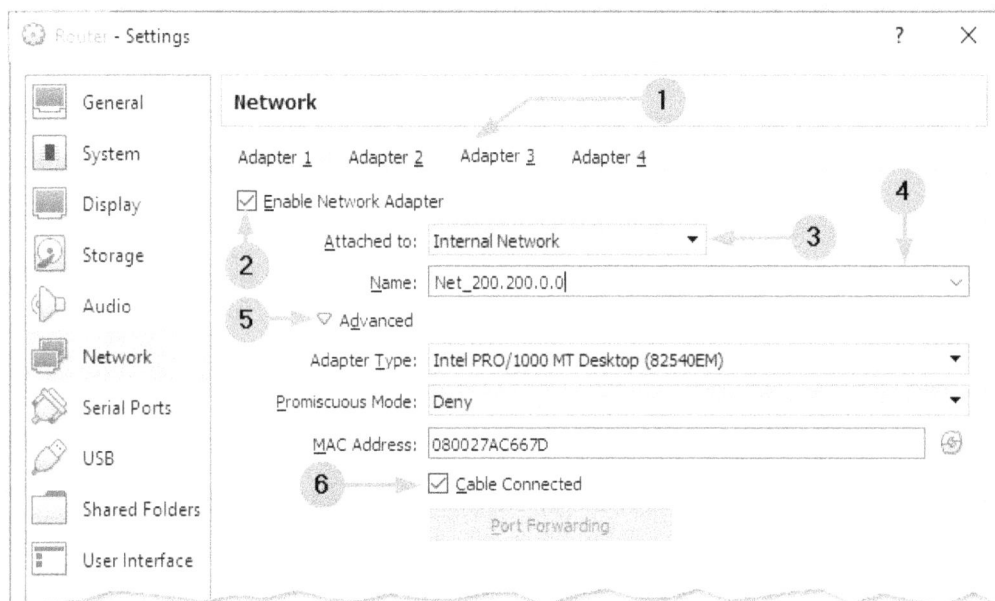

Figure 3.26 – Router VM networking configuration, Adapter 3

With VyOS router VM created, and with its network interfaces defined, follow this sequence to complete the installation from the ISO image:

1. In the **Router - Settings** options listed in the left column, select **Storage** [1]. Click to select the **Empty** disk icon in **Storage Devices** [2]. Click on the corresponding disk icon to the right of the **Optical Drive** field [3] and click on the **Choose a disk file...** option [4]:

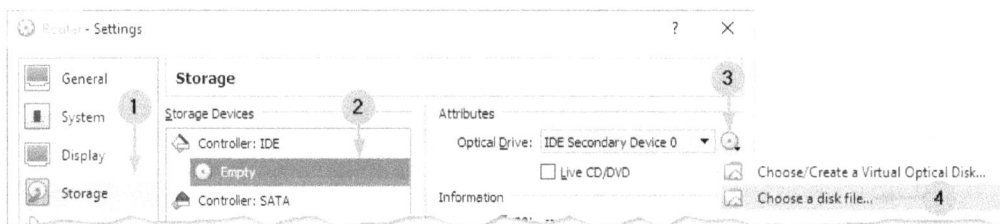

Figure 3.27 – Choosing installation media

2. Browse to the `LabShare\ISOs_and_OVAs` folder [1], select `vyos-rolling-latest.iso`, and click **OK**:

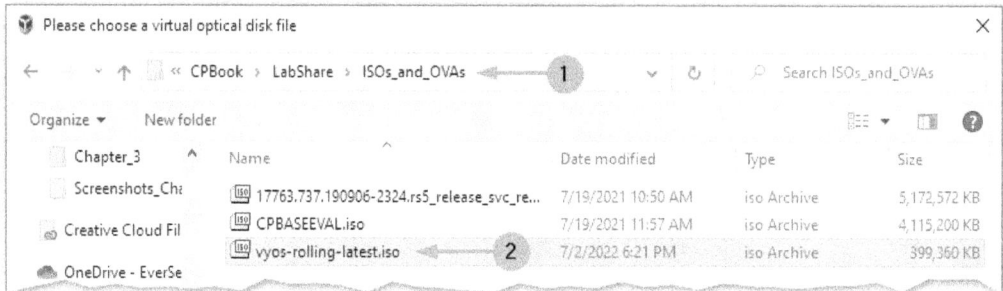

Figure 3.28 – Mounting VyOS installation image

3. In the Oracle VM VirtualBox Manager, right-click on the VyOS VM and click **Start | Normal Start**.

4. After the initial boot from the mounted ISO image is complete, we must install the router OS on the hard drive. Follow these steps to do so:

    A. At the `vyos :` log in prompt, type `vyos` and `vyos` for username and password.

    B. At the `vyos@vyos:~$` prompt, type `install image`.

5. We will then accept all defaults, answering a single question in the affirmative:

    A. At `Would you like to continue? (Yes/No) [Yes] :`, press *Enter*.

    B. At `Partition (Auto/Parted/Skip) [Auto] :`, press *Enter*.

    C. At `Install the image on? [sda] :`, press *Enter*.

    D. At `This will destroy all data on /dev/sda. Continue? (Yes/No) [No] :`, type `Yes` and press *Enter*.

    E. At `How big of a root partition should I create? (2000MB - 8589MB) [8589] MG:`, press *Enter*.

    F. At `What would you like to name this image? [...] :`, press *Enter*.

    G. At `Which one should I copy to sda? [...] :`, press *Enter*.

    H. At `Enter password for user 'vyos' :`, type `vyos` and press *Enter*.

    I. At `Retype password for user 'vyos' :`, type `vyos` and press *Enter*.

    J. At `Which drive should GRUB modify the boot partition on? [sda] :`, press *Enter*.

The same actions are depicted in the following screenshot:

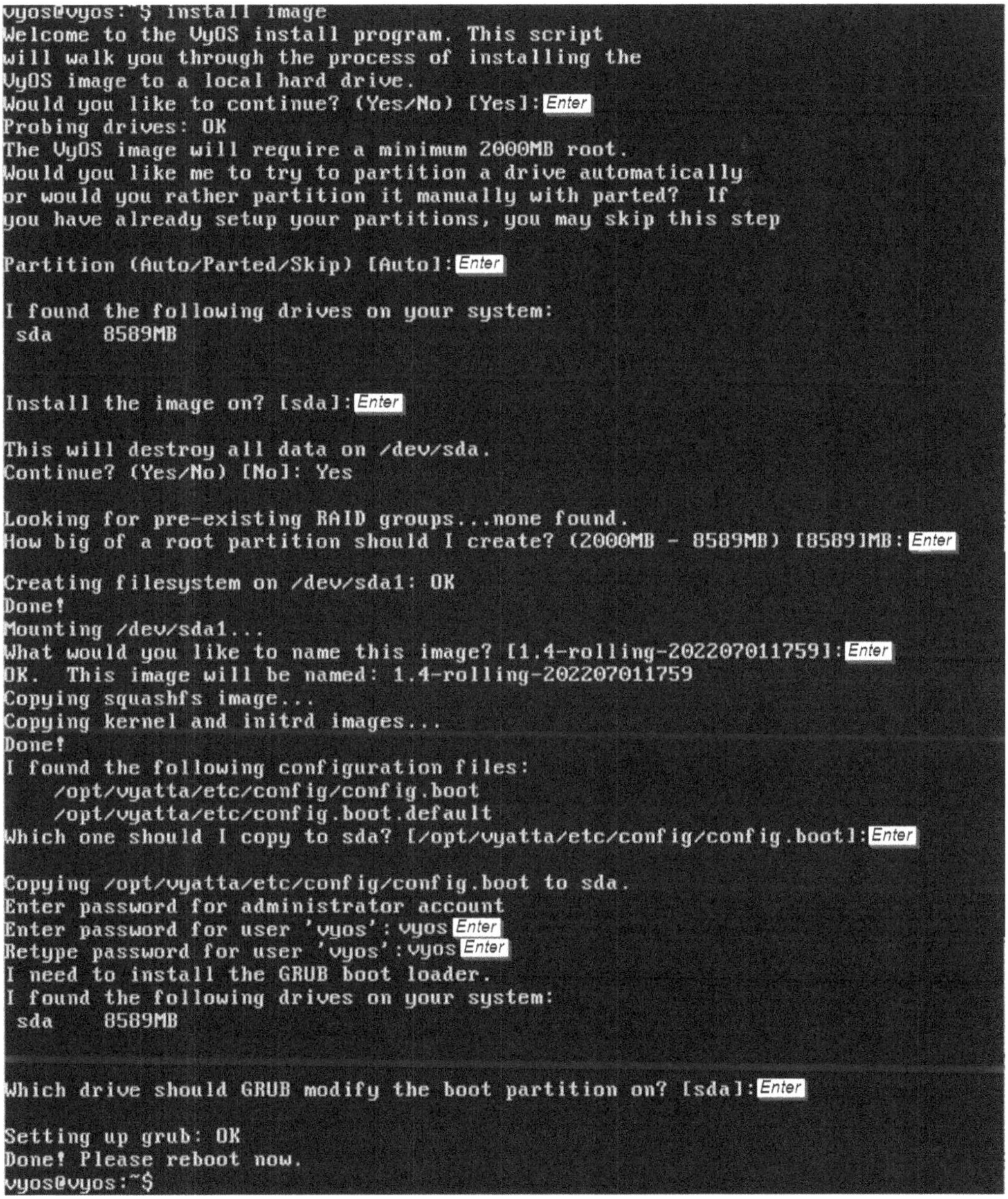

```
vyos@vyos:~$ install image
Welcome to the VyOS install program. This script
will walk you through the process of installing the
VyOS image to a local hard drive.
Would you like to continue? (Yes/No) [Yes]: Enter
Probing drives: OK
The VyOS image will require a minimum 2000MB root.
Would you like me to try to partition a drive automatically
or would you rather partition it manually with parted?  If
you have already setup your partitions, you may skip this step

Partition (Auto/Parted/Skip) [Auto]: Enter

I found the following drives on your system:
 sda     8589MB


Install the image on? [sda]: Enter

This will destroy all data on /dev/sda.
Continue? (Yes/No) [No]: Yes

Looking for pre-existing RAID groups...none found.
How big of a root partition should I create? (2000MB - 8589MB) [8589]MB: Enter

Creating filesystem on /dev/sda1: OK
Done!
Mounting /dev/sda1...
What would you like to name this image? [1.4-rolling-202207011759]: Enter
OK.  This image will be named: 1.4-rolling-202207011759
Copying squashfs image...
Copying kernel and initrd images...
Done!
I found the following configuration files:
    /opt/vyatta/etc/config/config.boot
    /opt/vyatta/etc/config.boot.default
Which one should I copy to sda? [/opt/vyatta/etc/config/config.boot]: Enter

Copying /opt/vyatta/etc/config/config.boot to sda.
Enter password for administrator account
Enter password for user 'vyos': vyos Enter
Retype password for user 'vyos': vyos Enter
I need to install the GRUB boot loader.
I found the following drives on your system:
 sda     8589MB


Which drive should GRUB modify the boot partition on? [sda]: Enter

Setting up grub: OK
Done! Please reboot now.
vyos@vyos:~$
```

Figure 3.29 – VyOS router installation procedure

With installation completed, let's remove the ISO image from the virtual **Optical Drive** following this short sequence:

1. When you see ⏻one! Please reboot now, right-click on the **Router [Running]** VM, click **Close**, and click **ACPI Shutdown**.

2. Confirm **ACPI Shutdown**.

3. When the VM state changes from **Running** to **Powered off**, right-click on it and click **Settings** [1]. In the VyOS settings, click **Storage** [2]. Click on vyos-rolling-latest.iso [3]. Click on the corresponding disk icon to the right of the **Optical Drive** field [4] and click on the **Remove Disk from Virtual Drive** option [5]:

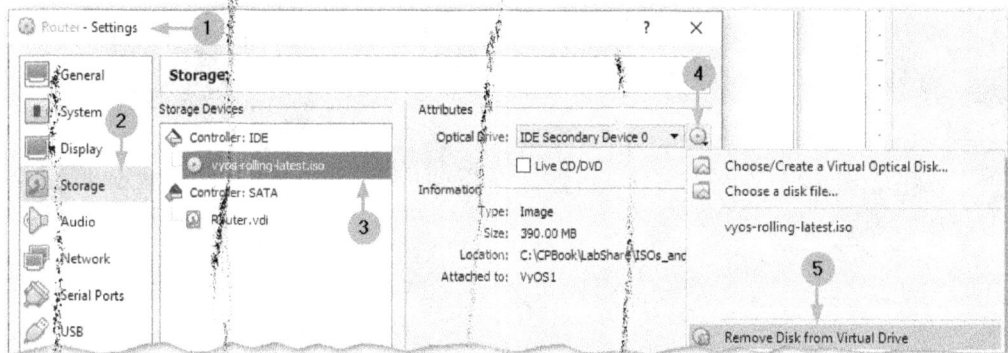

Figure 3.30 – Removing the VyOS ISO from Virtual Drive

4. Click **OK** (not shown).

This completes the VyOS router VM creation and installation sequences.

Let's power it up to configure:

1. Right-click on the **Router** VM and click **Start | Normal Start** (if you do not see the VM console, double-click on the VM icon again and you'll see the boot process of the installed Router VM).

2. Once the boot process is completed, at the login prompt, use vyos for both username and password.

3. After you are logged on, at the vyos@vyos:~$ prompt, type configure, and press *Enter*.

4. Type these commands, pressing *Enter* after each line's completion:

A. set interfaces ethernet eth0 address dhcp

B. set service ssh

C. commit

D. save

5. Type `show interfaces ethernet eth0 brief` and note the IP address your DHCP server has assigned to the Router VM `eth0` interface.

6. On your LabHost PC, launch PuTTY and create a new session using the IP address obtained in the preceding step. Name the session `Router`, save, and open.

7. Repeat the login process in your SSH session using the same credentials (vyos/vyos).

8. Copy and paste the following block of code:

```
set interfaces ethernet eth0 description 'OUTSIDE'
set interfaces ethernet eth1 address '200.100.0.254/24'
set interfaces ethernet eth1 description
'Net_200.100.0.0'
set interfaces ethernet eth2 address '200.200.0.254/24'
set interfaces ethernet eth2 description
'Net_200.200.0.0'
set nat source rule 100 outbound-interface 'eth0'
set nat source rule 100 source address '200.100.0.0/24'
set nat source rule 100 translation address 'masquerade'
set nat source rule 200 outbound-interface 'eth0'
set nat source rule 200 source address '200.200.0.0/24'
set nat source rule 200 translation address 'masquerade'
set system host-name 'router'
set system name-server '9.9.9.9'
```

9. Press *Enter* to execute the last command.

10. Copy and paste the following command, omitting `US/Eastern` if you are in a different time zone: `set system time-zone US/Eastern`.

11. Instead, start typing your common locale and press the *Tab* key after the first few letters, for example, `set system time-zone Asi` [*Tab*] to see all choices available for Asia. Finish typing yours and press *Enter* to complete the time zone configuration.

12. Once done, type each of the following commands followed by *Enter*:

    A. `commit`

    B. `save`

    C. `exit`

13. Use `ping` to a common website to verify external connectivity.

We have now completed the deployment of the virtual router that will be used by the Check Point components of the lab to route the traffic between the *left* and *right* segments of the lab, as well as outside of the LabHost. We can now move on to the preparation of the templates of VMs (referred to later as *Base VMs*) for Windows hosts and Check Point management and gateways.

Summary

With Oracle VirtualBox, the extension pack, and the VyOS router installed and configured on the LabHost, we can move on to the installation of the Check Point and Windows guest VMs in *Chapter 4*, *Building a Check Point Lab Environment – Part 2*, which is a continuation of this chapter.

4

Building
a Check Point
Lab Environment
– Part 2

In this chapter, we will continue working on creating our **Check Point** virtual lab environment. With **VirtualBox** and **VirtualBox extensions** installed, a virtual router deployed, and the prerequisite software and installation media downloaded and placed in the predefined directories, we will proceed with the creation of the **virtual machines** (**VMs**). Both GUI and command-line deployment options will be presented.

We are going to cover the following main topics:

- Creating a Windows base VM
- Creating a Check Point base VM
- Creating and preparing linked clones

Technical requirements

Download all the scripts from `https://github.com/PacktPublishing/Check-Point-Firewall-Administration-R81.10-/tree/main/Chapter04/Scripts` to your `C:\CPBook\LabShare\Scripts` folder.

If you are using an OS other than Windows on your LabHost, you will have to convert *only the scripts you intend to run on it* for compatibility from Windows batch files to Bash scripts. These scripts are as follows:

- `create_Router.bat`
- `unattended_swap.bat`
- `create_WINBASE.bat`
- `create_CPBASE.bat`
- `eject_and_snapshot_WINBASE.bat`
- `eject_and_snapshot_CPBASE.bat`
- `clonevms.bat`

Creating a Windows base VM

To simplify and speed up the deployment of lab components, as well as in the interest of saving space, we will prepare a **Windows Server base VM** and use linked clones to create the rest.

If you are interested in the scripted creation of the Windows base VM, skip to the *Windows Server base image scripted* section; otherwise, proceed with the following section.

Creating a Windows Server base VM in the GUI

The following steps will help us create our Windows Server template:

1. In **Oracle VM VirtualBox Manager** [1], click **New** [2]:

Figure 4.1 – Creating a new VM

2. In the **Create Virtual Machine** [1] window, enter WINBASE in the **Name** field [2], select **Microsoft Windows** as the type [3], and in **Version**, choose **Windows 2019 (64-bit)** [4], then click **Next** [5]:

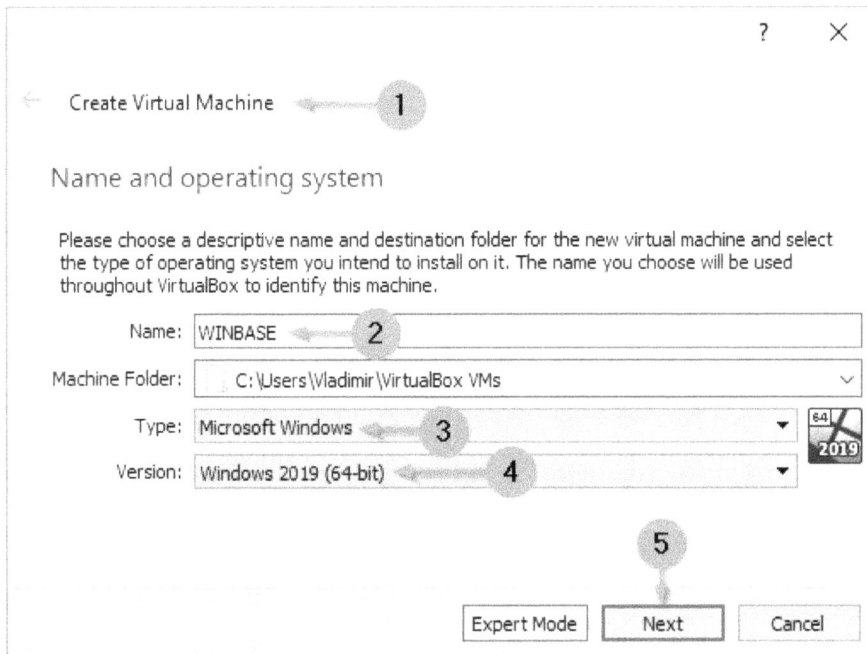

Figure 4.2 – Choosing the name and OS

3. Leave **Memory size** at the default of **2048 MB** and click **Next** [1]:

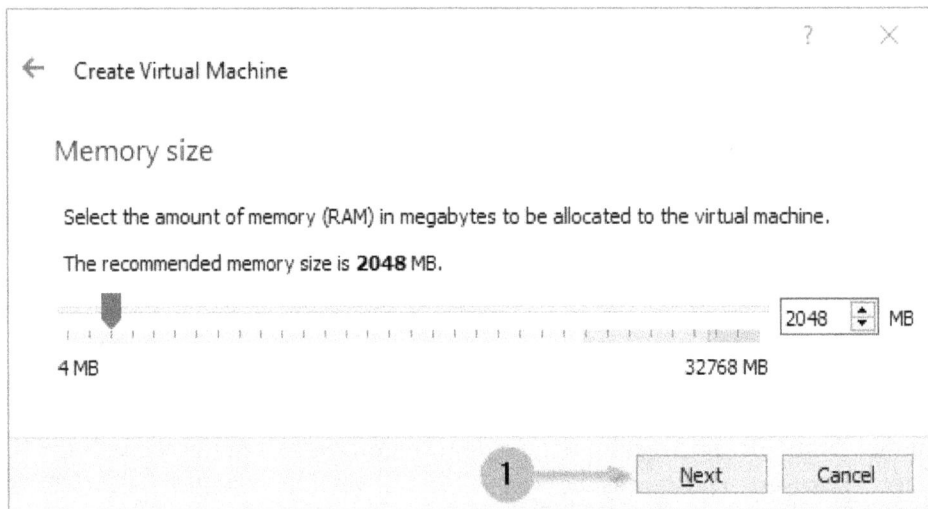

Figure 4.3 – Allocating memory

4. Leave the hard disk size at the recommended 50 GB [1] and click **Create** [2]:

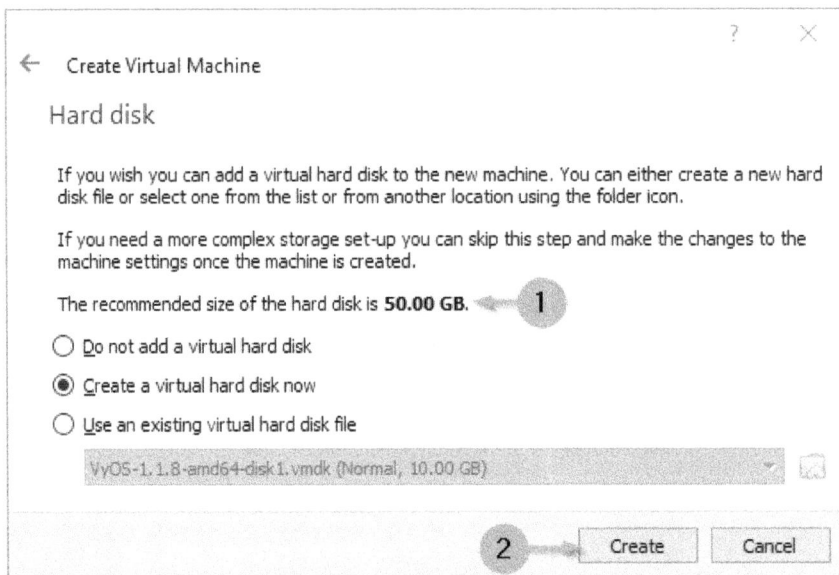

Figure 4.4 – Creating a hard disk

5. In **Hard disk file type**, leave it as **VDI (VirtualBox Disk Image)** [1] and click **Next** [2]:

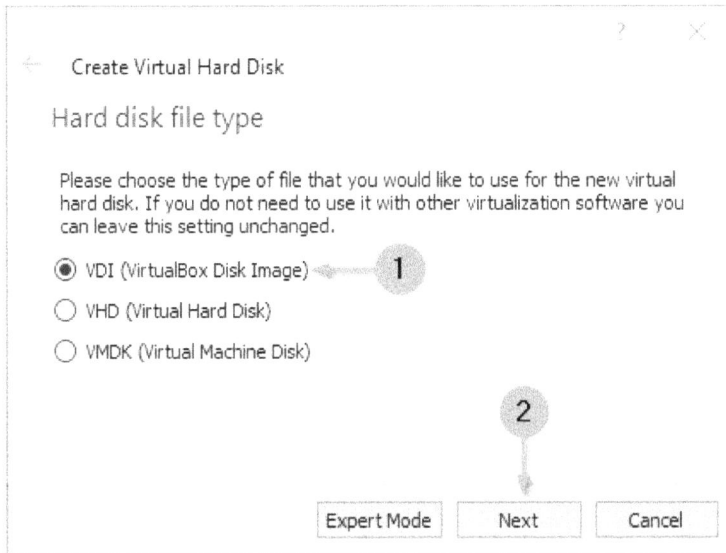

Figure 4.5 – Defining the file type

6. In **Storage on physical hard disk**, leave **Dynamically allocated** [1] and click **Next** [2]:

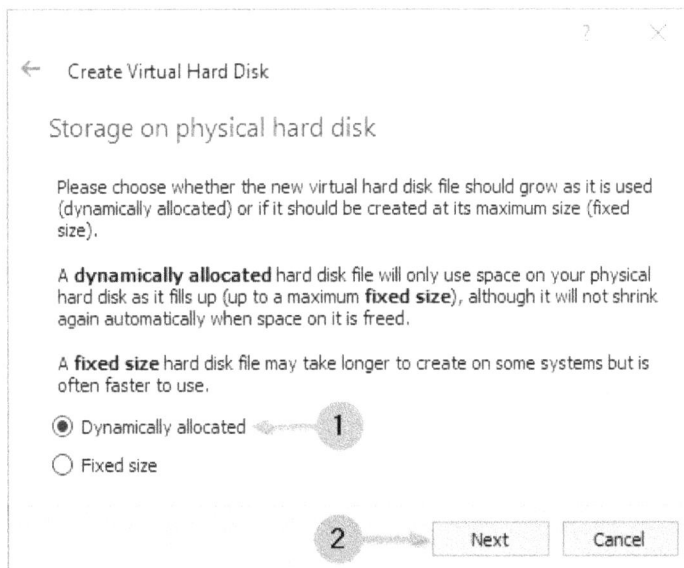

Figure 4.6 – Choosing dynamic allocation of space

7. In **File location and size**, change the value to **40.00 GB** [1] and click **Create** [2]:

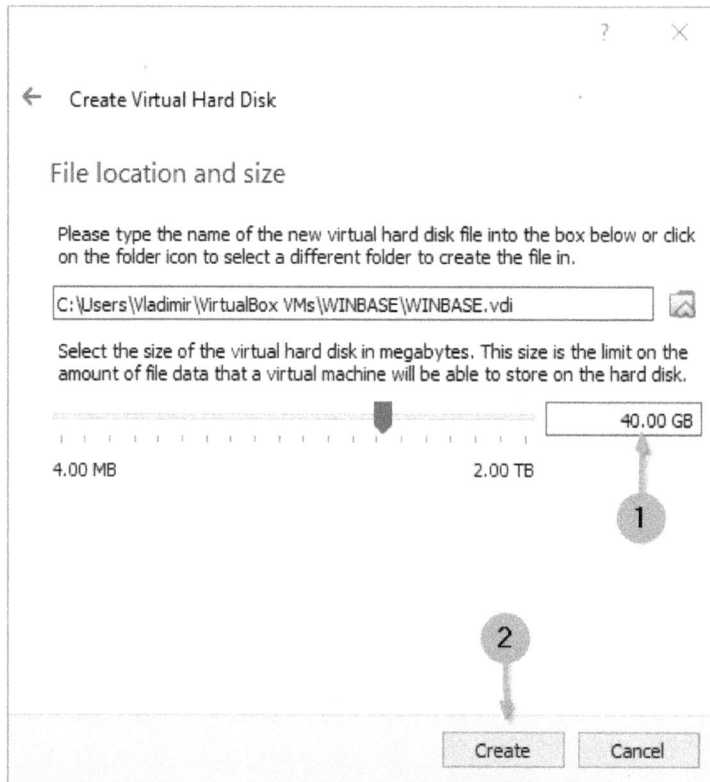

Figure 4.7 – Configuring the file location and size

8. With the VM template created, select it in VirtualBox Manager [1] and click **Settings** [2]:

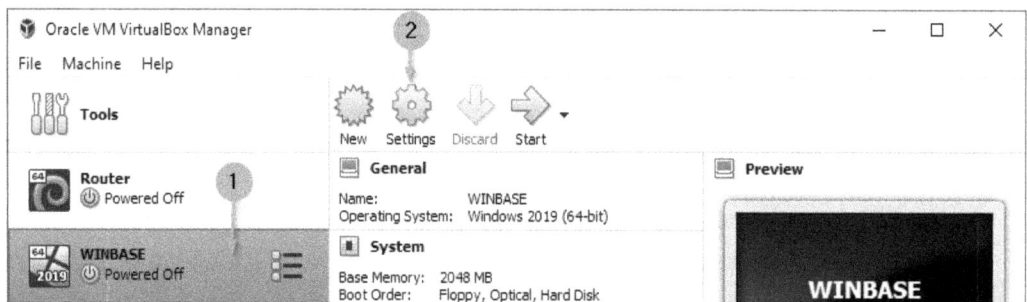

Figure 4.8 – WINBASE VM settings

9. In the **WINBASE - Settings** window, click on **Storage** [1], then on the **Empty** DVD icon [2], and once more on the DVD icon to the right of **SATA Port 1** [3]. Click on **Choose a disk file…** [4]:

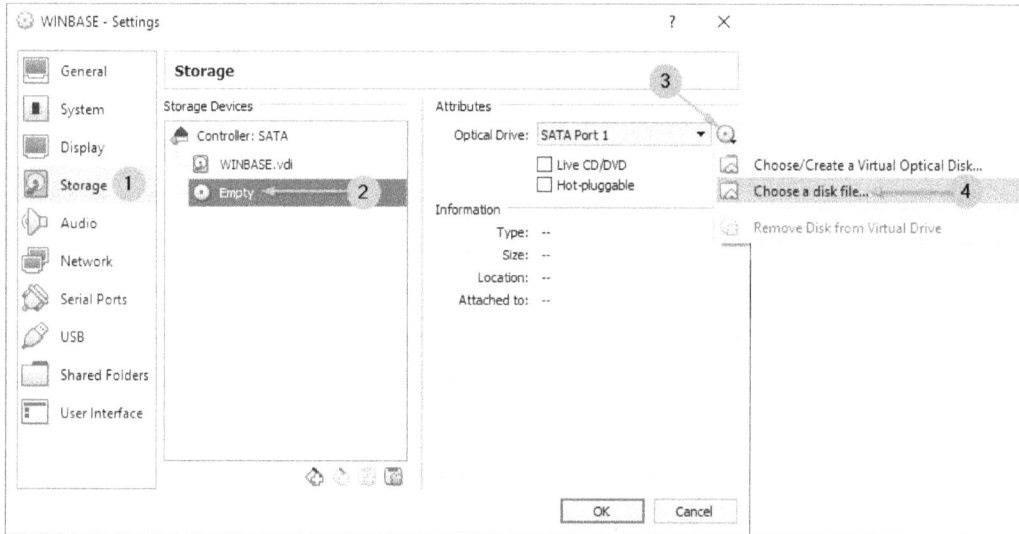

Figure 4.9 – WINBASE installation media part 1

10. Browse to the location of the files downloaded for the lab [1], click on the Windows Server 2019 ISO file [2], and click **Open** [3]:

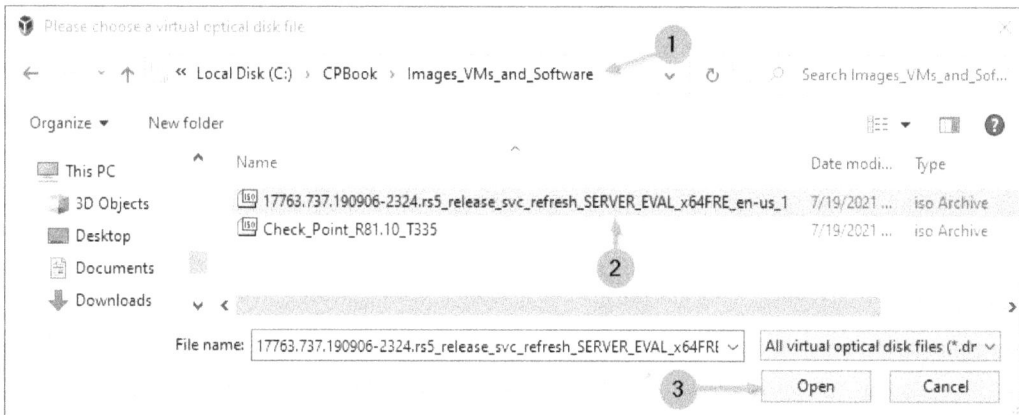

Figure 4.10 – WINBASE installation media part 2

11. Click on **Network** [1] and in the **Adapter 1** tab [2], in the **Attached to** field, choose **Internal Network** [3]. In the **Name** field, enter Net_DeadEnd [4]:

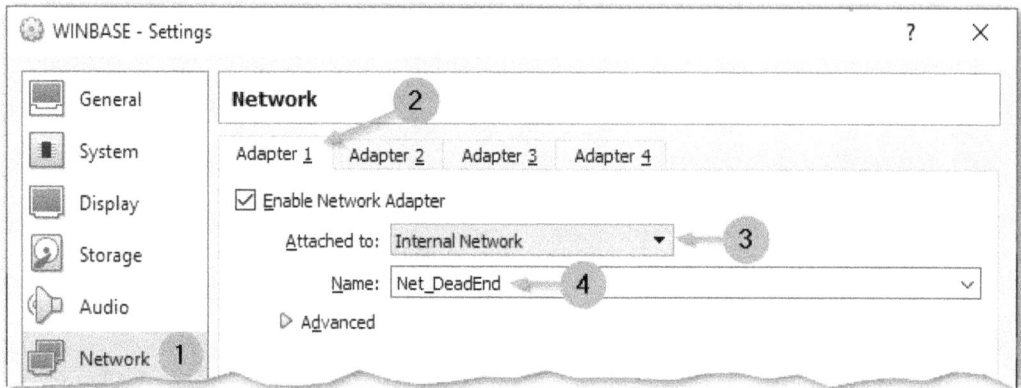

Figure 4.11 – WINBASE network settings

12. In **General** pane, click the **Advanced** tab [2], and for **Shared Clipboard** [3] and **Drag'n'Drop** [4], choose **Bidirectional**. Click **OK** [5]:

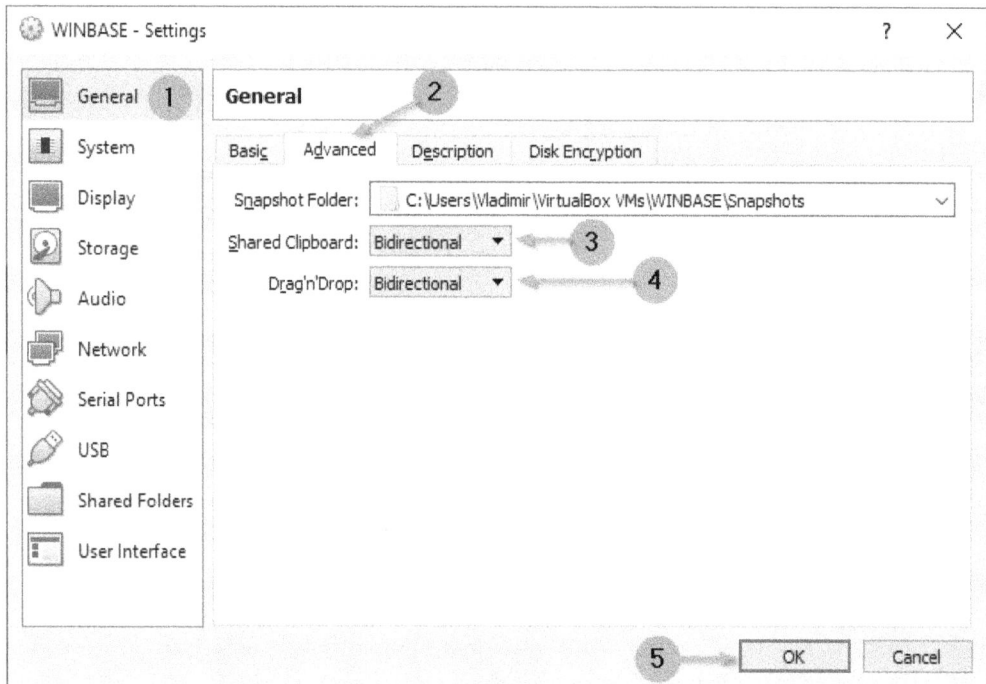

Figure 4.12 – WINBASE clipboard and drag and drop

13. In **Shared Folders** [1], click the blue folder icon with the green plus sign [2], and in the **Add Share** popup's **Folder Path** setting [3], browse to the LabHost's folder with the downloaded files, select it, and click **Select Folder**. Once it is shown in **Folder Path**, check the **Auto-mount** checkbox [4], enter F: into the **Mount point** field [5], check **Make Permanent** [6], and click **OK** twice [7 and 8]:

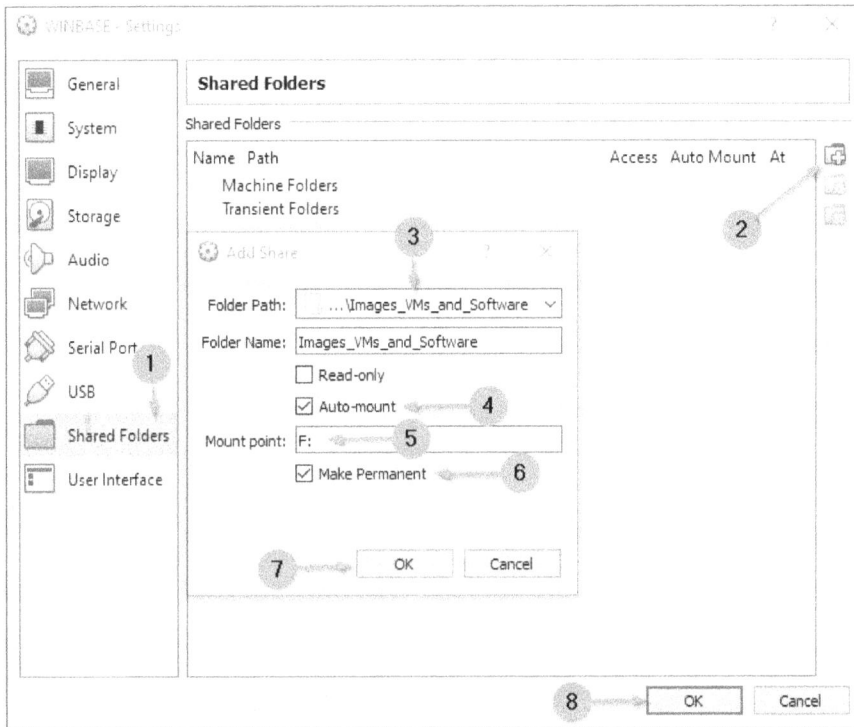

Figure 4.13 – WINBASE Shared Folders

14. In VirtualBox Manager, select the **WINBASE** VM [1] and click **Start** [2]:

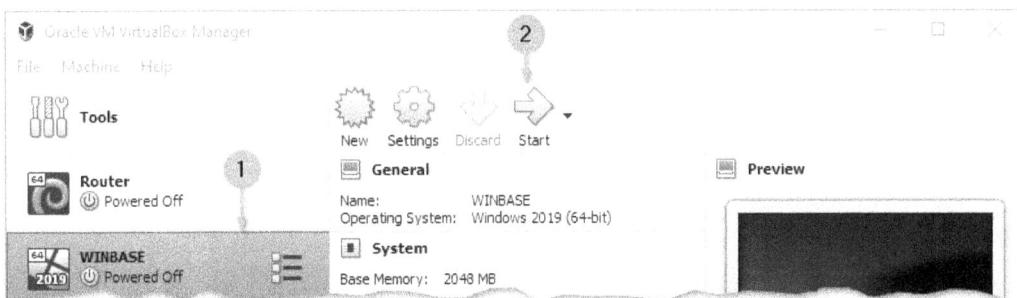

Figure 4.14 – Starting the WINBASE VM

15. Go through the initial Windows Server installation procedure, choosing the
Windows Server 2019 Standard Evaluation (Desktop Experience) [1] and
Custom: Install Windows only (advanced) [2] choices when prompted. Log on to
it, once it is completed:

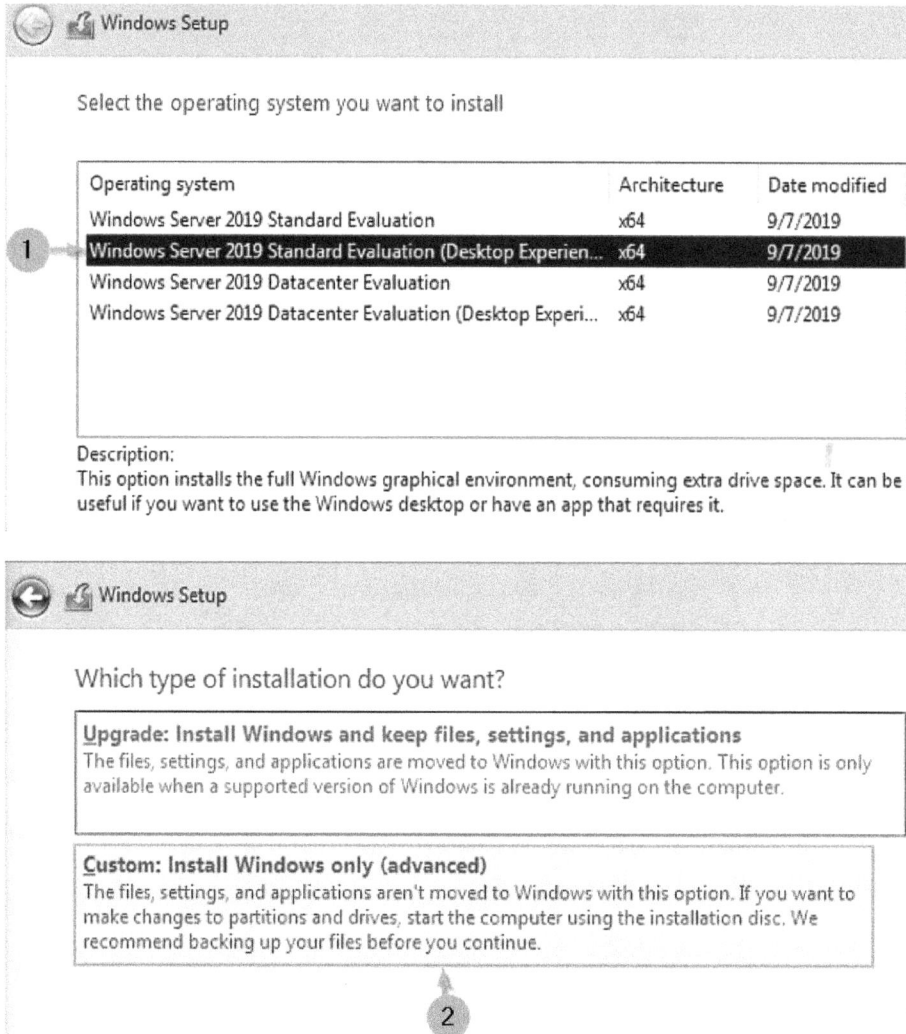

Figure 4.15 – Choosing the Windows Server installation type

16. Once logged on, close the Server Manager pop-up window and Server
Manager itself.

17. In the **WINBASE** VM interface [1], click on **Devices** [2] and then on **Insert Guest Additions CD image…** [3]:

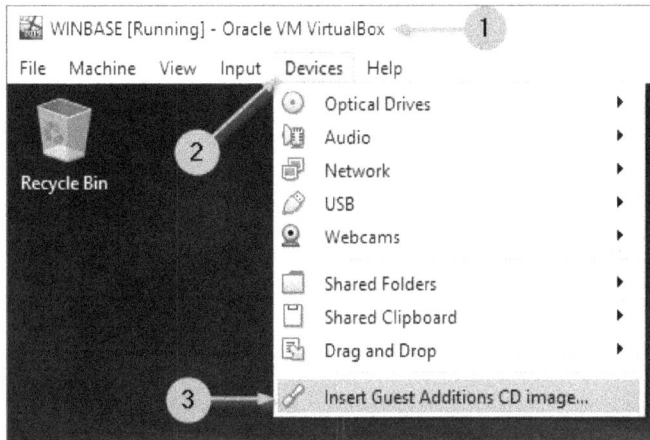

Figure 4.16 – Insert Guest Additions CD image

18. Open Windows File Explorer [1], click on **This PC** [2], and double-click on **CD Drive (D:) VirtualBox Guest Additions** [3]:

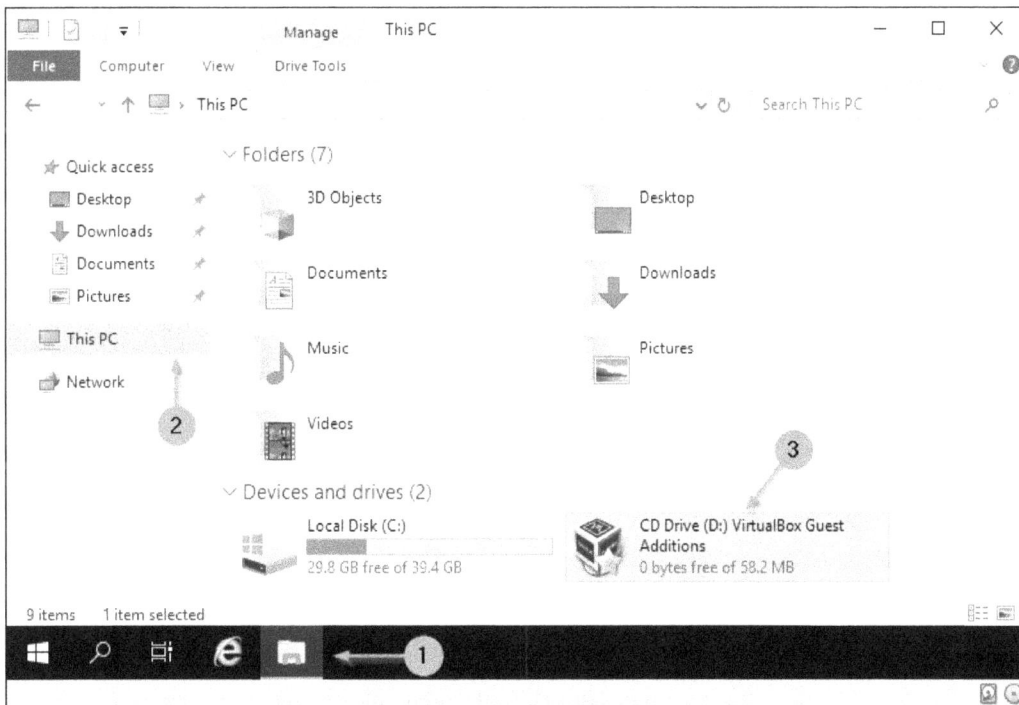

Figure 4.17 – Guest Additions

19. When presented with **Welcome to Oracle VM VirtualBox Guest Additions #.#.##
Setup** dialog, click **Next | Next | Install**:

20. When prompted by Windows Security [1], verify that **Always trust software from
"Oracle Corporation".** is checked [2] and click **Install** [3]:

Figure 4.18 – Windows Security prompt confirmation

21. Verify that **Reboot now** is selected [1] and click **Finish** [2]:

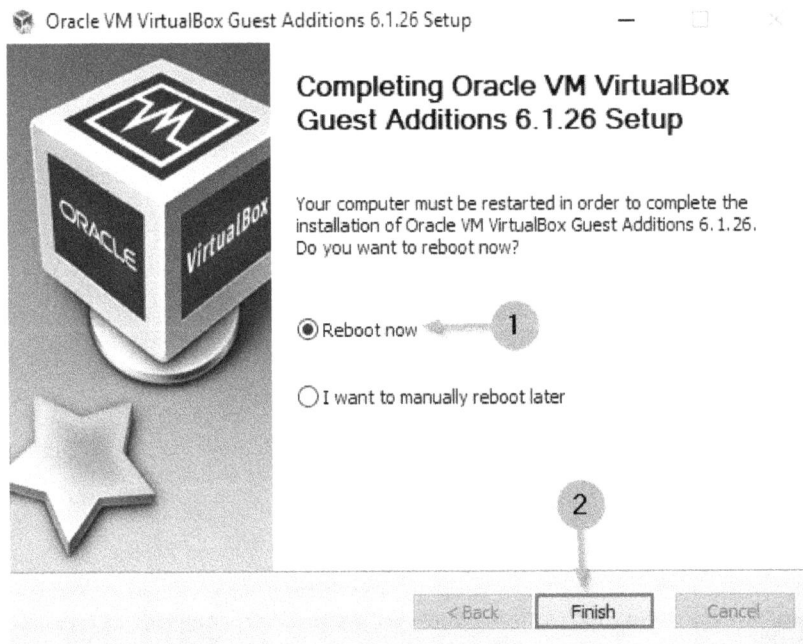

Figure 4.19 – Completing Guest Additions installation

22. Once the reboot is completed, using Windows Explorer [1] open the `Software` folder located in your PC's [2] mounted disk (`F:`) [3] and install PuTTY, WinSCP, Chrome, and Notepad++:

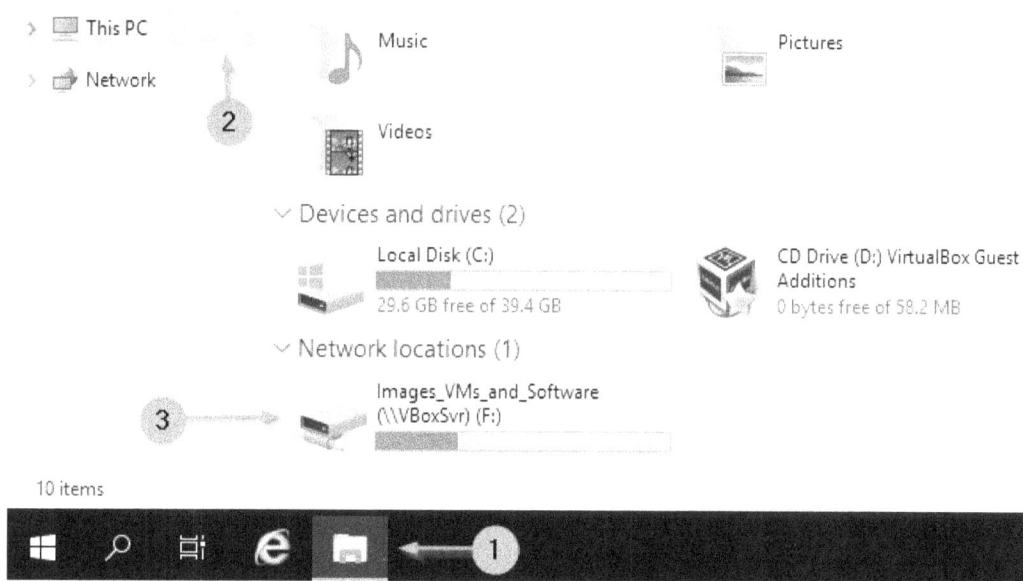

Figure 4.20 – LabShare mapped drive (F:)

> **Note**
>
> If you are not seeing the `F:` drive in **Network locations**, reboot the VM one more time for mapping to take effect.

The next section shows how the same actions could be accomplished using the CLI.

Windows Server base image scripted

If you are pursuing scripted lab implementation, you can either use the commands depicted as follows or execute referenced batch files.

> **Important Note!**
>
> If you have already performed the GUI-based steps previously, skip this section and go to the *Finalizing the Windows Server base VM installation* section.

The steps to create the Windows Server base using the CLI are as follows:

1. Modify the answer file for unattended installation. From an elevated Command Prompt, create a backup copy of an existing unattended Windows answer file and copy a modified one into the same directory:

   ```
   RENAME "C:\Program Files\Oracle\VirtualBox\
   UnattendedTemplates\win_nt6_unattended.xml" win_nt6_
   unattended_backup1.xml
   COPY C:\CPBook\LabShare\Scripts\win_nt6_unattended.xml
   "C:\Program Files\Oracle\VirtualBox\UnattendedTemplates\
   win_nt6_unattended.xml"
   ```

   Or execute the `C:\CPBook\LabShare\Scripts\unattended_swap.bat` file.

2. Create a Windows Server 2019 base VM using the CLI. In the process of doing that, we will rename the ISO file downloaded from Microsoft to a shorter one using wildcards to avoid errors with future versions. Copy and paste the following code lines into the Command Prompt of your LabHost:

   ```
   SET PATH=%PATH%;C:\Program Files\Oracle\VirtualBox
   SET VM=WINBASE
   REN C:\CPBook\LabShare\ISOs_and_OVAs\*SERVER_EVAL_
   x64FRE*.iso WINBASEEVAL.iso
   VBoxManage createhd --filename %VM%.vdi --size 40960
   VBoxManage createvm --name %VM% --ostype "Windows2019_64"
   --register
   VBoxManage storagectl %VM% --name "SATA Controller" --add
   sata --controller IntelAHCI
   VBoxManage storageattach %VM% --storagectl "SATA
   Controller" --port 0 --device 0 --type hdd --medium %VM%.
   vdi
   VBoxManage storageattach %VM% --storagectl "SATA
   ```

```
Controller" --port 1 --device 0 --type dvddrive --medium
emptydrive
VBoxManage modifyvm %VM% --ioapic on
VBoxManage modifyvm %VM%  --boot2 disk --boot1 dvd
--boot3 none --boot4 none
VBoxManage modifyvm %VM% --memory 2048 --vram 128
VBoxManage modifyvm %VM% --nic1 intnet
VBoxManage modifyvm %VM% --nictype1 82540EM
VBoxManage modifyvm %VM% --macaddress1 auto
VBoxManage modifyvm %VM% --cableconnected1 on
VBoxManage modifyvm %VM% --intnet1 Net_DeadEnd
VBoxManage modifyvm %VM% --graphicscontroller vboxsvga
VBoxManage modifyvm %VM% --usbxhci on
VBoxManage modifyvm %VM% --mouse usbtablet
VBoxManage sharedfolder add %VM% --name "Labshare"
--hostpath "C:\CPBook\LabShare\" --automount --auto-
mount-point F:
VBoxManage modifyvm %VM% --clipboard-mode=bidirectional
VBoxManage modifyvm %VM% --draganddrop=bidirectional
VBoxManage modifyvm %VM% --nested-hw-virt on
VBoxManage modifyvm %VM% --pae off
VBoxManage unattended install %VM% --iso=C:\CPBook\
LabShare\ISOs_and_OVAs\WINBASEEVAL.iso --image-index=2
--user=Administrator --password=CPL@b8110 --install-
additions
VBoxManage startvm %VM% --type gui
```

Or, in the Command Prompt of your LabHosts, execute the `C:\CPBook\`
`LabShare\Scripts\create_WINBASE.bat` file.

> **Important Note**
> When running this script, the name of the original Windows Server
> installation ISO will be overwritten.

The Windows VM is now created and is booting up. Once it is up and running, we will be
preparing it to serve as a template for the rest of the Windows VMs in our lab.

Finalizing the Windows Server base VM installation

To prepare the Windows template VM for cloning, we must perform generalization to remove the computer **security identifier** (**SID**):

1. After you are presented with the Windows Server UI running Server Manager, in your guest VM, click **Start** | **Power** | **Restart** and choose **Other (Planned)** to complete the user environment variable settings and finish mounting the shared folder as the F: drive.

2. Once the reboot is complete, in the guest WINBASE VM, close Server Manager and run this line in PowerShell:

   ```
   Get-ScheduledTask -TaskName ServerManager | Disable-
   ScheduledTask; cmd /k %WINDIR%\System32\sysprep\sysprep.
   exe /oobe /generalize /shutdown
   ```

   Or in the guest WINBASE VM, click **Start**, type run, and press *Enter*. In the **Run** prompt, paste this line:

   ```
   powershell.exe -NoProfile -InputFormat None
   -ExecutionPolicy Bypass -Command (F:\Scripts\WINBASE_
   prep.ps1)"
   ```

   Press *Enter* to prevent Server Manager from starting automatically and to initiate the sysprep for creation of the rapidly deployable Windows template. Wait for the VM to shut down.

3. Once the VM is shut down, on your LabHost, in Command Prompt, type or copy/paste the following lines, pressing *Enter* when done:

   ```
   SET PATH=%PATH%;C:\Program Files\Oracle\VirtualBox
   SET VM=WINBASE
   VBoxManage storageattach %VM% --storagectl Floppy --port
   0 --device 0 --medium emptydrive
   VBoxManage storageattach %VM% --storagectl "SATA
   Controller" --port 1 --device 0 --medium emptydrive
   VBoxManage storageattach %VM% --storagectl "SATA
   Controller" --port 2 --device 0 --medium emptydrive
   VBoxManage snapshot %VM% take "Snapshot 1"
   ```

   Or, in the Command Prompt of your LabHost, execute the C:\CPBook\ LabShare\Scripts\eject_and_snapshot_WINBASE.bat file.

This completes the creation of the WINBASE Windows Server template VM. Let's move on to the creation of the Check Point template VM.

Creating a Check Point base VM

We are going to have two different types of Check Point VMs in our environment: the **Security Management Server** (**SMS**) and the gateways (CPCM1 and CPCM2 in a **High Availability** (**HA**) cluster and CPGW as a single gateway), but in the interest of saving time, we will be using the same base VM for all of them.

> Note
>
> Keep in mind that the actual hardware (or virtual hardware requirements) for Check Point gateways and servers are quite higher than the values we are using in the lab. These could be found at `https://sc1.checkpoint.com/documents/R81.10/WebAdminGuides/EN/CP_R81.10_RN/Topics-RN/Hardware-Requirements.htm?tocpath=Open%20Server%20Hardware%20Requirements%7C_____1#Minimal_Hardware_Requirements`. If space is not an issue in your lab environment, for HDD, allocate 80 GB and leave the partition sizes at suggested default values, when prompted. Otherwise, use the values specified in the following steps.

If you are interested in scripted creation of a Check Point base VM, skip to the *Check Point base image scripted* section; otherwise, proceed as follows:

1. In **Oracle VM VirtualBox Manager** [1], click **New** [2]:

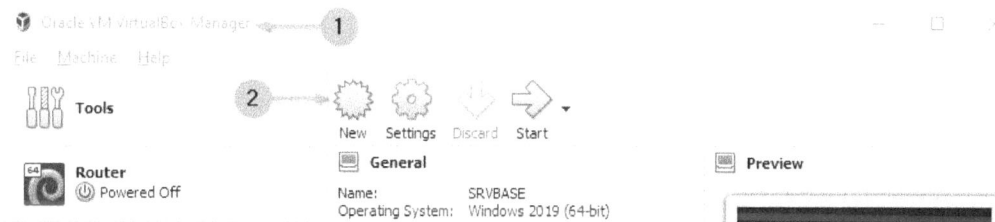

Figure 4.21 – Creating a new VM

2. In the **Create Virtual Machine** window [1], in the **Name** field, type CPBASE [2], set **Type** as **Linux** [3] and **Version** as **Red Hat (64-bit)** [4], and click **Next** [5]:

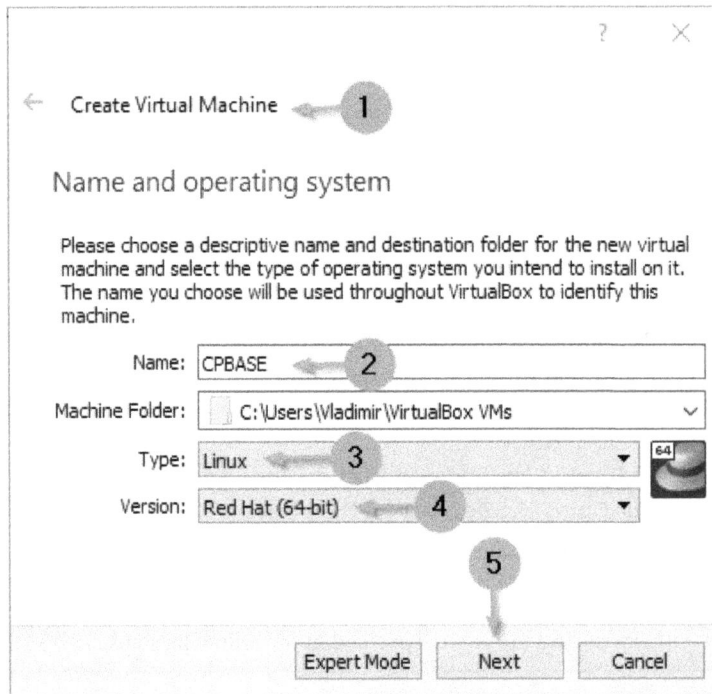

Figure 4.22 – Choosing the OS type and version

3. In **Memory size** [1], adjust the allocated amount of RAM to 4 GB [2] and click **Next** [3]:

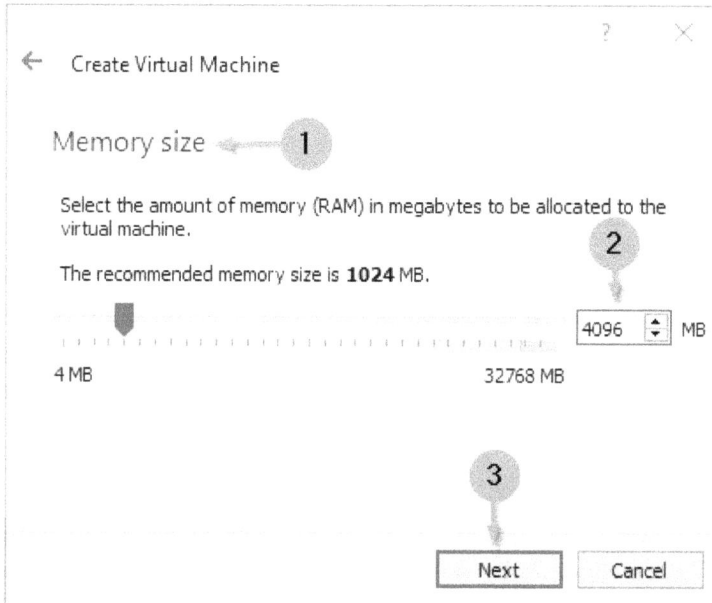

Figure 4.23 – Allocating memory

4.   In **Hard disk** [1], choose **Create a virtual hard disk now** [2] and click **Create** [3]:

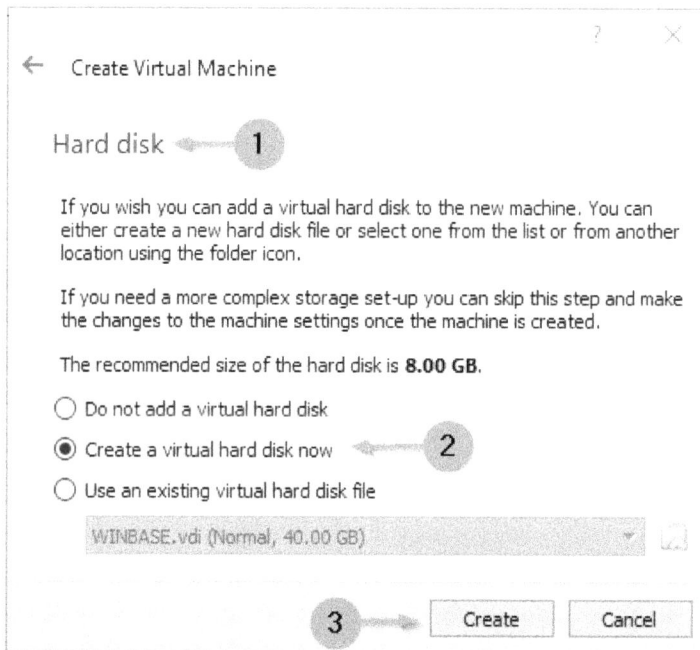

Figure 4.24 – Creating a virtual hard disk

5. In **Hard disk file type** [1], choose **VDI (VirtualBox Disk Image)** [2] and click **Next** [3]:

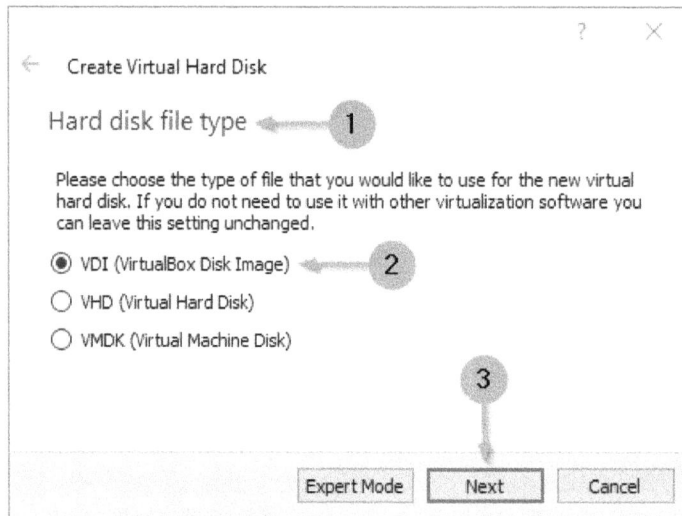

Figure 4.25 – Selecting a hard disk file type

6. In **Storage on physical hard disk** [1], choose **Dynamically allocated** [2] and click **Next** [3]:

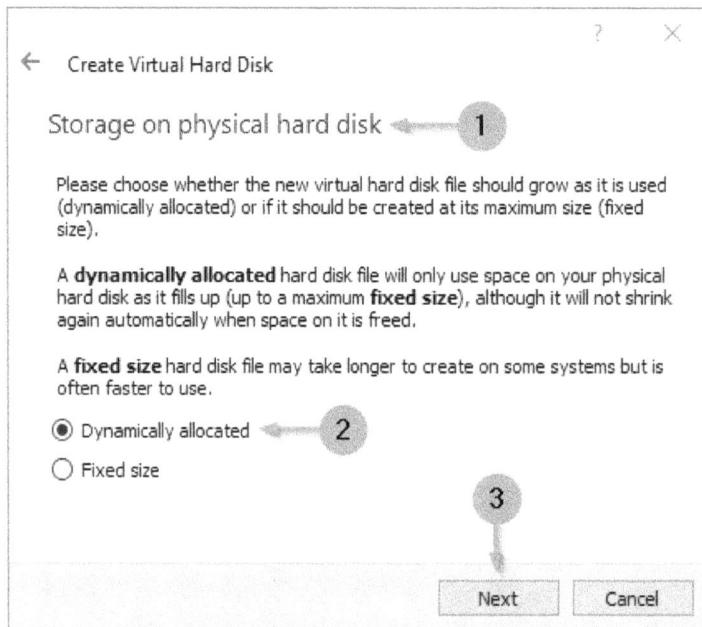

Figure 4.26 – Choosing dynamic space allocation

7. In **File location and size** [1], type 60.00 GB [2] and click **Create** [3]:

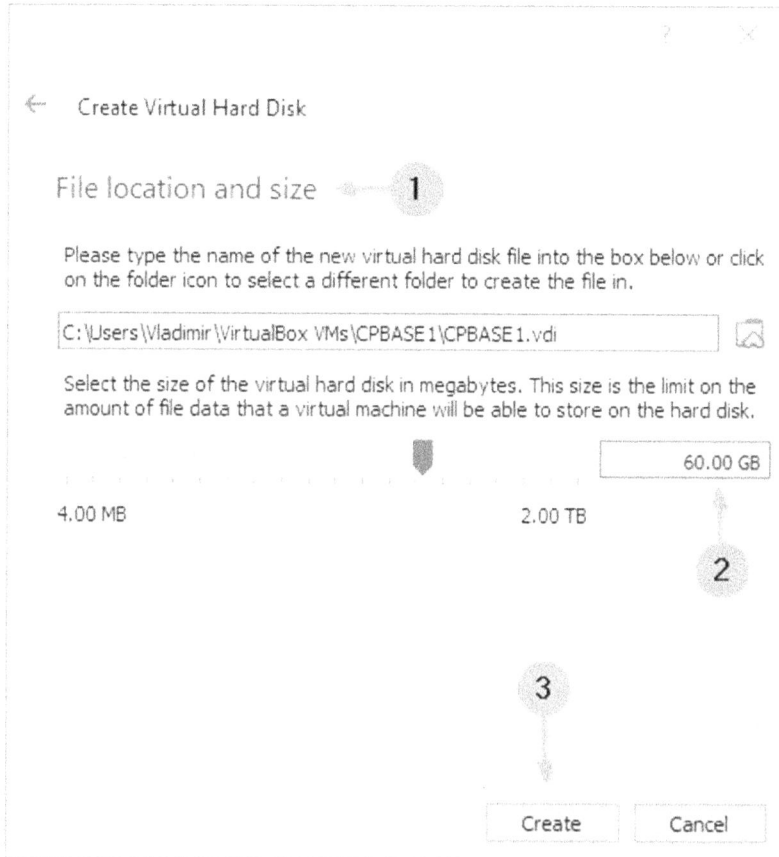

Figure 4.27 – Configuring the file location and size

In the newer versions of VirtualBox, the previous two steps are combined into one, as shown in the following screenshot:

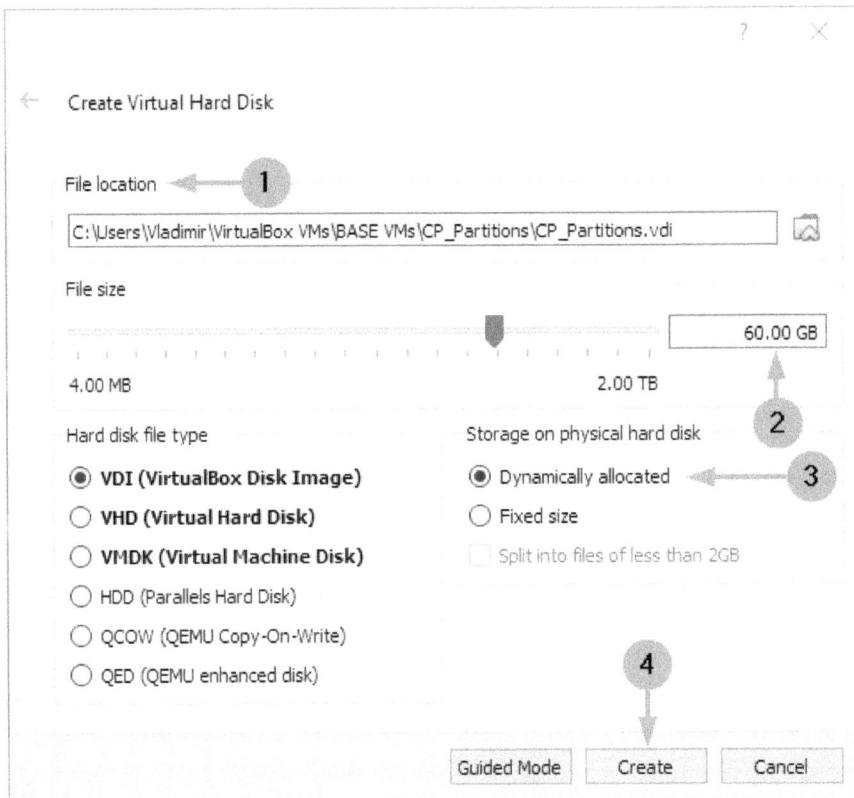

Figure 4.28 – Configuring dynamic space allocation, file location, and size

8. In **CPBASE - Settings** [1], click on **System** [2], then on the **Processor** tab [3], and type 4 in **Processor(s)** [4]. Uncheck **Enable PAE/NX** [5] and check **Enable Nested VT-x/AMD-V** [6]:

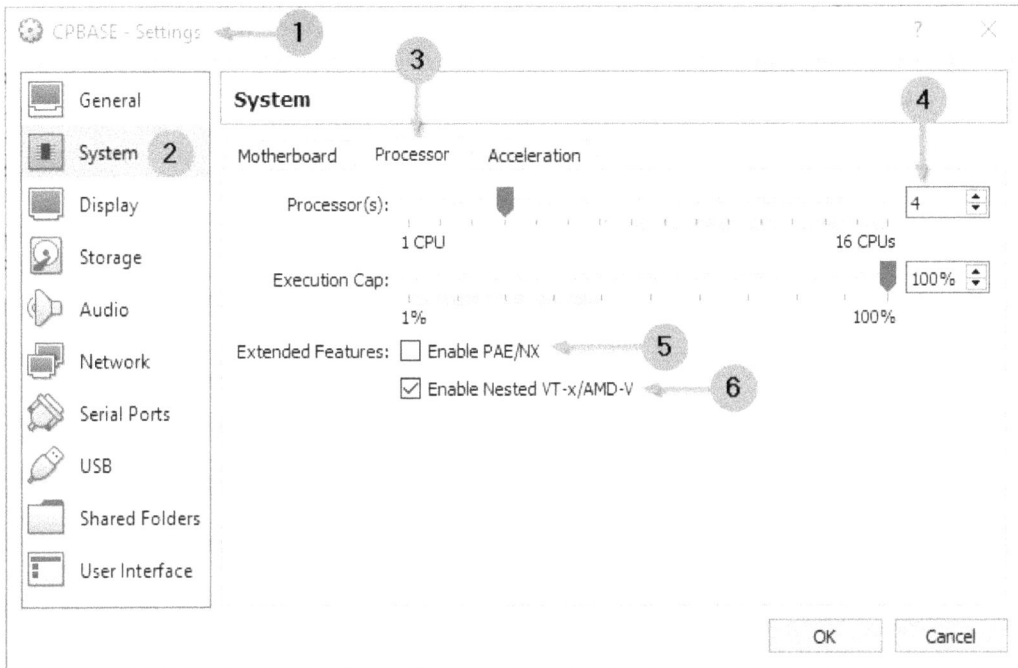

Figure 4.29 – Specifying the number of CPUs

9. While still in **System** [1], click on the **Acceleration** tab [2], choose **None** for
    Paravirtualization Interface, and verify that **Enable Nested Paging** is checked [3]:

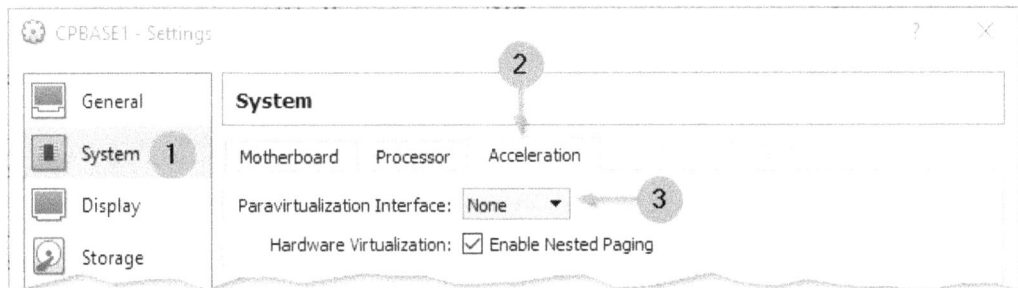

Figure 4.30 – Selecting virtualization parameters

10. In **Storage**, click on the **Empty** DVD icon [2]. Verify that **Optical Drive** is set as **IDE Secondary Device 0** and click on the blue DVD icon to the right of it [3]. In the drop-down box, click on **Choose a disk file...** [4]:

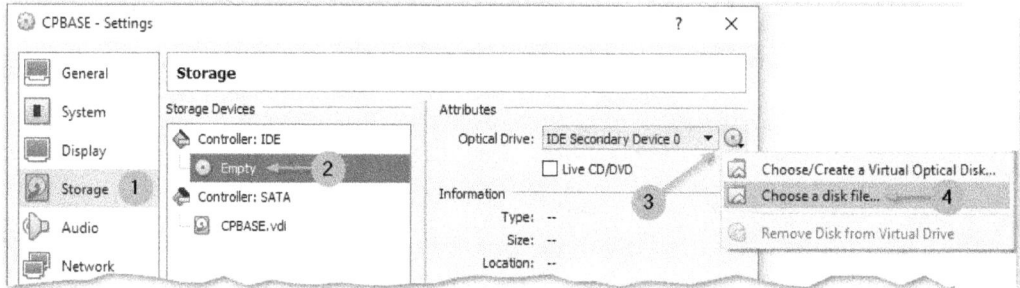

Figure 4.31 – Selecting installation media part 1

11. In the **Please choose a virtual optical disk file** popup [1], navigate to the C:\ CPBook\LabShare\ISOs_and_OVAs\ folder [2], click once on the Check_ Point_R81.1#_T###.iso file [3], and click **Open** [4]:

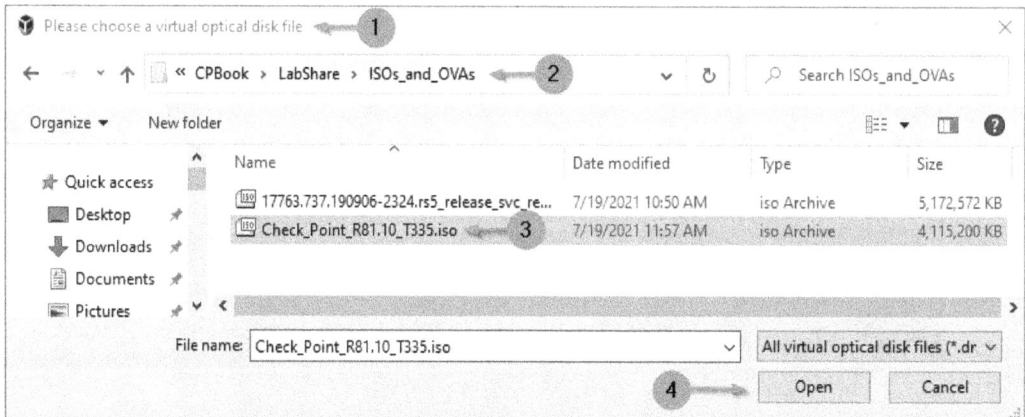

Figure 4.32 – Selecting installation media part 2

12. While still in **Storage** [1], click on the second controller [2]. Define it as SATA [3] and **LsiLogic SAS** [4] and verify that **Use Host I/O Cache** is checked [5]:

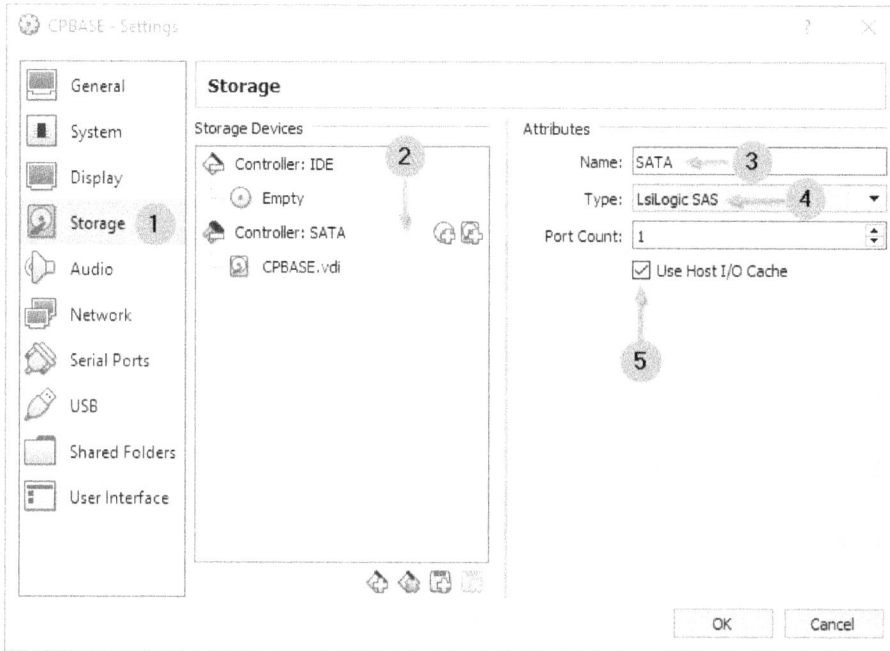

Figure 4.33 – Specifying storage controller configuration

13. In **Audio** [1], uncheck **Enable Audio** [2]:

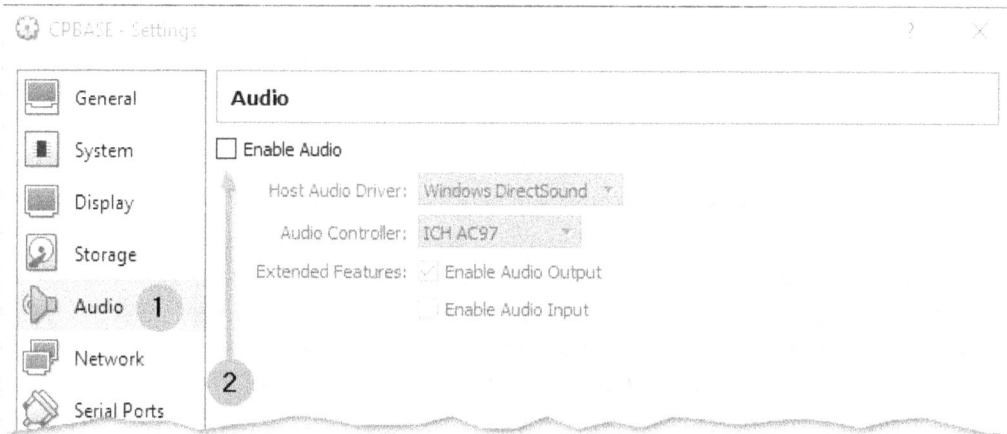

Figure 4.34 – Disabling audio

14. In the **Network** [1] pane | **Adapter 1** tab, choose **Internal Network** [2], and in the **Name** field dropdown, choose **Net_DeadEnd** [3]:

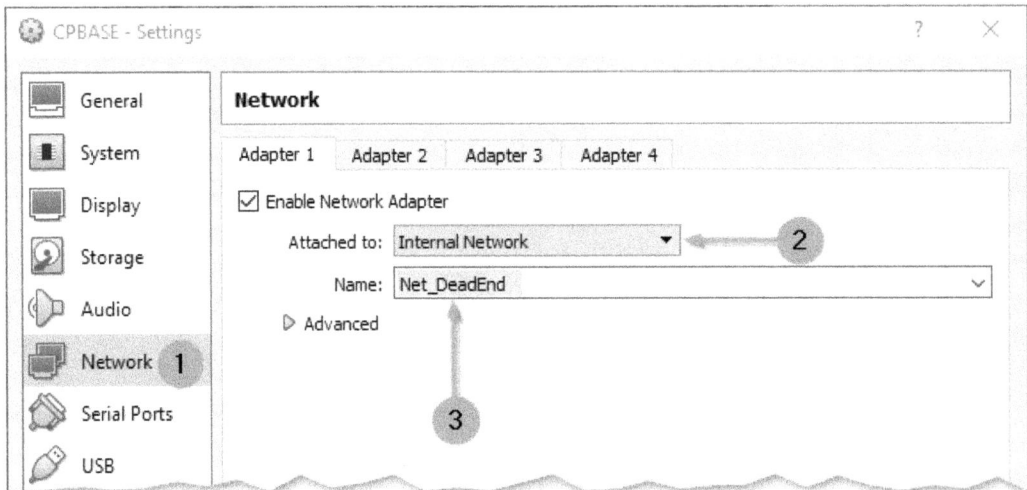

Figure 4.35 – Defining the network

15. In **USB**, uncheck **Enable USB Controller** [1] and click **OK** [2]:

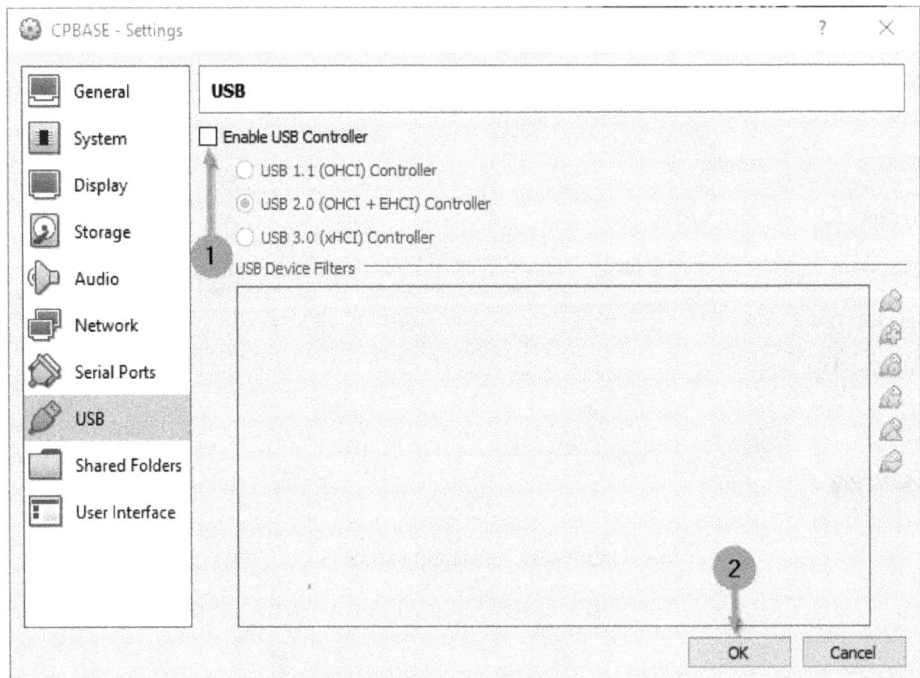

Figure 4.36 – Disabling USB

16. Now, in **Oracle VM VirtualBox Manager** [1], with **CPBASE** selected [2], click **Start** [3]:

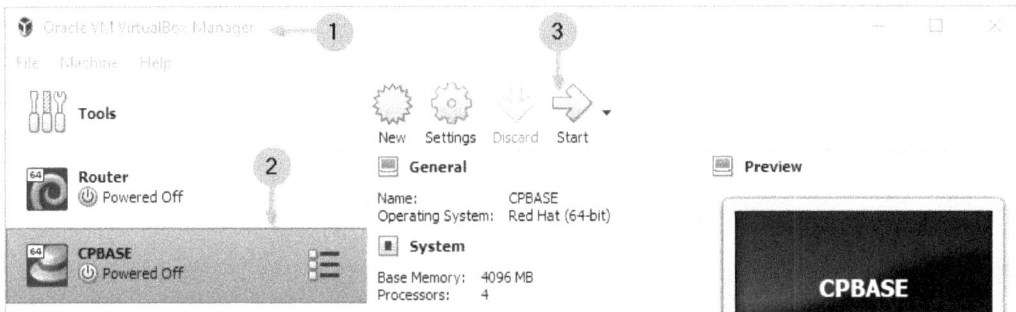

Figure 4.37 – Starting the CPBASE VM

The Check Point template VM called CPBASE is now created and running. We may proceed to the *Finalizing the Check Point base VM installation* section. The following section is optional for those interested in scripted VM creation.

Check Point base image scripted

If you prefer to use scripting, the creation of this VM in Oracle VirtualBox could be accomplished via the CLI.

> **Important Note**
>
> If you have already performed the GUI-based steps previously, skip this section and go to the *Finalizing the Check Point base VM installation* subsection. This script will overwrite the original name of the Check Point installation media ISO.

On your LabHost, open Command Prompt and paste the following code snippet in it:

```
SET PATH=%PATH%;C:\Program Files\Oracle\VirtualBox
SET VM=CPBASE
REN C:\CPBook\LabShare\ISOs_and_OVAs\Check_Point_R81.10_*.iso
CPBASEEVAL.iso
VBoxManage createhd --filename %VM%.vdi --size 61,440
VBoxManage createvm --name %VM% --ostype RedHat_64 --register
VBoxManage storagectl %VM% --name "IDE" --add IDE --controller
PIIX4
VBoxManage storageattach %VM% --storagectl "IDE" --port 0
--device 0 --type dvddrive --medium "C:\CPBook\LabShare\ISOs_
```

```
and_OVAs\CPBASEEVAL.iso"
VBoxManage storagectl %VM% --name "SATA" --add SAS --controller
LsiLogicSas
VBoxManage storageattach %VM% --storagectl "SATA" --port 0
--device 0 --type hdd --medium %VM%.vdi
VBoxManage modifyvm %VM% --cpus 4
VBoxManage modifyvm %VM% --ioapic on
VBoxManage modifyvm %VM% --paravirtprovider none
VBoxManage modifyvm %VM% --nestedpaging on
VBoxManage modifyvm %VM%  --boot1 disk --boot2 DVD --boot3 none
--boot4 none
VBoxManage modifyvm %VM% --memory 4096 --vram 128
VBoxManage modifyvm %VM% --keyboard ps2
VBoxManage modifyvm %VM% --nic1 intnet
VBoxManage modifyvm %VM% --nictype1 82540EM
VBoxManage modifyvm %VM% --macaddress1 auto
VBoxManage modifyvm %VM% --cableconnected1 on
VBoxManage modifyvm %VM% --intnet1 Net_DeadEnd
VBoxManage modifyvm %VM% --graphicscontroller vmsvga
VBoxManage modifyvm %VM% --acpi on
VBoxManage modifyvm %VM% --pae off
VBoxManage startvm %VM% --type gui
```

Press *Enter* after the last line to execute. Or, from LabHost Command Prompt, execute the
C:\CPBook\LabShare\Scripts\create_CPBASE.bat file.

This completes the creation of the Check Point template VM via the CLI. We may now
finalize the installation.

Finalizing the Check Point base VM installation

In a previous step, this VM was created using either the GUI or CLI. Now, after it is
powered up, we must complete the installation process by following these instructions:

1. Click once inside **CPBASE [Running]** window, press the *Tab* key to select **Install
    Gaia on this system** [1], and press *Enter*:

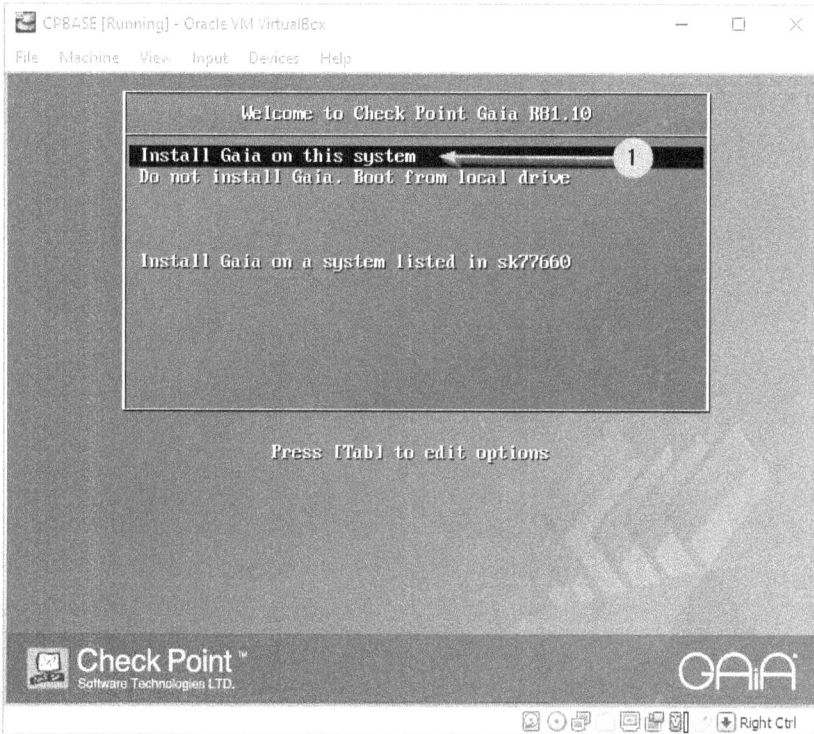

Figure 4.38 – Choosing to install Gaia

2.   Click **OK** [1] to proceed with the installation:

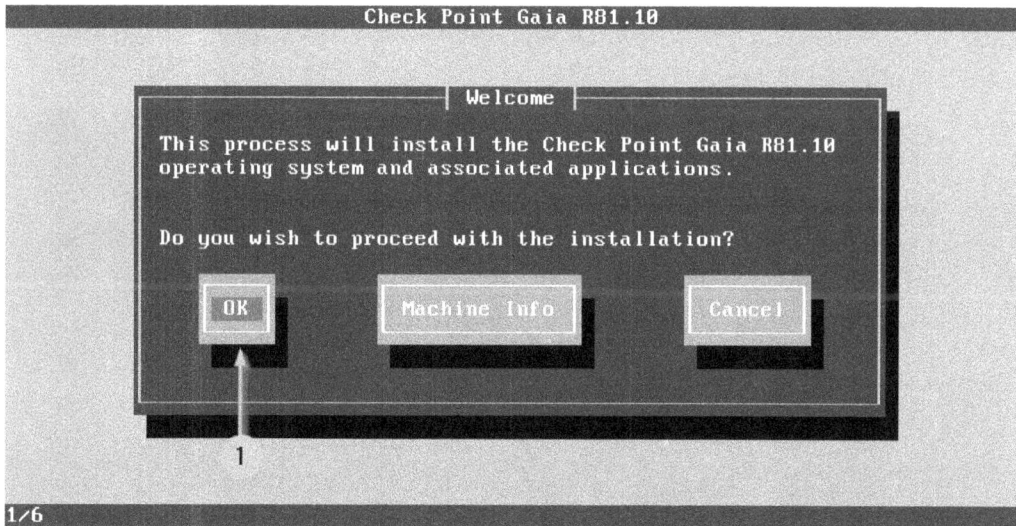

Figure 4.39 – Gaia installation prompt 1

3. Choose your keyboard locale (in our case, US) [1] and click **OK** [2]:

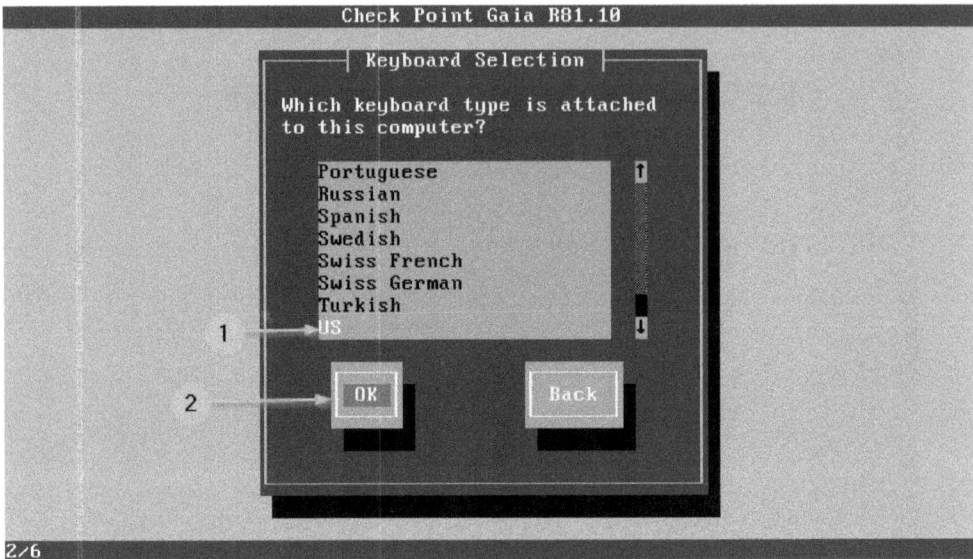

Figure 4.40 – Gaia installation prompt 2

4. Change the partition configuration using the *Tab* key to move between the fields and set **System-root (GB)** to 15 [1] and **Logs (GB)** to 11 [2]. Click **OK** [3] (recall note at the beginning of this section regarding HDD and partition sizes)

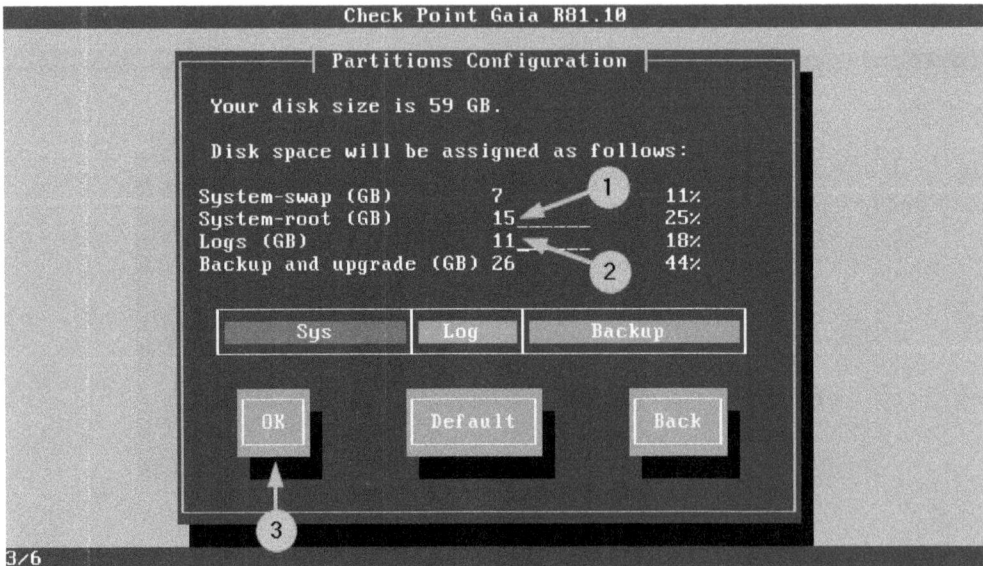

Figure 4.41 – Gaia installation prompt 3

5. Enter our lab password, CPL@b8110, into both fields [1] and click **OK** [2]:

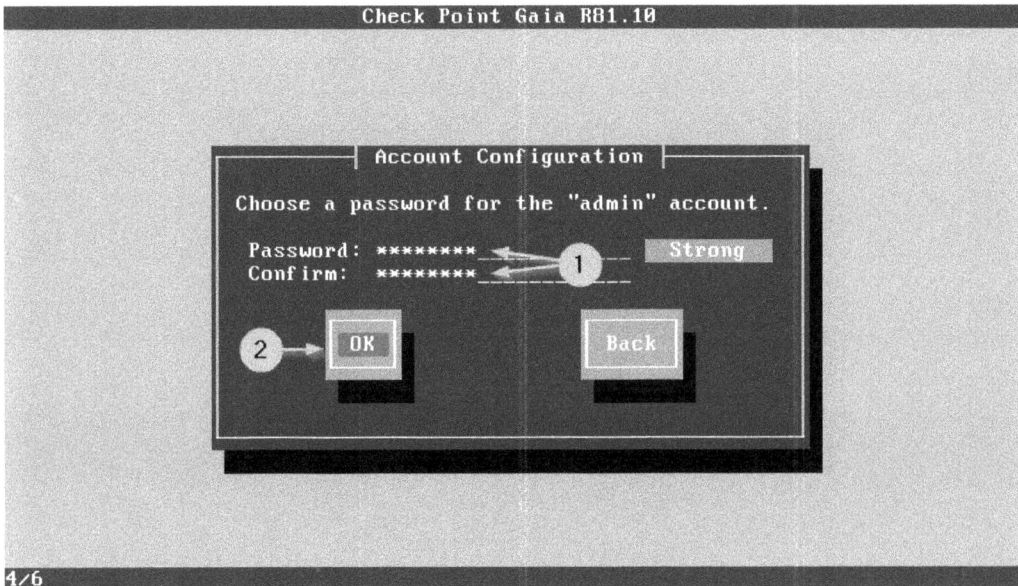

Figure 4.42 – Gaia installation prompt 4

6. Enter a temporary IP address [1] and netmask [2] and delete the automatically populated entry for the default gateway. Click **OK** [3]:

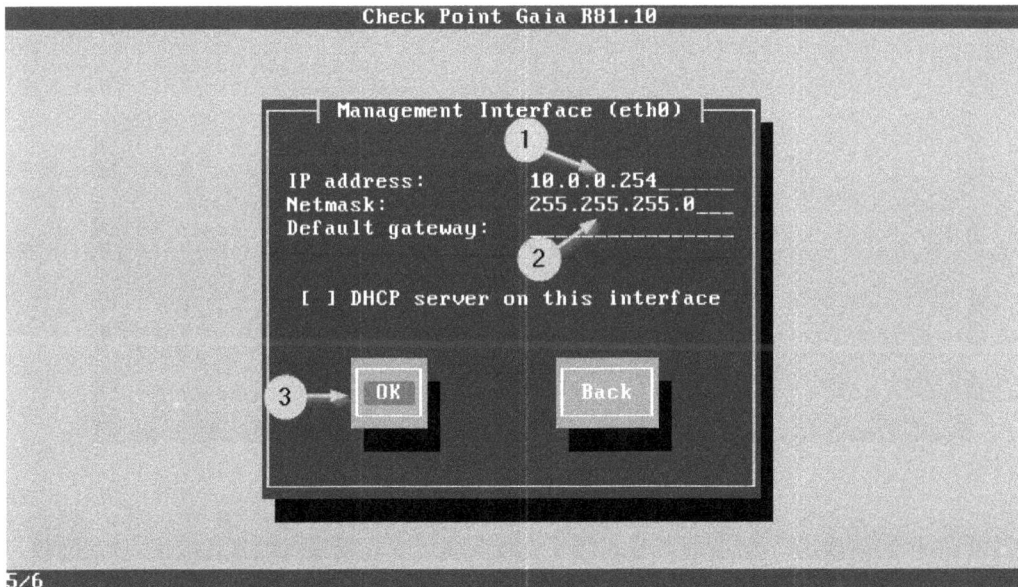

Figure 4.43 – Gaia installation prompt 5

7.   Click **OK** [1] to agree to the formatting of your VM's hard disk:

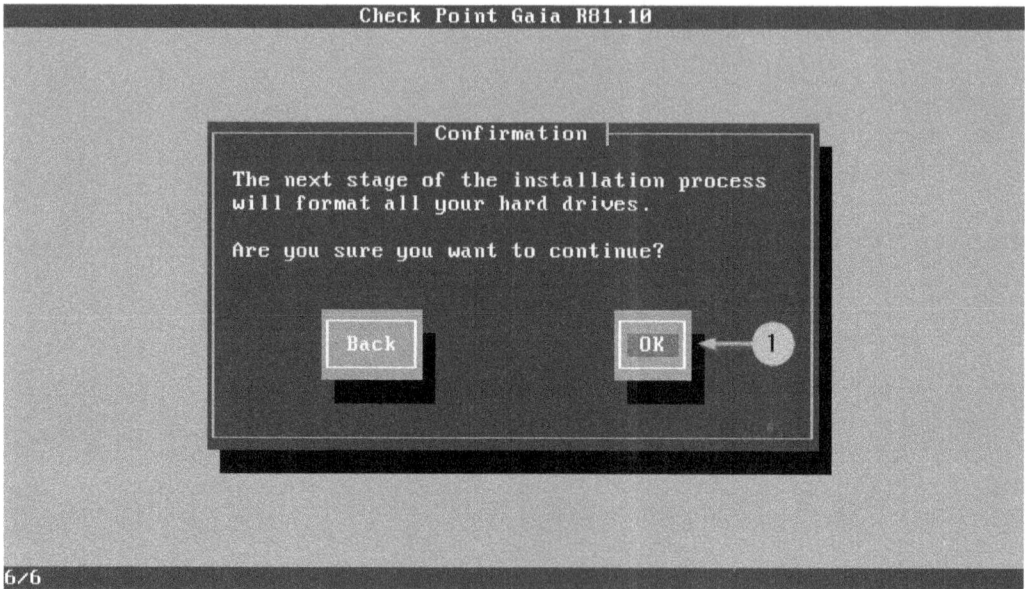

Figure 4.44 – Gaia installation prompt 6

8.   Click **Reboot** [1] to complete the installation process:

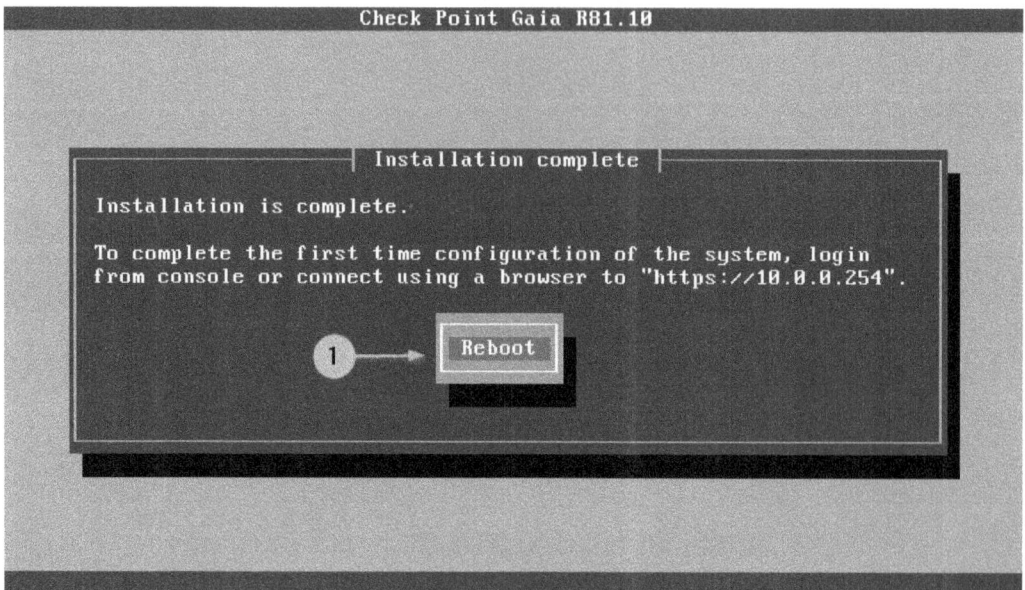

Figure 4.45 – Reboot

9. Observe the boot process, ignoring warnings that may show up due to your choice of hardware. These are typically harmless and will allow you to continue:

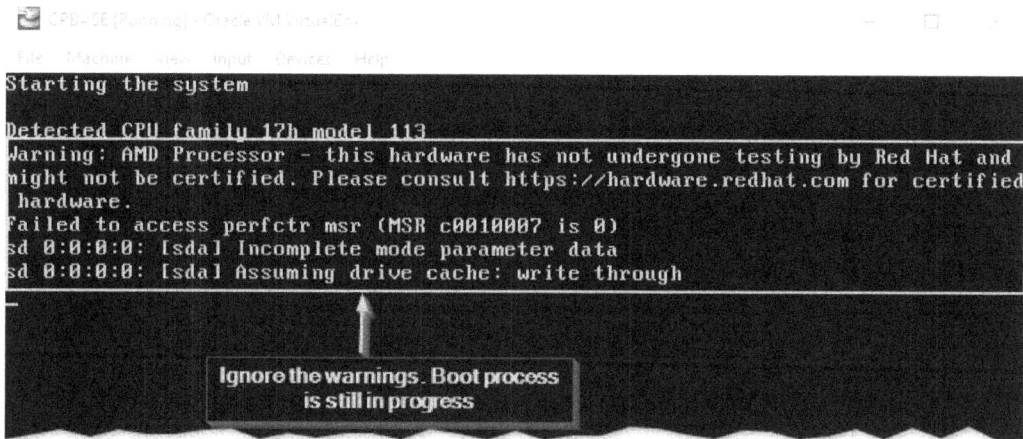

Figure 4.46 – Post-installation boot process

10. Once you see the login prompt, log on as `admin` using `CPL@b8110` as your password. You should see the invitation to connect to the web UI [1]. Ignore that for the moment and instead, change the randomly assigned hostname to `CPBASE` [2], save the configuration [3], and shut down the VM using the `halt` command [4] with confirmation, `Y` [5], and press *Enter*:

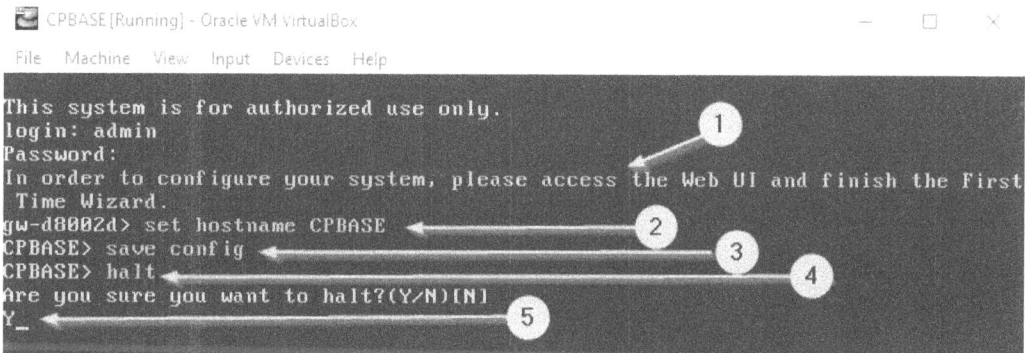

Figure 4.47 – Post-installation CPBASE configuration

11. Once the VM is shut down, on your LabHost, paste these lines in Command Prompt:

```
SET PATH=%PATH%;C:\Program Files\Oracle\VirtualBox
SET VM=CPBASE
VBoxManage storageattach %VM% --storagectl "IDE" --port 0
--device 0 --medium emptydrive
```

```
VBoxManage snapshot %VM% take "Snapshot 1" --description
"Do NOT delete!!! This snapshot is used to spawn the
linked clones of all Check Point components in the lab."
```

Or, in your LabHost Command Prompt, execute the `C:\CPBook\LabShare\Scripts\eject_and_snapshot_CPBASE.bat` file.

Now, that we have prepared the base images, we can create linked clones, that is, VMs that will be sharing the same common storage provided by the snapshots of base VMs and will only consume additional space for differential disks, established after their creation.

Creating linked clones

Now, with the base VMs prepared, let's create linked clones for the rest of the lab's infrastructure:

1. On your LabHost, open Command Prompt.
2. Change your current directory to `C:\CPBook\CPLab\Scripts`.
3. To see what the script is doing, I recommend opening it in a text editor, such as Notepad on Windows. This script will create linked clones of our base VMs, attach additional network interfaces, when necessary, and connect them to the corresponding network segments.

    You have an option of either executing the `clonevms.bat` script or copying and pasting the code contained in it into the CMD.

    The following is an excerpt from the batch file demonstrating its operation. Note that we must use the CLI or a batch file to accomplish the cloning process, since the Oracle VirtualBox GUI allows the creation of VMs with a maximum of four network interfaces. The CLI options allow the creation of up to eight:

    ```
    SET PATH=%PATH%;C:\Program Files\Oracle\VirtualBox
    VBoxManage clonevm WINBASE --snapshot "Snapshot 1"
    --options link --name LeftHost --register
    VBoxManage modifyvm "LeftHost" --nic1 intnet
    VBoxManage modifyvm "LeftHost" --nictype1 82540EM
    VBoxManage modifyvm "LeftHost" --macaddress1 auto
    VBoxManage modifyvm "LeftHost" --cableconnected1 on
    VBoxManage modifyvm "LeftHost" --intnet1 Net_10.10.10.0
    VBoxManage clonevm CPBASE --snapshot "Snapshot 1"
    --options link --name CPCM1 --register
    VBoxManage modifyvm "CPCM1" --nic1 intnet
    ```

```
VBoxManage modifyvm "CPCM1" --nictype1 82540EM
VBoxManage modifyvm "CPCM1" --macaddress1 auto
VBoxManage modifyvm "CPCM1" --cableconnected1 on
VBoxManage modifyvm "CPCM1" --intnet1 Net_10.0.0.0
VBoxManage modifyvm "CPCM1" --nic2 intnet
VBoxManage modifyvm "CPCM1" --nictype2 82540EM
VBoxManage modifyvm "CPCM1" --macaddress2 auto
VBoxManage modifyvm "CPCM1" --cableconnected2 on
VBoxManage modifyvm "CPCM1" --intnet2 Net_10.10.10.0
VBoxManage modifyvm "CPCM1" --nic3 intnet
VBoxManage modifyvm "CPCM1" --nictype3 82540EM
VBoxManage modifyvm "CPCM1" --macaddress3 auto
VBoxManage modifyvm "CPCM1" --cableconnected3 on
VBoxManage modifyvm "CPCM1" --intnet3 Net_10.20.20.0
VBoxManage modifyvm "CPCM1" --nic4 intnet
VBoxManage modifyvm "CPCM1" --nictype4 82540EM
VBoxManage modifyvm "CPCM1" --macaddress4 auto
VBoxManage modifyvm "CPCM1" --cableconnected4 on
VBoxManage modifyvm "CPCM1" --intnet4 Net_192.168.255.0
VBoxManage modifyvm "CPCM1" --nic5 intnet
VBoxManage modifyvm "CPCM1" --nictype5 82540EM
VBoxManage modifyvm "CPCM1" --macaddress5 auto
VBoxManage modifyvm "CPCM1" --cableconnected5 on
VBoxManage modifyvm "CPCM1" --intnet5 Net_200.100.0.0
VBoxManage modifyvm "CPCM1" --nic6 intnet
VBoxManage modifyvm "CPCM1" --nictype6 82540EM
VBoxManage modifyvm "CPCM1" --macaddress6 auto
VBoxManage modifyvm "CPCM1" --cableconnected6 on
VBoxManage modifyvm "CPCM1" --intnet6 Net_10.30.30.0
```

4. As the script is being executed, you will see nine new VMs appearing in **Oracle VM VirtualBox Manager** [1 to 9]:

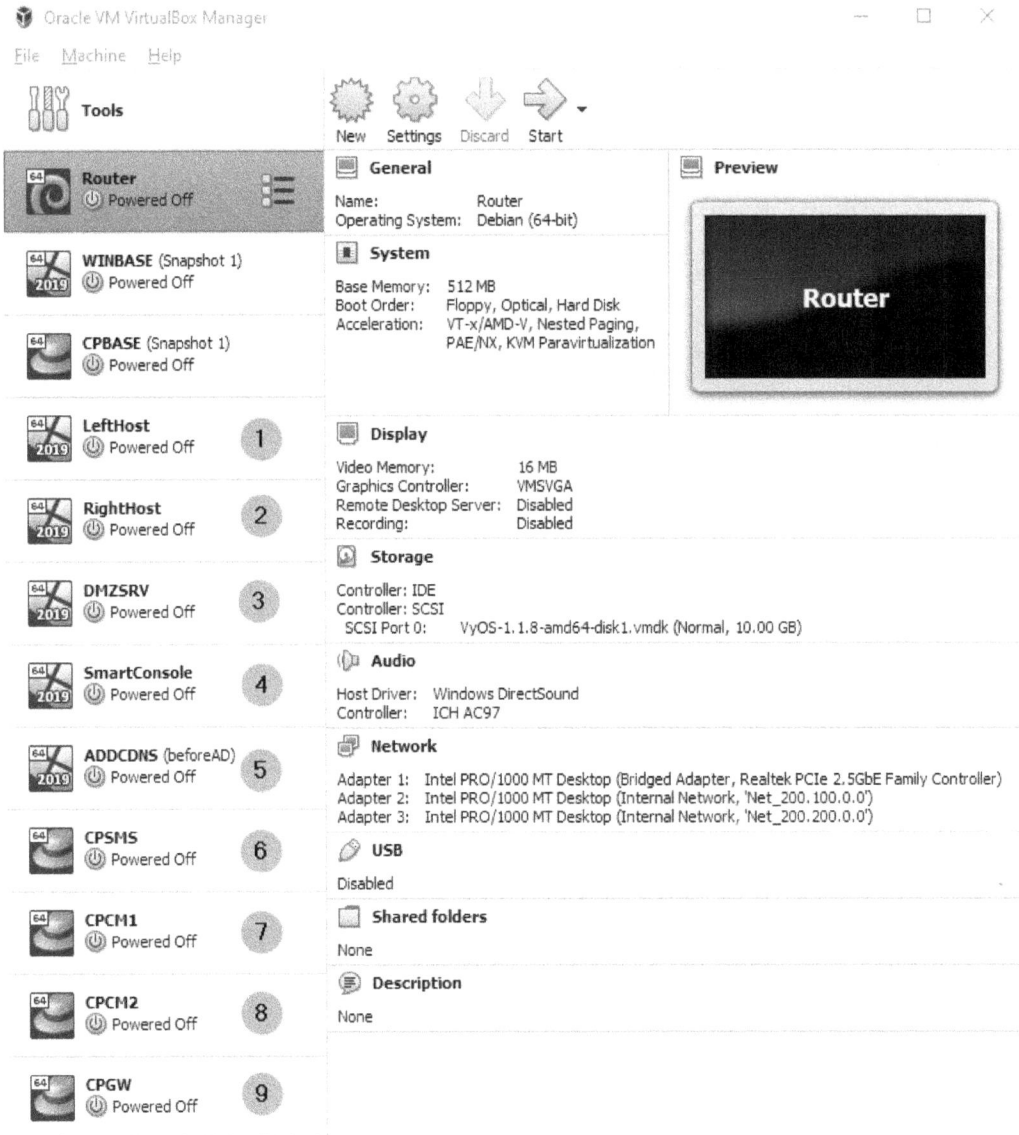

Figure 4.48 – Cloned VMs

5. To avoid inadvertently deleting or changing the configuration of the base VMs and their snapshots used for cloned VMs, in Oracle VM VirtualBox Manager, select both base VMs by holding *Ctrl* and clicking on each, so that both of them are blue [1], right-click anywhere in the blue area, and click **Group** [2]:

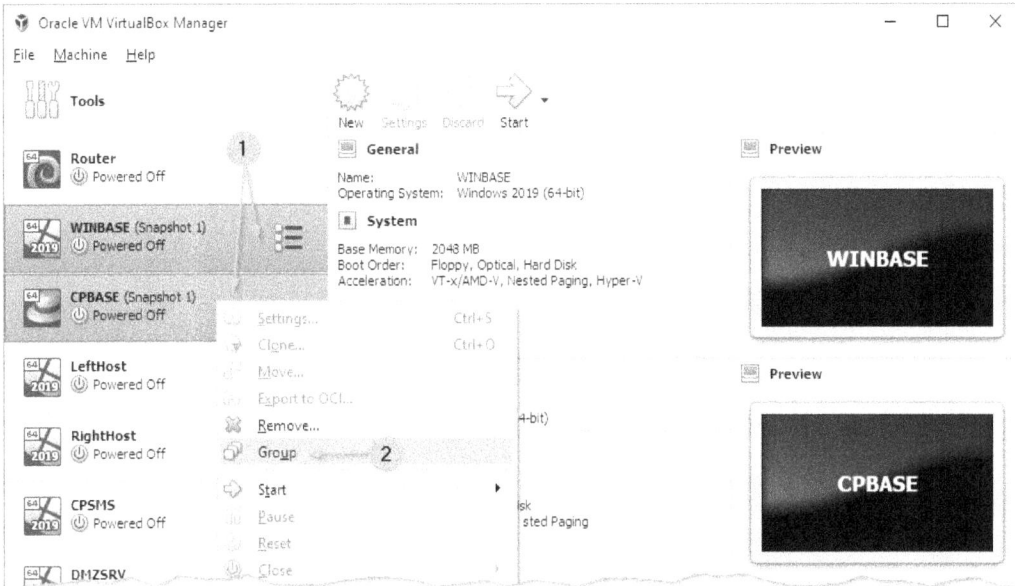

Figure 4.49 – Grouping base VMs

6. Once the VMs are grouped, click on **New group** [1] and click **Rename Group…** [2]. Type in the new group name, BASE VMs, and press *Enter*. Click on the expansion arrow on the left side of the group name [3] to hide the group member VMs:

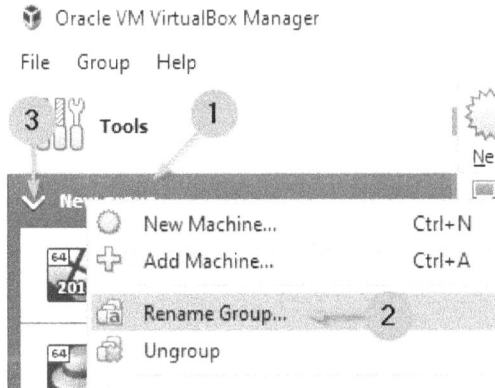

Figure 4.50 – Hiding base VMs

With the creation of the linked clones complete, we can finalize the preparation of individual cloned VMs.

Preparing cloned Windows hosts

Now we must finish the configuration of the freshly cloned Windows VMs. For each Windows VM *except* ADDCDNS, do the following:

1. Power them up one at a time.

2. Once the new Windows VM is powered up, complete choosing the locale and language.

3. Accept the **End User License Agreement** (**EULA**).

4. Set the lab password for the local administrative account(s) to CPL@b8110.

5. Log on to the cloned Windows VM.

6. Open File Explorer and click on **This PC**.

7. Under **Network locations**, double-click on LabShare (\\VBoxSrv) (F:).

8. Double-click on the Scripts folder.

9. Click **View** [1] and check the **File name extensions** box [2]:

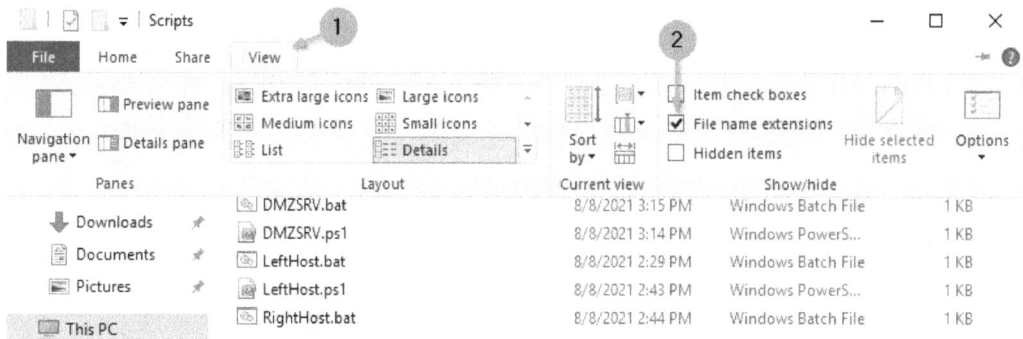

Figure 4.51 – Filename extensions visibility

10. You can execute the batch file [1] named after the VM in which you are working [2] in that VM's Command Prompt (by specifying the full path to it, that is, F:\ LabShare\Scripts\<vmname.bat>). The batch files are running associated PowerShell scripts bypassing the localhost execution policy.

Or, right-click the .ps1 script with the same name [3] and click **Edit with Notepad++** [4] to examine the script and execute it manually:

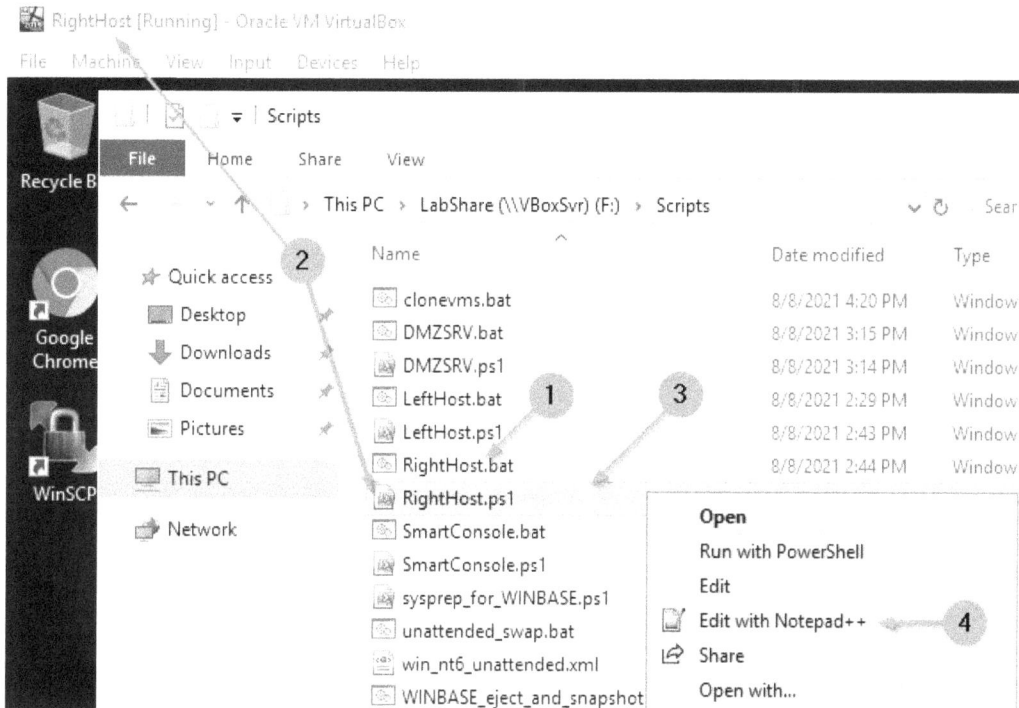

Figure 4.52 – Batch files to VM name correlation

11. If you have chosen to execute the `.bat` file, it will call the PowerShell script with the same name, apply necessary changes, and restart your VM with a new hostname and IP address. Additionally, automatic Windows Update services will be disabled on this VM from here on to conserve the space on your hard drive. If you have ample space, follow the comments inside the `.ps1` files to comment out the indicated lines in the script using # in front of the lines, and save the script before running it.

12. If you have chosen to run the PowerShell commands manually, in Notepad++, copy all lines containing text (do not copy the last empty line: if it is pasted in the PowerShell prompt, the script will cause the VM to reboot immediately upon completion and you will not be able to observe its output).

Note: If you do not want to download the scripts and would rather copy code from the book, the following are the PowerShell code snippets for each of the Windows VMs except ADDCDNS, which will be provided later in this chapter:

- LeftHost:

```
Get-ScheduledTask -TaskName ServerManager | Disable-
ScheduledTask -Verbose
Rename-Computer -NewName LeftHost
New-NetIPAddress -IPAddress 10.10.10.10 -DefaultGateway
10.10.10.1 -PrefixLength 24 -InterfaceIndex
(Get-NetAdapter).InterfaceIndex
$WindowsUpdate = "Scheduled Start"
Get-ScheduledTask -TaskName $WindowsUpdate | Disable-
ScheduledTask  -Verbose
Get-ScheduledTask -TaskName StartComponentCleanUp |
Disable-ScheduledTask  -Verbose
Get-Service -Name wuauserv | Set-Service -StartupType
Disabled -Confirm:$false
Stop-Service wuauserv -Force
Restart-Computer
```

- RightHost:

```
Get-ScheduledTask -TaskName ServerManager | Disable-
ScheduledTask -Verbose
Rename-Computer -NewName RightHost
New-NetIPAddress -IPAddress 172.16.16.10 -DefaultGateway
172.16.16.1 -PrefixLength 24 -InterfaceIndex
(Get-NetAdapter).InterfaceIndex
$WindowsUpdate = "Scheduled Start"
Get-ScheduledTask -TaskName $WindowsUpdate | Disable-
ScheduledTask  -Verbose
Get-ScheduledTask -TaskName StartComponentCleanUp |
Disable-ScheduledTask  -Verbose
Get-Service -Name wuauserv | Set-Service -StartupType
Disabled -Confirm:$false
Stop-Service wuauserv -Force
Restart-Computer
```

- DMZSRV:

  ```
  Get-ScheduledTask -TaskName ServerManager | Disable-
  ScheduledTask -Verbose
  Rename-Computer -NewName DMZSRV
  New-NetIPAddress -IPAddress 10.30.30.5 -DefaultGateway
  10.30.30.1 -PrefixLength 24 -InterfaceIndex
  (Get-NetAdapter).InterfaceIndex
  $WindowsUpdate = "Scheduled Start"
  Get-ScheduledTask -TaskName $WindowsUpdate | Disable-
  ScheduledTask  -Verbose
  Get-ScheduledTask -TaskName StartComponentCleanUp |
  Disable-ScheduledTask  -Verbose
  Get-Service -Name wuauserv | Set-Service -StartupType
  Disabled -Confirm:$false
  Stop-Service wuauserv -Force
  Restart-Computer
  ```

- SmartConsole:

  ```
  Get-ScheduledTask -TaskName ServerManager | Disable-
  ScheduledTask -Verbose
  Rename-Computer -NewName SmartConsole
  New-NetIPAddress -IPAddress 10.0.0.20 -DefaultGateway
  10.0.0.1 -PrefixLength 24 -InterfaceIndex
  (Get-NetAdapter).InterfaceIndex
  $WindowsUpdate = "Scheduled Start"
  Get-ScheduledTask -TaskName $WindowsUpdate | Disable-
  ScheduledTask  -Verbose
  Get-ScheduledTask -TaskName StartComponentCleanUp |
  Disable-ScheduledTask  -Verbose
  Get-Service -Name wuauserv | Set-Service -StartupType
  Disabled -Confirm:$false
  Stop-Service wuauserv -Force
  Restart-Computer
  ```

13. Click **Start** and click **Windows PowerShell**.

14. Wait for PowerShell to load and, at the prompt, paste the lines you have copied in *step 12*.

15. Scroll through the output to see how each command was executed and, when done, hit *Enter* to execute the last pasted command (Restart-Computer).

16. Shut down the Windows VMs you are done with, until they are needed, to conserve LabHost's resources.

An example of the PowerShell script running on LeftHost is as follows:

```
Administrator: Windows PowerShell                                        —    □    ×

Windows PowerShell
Copyright (C) Microsoft Corporation. All rights reserved.

PS C:\Users\Administrator> #Caution!!! This script disables the Windows Update Services.
PS C:\Users\Administrator> #It is intended for the use in isolated lab environments for space saving purposes only.
PS C:\Users\Administrator> #If you do not have space constraints, comment-out five lines before the last one and save th
e script!!!
PS C:\Users\Administrator> Get-ScheduledTask -TaskName ServerManager | Disable-ScheduledTask -Verbose

TaskPath                                    TaskName                    State
--------                                    --------                    -----
\Microsoft\Windows\Server Manager\          ServerManager               Disabled


PS C:\Users\Administrator> Rename-Computer -NewName SmartConsole
WARNING: The changes will take effect after you restart the computer WIN-Q3I97U5JFD6.
PS C:\Users\Administrator> New-NetIPAddress -IPAddress 10.30.30.5 -DefaultGateway 10.30.30.1 -PrefixLength 24 -Interface
Index (Get-NetAdapter).InterfaceIndex


IPAddress         : 10.30.30.5
InterfaceIndex    : 6
InterfaceAlias    : Ethernet
AddressFamily     : IPv4
Type              : Unicast
PrefixLength      : 24
PrefixOrigin      : Manual
SuffixOrigin      : Manual
AddressState      : Tentative
ValidLifetime     : Infinite ([TimeSpan]::MaxValue)
PreferredLifetime : Infinite ([TimeSpan]::MaxValue)
SkipAsSource      : False
PolicyStore       : ActiveStore

IPAddress         : 10.30.30.5
InterfaceIndex    : 6
InterfaceAlias    : Ethernet
AddressFamily     : IPv4
Type              : Unicast
PrefixLength      : 24
PrefixOrigin      : Manual
SuffixOrigin      : Manual
AddressState      : Invalid
ValidLifetime     : Infinite ([TimeSpan]::MaxValue)
PreferredLifetime : Infinite ([TimeSpan]::MaxValue)
SkipAsSource      : False
PolicyStore       : PersistentStore



PS C:\Users\Administrator> $WindowsUpdate = "Scheduled Start"
PS C:\Users\Administrator> Get-ScheduledTask -TaskName $WindowsUpdate | Disable-ScheduledTask  -Verbose

TaskPath                                    TaskName                    State
--------                                    --------                    -----
\Microsoft\Windows\WindowsUpdate\           Scheduled Start             Disabled


PS C:\Users\Administrator> Get-ScheduledTask -TaskName StartComponentCleanUp | Disable-ScheduledTask  -Verbose

TaskPath                                    TaskName                    State
--------                                    --------                    -----
\Microsoft\Windows\Servicing\               StartComponentCleanup       Disabled


PS C:\Users\Administrator> Set-Service wuauserv -Startup Disabled
PS C:\Users\Administrator> Stop-Service wuauserv -Force
PS C:\Users\Administrator> Restart-Computer_
```

Figure 4.53 – PowerShell script execution example

Now, as the last step in preparing Windows VMs, we will work on ADDCDNS.

Preparing Active Directory and the domain controller

Now that all the Windows VMs except ADDCDNS are configured, let's finish with this VM:

1. Power up the ACDCDNS VM and complete the locale and language settings, accept the EULA, and set the password to `CPL@b8110`.

2. Log on to the VM.

3. Click **Start** and click **Windows PowerShell**.

4. Copy and paste these two lines at the prompt to rename and reboot the server:

   ```
   Rename-Computer -NewName ADDCDNS
   Restart-Computer
   ```

5. After the computer has rebooted, log on.

6. Click **Start** and click **PowerShell**.

7. At the prompt, copy and paste the following code snippet to do the following:

   - Disable automatic Windows Update services.

   - Disable the Google update service.

   - Disable Microsoft MapsBroker.

   - Change the IP address.

   - Set the local DNS client to point to itself.

   - Install Active Directory and its management tools.

   - Configure this server as a domain controller and a DNS server.

   Paste the following code snippet:

   ```
   Get-ScheduledTask -TaskName ServerManager | Enable-
   ScheduledTask -Verbose
   $WindowsUpdate = "Scheduled Start"
   Get-ScheduledTask -TaskName $WindowsUpdate | Disable-
   ScheduledTask  -Verbose
   Get-ScheduledTask -TaskName StartComponentCleanUp |
   Disable-ScheduledTask  -Verbose
   Get-Service -Name wuauserv | Set-Service -StartupType
   Disabled -Confirm:$false
   Get-Service -Name MapsBroker | Set-Service -StartupType
   Disabled -Confirm:$false
   Get-Service -Name gupdate | Set-Service -StartupType
   ```

```
Disabled -Confirm:$false
Stop-Service wuauserv -Force
New-NetIPAddress -IPAddress 10.20.20.10 -DefaultGateway
10.20.20.1 -PrefixLength 24 -InterfaceIndex
(Get-NetAdapter).InterfaceIndex
Set-DNSClientServerAddress -InterfaceIndex
(Get-NetAdapter).InterfaceIndex -ServerAddresses
10.20.20.10
Install-WindowsFeature -name AD-Domain-Services
-IncludeManagementTools
Install-ADDSForest -DomainName mycp.lab
-DomainNetBIOSName AD -InstallDNS
```

8. Enter a password for `SafeModeAdministratorPassword` (also known as **Directory Services Restore Mode** (**DSRM**)) and confirm it (`CPL@b8110`).

9. Type y and press *Enter* to confirm that this server will be configured as a domain controller.

10. Observe the script in action and wait for the VM to reboot.

> **Note**
>
> After the VM has rebooted, it will go through a long **Applying Computer Settings** cycle, as it is getting ready to become the first domain controller.

11. Log on as `AD\administrator` with the `CPL@b8110` password.

12. When you log on, Server Manager will automatically start. Dismiss the **Try managing servers with Windows Admin Center** popup permanently by checking **Don't show this message again** and clicking on **X** in the top-right corner of the window.

13. Your **Server Manager | Dashboard** should look like this:

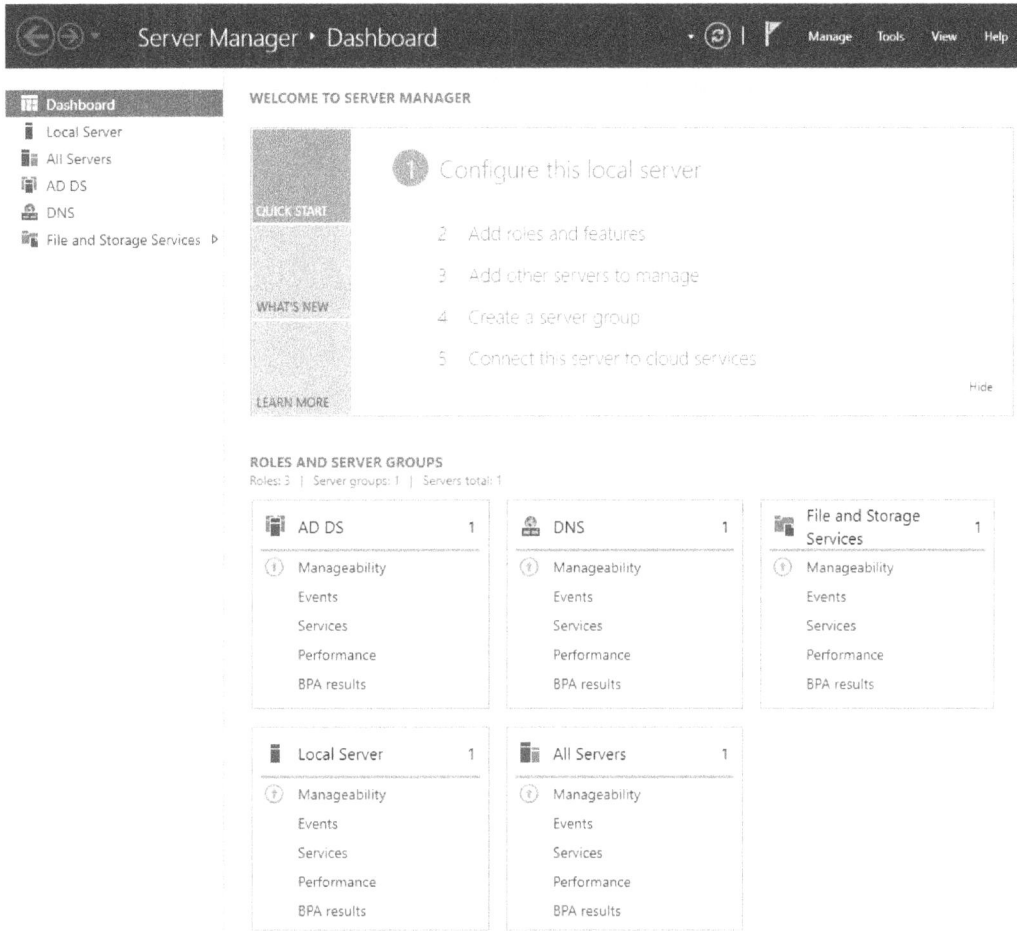

Figure 4.54 – Server Manager on ADDCDNS

14. To complete the configuration of subnets in **Sites and Services**, as well as adding the DNS forwarder (IBM's secure public DNS 9.9.9.9), creating reverse lookup zones, and adding host records for our lab infrastructure, you can either open PowerShell again to paste the following code snippets into it to do the following:

- Define AD subnets:

```
Import-Module ActiveDirectory
New-ADReplicationSubnet -Name "10.0.0.0/24"
New-ADReplicationSubnet -Name "10.10.10.0/24"
New-ADReplicationSubnet -Name "10.20.20.0/24"
New-ADReplicationSubnet -Name "10.30.30.0/24"
```

- Define DNS zones:

```
Add-DnsServerPrimaryZone -NetworkID "10.0.0.0/24"
-ReplicationScope "Forest"

Add-DnsServerPrimaryZone -NetworkID "10.10.10.0/24"
-ReplicationScope "Forest"

Add-DnsServerPrimaryZone -NetworkID "10.20.20.0/24"
-ReplicationScope "Forest"

Add-DnsServerPrimaryZone -NetworkID "10.30.30.0/24"
-ReplicationScope "Forest"
```

- Define the DNS forwarder:

```
Add-DnsServerForwarder -IPAddress 9.9.9.9 -PassThru
```

- Define DNS records:

```
Add-DnsServerResourceRecordA -Name "SmartConsole"
-ZoneName "mycp.lab" -AllowUpdateAny -IPv4Address
"10.0.0.20" -TimeToLive 01:00:00

Add-DnsServerResourceRecordPtr -Name "20" -ZoneName
"0.0.10.in-addr.arpa" -AllowUpdateAny -TimeToLive
01:00:00 -AgeRecord -PtrDomainName "SmartConsole.mycp.
lab"

Add-DnsServerResourceRecordA -Name "CPSMS" -ZoneName
"mycp.lab" -AllowUpdateAny -IPv4Address "10.0.0.10"
-TimeToLive 01:00:00

Add-DnsServerResourceRecordPtr -Name "10" -ZoneName
"0.0.10.in-addr.arpa" -AllowUpdateAny -TimeToLive
01:00:00 -AgeRecord -PtrDomainName "CPSMS.mycp.lab"

Add-DnsServerResourceRecordA -Name "CPCM1" -ZoneName
"mycp.lab" -AllowUpdateAny -IPv4Address "10.0.0.2"
-TimeToLive 01:00:00

Add-DnsServerResourceRecordPtr -Name "2" -ZoneName
"0.0.10.in-addr.arpa" -AllowUpdateAny -TimeToLive
01:00:00 -AgeRecord -PtrDomainName "CPCM1.mycp.lab"

Add-DnsServerResourceRecordA -Name "CPCM2" -ZoneName
"mycp.lab" -AllowUpdateAny -IPv4Address "10.0.0.3"
-TimeToLive 01:00:00

Add-DnsServerResourceRecordPtr -Name "3" -ZoneName
"0.0.10.in-addr.arpa" -AllowUpdateAny -TimeToLive
01:00:00 -AgeRecord -PtrDomainName "CPCM2.mycp.lab"

Add-DnsServerResourceRecordA -Name "CPCXL" -ZoneName
```

```
"mycp.lab" -AllowUpdateAny -IPv4Address "10.0.0.1"
-TimeToLive 01:00:00
Add-DnsServerResourceRecordPtr -Name "1" -ZoneName
"0.0.10.in-addr.arpa" -AllowUpdateAny -TimeToLive
01:00:00 -AgeRecord -PtrDomainName "CPCXL.mycp.lab"
Add-DnsServerResourceRecordA -Name "LeftHost" -ZoneName
"mycp.lab" -AllowUpdateAny -IPv4Address "10.10.10.10"
-TimeToLive 01:00:00
Add-DnsServerResourceRecordPtr -Name "10" -ZoneName
"10.10.10.in-addr.arpa" -AllowUpdateAny -TimeToLive
01:00:00 -AgeRecord -PtrDomainName "LeftHost.mycp.lab"
Add-DnsServerResourceRecordA -Name "DMZSRV" -ZoneName
"mycp.lab" -AllowUpdateAny -IPv4Address "10.30.30.5"
-TimeToLive 01:00:00
Add-DnsServerResourceRecordPtr -Name "5" -ZoneName
"30.30.10.in-addr.arpa" -AllowUpdateAny -TimeToLive
01:00:00 -AgeRecord -PtrDomainName "DMZSRV.mycp.lab"
```

Or, from the ADDCDNS VM's Command Prompt, execute the `F:\Scripts\Subnets_and_dns_ADDCDNS.bat` file.

15. Exit from PowerShell and CMD.

This concludes the preparation of Windows servers in the lab. With all Windows VMs configured, we can move on to the Check Point lab components.

Preparing cloned Check Point hosts

Each cloned Check Point VM has to be assigned an individual hostname and IP address in the corresponding network segment, as shown:

- **Preparing CPSMS**:

    A. In Oracle VM VirtualBox Manager, start the CPSMS VM.

    B. Wait for the login prompt.

    C. Log on using `admin` and `CPL@b8110` credentials.

    D. Change the hostname and the IP address of the `eth0` interface and save the configuration by typing these lines manually and pressing *Enter* after each line:

    ```
    set hostname CPSMS
    set interface eth0 ipv4-address 10.0.0.10 mask-length 24
    ```

```
set interface eth0 comments "Mgmt"
save config
```

- **Preparing CPCM1:**

  A. Start the CPCM1 VM.

  B. Wait for the login prompt.

  C. Log on using `admin` and `CPL@b8110` credentials.

  D. Change the hostname and IP address of the `eth0` interface and save the configuration by typing these lines manually and pressing *Enter* after each line:

  ```
  set hostname CPCM1
  set interface eth0 ipv4-address 10.0.0.2 mask-length 24
  set interface eth0 comments "Mgmt"
  save config
  ```

- **Preparing CPCM2:**

  A. Start the CPCM2 VM.

  B. Wait for the login prompt.

  C. Log on using `admin` and `CPL@b8110` credentials.

  D. Change the hostname and IP address of the `eth0` interface and save the configuration by typing these lines manually and pressing *Enter* after each line:

  ```
  set hostname CPCM2
  set interface eth0 ipv4-address 10.0.0.3 mask-length 24
  set interface eth0 comments "Mgmt"
  save config
  ```

- **Preparing CPGW:**

  A. Start the CPGW VM.

  B. Wait for the login prompt.

  C. Log on using `admin` and `CPL@b8110` credentials.

  D. Change the hostname and IP address of the `eth0` interface and save the configuration by typing these lines manually and pressing *Enter* after each line:

  ```
  set hostname CPGW
  set interface eth0 ipv4-address 200.200.0.1 mask-length
  24
  ```

```
set interface eth0 comments "Mgmt"
save config
```

Preparing SmartConsole

1. Start and log on to the SmartConsole VM.

2. Click **Start**, type `putty`, and press *Enter*.

3. In the **PuTTY Configuration** window, in the **Host Name (or IP address)** field, enter `10.0.0.10` [1], and in **Saved Sessions**, enter `CPSMS` [2]. Click **Save** [3]:

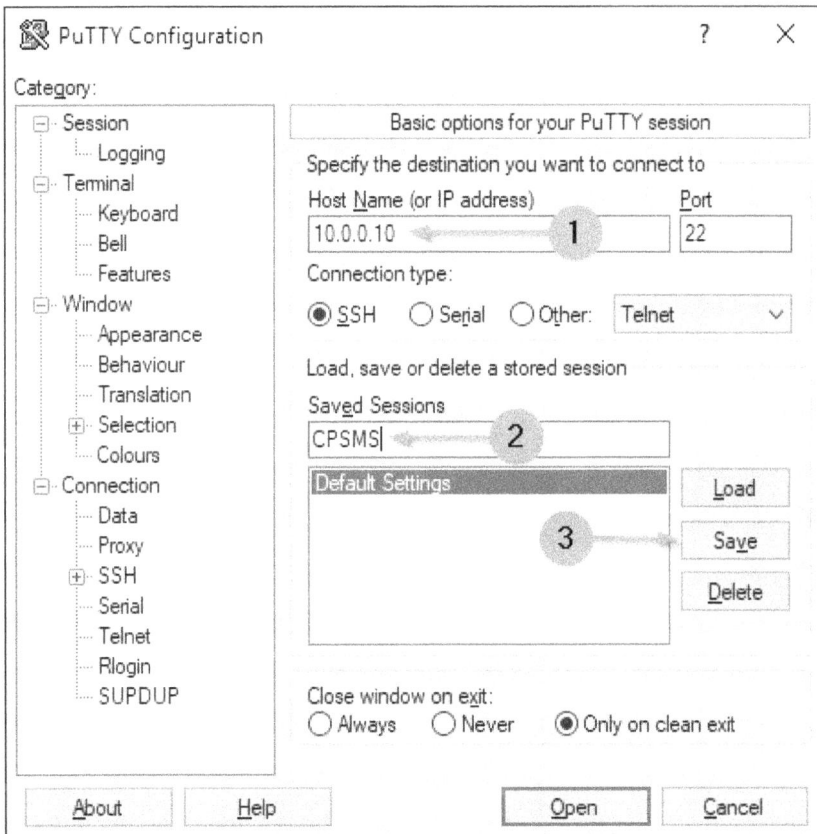

Figure 4.55 – Creating PuTTY sessions

4. Repeat this process, creating sessions for CPCM1 (`10.0.02`) and CPCM2 (`10.0.0.3`).

5. Once all three hosts' sessions are saved, double-click on **CPSMS**.

6. When prompted with **PuTTY Security Alert**, click **Accept** [1]:

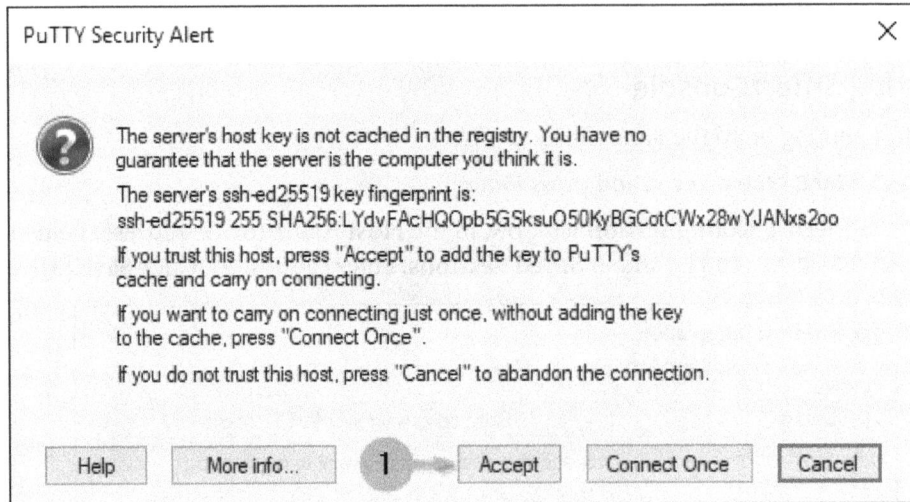

Figure 4.56 – Accepting hosts' keys

7. You should see similar screens for all three hosts on the left side of your topology:

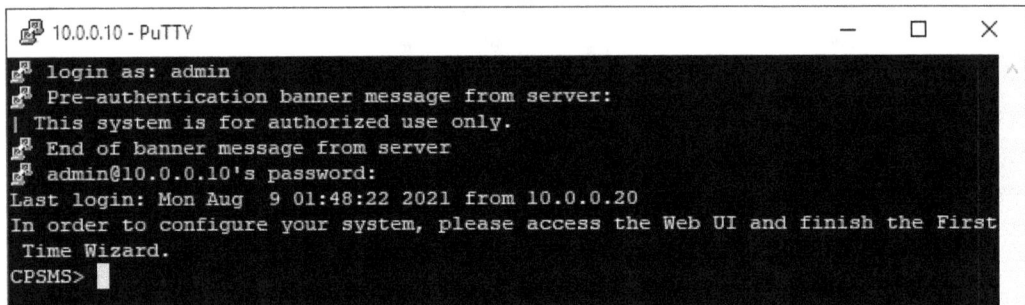

Figure 4.57 – SSH session to the Check Point host

8. While PuTTY is running, right-click on its icon in the Windows taskbar and click on **Pin to taskbar** for convenience.

This concludes the preparation of the Check Point components in the lab.

Summary

Over the course of the last two chapters, we have created our lab environment containing typical elements of the production infrastructure. We were able to take advantage of the VirtualBox scripting and batch files to accomplish some of these tasks and we have used PowerShell scripts to create or modify Windows components. You now have the necessary skills for building your own labs with different topologies in the future and can modify and reuse your scripts and commands to accomplish that.

Now that we have all the lab components in place and understand how all of them are interconnected, we are ready to start working on our Check Point component configuration using the Gaia OS web UI and CLI.

Part 2:
Introduction to Gaia, Check Point Management Interfaces, Objects, and NAT

In this portion of the book, you will be introduced to the Check Point Gaia, an underlying operating system used by Check Point gateways and management servers. You will learn how to navigate and use its WebUI and CLI. We'll progress to learn about the navigation and capabilities of SmartConsole, your primary administration interface. We then will cover different object types and network address translation specifics.

The following chapters will be covered in this section:

- *Chapter 5, Gaia OS, the First Time Configuration Wizard, and an Introduction to the Gaia Portal (WebUI)*

- *Chapter 6, Check Point Gaia Command-Line Interface; Backup and Recovery Methods; CPUSE*

- *Chapter 7, SmartConsole – Familiarization and Navigation*

- *Chapter 8, Introduction to Policies, Layers, and Rules*

- *Chapter 9, Working with Objects – ICA, SIC, Managed, Static, and Variable Objects*

- *Chapter 10, Working with Network Address Translation*

5

Gaia OS, the First Time Configuration Wizard, and an Introduction to the Gaia Portal (WebUI)

In this chapter, we will learn about Gaia OS, the operating system that all of the Check Point management servers and gateways are using. We will familiarize ourselves with its web interface and perform some of the common configuration tasks using the WebUI.

In this chapter, we are going to cover the following main topics:

- Learning about Gaia's roots – a historical note
- Using the First Time Configuration Wizard
- An introduction to the Gaia Portal (WebUI)

Technical requirements

For this chapter, you will need a PC with at least 4 CPU cores, 24 to 32 GB of RAM, 200 GB of HDD space (preferably SSD), and a monitor with at least a 1,024 x 768 resolution (1,920 x 1,080 is preferable) running **Windows 10**. References to the software used in the lab are provided in later sections. You will also need **Google Chrome** or **Microsoft Edge** as a browser and an SSH terminal emulator.

A PC running either Windows or any Linux distribution supported by Oracle VM VirtualBox as a host operating system (a full list of compatible host operating systems can be found here: `https://docs.oracle.com/en/virtualization/virtualbox/6.0/user/hostossupport.html`)

> **Important Note**
>
> If you are experienced with, and have access to, different virtualization platforms, such as VMware Workstation or Fusion, ESXi, Hyper-V, or KVM, you can adapt the lab to any of those with minor modifications.
>
> If you do not have access to the hardware necessary to create the lab, simply read the chapter. It is illustrated well enough to follow the progress.

Learning about Gaia's roots – a historical note

Gaia is the operating system developed by Check Point. It is based on a heavily modified and fortified version of Red Hat Enterprise Linux. It was conceived after Check Point's 2009 acquisition of the Nokia IP appliance business and the IPSO operating system, a popular choice with Check Point customers at the time. After merging Nokia IPSO with Check Point's own **SecurePlatform** (**SPLAT**) operating system, Nokia's Network Voyager interface has evolved into the Gaia WebUI and its CLI shell has evolved into Gaia's command-line interface, **CLI Shell** (**CLISH**).

Using the First Time Configuration Wizard

If you are working with physical Check Point appliances, you are not going to encounter the installation prompts shown in *Chapter 4*, *Building a Check Point Lab Environment – Part 2*, under the *Finalizing Check Point BASE VM installation* section, since those come with Gaia OS already installed. Even if you are performing a complete re-installation of the operating system on Check Point devices, the mechanism for that is different than the ISO installation. The ISO installation is relevant to the virtual environments and open servers supported by Check Point. A list of supported hardware can be found at `https://www.checkpoint.com/support-services/hcl/`, where you can see quite a few models by multiple manufacturers, including Dell, HP, IBM, Lenovo, Cisco, Fujitsu, and others.

To define the role of the Check Point appliance (physical, virtual, or running on OpenServer) in your environment, you must use the **First Time Configuration Wizard** (**FTW**).

Using the FTW for the primary management server

Let's start by configuring our management server, CPSMS:

1. In your Oracle VM VirtualBox, verify that the following VMs are powered up:

    -   SmartConsole

    -   CPSMS

2. Open the SmartConsole VM's interface and log in with credentials that we have defined for the lab.

3. From the SmartConsole VM, verify that your SMS (CPSMS VM) is accessible by pinging its interface.

4. Once connectivity is confirmed, open the Chrome web browser, and in the address bar, enter `https://10.0.0.10`.

5. You will now be prompted with the **Your connection is not private** warning, followed by the **NET::ERR_CERT_AUTHORITY_INVALID** error description. This is expected, as a self-signed certificate is used for encryption of traffic between your browser and the newly installed Check Point virtual appliance. Click on **Advanced** [1] and **Proceed to 10.0.0.10 (unsafe)** [2] at the bottom of the page:

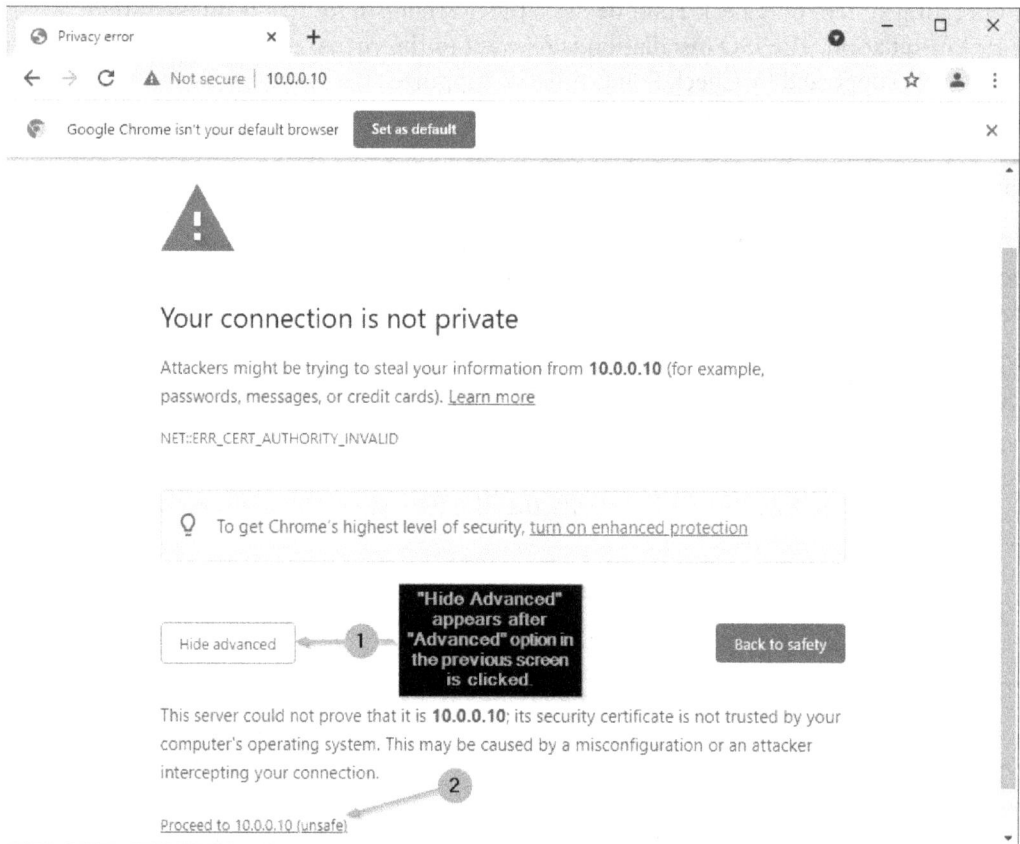

Figure 5.1 – Privacy warning when connecting to the Check Point WebUI

6. You are now presented with the Gaia WebUI login prompt (as credentials, use `admin` and `CPL@b8110`):

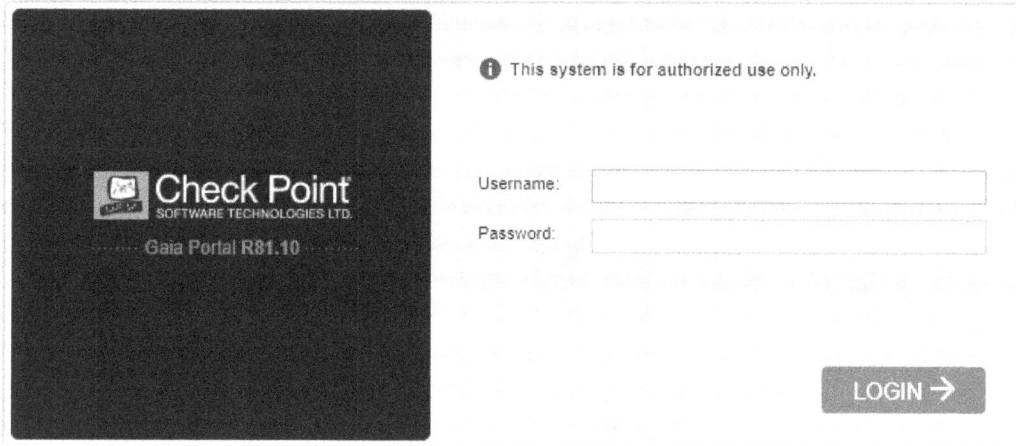

Figure 5.2 – Gaia WebUI login prompt

7. You are now prompted with **Check Point First Time Configuration Wizard** [1]. You can see that Gaia has recognized our platform as **Open Server** [2]. Click **Next** [3]:

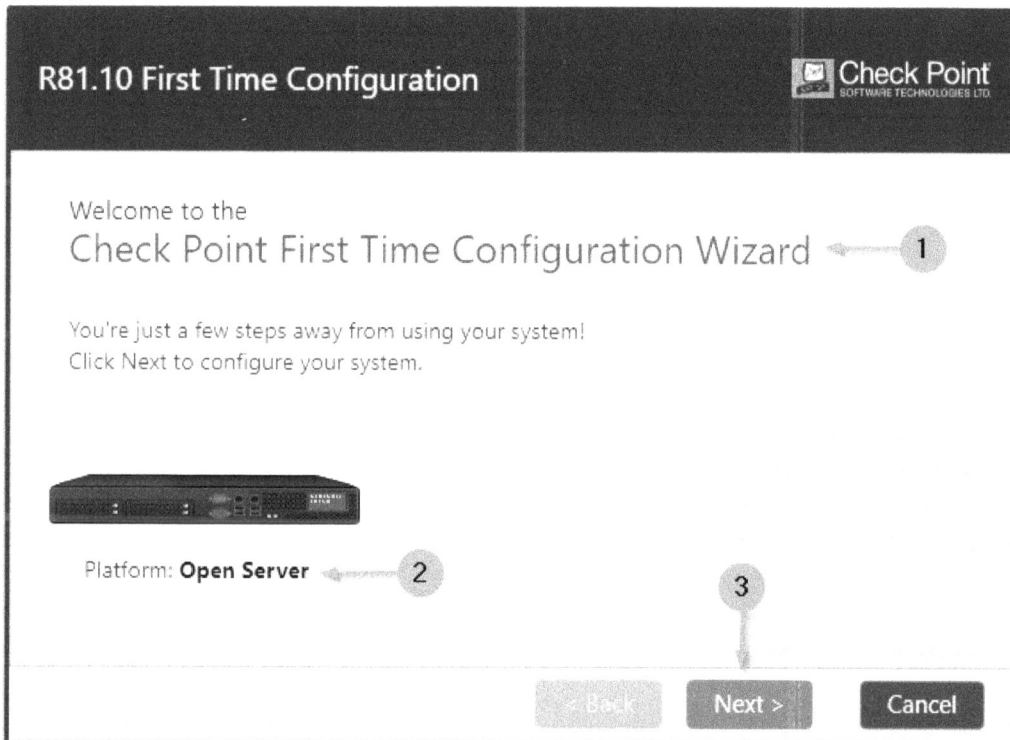

Figure 5.3 – The FTW

8. On the **Deployment Options** screen [1], we are choosing (or leaving the default selection as) **Continue with R81.10 configuration** [2]. Note the other available **Installation** and **Recovery** options [3], as you may use those in production environments. Click **Next** [4]:

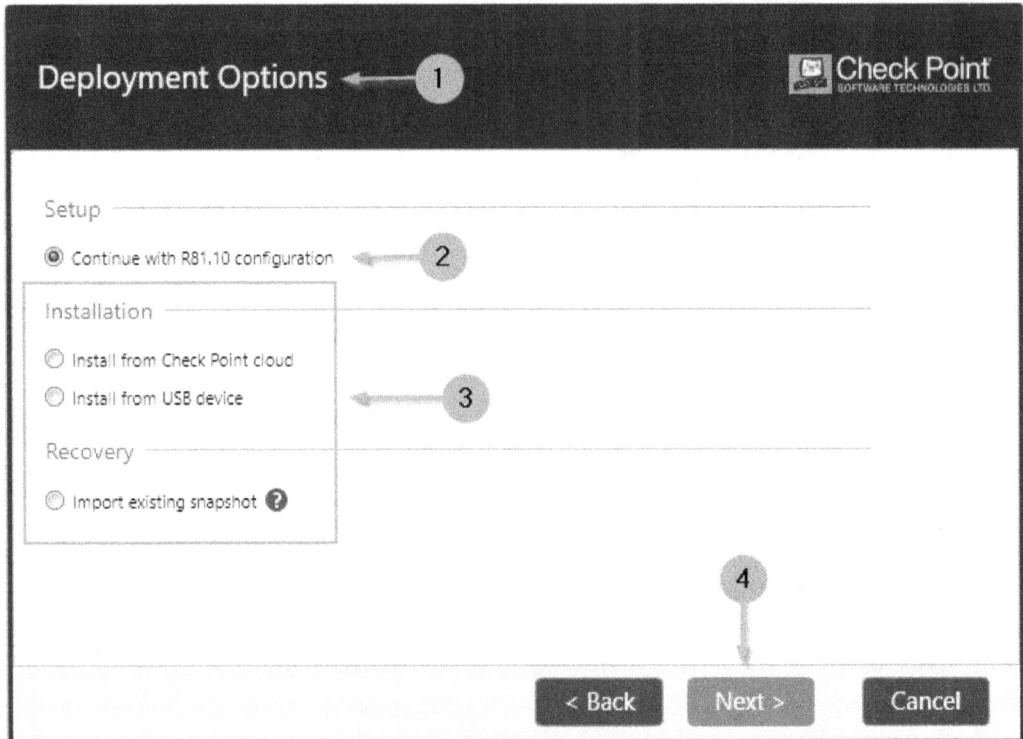

Figure 5.4 – Deployment Options

9. On the **Management Connection** screen [1], you can see that the IP address of the **eth0** interface is preconfigured. This was done by us in the previous chapter's *Preparing cloned Check Point hosts* section. Configure **Default Gateway:** [2] here as the future **virtual IP (vIP)** address of our cluster. Click **Next** to continue [3]:

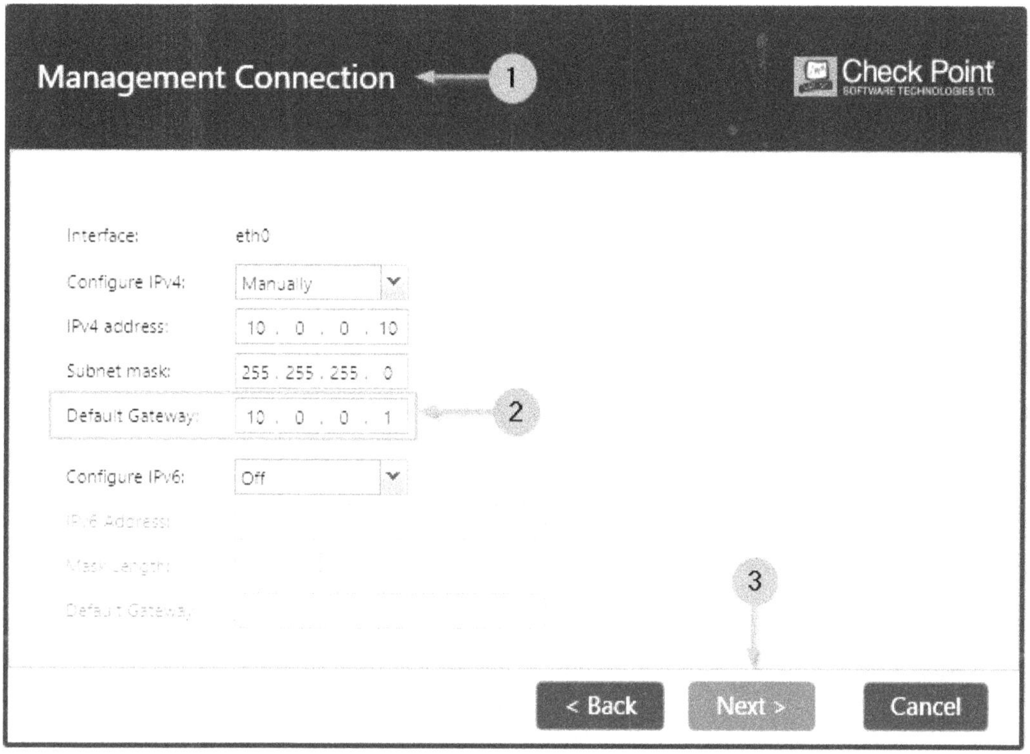

Figure 5.5 – Management Connection

10. On the **Device Information** screen [1], type the domain name we used when creating our `mycp.lab` Active Directory and the IP address of the ADDCDNS host [2]. Note the following **Proxy Settings** options for possible use in production environments [3]. Click **Next** [4]:

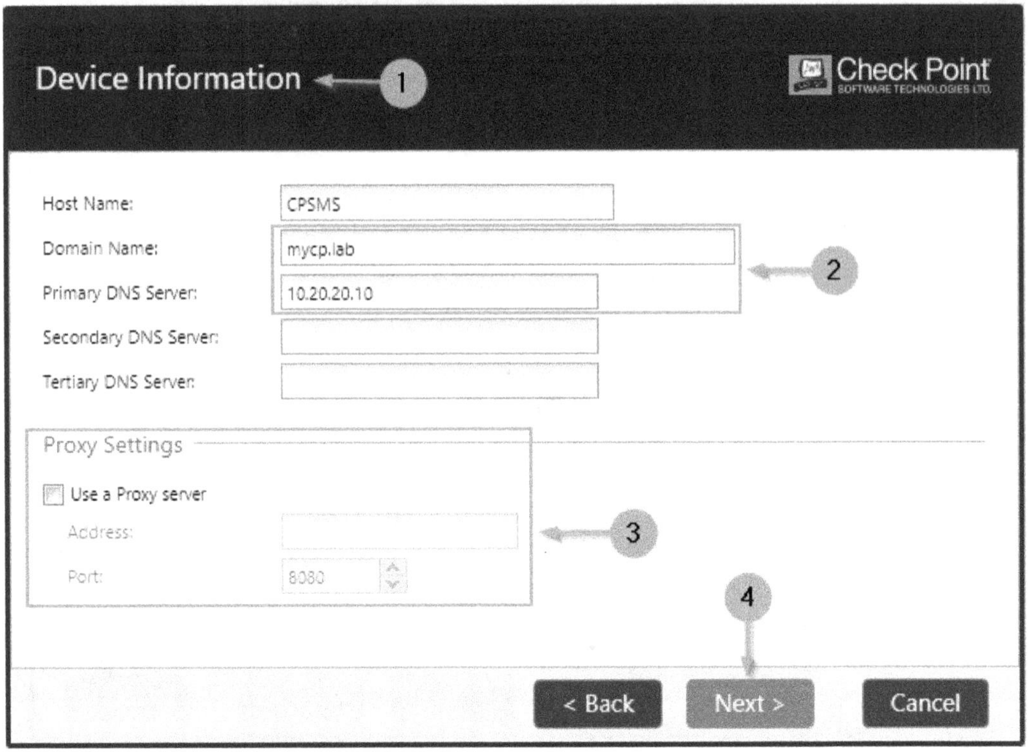

Figure 5.6 – Device Information

11. In **Date and Time Settings** [1], for now, we will configure the time and the time zone manually [2], until our management server can connect to the internet. In production environments, where you have a local time server accessible from the management network segment, use the labeled settings [3]. Click **Next** [4] to continue:

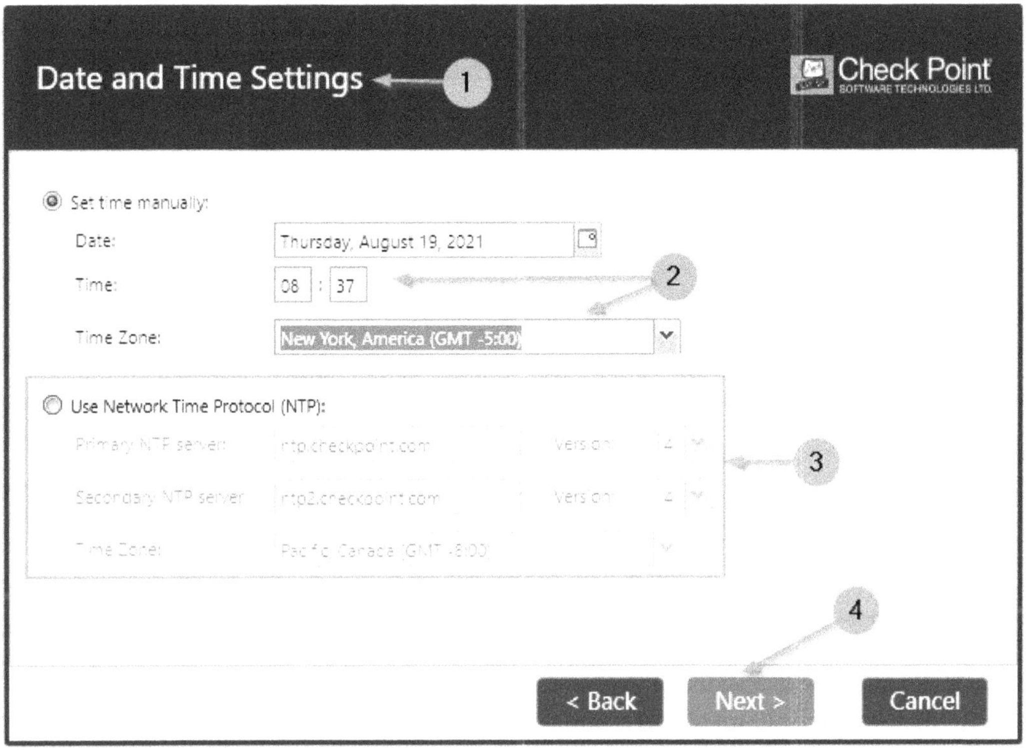

Figure 5.7 – Date and Time Settings

12. On the **Installation Type** screen [1], we are choosing the **Security Gateway and/or Security Management** option [2]. Note the **Multi-Domain Server** option [3] for future reference. Click **Next** [4]:

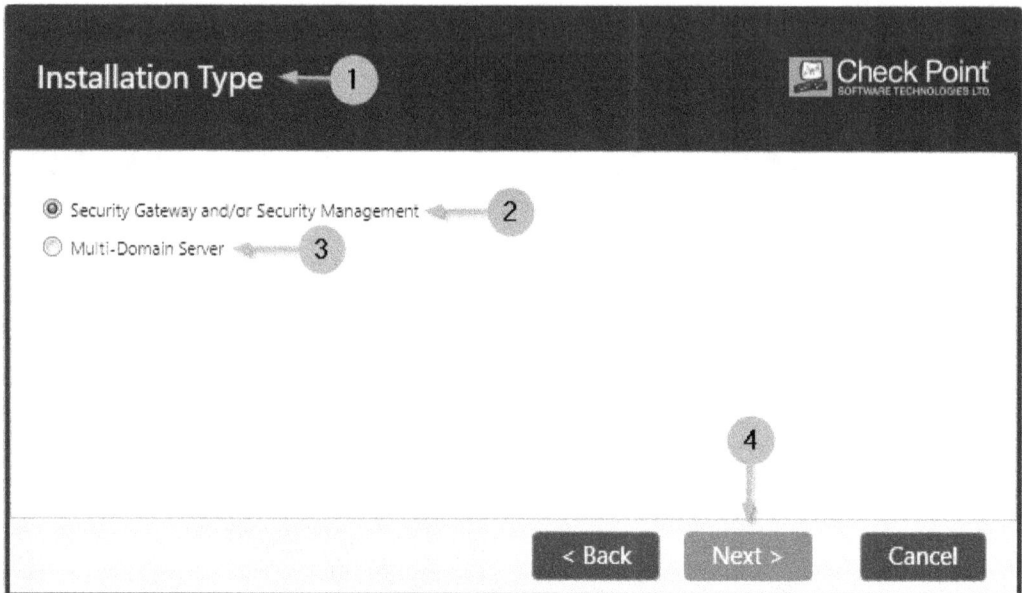

Figure 5.8 – Installation Type

13. On the **Products** screen [1], choose *only* **Security Management** [2]; in the **Define Security Management as:** field, leave **Primary** [3], but click on the dropdown to see other available options [4]. Leave the **Automatically download and install Blade Contracts, new software, and other important data (highly recommended)** option checked [5]. Click **Next** to continue:

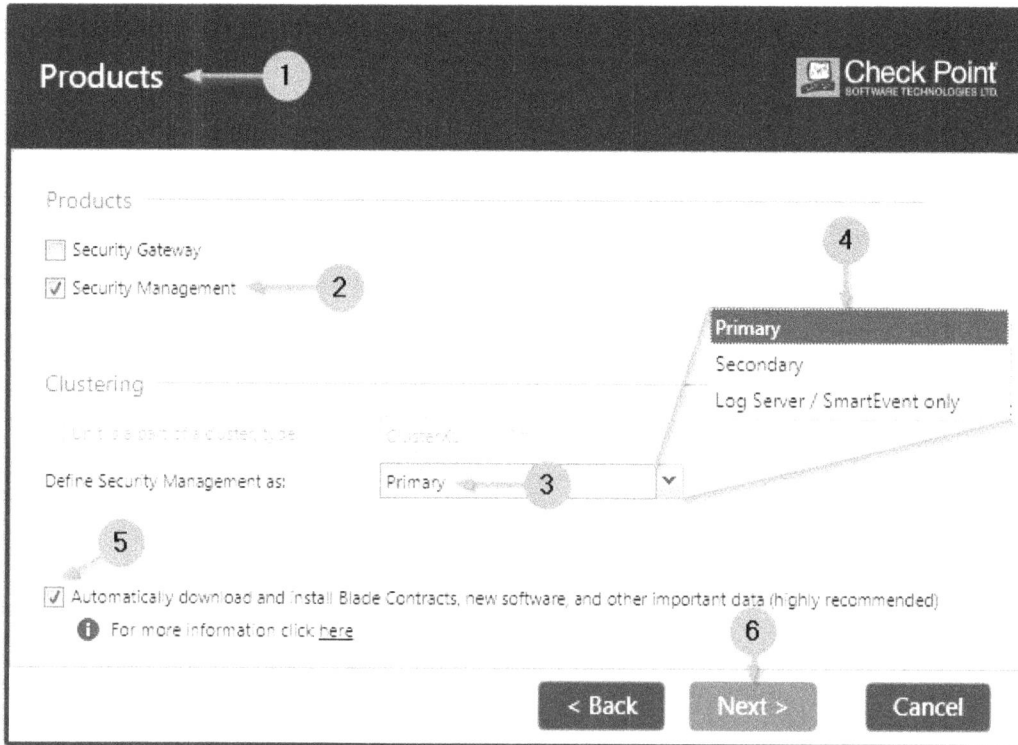

Figure 5.9 – Products (Security Management, Primary)

> **Important Note**
>
> See the *First Time Configuration Wizard for Gateways* section for **Products (Cluster Member)**.

14. In **Security Management Administrator** [1], choose **Define a new administrator** and use the account secadmin with the password CPL@b8110 [2]. Click **Next** [3] to continue:

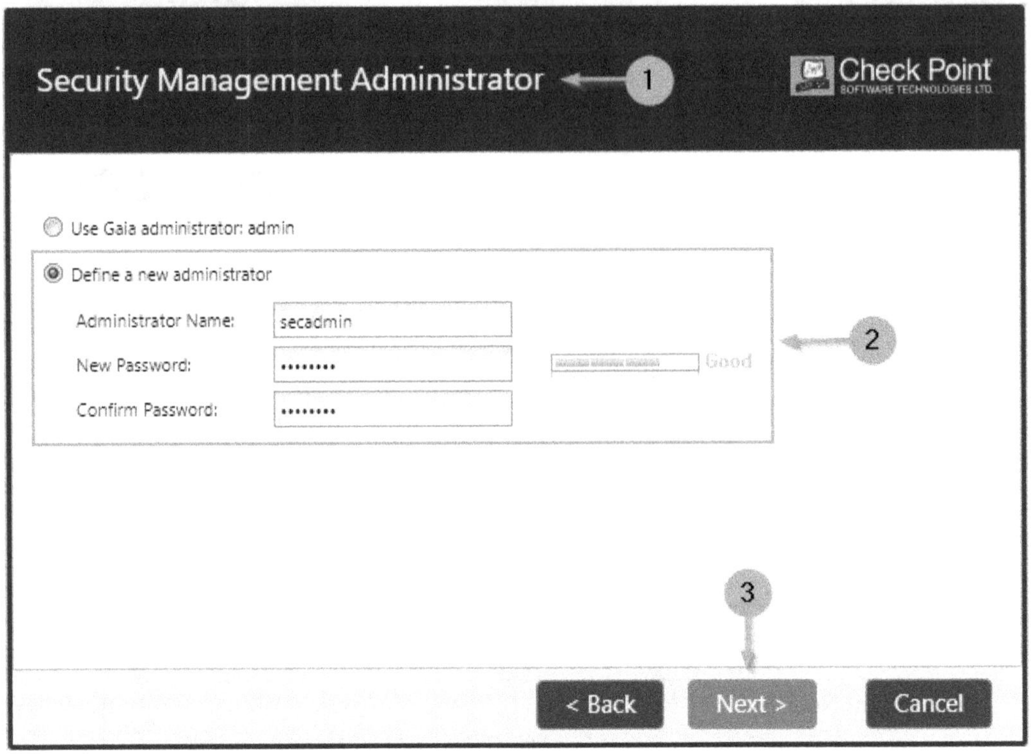

Figure 5.10 – Security Management Administrator

15. On the **Security Management GUI Clients** screen [1], in the **GUI clients can log into Security Management from:** section [2], select the **This machine** [3] option, and observe all available options. You will be able to add or modify these settings after the initial configuration is completed:

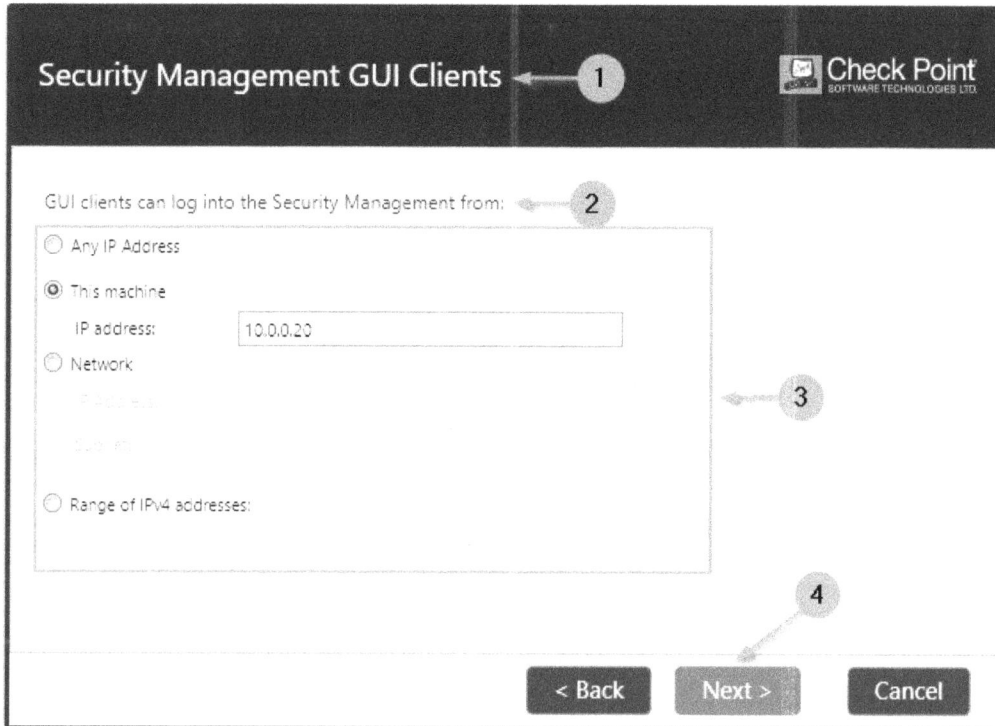

Figure 5.11 – Security Management GUI Clients

16. On the **First Time Configuration Wizard Summary** screen [1], leave **Send data to Check Point** checked [2] (click **more** > there to read about the data being sent to Check Point), and click **Finish** [3]:

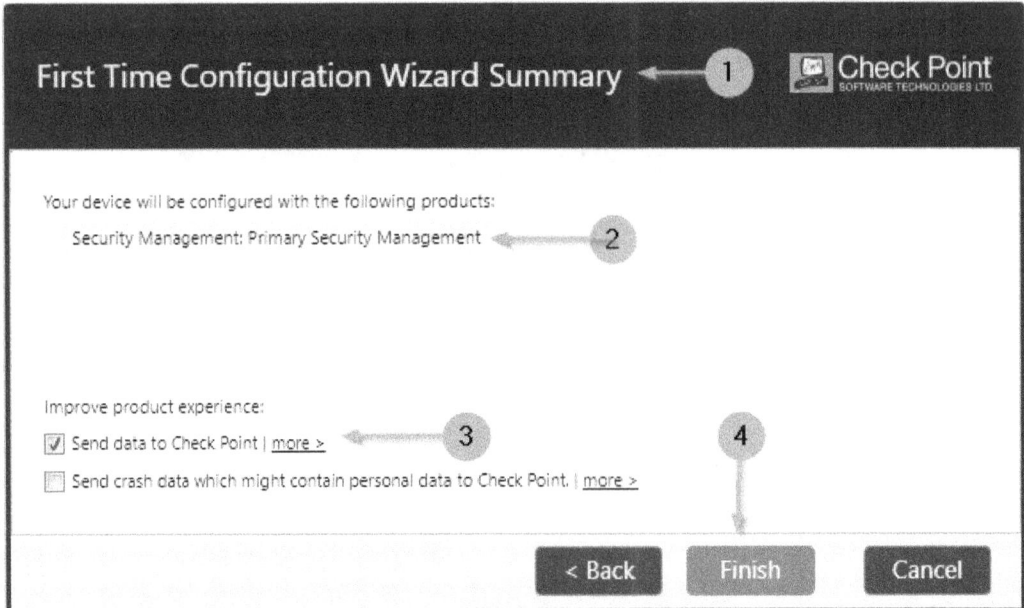

Figure 5.12 – First Time Configuration Wizard Summary

> **Important Note**
> Primary Security Management Server is the only component of your Check Point infrastructure in a single-domain environment that does not require you to configure the **Secure Internal Communication Activation Key** during the FTW.

17. Confirm the start of the configuration process by clicking **Yes** [1]:

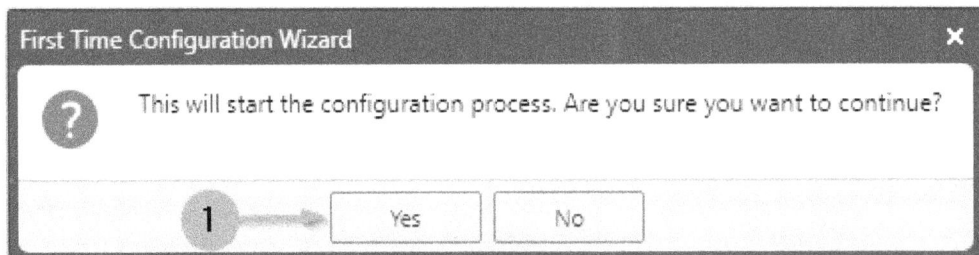

Figure 5.13 – Configuration start confirmation dialog box

18. Observe the progress of the installation process:

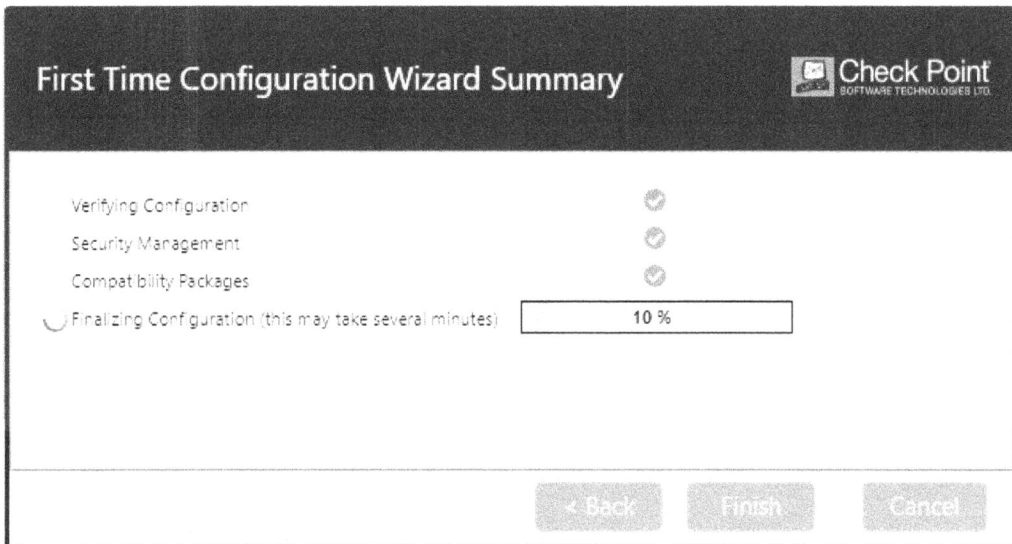

Figure 5.14 – The FTW progress screen

19. Once the process is complete, you are prompted with a report of the successful completion of the configuration. Click **OK** [1]:

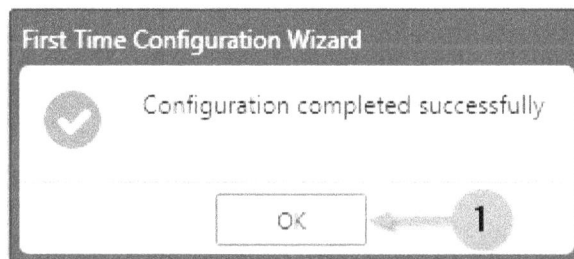

Figure 5.15 – Successful configuration completion

20. At this point, you will be redirected to the Gaia WebUI **System Overview** page of your management server [1] and will have a configuration lock on the database [2]:

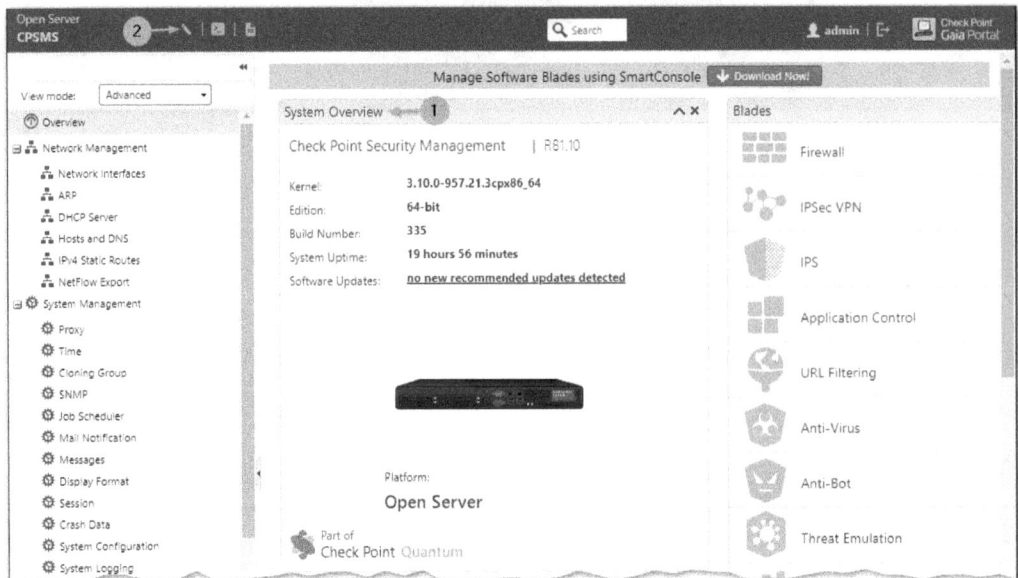

Figure 5.16 – WebUI overview

We will go over the WebUI in detail later in the chapter, but, for now, a few words about the database configuration lock: your WebUI session may time out, and when you log in again, you may find the icon of the pen replaced with that of the lock, indicating that you are now in read-only mode:

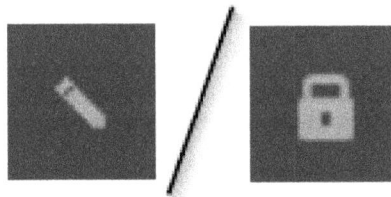

Figure 5.17 – Database lock

To unlock the database, click the lock icon and confirm your choice.

First Time Configuration Wizard for gateways

Repeat the entire process for the CPCM1 cluster member by powering it up and following the same configuration steps with these changes:

1. Unlike the management server, you will not be able to ping the gateways, because they are, by default, created with a restrictive security policy. You should only be able to connect to them via SSH or HTTPS to complete the FTW.

2. In the **Management Connection** [1] | eth0 interface [2], do *not* specify **Default Gateway**; instead, configure the eth0 interface of CPCM1 according to our diagrams and the *Hosts, interfaces, IP addresses, and networks* table in *Chapter 3, Building a Check Point Lab Environment – Part 1*. For CPCM1, it is 10.0.0.2/24 [3]. Click **Next** [4]:

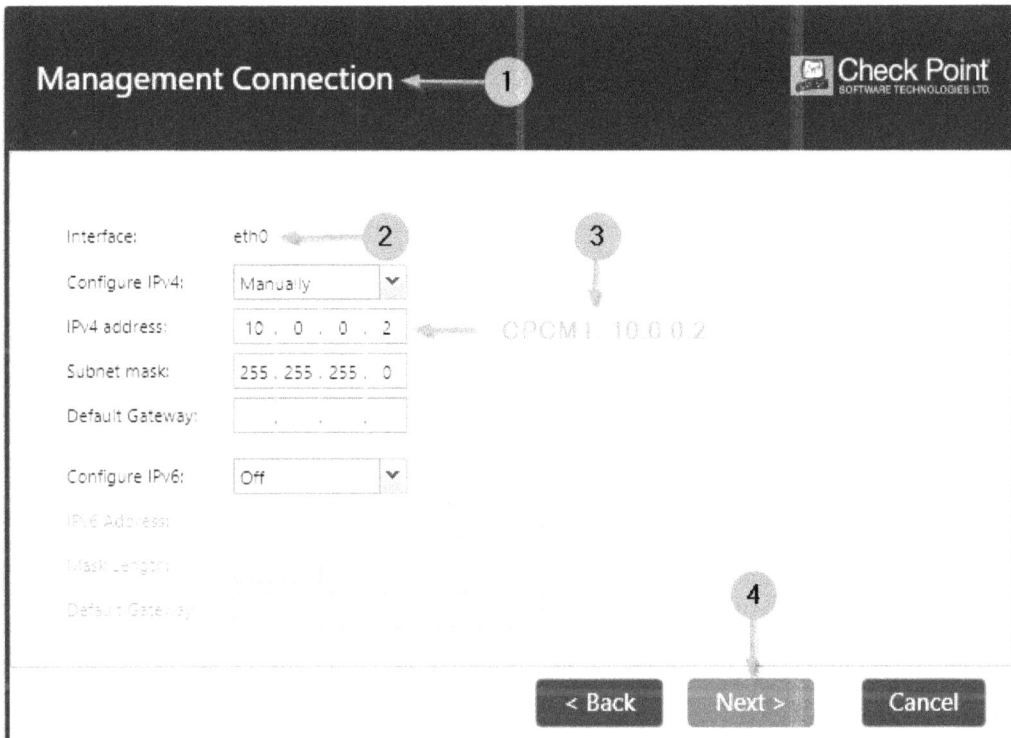

Figure 5.18 – Cluster member Management Connection interface IP and subnet mask

3. You will advance to the **Internet Connection** configuration step [1]. Although this step is optional, let's configure it here. In **Interface:**, choose **eth4** [2]. For CPCM1, we will use 200.100.0.2/24 [3]. Click **Next** [4]:

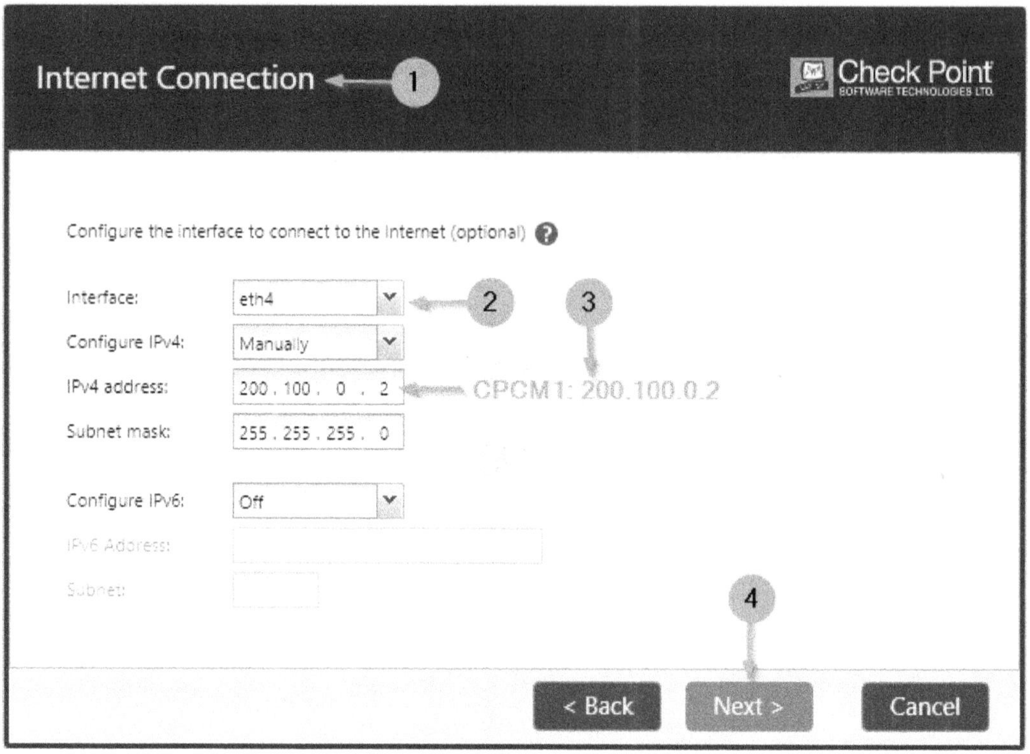

Figure 5.19 – Cluster member Internet Connection interface IP and subnet mask

Important Note

You will see the **Internet Connection** screen whenever any Check Point appliance, physical or virtual, has more than one network interface. It is described as optional, and you do not have to configure it. This connection could be defined later in the Gaia configuration process.

4. Next, in **Products** [1], we will uncheck the **Security Management** option and leave only **Security Gateway** checked [2]. We will check **Unit is a part of a cluster, type:** [3] and verify that **ClusterXL** [4] is selected. Click **Next** [5]:

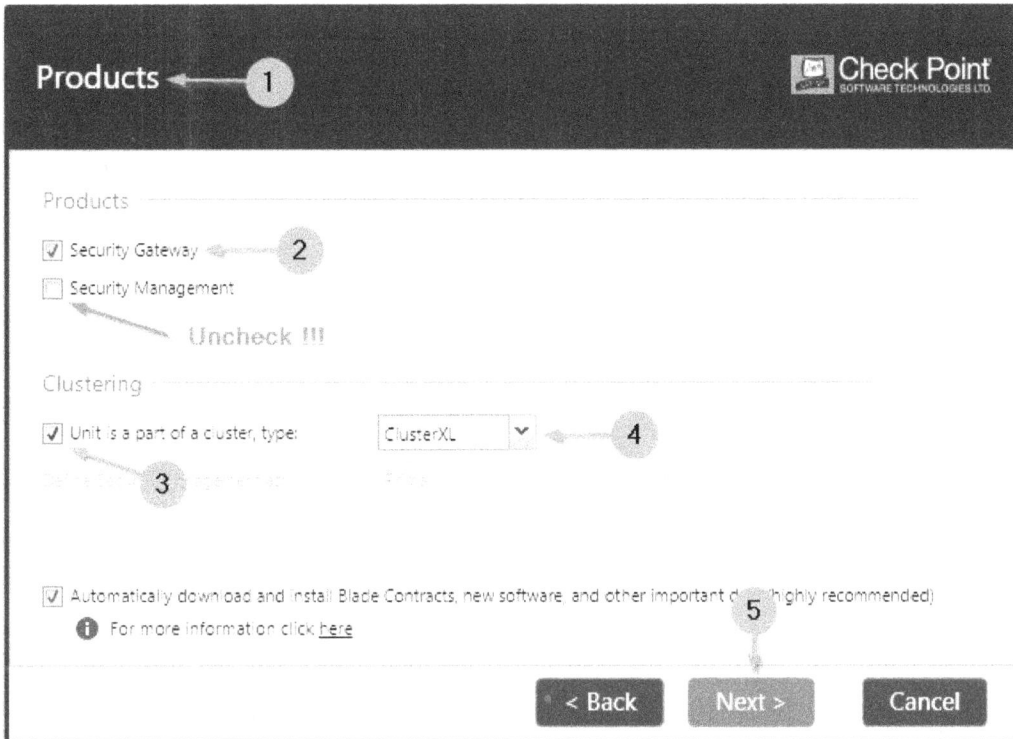

Figure 5.20 – Product configuration for ClusterXL member gateways

5. We now see a new screen, **Secure Communication to Management Server**. You will see it during the manual configuration of any appliance, physical or virtual, that is *not* a primary management server. For **Activation Key** [2], use the same password that we are using throughout the lab, CPL@b8110. We will discuss this subject, **Secure Internal Communication (SIC)**, in later chapters.

Also, note the **Connect to your Management as a Service** section of the screen [3]. This is the connectivity configuration for the Check Point Smart-1 Cloud service, which was briefly described in *Chapter 1, Introduction to Check Point Firewalls and Threat Prevention Products*. Click **Next** [4]:

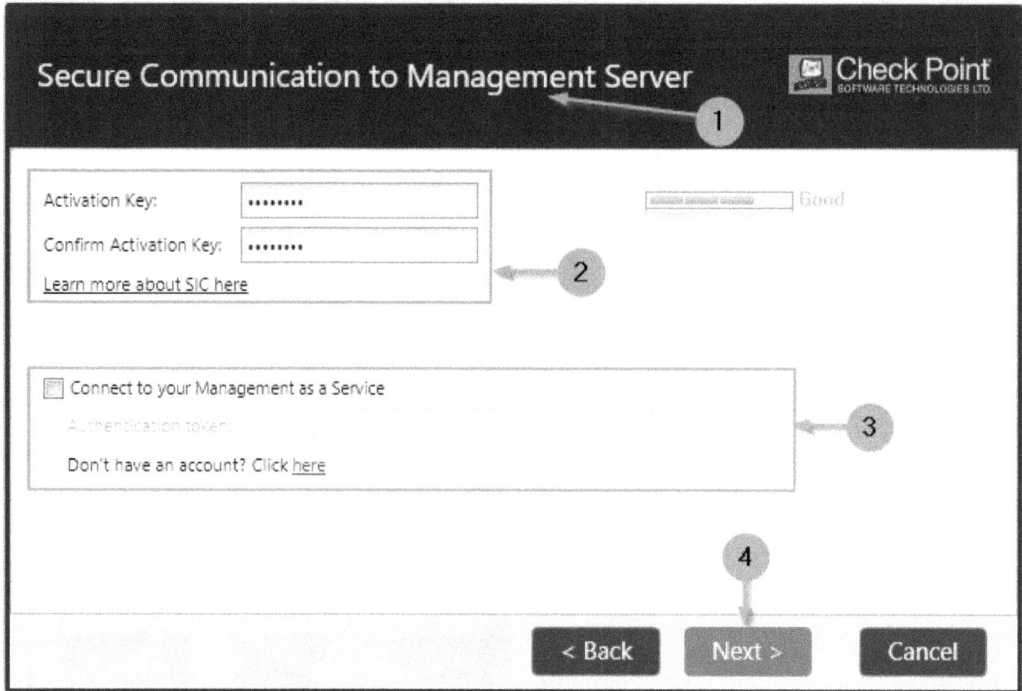

Figure 5.21 – Secure internal communication activation key

6. On the next screen, **First Time Configuration Wizard Summary** [1], you will see that this device will be configured as **Security Gateway** [2]. Click **Finish** [3] to complete the process:

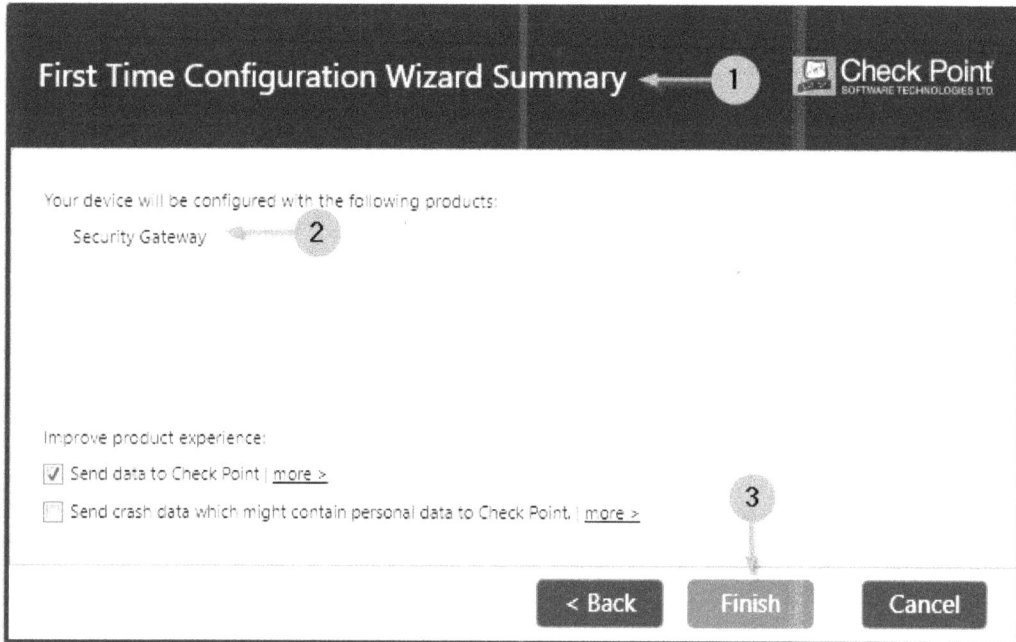

Figure 5.22 – First Time Configuration Wizard Summary for gateways and cluster members

After a display of progress bars for different configuration stages, you are presented with a **Configuration completed successfully** popup:

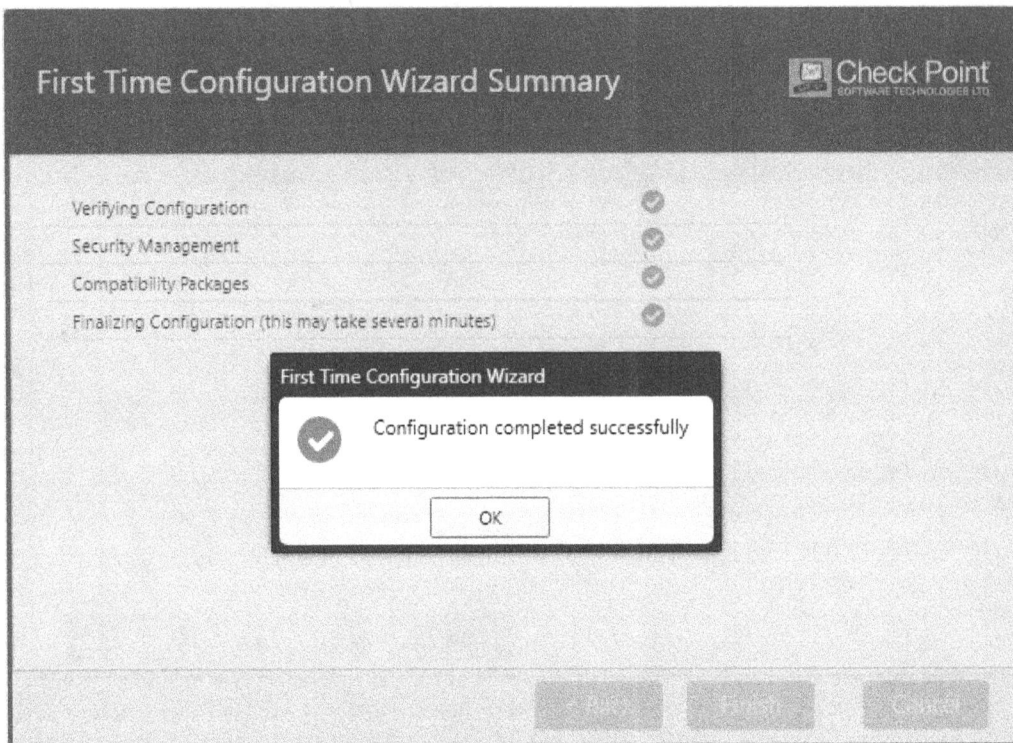

Figure 5.23 – FTW completed successfully

Now that we have learned how to perform the first-time configuration using the graphical UI, let's see how the same can be achieved using a command line.

First-time configuration using the CLI

All operations described in the previous section that rely on a rich visual user interface can also be performed using a command line. Unlike the UI FTW configuration option, which is interactive, the CLI option is not, thus a predefined collection of key-value pairs is used to complete the process.

> **Important Note**
>
> If you are reading this section for the first time and have no prior introduction to Check Point's Command-Line Interface then, skip this section and move to the *An Introduction to the Gaia Portal (WebUI)* section. You can return to this section for instructions after reading the *Introduction to Command-Line Interface (CLISH)* section of *Chapter 6, Check Point Gaia Command-Line Interface; Backup and Recovery Methods; CPUSE.*

To perform a command-line operation with the FTW, follow these steps:

1. SSH into CPCM2 from the SmartConsole VM.

2. Log in using our Gaia credentials, `admin` and `CPL@b8110`.

3. Set an expert password, if it is not configured yet, by executing the following:

    ```
    set expert-password
    ```

    Enter a new expert password twice (use the same password, `CPL@b8110`) and `save config`.

> **Important Note**
>
> Note that while you are typing `expert-password`, you do not see the cursor move. You cannot use *backspace* to edit your expert password. If you do that, an `Invalid characters detected` error will be returned. Simply retype the password if you see that error.

4. Log in to the expert mode by typing `expert` and entering the expert password.

From here on, we have two ways of moving forward: either using a configuration file created from a template or using a single CLI command, a *one-liner*.

Using a configuration file

This is the first option:

5. Type `config_system -t CPCM2_FTCW` and press *Enter* to create a CPCM2_FTCW FTW template file (this filename and its format are arbitrary; you can use anything sufficiently descriptive instead).

    You now have two options: you can choose to download the template file for editing and upload the completed version to the target Check Point appliance, or you can proceed to the following steps and edit the file locally.

> **Important Note**
>
> Before editing the First Time Configuration Wizard template on the existing cluster member or any other Check Point appliance where you have the administrative password configured, using expert mode, execute `dbget passwd:admin:passwd` or retrieve `admin_hash value`:
>
> `(Expert@CPCM1:0)# dbget passwd:admin:passwd`
>
> `$6$rounds=10000$Ra85m9P9$Q08xjxqqyb6z/QDY.`
> `GGphCBC5jWNtDt2akqFdUqtGosolS27PCQN9MZ5BsZdc1yMI`
> `4HIKfQ/Wqx0iPqSpXGZw/`
>
> `(Expert@CPCM1:0)#`

6. Open that file in the **vi** editor by typing `vi CPCM2_FTCW` and pressing *Enter*.

7. Change the mode to `- -INSERT - -` by pressing `i`.

8. Use the arrow keys to get to the relevant fields and set them up as follows (keep an eye out for details, as *some values must be defined without quotes, some with single, and some with double*):

```
install_security_gw="true"
gateway_daip="false"
gateway_cluster_member="true"
install_security_managment="false"
download_info="true"
upload_info="true"
upload_crash_data="false"
ftw_sic_key="CPL@b8110"
admin_hash=' $6$rounds=10000$Ra85m9P9$Q08xjxqqyb6z/QDY.
GGphCBC5jWNtDt2akqFdUqtGosolS27PCQN9MZ5BsZdc1yMI4HIKfQ/
Wqx0iPqSpXGZw/'
iface=eth0
ipstat_v4=manually
ipaddr_v4=10.0.0.3
masklen_v4=24
default_gw_v4=200.100.0.254
ipstat_v6=off
hostname=CPCM2
domainname=mycp.lab
timezone='America/New_York'
```

```
primary=10.20.20.10
reboot_if_required="true"
```

9. Once all of the values are defined in the vi editor, press *Esc*, type `:wq!`, and press *Enter*.

10. Now, execute `config_system -f CPCM2_FTCW --dry-run` to validate the configuration file.

11. If the dry run succeeds, execute `config_system -f CPCM2_FTCW`.

Using a single command one-liner

As a second option, the same task of first-time configuration via the CLI can be accomplished by chaining these key-value pairs using the & sign and executing a one-liner command by enclosing all key-value pairs in double quotes (`"  "`), omitting the `"admin_hash"` option:

```
(Expert@CPCM2:0)# config_system -s "install_
security_gw="true"&gateway_daip="false"&gateway_
cluster_member="true"&install_security_
managment="false"&download_info="true"&upload_
info="true"&upload_crash_data="false"&ftw_sic_key="CPL@
b8110"&iface=eth0&ipstat_v4=manually&ipaddr_v4=10.0.0.3&masklen_
v4=24&default_gw_v4=200.100.0.254&ipstat_
v6=off&hostname=CPCM2&domainname=mycp.lab&timezone='America/
New_York'&primary=10.20.20.10&reboot_if_required="true""


Validating configuration file:   Done
Configuring OS parameters:       Done
Configuring products:            Done
Verifying installation...


First time configuration was completed!


Reboot will be performed in order to complete the installation


Going to load initial security policy.
Your remote session will be disconnected now.
Log in again to continue the configuration.
```

We have now learned how to complete the FTW using two CLI methods.

Rerunning the FTW

Should you determine that the FTW was executed with incorrect parameters and must be rerun again, using expert mode, delete the two flag files by running the following:

```
rm -i /etc/.wizard_accepted
rm -i /etc/.wizard_started
```

Confirm your choice to delete them.

Note that if you intend to rerun the FTW on a management server, in addition to removing the `wizard_accepted` and `wizard_started` files, two more files must be removed:

```
rm -i $FWDIR/conf/ICA.crl
rm -i $FWDIR/conf/InternalCA.*
```

The next time you access the system via a web browser, you will be greeted with the FTW; or, after you exit a currently running session, SSH into the appliance again, and, after logging on, you will see the notice that the FTW must be completed:

```
login as: admin
Pre-authentication banner message from server:
| This system is for authorized use only.
End of banner message from server
admin@10.0.0.3's password:
Last login: Sat Aug 21 02:58:16 2021 from 10.0.0.20
In order to configure your system, please access the Web UI and
finish the First Time Wizard.
CPCM2>
```

> **Important Note**
> Although you are prompted to use the WebUI, you may choose to use CLI FTW configuration options instead.

We have learned about the function and different means of completing the FTW. Now that we have defined which product our appliances will be running (management or gateway) and their cluster membership in our Check Point infrastructure, let's proceed with the configuration.

Introduction to the Gaia Portal (WebUI)

Let's log in to our CPCM1 virtual gateway appliance to continue its configuration via the WebUI.

The main view is divided into the toolbar [1], the navigation tree [2], and the widgets and status bar pane [3]. The latter one contains an *at a glance* **System Overview**, current **Network Configuration**, and **Link Status**, as well as currently active gateway functions (aka **Blades**):

Figure 5.24 – WebUI main page overview

Let's discuss each of these elements in detail in the following subsections.

Toolbar

You can see the different elements of the toolbar highlighted in the following screenshot:

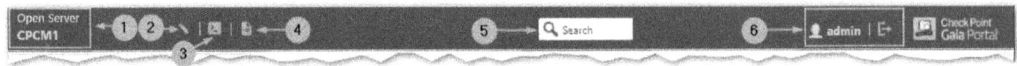

Figure 5.25 – WebUI navigation – toolbar

These are as follows:

1. **Platform and hostname**.

2. **Configuration lock**: Only a single administrator at a time is permitted to maintain the lock on the configuration database. A pen icon indicates that you presently have a lock and a write permission in this session. If another administrator is logged on and is editing the system, or if your earlier Gaia WebUI session has timed out, the pen icon is replaced with a lock icon. When you hover the mouse over the lock icon (A), you can see which administrator has a lock on the configuration database (B):

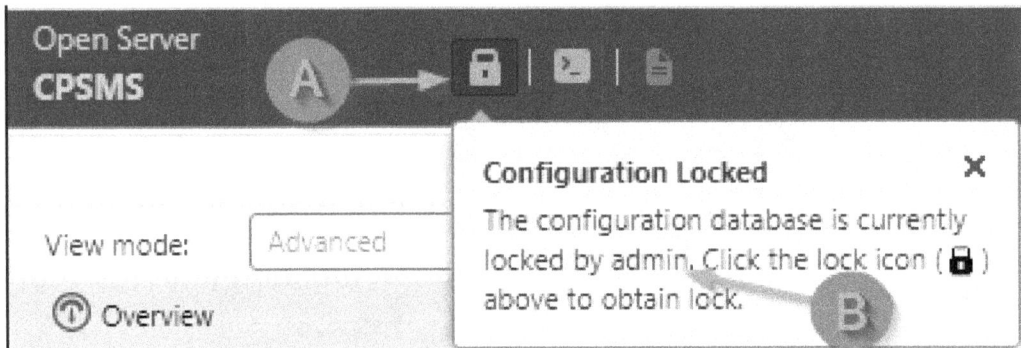

Figure 5.26 – WebUI session locked

If you see that it is your session maintaining the lock, you can click on the lock icon to unlock it. When you do so, the icon will revert to that of a pen.

3. **Virtual terminal**: If you need to perform a quick command-line operation that does not warrant having a dedicated SSH session open, you can launch the virtual terminal from the WebUI. You will have to authenticate again to access the shell. This option is of limited value, as you cannot copy and paste to or from that window, but it is nice to have it. This option is accessible when you maintain the lock on the database.

4. **Scratchpad**: This is a persistent text for running comments or temporary storage for values you may need to look up. The scratchpad does support copy and paste functionality and its content survives reboots. This option is accessible when you maintain the lock on the database.

5. **Search**: If you are unsure where in the WebUI certain options are present, begin typing the relevant term into the search box and, as you type, the available options that contain the string will be presented in the **Found # items** window, as shown here:

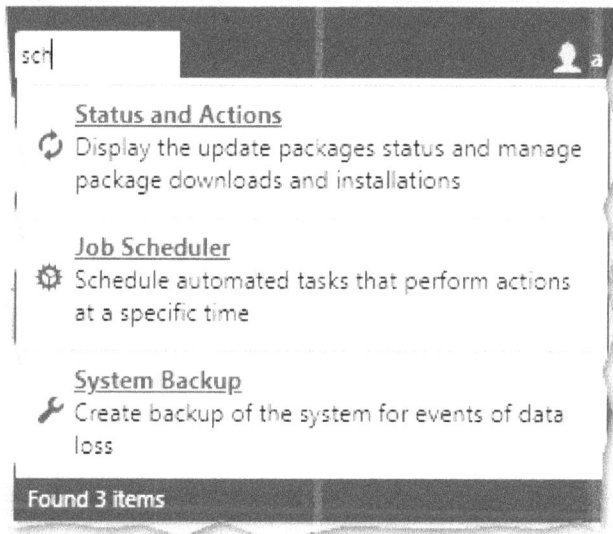

Figure 5.27 – WebUI search

6. Currently logged-in administrator and a logout option.

Next, let's get an overview of the navigation tree.

Navigation tree

The navigation tree comprises either six parts (in basic view mode) or seven (in advanced view mode), including the **Advanced Routing** section. Each section in the advanced view includes more configuration and action options.

For those reading this book without access to the lab, these are the sections and the components visible in the advanced view of the navigation tree:

□ 🔠 Network Management
    🔠 Network Interfaces
    🔠 ARP
    🔠 DHCP Server
    🔠 Hosts and DNS
    🔠 IPv4 Static Routes
    🔠 NetFlow Export

□ ⚙ System Management
    ⚙ Proxy
    ⚙ Time
    ⚙ Cloning Group
    ⚙ SNMP
    ⚙ Job Scheduler
    ⚙ Mail Notification
    ⚙ Messages
    ⚙ Display Format
    ⚙ Session
    ⚙ Crash Data
    ⚙ System Configuration
    ⚙ System Logging
    ⚙ Network Access
    ⚙ Host Access
    ⚙ LLDP

□ 🔁 Advanced Routing
    🔁 DHCP Relay
    🔁 BGP
    🔁 IGMP
    🔁 IP Broadcast Helper
    🔁 PIM
    🔁 Static Multicast Routes
    🔁 RIP
    🔁 IP Reachability Detection
    🔁 IPsec Routing
    🔁 OSPF
    🔁 Route Aggregation
    🔁 Inbound Route Filters
    🔁 Route Redistribution
    🔁 Routing Options
    🔁 Router Discovery
    🔁 Policy Based Routing
    🔁 NAT Pools
    🔁 Routing Monitor

□ 👥 User Management
    👥 Change My Password
    👥 Users
    👥 Roles
    👥 Password Policy
    👥 Authentication Servers
    👥 System Groups

□ ☁ High Availability
    ☁ VRRP
    ☁ Advanced VRRP

□ 🔧 Maintenance
    🔧 License Status
    🔧 Snapshot Management
    🔧 System Backup
    🔧 Download SmartConsole
    🔧 Shut Down

□ 🔃 Upgrades (CPUSE)
    🔃 Status and Actions
    🔃 Software Updates Policy

Figure 5.28 – Navigation tree components

We will be using a small subset of the options available in the navigation tree in our lab.

> **Important Note**
>
> The official Gaia administration guide (roughly a 500-page document for version R81.10) can be found at `https://sc1.checkpoint.com/documents/R81.10/WebAdminGuides/EN/CP_R81.10_Gaia_AdminGuide/Default.htm`.

Let's configure the necessary components. As we go through each, various features of the WebUI will be highlighted.

Network Interfaces

In the **Network Management** section of the navigation tree, click on **Network Interfaces**. You are now presented with the following screen:

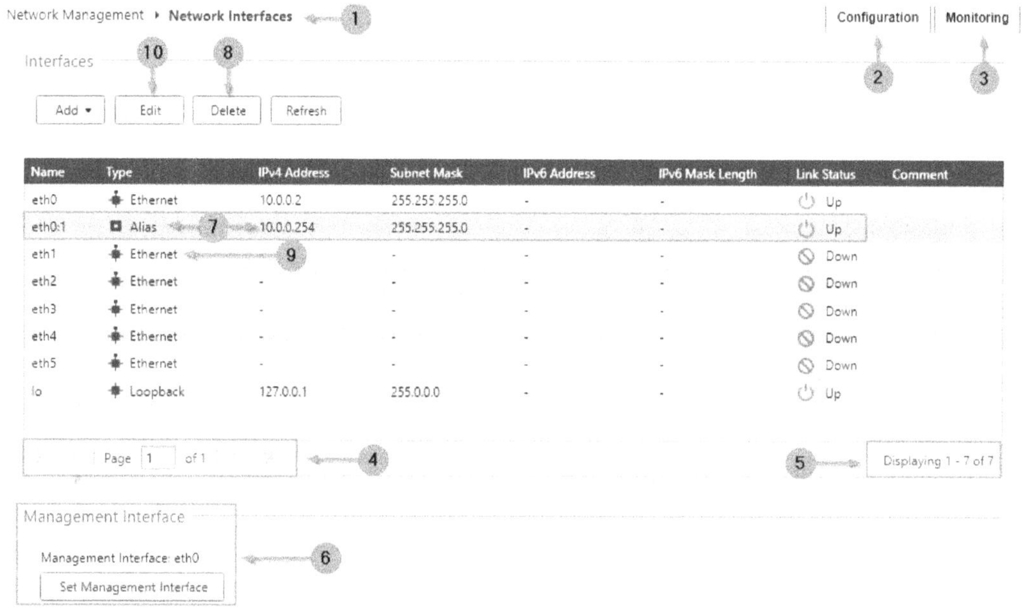

Figure 5.29 – Network Interfaces

Let's look over the components of the screen, as some of them will be encountered in other options of the navigation tree:

1. Your current location in the navigation tree.

2. The **Configuration** tab (presently active).

3. The **Monitoring** tab (present in all sections where a dynamic state change is possible).

4. Multi-page navigation (if there are too many items to fit on one screen, you can flip through the pages).

5. The current page in multi-page navigation.

6. The currently defined **Management Interface** (this is just a tag).

    In hardware appliances, there are two predefined network interfaces labeled as **Mgmt** and **Sync**. The MAC address of the **Mgmt** interface is used as a device ID in Check Point User Center for licensing purposes. In more complex implementations, where management plane separation may be warranted, interfaces used for management and cluster synchronization must be assigned to it. In such a case, these tags serve as convenient identifiers but, in principle, it could be any interface.

    The alias of the interface that we have configured with the IP address is different from that assigned to the BASECP VM (the origin of the clone).

7. In hardware appliances, the interface labeled **Mgmt** is pre-configured with the IP address of 192.168.1.1/24. In order to not lose connectivity to your current session, the default or, in our case, the existing IP address, is reassigned to **Alias**.

8. After exiting the existing WebUI session and logging into Gaia via its newly assigned interface IP of 10.0.0.2, select the line containing **Alias** and click **Delete**.

9. Either double-click the **eth1** interface or click it once to select.

10. Click **Edit** if you selected the interface in the previous step.

Now, open the interface editing window:

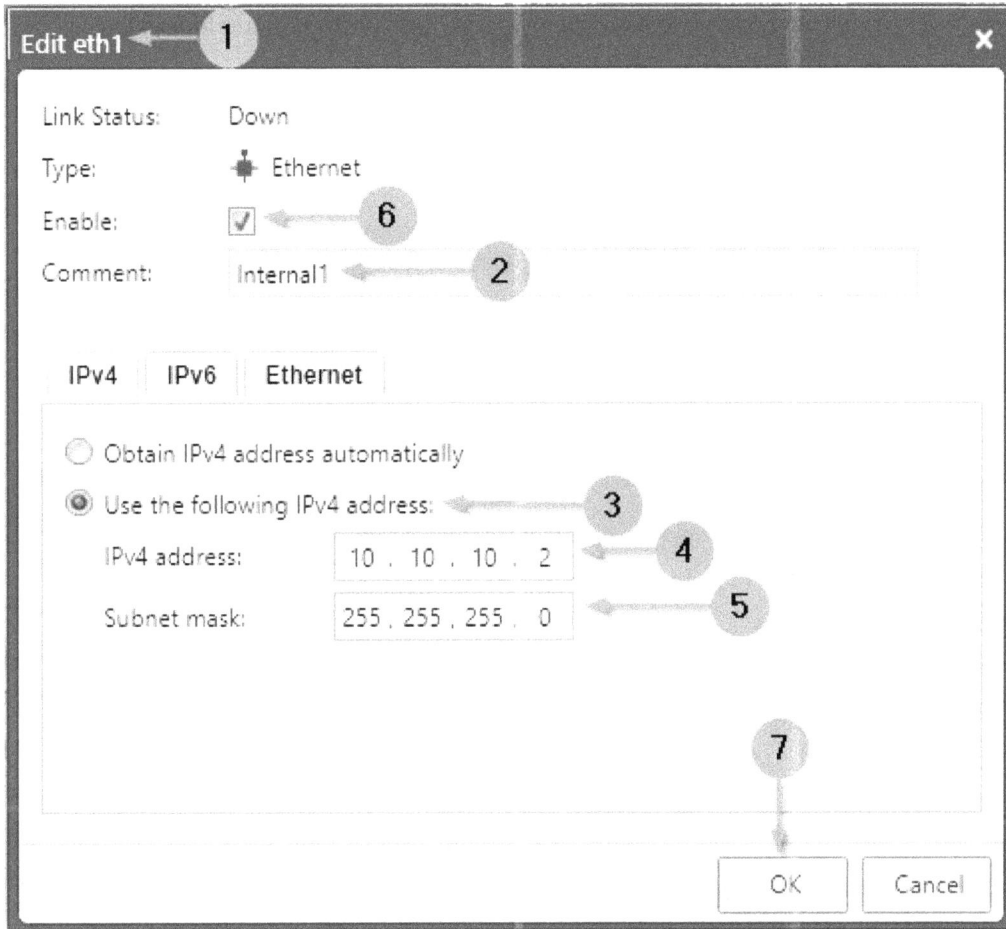

Figure 5.30 – Editing network interface properties

We perform the following steps in this window:

1. Verify that you are working on the correct interface.

2. In the comment field, type the identifier to the network segment it is connected to. In our case, we are following the lab topology diagram that identifies cluster interfaces, which is `Internal1`.

3. Verify that the **Use following IP address:** radio button is selected.

4. Enter the IP address defined in our lab topology diagram or table, `10.10.10.2`.

5. Enter the subnet mask for this interface, `/24` or `255.255.255.0`.

6. Check the **Enable:** checkbox.

7. Click **OK** to complete the interface configuration.

After a few seconds, you will see the indicator in the **Link Status** column for **eth1** changing from a gray **Down** to a green **Up**.

Repeat the interface editing process for the remaining interfaces and, in the case of **eth0**, when you are prompted with the following, you can proceed safely since the only thing you are changing is the comment:

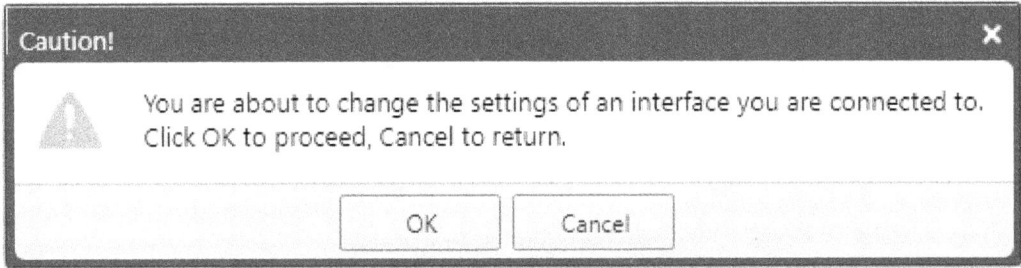

Caution!	**✕**
⚠	You are about to change the settings of an interface you are connected to. Click OK to proceed, Cancel to return.
	OK Cancel

Figure 5.31 – Changing the properties of the interface used for the current management session

In the end, for the CPCM1 cluster member gateway, your network interfaces should be configured as follows:

Name	Type	IPv4 Address	Subnet Mask	IPv6 Address	IPv6 Mask Length	Link Status	Comment
eth0	Ethernet	10.0.0.2	255.255.255.0	-	-	⏻ Up	Mgmt
eth1	Ethernet	10.10.10.2	255.255.255.0	-	-	⏻ Up	Internal1
eth2	Ethernet	10.20.20.2	255.255.255.0	-	-	⏻ Up	Internal2
eth3	Ethernet	192.168.255.1	255.255.255.0	-	-	⏻ Up	Sync
eth4	Ethernet	200.100.0.2	255.255.255.0	-	-	⏻ Up	External
eth5	Ethernet	10.30.30.2	255.255.255.0	-	-	⏻ Up	DMZ
lo	Loopback	127.0.0.1	255.0.0.0	-	-	⏻ Up	

Figure 5.32 – CPCM1 cluster member gateway network interfaces configuration and state

Before we move on to the configuration of the next items for the lab, let's look at the **Network Interface** options available to us by clicking on the **Add** button in the **Network Interfaces** section. Doing that will present you with the following choices:

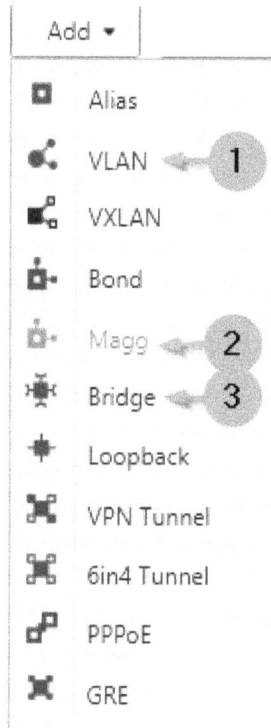

Figure 5.33 – Interface options

To those with a networking background, most of these are self-explanatory, but I'd like to point out a few particulars:

1. **VLAN** interfaces are defined as **Members Of: Interfaces** or **Bonds**. A VLAN interface cannot be assigned a VLAN ID of 1 (or default VLAN).

    > **Important Note**
    > A parent interface that has VLAN subinterfaces must *not* have an IP address configured, even though you are not prevented from doing it by Gaia.

2. The **Magg** interface is specific to Check Point physical appliances and is used for interconnects with Maestro Hyperscale Orchestrators.

3. **Bridge** interfaces are used in cases where you must protect the existing network segments while not relying on routing. This option allows for in-place drop-in of the Check Point gateways for access control and threat management without altering existing network topologies.

Hosts and DNS

When clicked, this section of the navigation tree will display the following window:

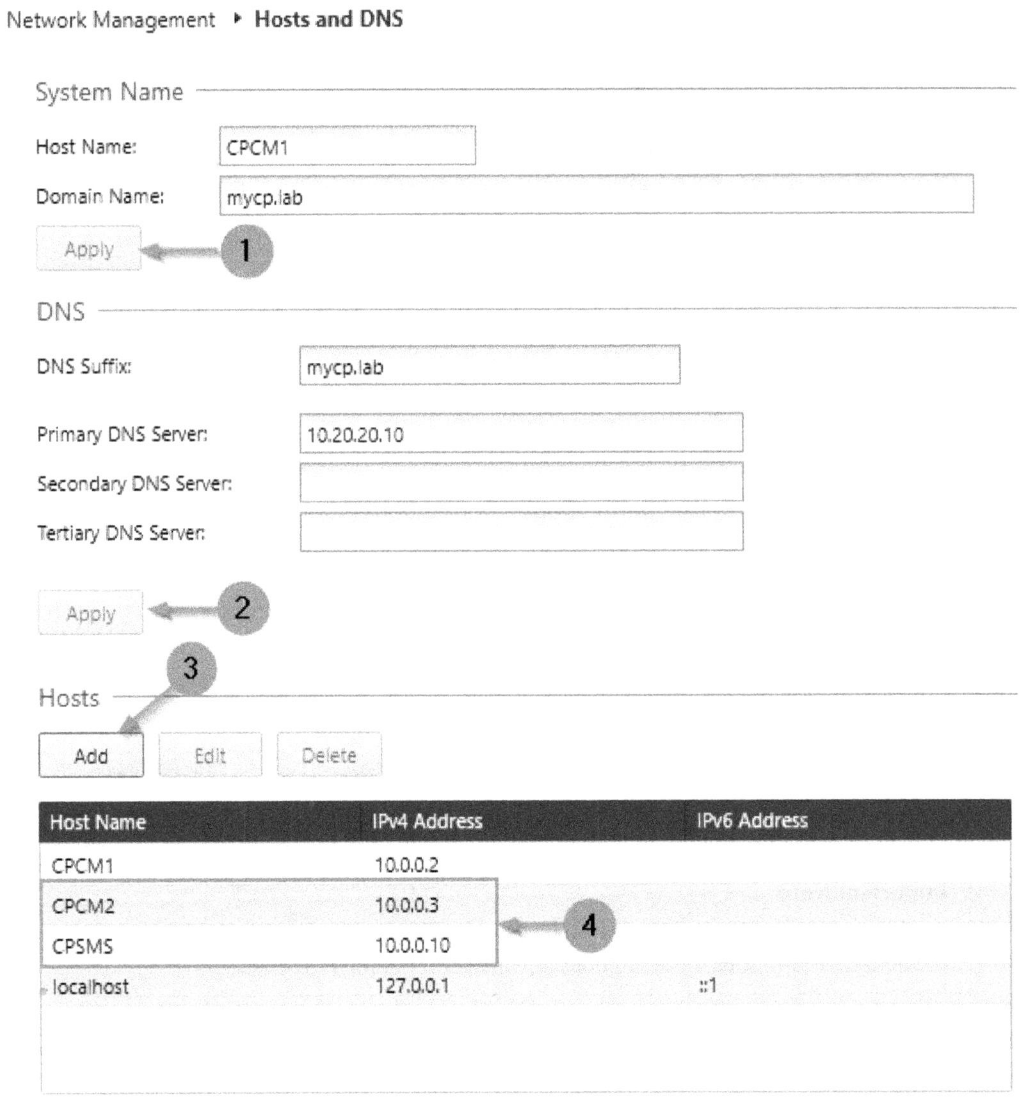

Figure 5.34 – Hosts and DNS properties

Let's go over the fields in this window:

1. Note that changes in **System Name** fields should *not* be performed casually, since it will affect communication between the gateways and management servers if it is done after the gateways are added as objects in SmartConsole. In order for the changes to take effect, click **Apply** before moving away from these fields.

2. It is imperative that the DNS servers configured in your Check Point infrastructure components are accurately defined. These should be the same internal servers used by the rest of your infrastructure. Your change management process for DNS should include a specific requirement to adjust these values. It is unfortunately a common occurrence to find the IP addresses of deprecated servers in production environments. If changes in DNS are made, click **Apply** before moving away from these fields.

3. In the **Hosts** subsection, click **Add**.

4. One of the entries of the **Hosts** table should be of the appliance you are working on (CPCM1 in our case). Verify that the IP address assigned to it is the one you intend to use for management. It is a best practice to have the rest of your Check Point infrastructure components' names and IP addresses defined in **Hosts**. Additionally, in a stable infrastructure, you may want to include entries for SMTP relays used for notifications and alerts, SFTP or SCP servers used for backups, as well as proxy servers and other hosts relying on frequent communication with your Check Point appliances that are not defined by IP addresses. Your change management processes should include reviews and adjustments of **Hosts** records in the Gaia configuration.

IPv4 Static Routes

When clicked, this section of the navigation tree displays the following window:

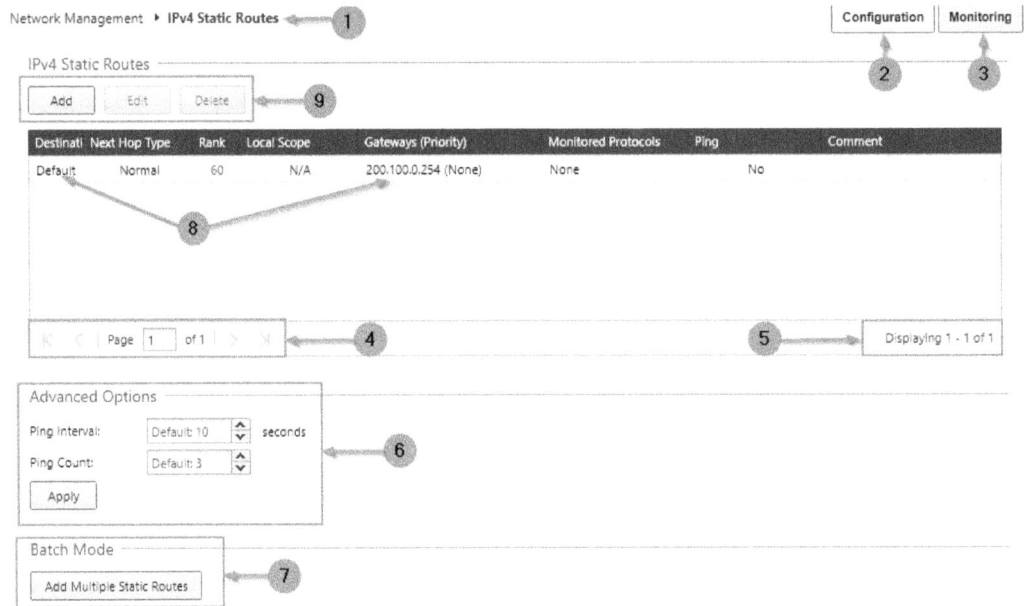

Figure 5.35 – IPv4 Static Routes

The elements visible on the screen are as follows:

1. The location in the navigation tree

2. The **Configuration** tab (default view)

3. The **Monitoring** tab

4. Static routes page navigation

5. Static route current page

6. Ping probing of the next hop

7. Batch addition of the static routes using space-delimited files

8. The default route configured by the FTW

9. The **IPv4 Static Routes** action menu

We do not need any additional static routes in our lab, but just to see the configuration options, let's go through the exercise of adding a route that we will later remove. In the screen shown previously, click on the **Add** option. You should be presented with the **Add Destination Route** window:

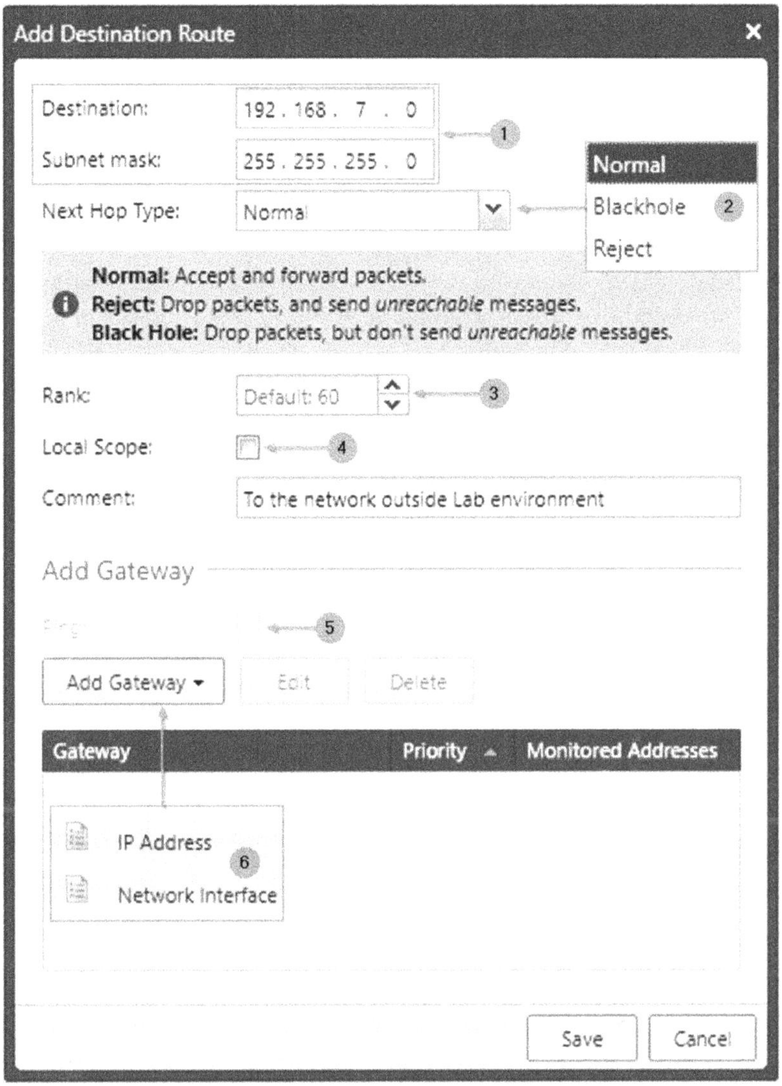

Figure 5.36 – Static route configuration options

Here we can configure the following:

1. The **Destination** network and **Subnet mask**.

2. The **Next Hop Type** options are **Normal**, **Blackhole**, and **Reject**.

3. **Rank** is set to a default static route value of **60**.

4. The **Local Scope** option is used for ClusterXL implementations where IP addresses of the physical interfaces are *not* in the same network as the vIP.

5. **Ping** is the ICMP probing of the next hop *only* if it is defined as the IP address. If there are multiple gateways defined for the same route and even one of them is defined as the interface, **Ping** becomes unchecked and is grayed out.

6. The gateway manipulation options. When adding gateway(s) for the static route (this action is only available for the **Normal** routes), you have two options: **IP Address** and **Network Interface**. Generally, **IP Address** is the preferred method. The interface is the viable option when you have no control over the IP addresses of the hosts/appliances connected to the network that this Check Point gateway's interface belongs to. This situation could be encountered with certain service providers that belong to your extranet.

> **Important Note**
> Do *not* use the **Network Interface** option for the **Default** routes. This will result in an ARP cache overflow!

When we click **Add Gateway** or **Edit** in this section, we are presented with the following child window:

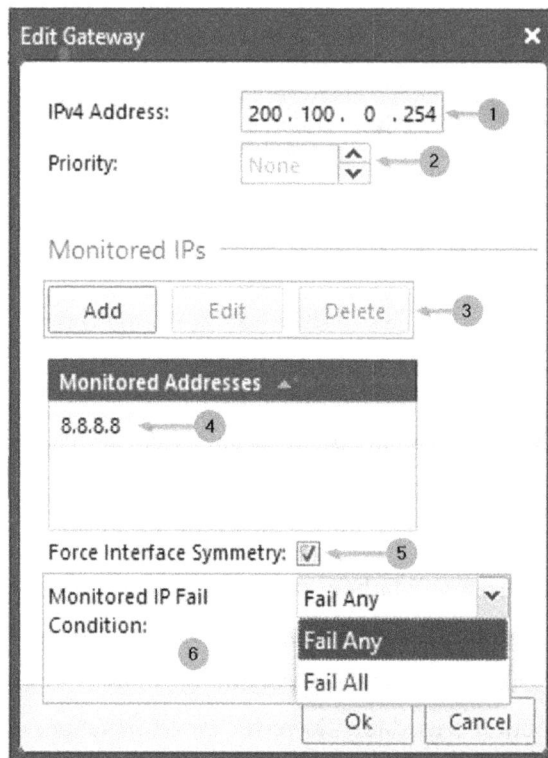

Figure 5.37 – Static route IP address gateway configuration

In it, we define the following:

1. The IP address of the next hop.

> **Important Note**
> The rest of the settings in this window are optional.

2. **Priority** can be left as **None** or assigned an integer value between **1** and **8**. Lower values have higher priority.

3. The **Monitored IPs** section. This option can only be used if the monitored addresses are already defined in a different section of the navigation tree, **IP Reachability Detection**. In that section, you define the IP addresses used for either single-hop or multi-hop probing, using either ICMP or **Bidirectional Forwarding Detection (BFD)** protocols.

4. The monitored address(es) selected in a previous step.

5. The **Force Interface Symmetry** checkbox should be selected if there is more than one possible route to the destination. It is checked when monitoring and should disable the route in case of asymmetric routing.

6. **Monitored IP Fail Condition** can be either **Fail Any** or **Fail All**. **Fail Any** would disable the route if any of the monitored IP addresses became unreachable. **Fail All** will do the same if all of the monitored IP addresses are not responding.

Display Format

Before we configure the **Time** settings in Gaia, it is preferable to change the **Display Format** values:

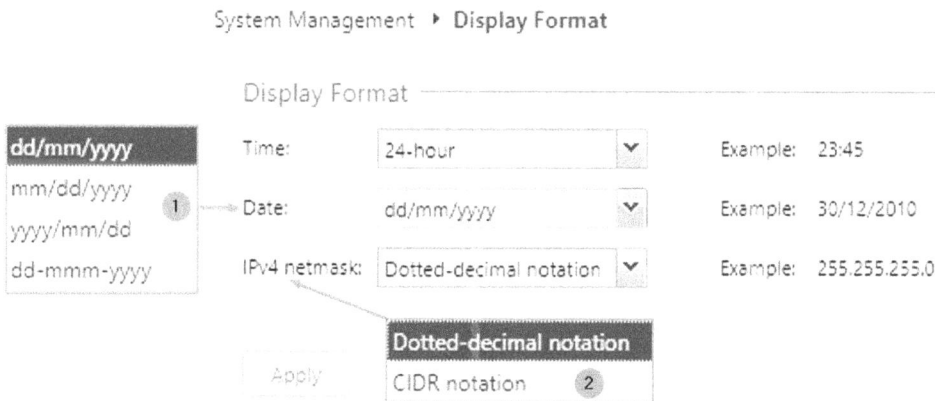

Figure 5.38 – Display Format for time, date, and IPv4 netmask notation

Change the **Date** format to what is acceptable in your region (1) and **IPv4 netmask** (2) to **CIDR notation** or **Dotted-decimal notation**, according to your preferences.

Time

When the **Time** option is clicked in the navigation tree, you are prompted with the following window:

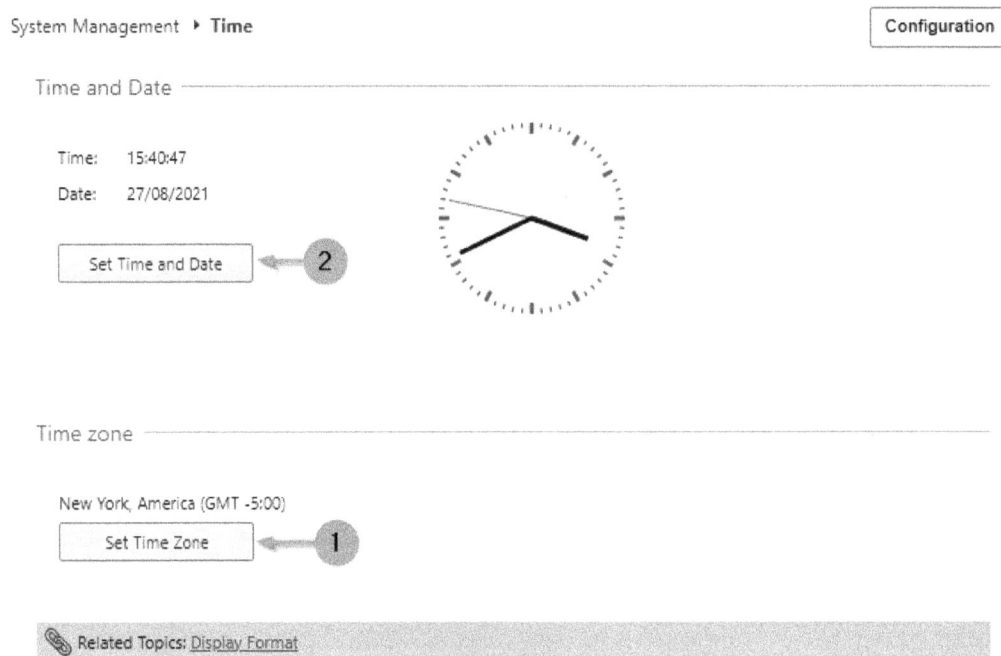

Figure 5.39 – Time configuration

In it, we should do the following:

1. Set **Time Zone** by choosing the zone where the appliance is residing or, if you are using common time for log correlation, the one your management server and SIEMs are using as a reference zone:

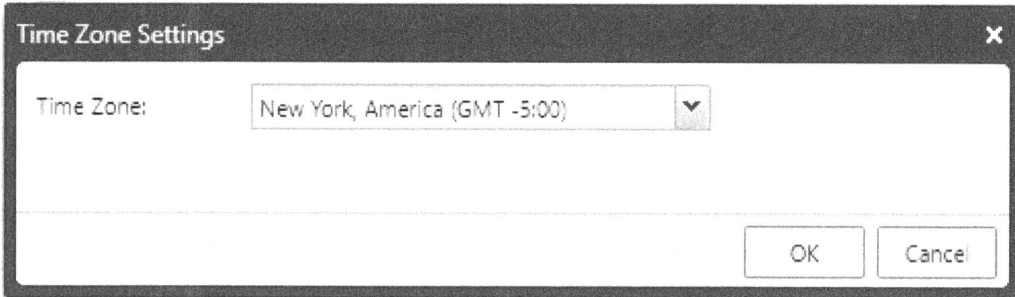

Figure 5.40 – Time Zone

2. Configure the **Network Time Protocol (NTP)** servers:

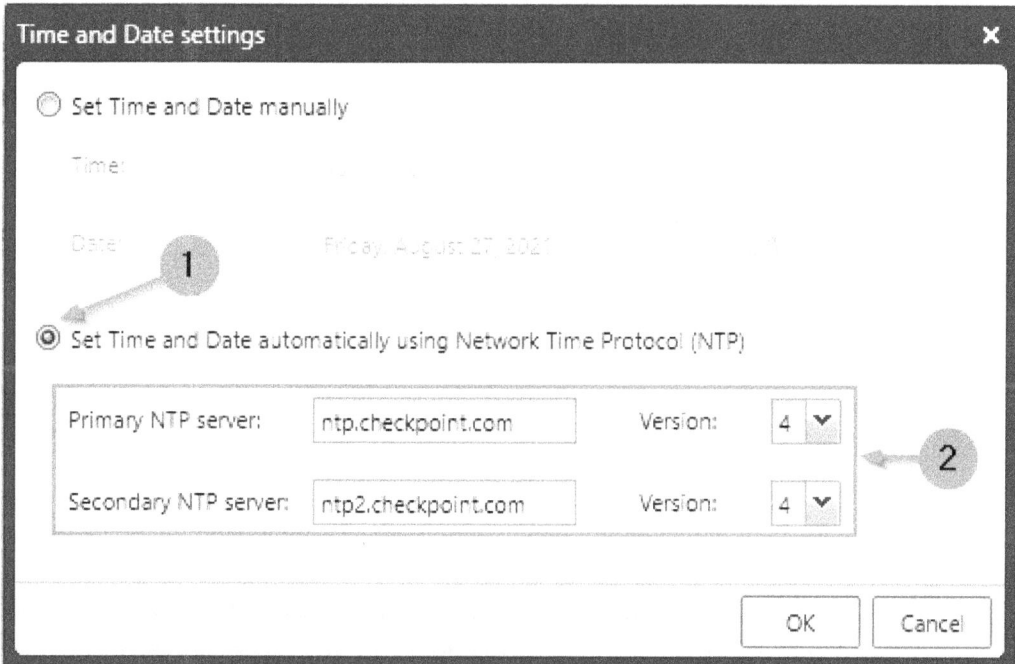

Figure 5.41 – NTP servers and protocol version

> **Important Note**
>
> While for our lab we are using external time servers, even for an infrastructure of moderate complexity, it is recommended to have dedicated internal time servers. In small- and medium-sized enterprises, it is common to see Windows domain controllers being used as NTP servers. If this is the case in your environment, consider enabling high-accuracy settings:
>
> ```
> https://docs.microsoft.com/en-us/windows-server/
> networking/windows-time-service/configuring-
> systems-for-high-accuracy
> ```
>
> Also, note that this is not a requirement for Check Point implementation and that the default accuracy settings are sufficient.

Mail Notification

When **Mail Notification** is clicked in the navigation tree, you are prompted to specify **Mail Server** and **User Name**:

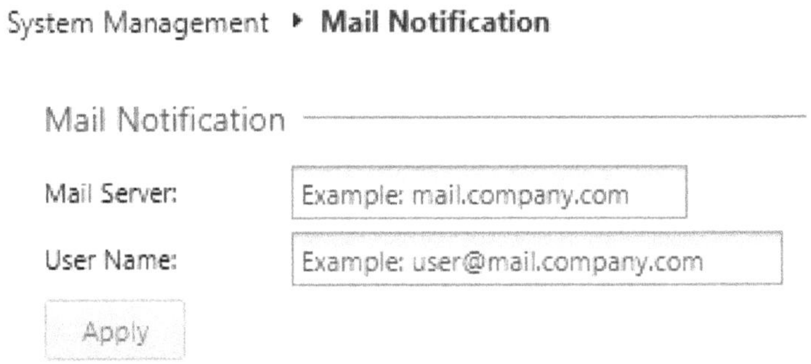

Figure 5.42 – Mail Notification

While we are not using this feature in the lab, it will allow you to use the alerting functionality of Check Point in production environments that already have mail relay servers in place.

Messages

In some environments, you are required to display a legal disclaimer or a warning when certain hosts are being accessed:

System Management ▸ **Messages**

Messages ──────────────────────────────────

☑ Banner message

This system is for authorized use only.

(1)

☑ Message of the day

You have logged into the system.

(2)

☐ Show hostname on login page ◂— (3)

Apply ◂——— (4)

Figure 5.43 – Banner message and Message of the day

If this is the case, the **Messages** option in the navigation tree is the place where these banners are configured [1]. Additionally, **Message of the Day** allows you to display different text after successful login. This option may be used to alert other administrators of the system about pending or completed changes, troubleshooting cases in progress, and so on, and will be visible in either the CLI or the WebUI upon login:

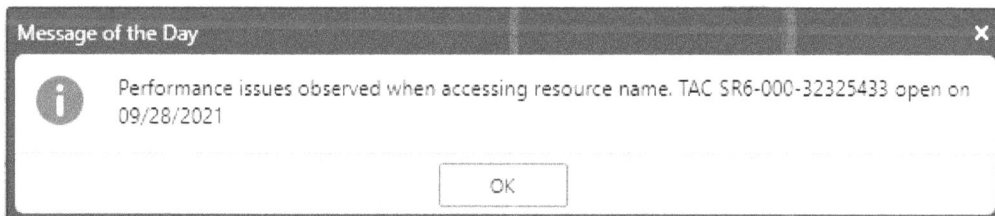

Message of the Day   ✕

ⓘ   Performance issues observed when accessing resource name. TAC SR6-000-32325433 open on
    09/28/2021

                              OK

Figure 5.44 – Message of the Day used for administration references

In addition to that, you also have the option of choosing whether the hostname is displayed before successful login [3]. When these options are configured or modified, click **Apply**.

Session

The **Session** option in the navigation tree allows for the adjustment of timeout parameters for the WebUI and CLI sessions [1] and specifies the refresh rate for the information tables in the WebUI [2]. Knowing its value (the default is **15** seconds), you should allow for the same before expected changes are reflected in the WebUI. If any changes are made here, click **Apply**:

> **Important Note**
> The administrative sessions' inactivity timeouts in your environment may be governed by your information security policy. If you need to adjust these parameters, which may be the case for maintenance operations, verify that the written provisions for this are included and that these parameters are reset to the normal operating values afterward.

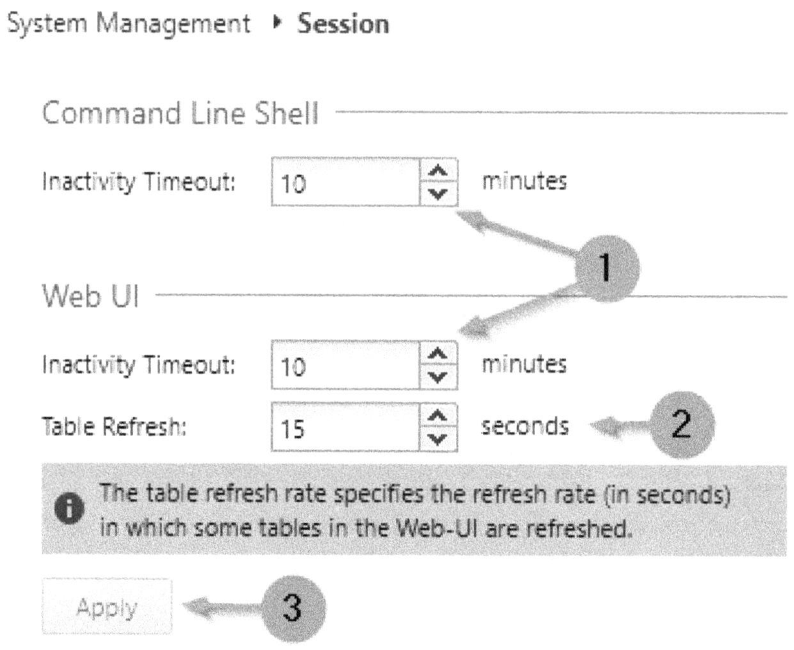

Figure 5.45 – Session settings

System Logging

When **System Logging** is clicked in the navigation tree, the following window is displayed:

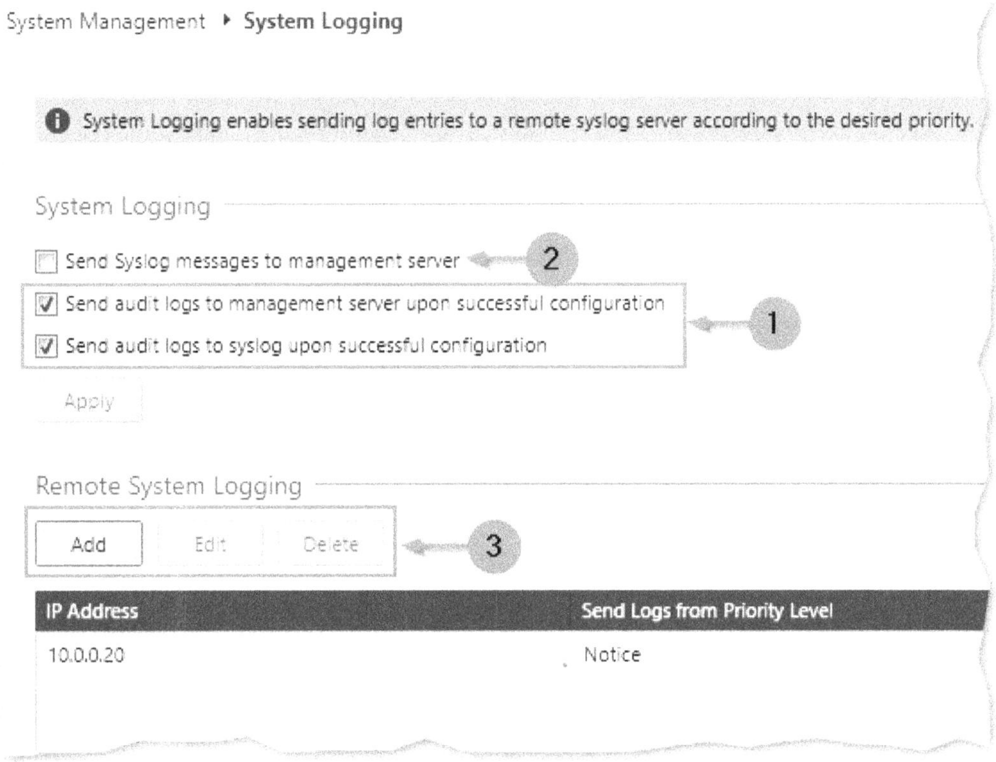

Figure 5.46 – System Logging options

Note that these configuration options are only relevant to the Gaia events logging and not the Check Point logs. The **Send audit logs to management server upon successful configuration** and **Send audit logs to syslog upon successful configuration** options [1] are selected by default. In this case, *syslog* refers to the local syslog facility.

The **Send Syslog messages to management server** option [2], is unchecked by default, but I recommend checking it once you have configured your Check Point infrastructure. Doing this will allow you to see the Gaia events in the Check Point logs and may aid in troubleshooting without relying on external event correlation facilities. Note that the Gaia syslog messages forwarded to Check Point logs contain the data in the information field, which is not indexed. You may be able to filter out the syslog messages, but you would have to search through them manually. That said, if you are planning to forward your Check Point logs to the same external syslog server as defined in **Remote System Logging** [3], you will have duplicate entries – some that will be sourced from individual Check Point appliances and all those that were sent to the management server.

The **Remote System Logging** subsection allows us to add external syslog server(s) used in your infrastructure. When you select the desired **Priority** level of the forwarded logs, all higher-level events will be automatically included, as depicted in the following screenshot (the escalation levels in the drop-down box are inverted):

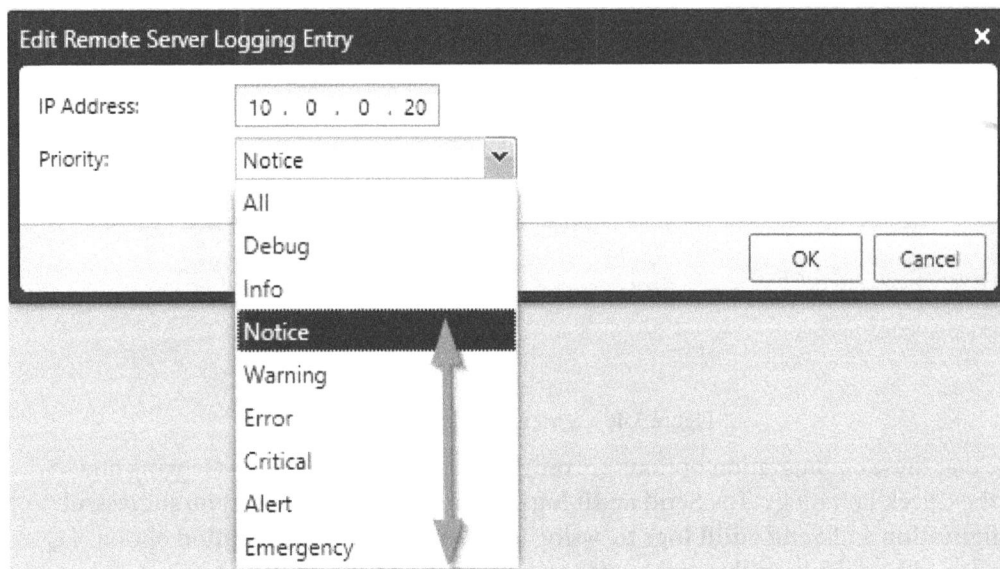

Figure 5.47 – Remote server logging entry and level

User Management

Note that users in this context are the administrators of Gaia and *not* Check Point users or administrators. It is possible to create identical local users or administrators for both, but their credentials will not be synchronized. The only possible exception is the Gaia default **admin** account [1], if it is chosen as **Security Management Administrator** [2] during the FTW:

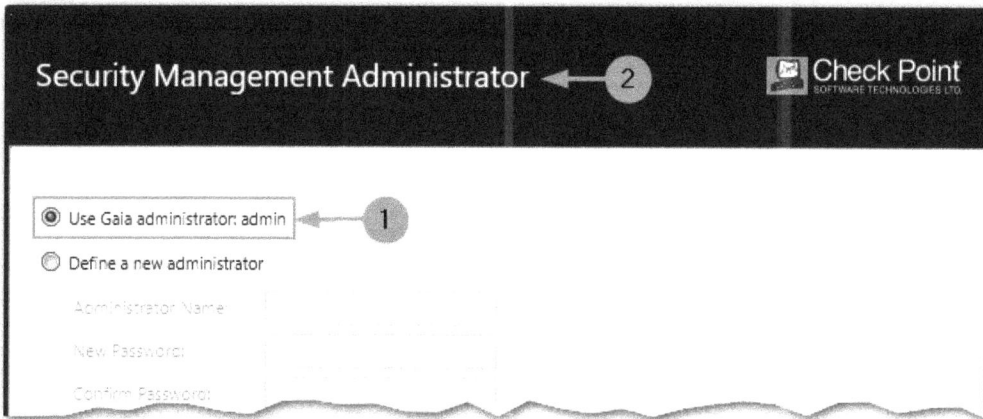

Figure 5.48 – Gaia admin account as Security Management Administrator

Roles and **Users** are two of the subsections of **User Management** that we will look at next.

Roles

Gaia employs role-based administration; that is, either predefined or newly created roles must be assigned to users for them to have rights on the system.

By default, three roles are created, **adminRole**, **cloningAdminRole**, and **monitorRole**, as well as two users, **admin** and **monitor**, with corresponding roles assigned to each.

When **Roles** in the **User Management** section is clicked, we are presented with the following window:

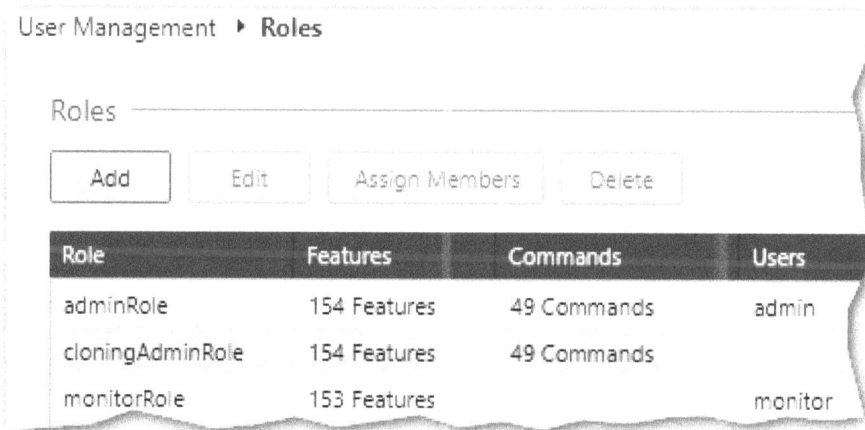

Figure 5.49 – User Management – Roles

If additional roles are needed (for instance, for dedicated routing administrators), click **Add**, and in the **Add Role** window [1], enter a new **Role Name** [2], filter the available list of features [3] (or scroll to select those you are interested in), and do either of the following:

- Click on the drop-down arrow next to each [4] and choose between **None**, **Read Only**, or **Read / Write** for each [5]. Click **OK** when done [6]:

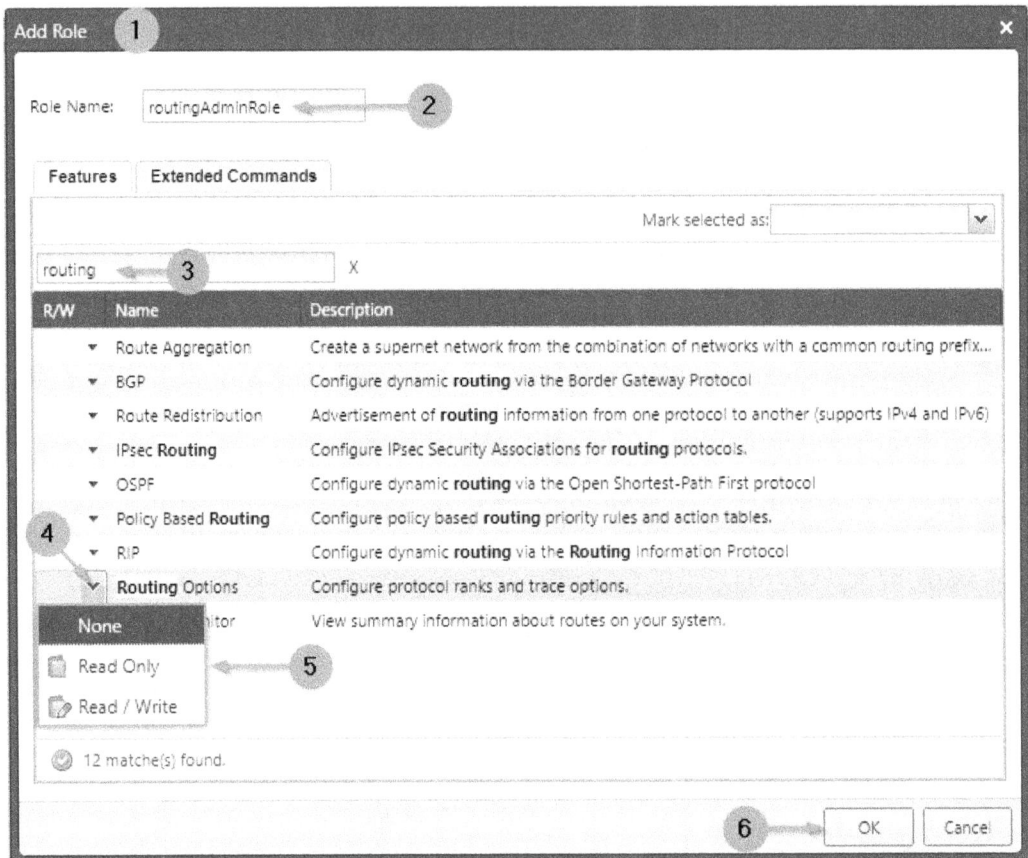

Figure 5.50 – User Management –individual feature rights of roles

- Repeat the first three steps and then use bulk selection (*Ctrl* + click, or click the top, scroll down, and *Shift* + click on the bottom) [4], click on **Mark selected as:**, and choose the rights you would want to grant for all of the selected features [6]. Click **OK** when done [7]:

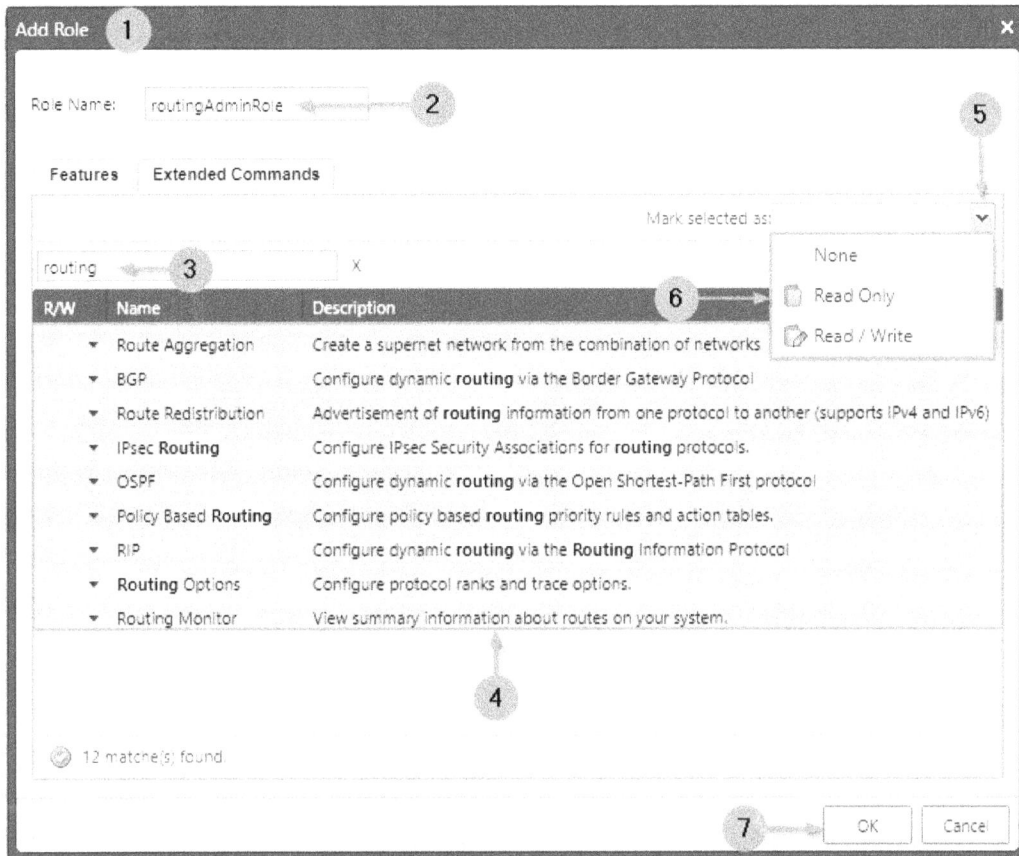

Figure 5.51 – User Management – bulk assignment of roles

Important Note

Note that roles cannot be renamed, but the assigned features and the rights can be edited. If you must rename a role, remove it from the users it was previously assigned to, delete, recreate the role, and assign it to the users again.

In larger environments, where administrators rely on **Authentication, Authorization, and Accounting (AAA)** servers, the roles must still be defined on Check Point appliances, but there is no need to create local user accounts assigned to those roles.

Users

If there are no AAA servers in your environment, define users locally. When **Users** is clicked in the navigation tree, you are presented with the following window:

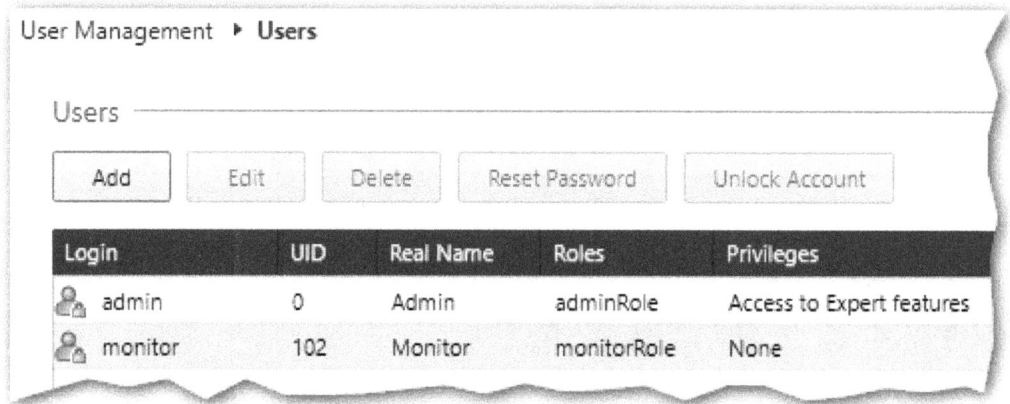

Figure 5.52 – User Management – Users

Click **Add** to open the following window:

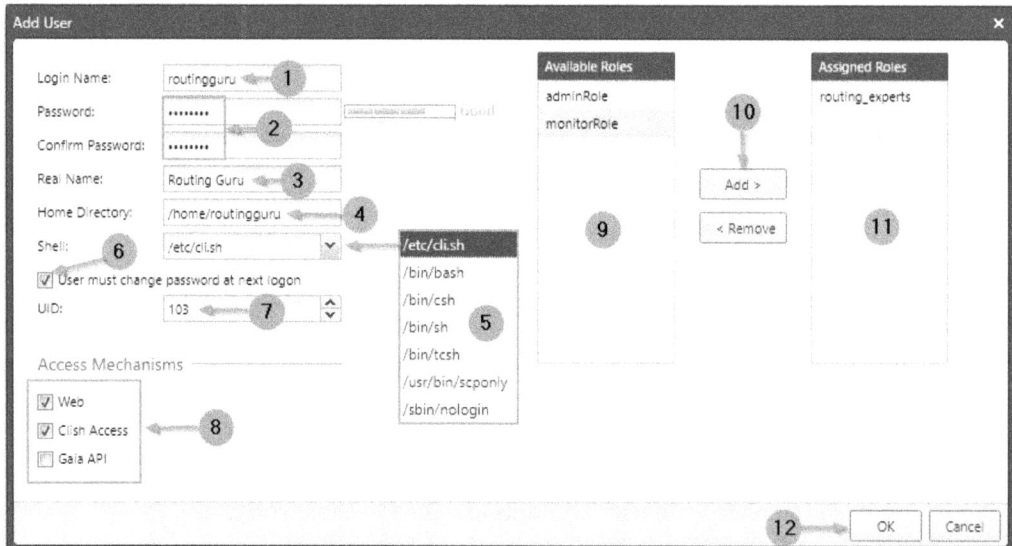

Figure 5.53 – User Management – Add User

Perform the following actions:

1. Define **Login Name**.

2. Set a temporary **Password**.

3. Adjust **Real Name** (the default is the same as **Login Name** with the first letter in uppercase).

4. Verify or change **Home Directory**.

5. Choose a **Shell** option (important – leave it as **/etc/cli.sh** unless a change is warranted).

6. Check **User must change password at next logon** if provisioning for others.

7. Specify **UID** (an integer between **103** and **65533**).

8. Choose **Access Mechanisms**.

9. Select one or more roles (*Ctrl* + click for multiples).

10. Click **Add** to move the selection to **Assigned Roles**.

11. Verify that the necessary roles are assigned.

12. Click **OK**.

Cloning groups

By now, after configuring a fraction of all available options for a single Check Point appliance, you may be wondering how to do this consistently over even a simple infrastructure without missing anything. Cloning groups is a feature that can significantly simplify the process.

This feature can be enabled once the Check Point infrastructure is defined in SmartConsole and a rule allowing communication between participating Check Point appliances on the 1129 TCP port is included in the installed security policy.

A description of cloning group functionality could take an entire chapter, but I will simply point out that you can use a multi-step approach by first creating cloning groups containing only the settings that are common throughout the infrastructure, such as **Roles**, **Hosts**, **DNS**, **Messages**, and **Notifications**, and synchronizing those.

You may then **Leave the cloning group** by performing this action on all of its members.

Once all members have left the cloning group, it disappears. You may then create another one for your cluster members using the **Cloning Group follows ClusterXL** option to enable automatic synchronization of a wider set of options that includes advanced routing, ARP, DHCP, and many more.

We have covered the options necessary for the configuration of our lab, but you saw that there are more options available in the navigation tree of the Gaia Portal for use in more complex environments, or for troubleshooting and maintenance.

Widgets and status bar

Scrolling to the bottom of the **Overview** page after logging into the Check Point appliance will bring you to the **Add Widget** button [1], which, when clicked, presents you with the option to select the widgets to be displayed [2], as well as a status bar that can be expanded when the arrow on its right side is clicked [3]. Doing that will show the status of the last configuration action executed in the WebUI:

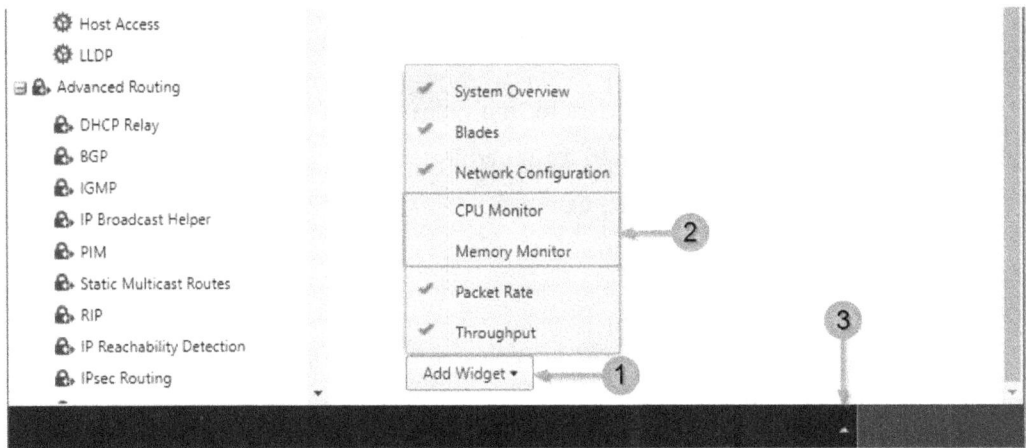

Figure 5.54 – Add Widget button and status bar

You may pick the relevant widgets for the type of appliance you are working on; that is, **Packet Rate**, **Throughput**, and **Blades** may not be relevant to your management server(s), but they are relevant to your gateways. You may also select all the widgets for each of your appliances and move them around in the **Overview** section to have the most relevant closer to the top.

This concludes the introduction to the WebUI, which we have used to perform the configuration of the first cluster member. We have learned how to navigate in this interface and explored some of its capabilities.

Summary

In this chapter, we have learned about the origins of the Gaia operating system. We have used the FTW to prepare the Check Point products on the management server and one of the cluster members. Additionally, we have used the CLI to achieve the same on the remaining cluster member.

We then completed the configuration of one of the cluster members using the WebUI.

In the next chapter, by way of introduction to the Gaia CLI shell, *CLISH*, we will perform the configuration of the second cluster member. We will learn how to save, modify, and load the Gaia configuration of the Check Point appliances.

6

Check Point Gaia Command-Line Interface; Backup and Recovery Methods; CPUSE

In this chapter, we will learn about the Check Point **command-line interface** (**CLI**). We will familiarize ourselves with its structure and operation and perform some of the common Gaia configuration tasks using **Command-Line-Interface Shell** (**CLISH**) and Expert mode shell. Later, we will go over Check Point backup and recovery options, offline Gaia configuration manipulation, and the **Check Point Upgrade Service Engine** (**CPUSE**) built-in upgrade and update tool.

In this chapter, we are going to cover the following main topics:

- Learning about the Check Point Gaia CLI
- Introduction to Expert mode

- Configuring Gaia using CLISH

- Saving Gaia configuration, backups, snapshots, and migration tools

- Saving and loading the configuration

- Offline configuration editing and comparison

- Using CPUSE

Learning about the Check Point Gaia CLI

The default Gaia shell (CLISH) allows us to perform the same Gaia configuration tasks as the WebUI. It is intuitive, is easy to learn, and can be accessed in a number of different ways: using a **Registered Jack - 45 (RJ45)** serial console, a **Universal Serial Bus (USB)** Type-C console, or a **HyperText Markup Language 5 (HTML5) Lights Out Management (LOM)** console emulator of the physical appliances; **Secure Shell (SSH)**, from within the Gaia WebUI virtual terminal, or from **Actions | Open Shell** of a selected appliance within SmartConsole.

Physical console connection parameters are listed here:

- 9600 **bits per second (bps)**

- 8 bits

- No parity

- 1 stop bit (8N1)

- **Flow control | None**

The USB Type-C connection has a higher priority, and if you are relying on the console server connected to an RJ45 port for remote administration, verify that the USB Type-C cable is disconnected.

USB Type-C console drivers should be downloaded from the appliance model's home pages, as denoted by numbers [1] and [2] in the following screenshot:

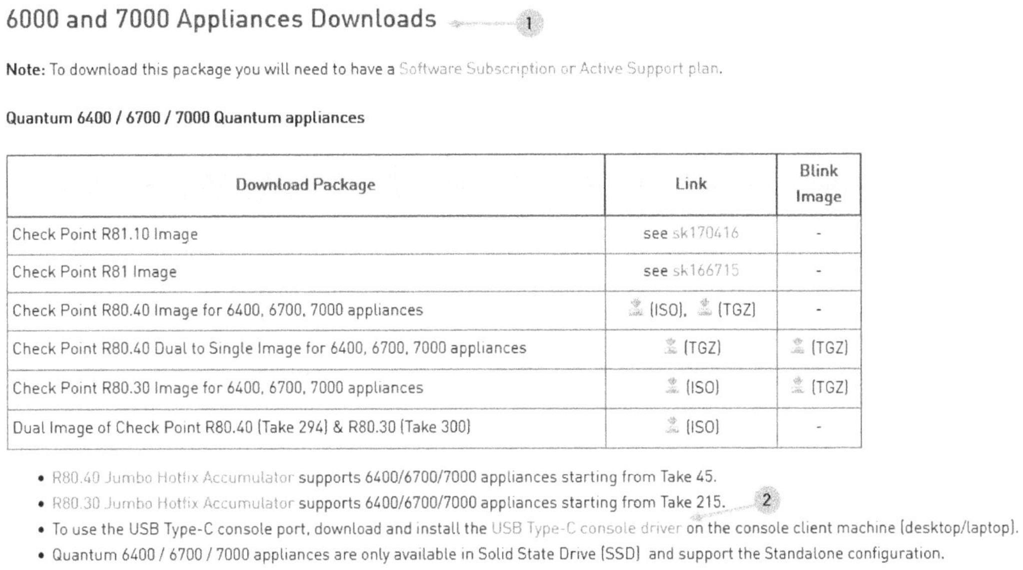

6000 and 7000 Appliances Downloads ⟵---①

Note: To download this package you will need to have a Software Subscription or Active Support plan.

Quantum 6400 / 6700 / 7000 Quantum appliances

Download Package	Link	Blink Image
Check Point R81.10 Image	see sk170416	-
Check Point R81 Image	see sk166715	-
Check Point R80.40 Image for 6400, 6700, 7000 appliances	⤓ (ISO), ⤓ (TGZ)	-
Check Point R80.40 Dual to Single Image for 6400, 6700, 7000 appliances	⤓ (TGZ)	⤓ (TGZ)
Check Point R80.30 Image for 6400, 6700, 7000 appliances	⤓ (ISO)	⤓ (TGZ)
Dual Image of Check Point R80.40 (Take 294) & R80.30 (Take 300)	⤓ (ISO)	-

- R80.40 Jumbo Hotfix Accumulator supports 6400/6700/7000 appliances starting from Take 45.
- R80.30 Jumbo Hotfix Accumulator supports 6400/6700/7000 appliances starting from Take 215. ⟵---②
- To use the USB Type-C console port, download and install the USB Type-C console driver on the console client machine (desktop/laptop).
- Quantum 6400 / 6700 / 7000 appliances are only available in Solid State Drive (SSD) and support the Standalone configuration.

Figure 6.1 – USB Type-C console driver download

CLISH is a restricted, role-based shell with only the options assigned to the role of the users/administrators available to them. If you recall our exercise in the previous chapter, where we defined a sample role for the routing administrators in WebUI, the same permissions are now available for this role in CLISH.

The most commonly used CLISH commands pertaining to Gaia configuration that you will encounter are these:

- `lock`
- `show`
- `set`
- `add`
- `save`
- `expert`

The *Tab* key provides both autocompletion (if there are no alternatives) and a list of additional options or parameters (if entered after the *spacebar* is pressed after the last command or option).

A question mark after the command either describes its intended functionality or prompts you to use *Tab* in order to narrow down your choices before using ? for a more detailed explanation.

Should you want to abandon the typed command without execution, use *Ctrl + C* to get back to the prompt.

The *up* and *down* arrow keys can cycle through previously entered commands. history will display a numbered list of previously executed commands, and you can call any of those by entering !## (where ## is the number of the command in the history list).

If the output of the command spans multiple pages, you can use *Enter* to advance by line, *spacebar* to scroll down a page, and either *Ctrl + C* or q to get back to the prompt.

If you simply type ? at the CLISH prompt, you will see a brief reminder of these options.

If you have the Gaia WebUI open and have logged on via a terminal session, a lock is maintained by the WebUI. This restricts you to only using the show command. If you attempt to execute any other command, you will encounter the following situation:

```
CPCM1> set time 11:24
CLINFR0519  Configuration lock present. Can not execute this
command. To acquire the lock use the command 'lock database
override'.
CPCM1>
```

If you suspect that the configuration is locked by someone else, execute the show config-lock command to determine if that is the case, as follows:

```
CPCM1> show config-lock
Configuration locked by admin (300 seconds to expiration)
CPCM1>
```

To override the lock, you will have to use the lock database override command.

If you have built the lab that was described in the previous chapters, I suggest trying each of the commonly used commands with *Tab* then *spacebar* then *Tab* to see lists of available options.

For the `lock database override` command in particular, no options are available, so after typing `lock` followed by a space, simply press *Tab* twice to autocomplete the command and press *Enter* to execute.

> **Important Note**
>
> The preceding example is also an important reminder that the Gaia configuration is not a static configuration file, but a database.

To showcase the autocomplete/show options behavior better, type `show rout` (an intentionally incomplete word), as illustrated in the following code snippet, and press *Tab*:

```
CPCM1> show rout
route           - Show routing table information
routed          - Show version, state of the Routing Daemon
routemap        - Show Route Map configuration
routemaps       - Show configuration of all Route Maps
router-id       - Show the Router ID
router-options - Show Router Options configuration state
CPCM1>
```

You are presented with all available options and their descriptions.

We have already encountered CLISH when we were changing the hostnames of cloned Check Point **virtual machines** (**VMs**) by logging in to Gaia and executing `set hostname` and `save config` commands.

To see the configuration of the appliance, execute `show configuration` on the CPCM1 VM in our lab. The resultant output is approximately 200 lines long. Press *spacebar* repeatedly to advance by a single page to the end of the configuration output. Scrolling through it, you will recognize some of the settings we have configured either during the **First Time Configuration Wizard** (**FTW**) or afterward in the Gaia WebUI. The rest are default prerequisite settings. Take a look at the top section of the following output:

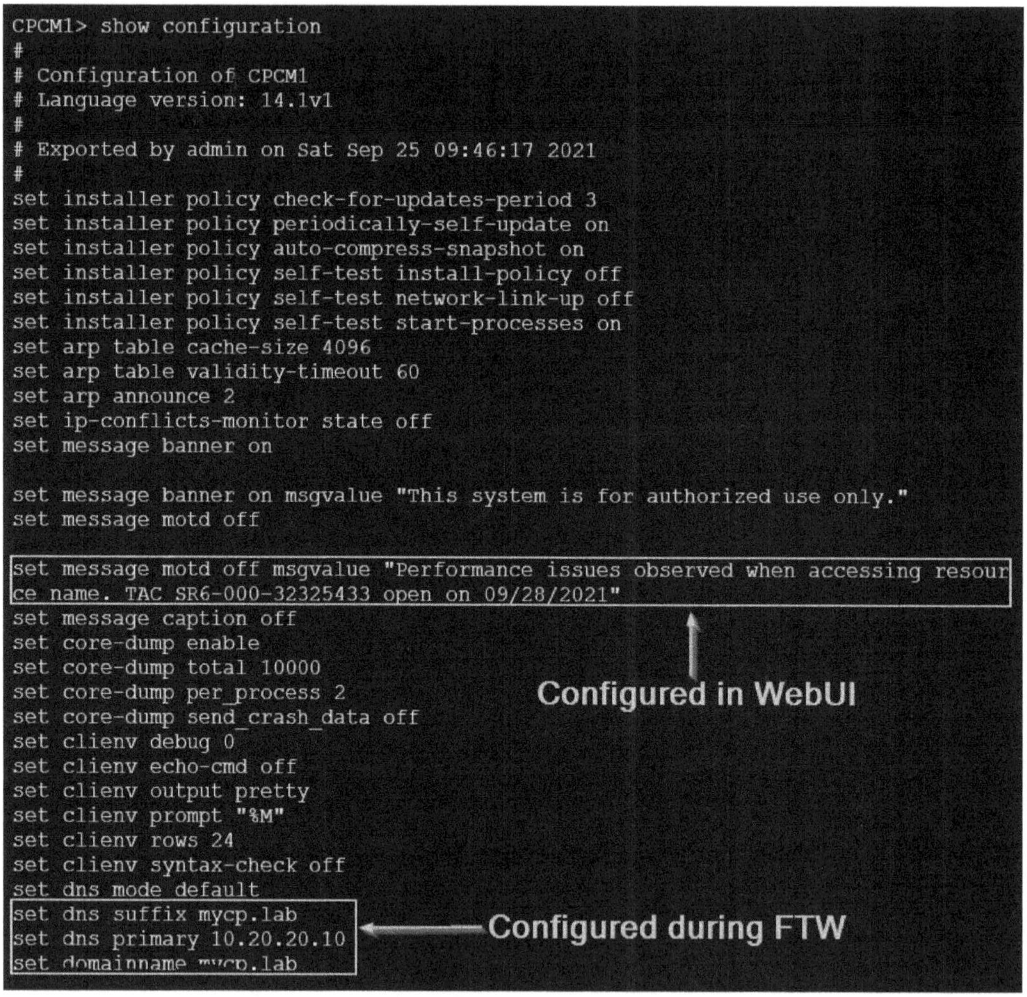

```
CPCM1> show configuration
#
# Configuration of CPCM1
# Language version: 14.1v1
#
# Exported by admin on Sat Sep 25 09:46:17 2021
#
set installer policy check-for-updates-period 3
set installer policy periodically-self-update on
set installer policy auto-compress-snapshot on
set installer policy self-test install-policy off
set installer policy self-test network-link-up off
set installer policy self-test start-processes on
set arp table cache-size 4096
set arp table validity-timeout 60
set arp announce 2
set ip-conflicts-monitor state off
set message banner on

set message banner on msgvalue "This system is for authorized use only."
set message motd off

set message motd off msgvalue "Performance issues observed when accessing resour
ce name. TAC SR6-000-32325433 open on 09/28/2021"
set message caption off
set core-dump enable
set core-dump total 10000
set core-dump per_process 2                    Configured in WebUI
set core-dump send_crash_data off
set clienv debug 0
set clienv echo-cmd off
set clienv output pretty
set clienv prompt "%M"
set clienv rows 24
set clienv syntax-check off
set dns mode default
set dns suffix mycp.lab
set dns primary 10.20.20.10    ◄────────Configured during FTW
set domainname mycp.lab
```

Figure 6.2 – CLISH show configuration output

Scrolling through the output, you will encounter another section—the lines containing the `secadmin` administrator that we have created in WebUI, as illustrated in the following screenshot:

```
add user secadmin uid 0 homedir /home/secadmin
add rba user secadmin roles adminRole
set user secadmin gid 100 shell /etc/cli.sh
set user secadmin realname "Secadmin"
set user secadmin password-hash $6$rounds=10000$OLkURgPY$6WN1Bd3xx3t6IzwyW/kKr3Z
OXuLotBWN153JCsoz6MK7fN9vTV8K8k9DfWUJuAnvgscXQYJ8HKwx3O9.WakNe/
```

Figure 6.3 – CLISH add and set sequence

Note the sequence of add and set commands in this section used to create a user account, assign it to a role, and define its parameters.

If performed out of sequence or not accurately completed, the account will not be created or active. That is, simply adding the user/administrator results in the following:

```
CPCM1>  add user testuser uid 0 homedir /home/testuser
WARNING Must set password and a role before user can login.
- Use 'set user USER password' to set password.
- Use 'add rba user USER roles ROLE' to set a role.

CPCM1>
```

Figure 6.4 – Configuration dependencies

The reason I am focusing your attention on this is that it is a common enough mistake to see administrators performing copy-paste of blocks of code while omitting some of the lines, which leads to unexpected results later on.

Delete this user account by executing delete testuser.

Now, scroll to the end of the configuration output and take a look at these two lines of code:

```
set snmp traps advanced coldStart reboot-only off ◄——— ①
set static-route default nexthop gateway address 200.100.0.254 on
CPCM1>                                                            ②
```

Figure 6.5 – CLISH "off" and "on" parameters in a different context

The on and off parameters result in a different behavior based on the context—that is, for the **Simple Network Management Protocol (SNMP)** and a majority of other settings, it will perform a toggle action and display its current state in the configuration output. For routes, off is equivalent to delete and will remove the record from the configuration.

To demonstrate, copy-paste the last line, changing the `nexthop gateway address` **Internet Protocol version 4 (IPv4)** value to `200.100.0.250`, press *Enter*, execute `show configuration`, and press *spacebar* repeatedly to page down to the end of the output. You will see something like this:

```
set static-route default nexthop gateway address 200.100.0.250 on
set static-route default nexthop gateway address 200.100.0.254 on
CPCM1>
```

Figure 6.6 – "on " and "off" usage for routes

Now, press the up arrow key twice to show the `set static-route` command again, *Backspace* to replace on with `off`, and press *Enter*. Execute `show configuration` and page down to the bottom. The route we have set in the previous step is no longer present.

Introduction to Expert mode

In addition to CLISH, which is the default restrictive shell, there is also an **Expert** shell (also referred to as **Expert mode**) that grants access to the advanced system and Linux functions. Before we can gain access to it, the expert password must be configured.

When logged in to CPCM1, execute the `set expert-password` command. When prompted, enter and confirm the expert password (use the same `CPL@b2021` password that we adopted for the lab).

Once your expert password is configured, execute `save configuration` or its value will not survive the reboot.

We will not focus too much on Expert mode functionality in this book, but it will be required to perform some of the tasks in this and subsequent chapters. The Expert shell is also the primary shell for troubleshooting.

While in CLISH, type `expert` and press *Enter*. Type your expert password and press *Enter* to switch to Expert mode, as illustrated in the following screenshot:

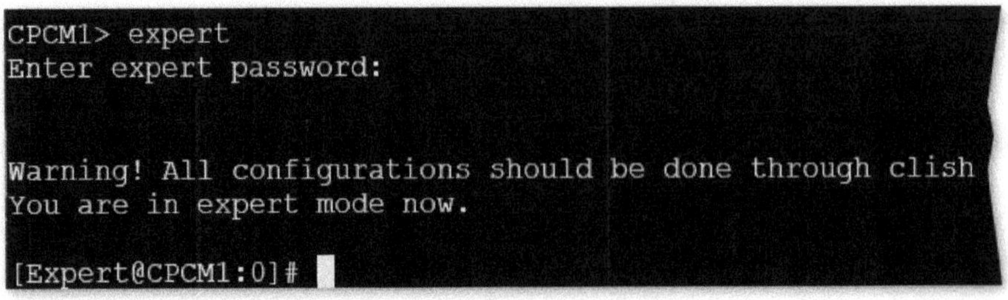

```
CPCM1> expert
Enter expert password:


Warning! All configurations should be done through clish
You are in expert mode now.

[Expert@CPCM1:0]#
```

Figure 6.7 – Entering Expert mode shell

Since CLISH does not have a search or filter functionality, this could be accomplished in Expert mode. As it allows us to use Linux functions, one of the interesting options is the ability to execute multiple CLISH commands and pipe their output for further processing, as shown in the following screenshot:

```
[Expert@CPCM1:0]# clish -c "show time";clish -c "show configuration" | grep eth0
;clish -c "show interface eth0" | grep errors
Time 12:55:49
set interface eth0 comments "Mgmt"
set interface eth0 link-speed 1000M/full
set interface eth0 state on
set interface eth0 auto-negotiation on
set interface eth0 mtu 1500
set interface eth0 ipv4-address 10.0.0.2 mask-length 24
set management interface eth0
TX bytes:16887913 packets:17678 errors:0 dropped:0 overruns:0 carrier:0
RX bytes:1178000 packets:9859 errors:0 dropped:0 overruns:0 frame:0
[Expert@CPCM1:0]#
```

Figure 6.8 – Chaining CLISH commands and filtering output in Expert mode

We also have to be in Expert mode to switch shells in order to permit **Secure Copy Protocol (SCP)** or **SSH File Transfer Protocol (SFTP)** file transfers. If you attempt to connect to the Check Point appliance using these protocols and valid credentials while your user's shell is defined as CLISH, you will encounter the following error:

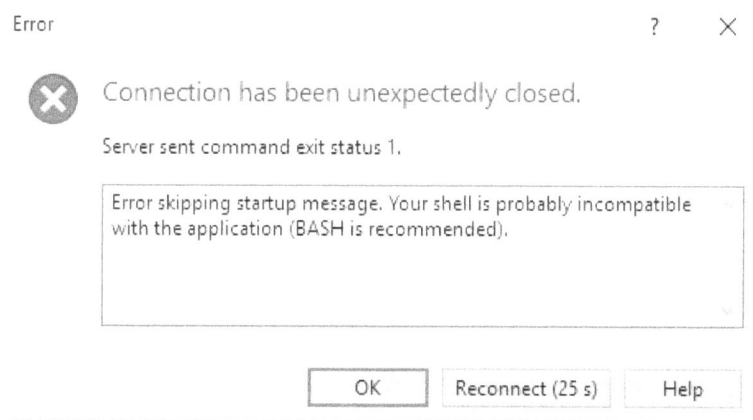

Error   ?   ✕

Connection has been unexpectedly closed.

Server sent command exit status 1.

Error skipping startup message. Your shell is probably incompatible with the application (BASH is recommended).

OK   Reconnect (25 s)   Help

Figure 6.9 – Incompatible shell message for users with CLISH as their shell

To change shells in Expert mode, execute the chsh -s /bin/bash command, perform file transfer(s), and change back to CLISH by executing chsh -s /etc/cli.sh.

It is possible to create a dedicated user account for file transfers with /usr/bin/scponly as its shell, to restrict logons to SCP only, and to prevent execution of unrelated Gaia commands, but this is more suitable for scripting and automation tasks.

> **Note**
>
> Now that you are familiarized with the basics of CLISH and Expert mode, return to *Chapter 5, Gaia OS, the First Time Configuration Wizard, and an Introduction to the Gaia Portal (WebUI)*, under the *First-Time Configuration using the CLI* section to complete the FTW for CPCM2 before continuing to the next section of this chapter.

This concludes our introduction to the Expert mode.

Configuring Gaia using CLISH

You should be reading this section after completing the *First-Time Configuration using the CLI* section of *Chapter 5, Gaia OS, the First Time Configuration Wizard, and an Introduction to the Gaia Portal (WebUI)*.

The interactive session segments depicted next illustrate the CLI configuration of Gaia. I am explicitly showing it this way instead of simply listing commands to highlight the interactive aspects of some of the commands. For example, the `add user` command produces a warning about additional steps that are required to define the user account, and `set user <username> password` prompts you for a password and verification. It is also a good idea to start typing commands by hand using the *Tab* key for autocompletion, where appropriate, to develop a muscle memory and to increase your comfort level with CLISH.

The steps to configure Gaia for our lab using CLISH are listed here:

1. Delete the alias interface (the one carrying the IP address from the clone source), as follows:

   ```
   CPCM2> delete interface eth0 alias eth0:1
   ```

2. Create a user/administrator `secadmin` account, assign `adminRole` to it, and set its password, as follows:

   ```
   CPCM2> add user secadmin uid 0 homedir /home/secadmin
   WARNING Must set password and a role before user can
   login.
   - Use 'set user USER password' to set password.
   - Use 'add rba user USER roles ROLE' to set a role.

   CPCM2> add rba user secadmin roles adminRole
   CPCM2> set user secadmin gid 100 shell /etc/cli.sh
   ```

```
CPCM2> set user secadmin realname "Secadmin"
CPCM2> set user secadmin password
New password:CPL@b2021
Verify new password:CPL@b2021
```

3. Configure and activate the interfaces, as follows:

```
CPCM2> set interface eth0 comments "Mgmt"
CPCM2> set interface eth1 comments "Internal1"
CPCM2> set interface eth1 state on
CPCM2> set interface eth1 ipv4-address 10.10.10.3 mask-
length 24
CPCM2> set interface eth2 comments "Internal2"
CPCM2> set interface eth2 state on
CPCM2> set interface eth2 ipv4-address 10.20.20.3 mask-
length 24
CPCM2> set interface eth3 comments "Sync"
CPCM2> set interface eth3 state on
CPCM2> set interface eth3 ipv4-address 192.168.255.2
mask-length 24
CPCM2> set interface eth4 comments "External"
CPCM2> set interface eth4 state on
CPCM2> set interface eth4 ipv4-address 200.100.0.3 mask-
length 24
CPCM2> set interface eth5 comments "DMZ"
CPCM2> set interface eth5 state on
CPCM2> set interface eth5 ipv4-address 10.30.30.3 mask-
length 24
```

4. Define entries for the localhost lookup, as follows:

```
CPCM2> add host name CPCM1 ipv4-address 10.0.0.2
CPCM2> add host name CPSMS ipv4-address 10.0.0.10
```

5. Save your configuration changes, like this:

```
CPCM2> save config
```

You may now execute the show configuration command, page down using the
spacebar key, and scroll through the output to see the entire configuration.

That's it—your second cluster member candidate is now preconfigured. Next, let's take a look at the various backup, recovery, and migration options.

Saving Gaia configuration, backups, snapshots, and migration tools

There are four different backup methods for Check Point appliances, each serving a different purpose. These are listed as follows:

- Gaia OS-level configuration backup (saving the configuration)
- System backup
- Snapshots
- Server migration tools

We will discuss each of these in detail in the following sections.

Gaia OS-level configuration backup

The Gaia OS-level configuration backup essentially produces the same content as a `show configuration` command output. It contains all Gaia configuration parameters necessary for the recovery on, or migration to, a new appliance with identical network interfaces and modules. If used to migrate to the new appliance with different interfaces or modules, it can be easily edited prior to loading it in the destination appliance.

> **Note**
>
> Keep in mind that this is only a Gaia OS configuration, so after it is loaded, there are additional steps required to make the appliance a part of your Check Point infrastructure. In the case of management servers, it may require a `migrate_server import` command, as described later in this section, or, when used for gateways, establishing **Secure Internal Communication** (**SIC**) and, if necessary, making adjustments to the gateway object's topology. We will address these subjects in a later chapter.

We will cover this option later in this chapter in a lot more detail, in the *Saving and loading the configuration* and *Offline configuration editing and comparison* sections.

Before we do, let's go over other backup options.

System backup

A system backup contains the same configuration as the Gaia OS-level configuration backup, with the addition of Check Point-specific data. It could be restored on the same appliance where the backup was performed or on a different appliance of an identical model.

Backups can be restored only on an appliance with the same version of Check Point software and the same version of the **Jumbo Hotfix Accumulator (JHFA)** as the source.

No additional configuration of the Check Point gateway object is required after restoring, as the appliance is recognized as part of the Check Point infrastructure.

There is a built-in mechanism for scheduled backups that could be performed locally, to the management server, or to the remote **SCP**, **File Transfer Protocol** (FTP), or **Trivial FTP (TFTP)** servers.

The backups interface is self-explanatory and is easy to use, as we can see here:

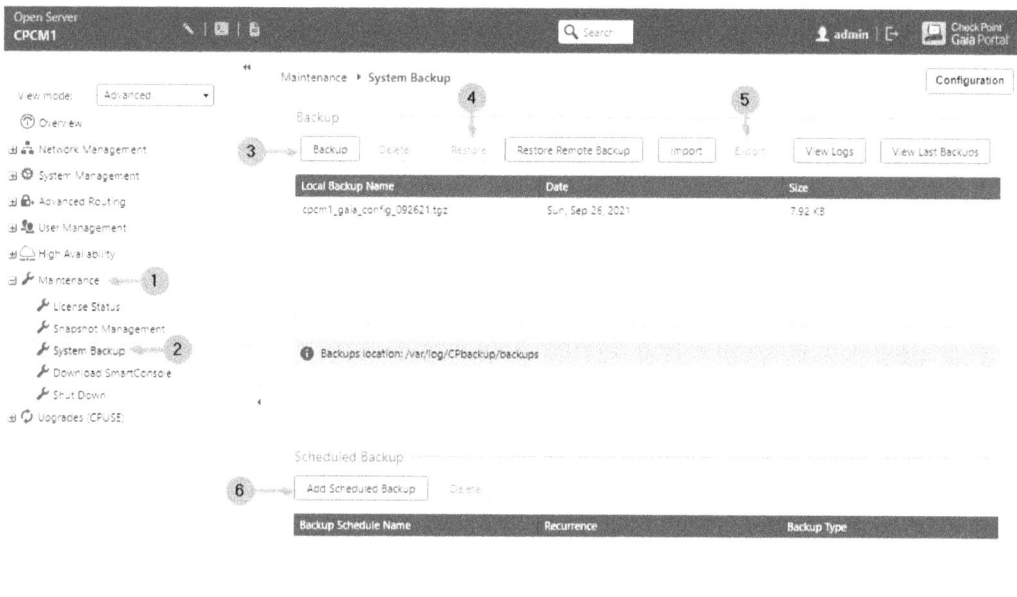

Figure 6.10 – Gaia System Backup and Restore

Located in the **Maintenance** [1] section of the navigation tree **System Backup** [2] option, the simplest way to perform backups is to click **Backup** [3].

To restore, click on the backup .tgz file to select and click **Restore** [4]. Click **Export** [5] to download the backup to your PC via the browser's download function.

To properly set up recurring backups, click the **Add Scheduled Backup** button [6].

Now, in the following screenshot, name the backup [1] and choose one of the remote SCP, FTP, or TFTP server options [2]. Define the IP address, credentials, and upload path [3]. Set the frequency and time [4] and click **Add** [5]:

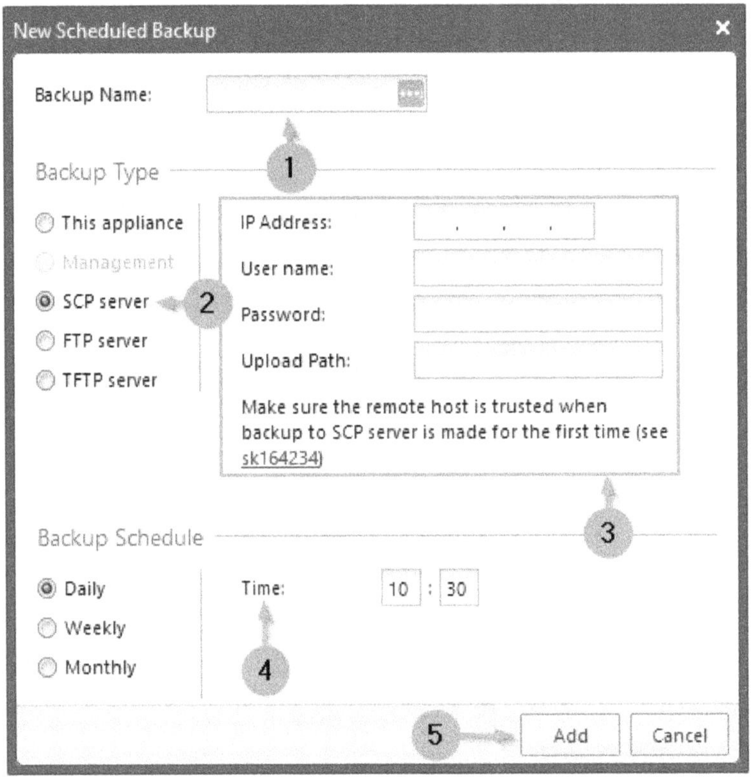

Figure 6.11 – Gaia scheduled backups

And that's it for system backups.

Snapshots

Snapshots are the most comprehensive backup method since they are preserving the binary image of the root disk partition. There is no need to pre-install the same version of Check Point products or hotfixes before recovering from snapshots on the same appliance where it was created, or to an identical appliance model.

We even have an option of initiating recovery from a snapshot during the FTW's **Deployment Options** functionality [1], by choosing the **Import existing snapshot** [2] option, as shown in the following screenshot:

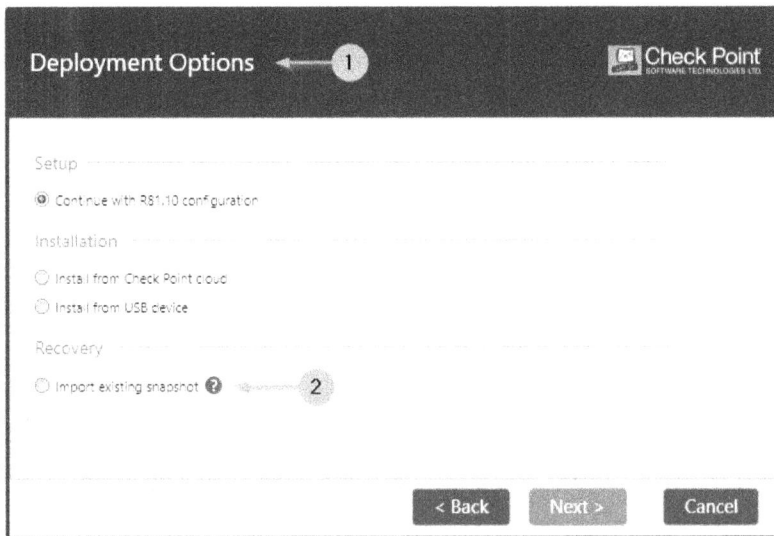

Figure 6.12 – FTW Deployment Options: Import existing snapshot

Snapshots, especially those of management servers, can be quite large and may take a while to create. It is recommended to create snapshots after a fresh installation of the product, before upgrades, and prior to the application of JHFAs.

A few important points about snapshots are noted here:

- Snapshots retain the **media access control** (**MAC**) addresses of configured interfaces. If you are performing recovery from a snapshot on a new appliance, the MAC addresses of corresponding interfaces will be overwritten with the old values. Technically, this is beneficial when recovering from hardware failure using replacement appliances. This may, however, require reaching out to Check Point support to update licensing information as, by default, appliances replaced using a **return merchandise authorization** (**RMA**) may be identified by their original MAC address in the licensing portal.

- Snapshot names should not contain spaces or special characters. Snapshot descriptions may contain underscores or dashes, but I recommend avoiding those as well and using uninterrupted alphanumeric strings in both fields.

- Snapshots created locally could and should be exported to network storage, or else you will not be able to use them to recover from hardware failures.

- The space requirement for the snapshot is 1.15 x the size of the root partition. The space required for the export of a snapshot is 2 x the size of the snapshot.

- Do not rename exported snapshots: if the exported files are renamed, you cannot recover from them.

Snapshot space requirements can be determined via WebUI by selecting the **Maintenance | Snapshot Management** [1] option in the navigation tree as shown in the following screenshot:

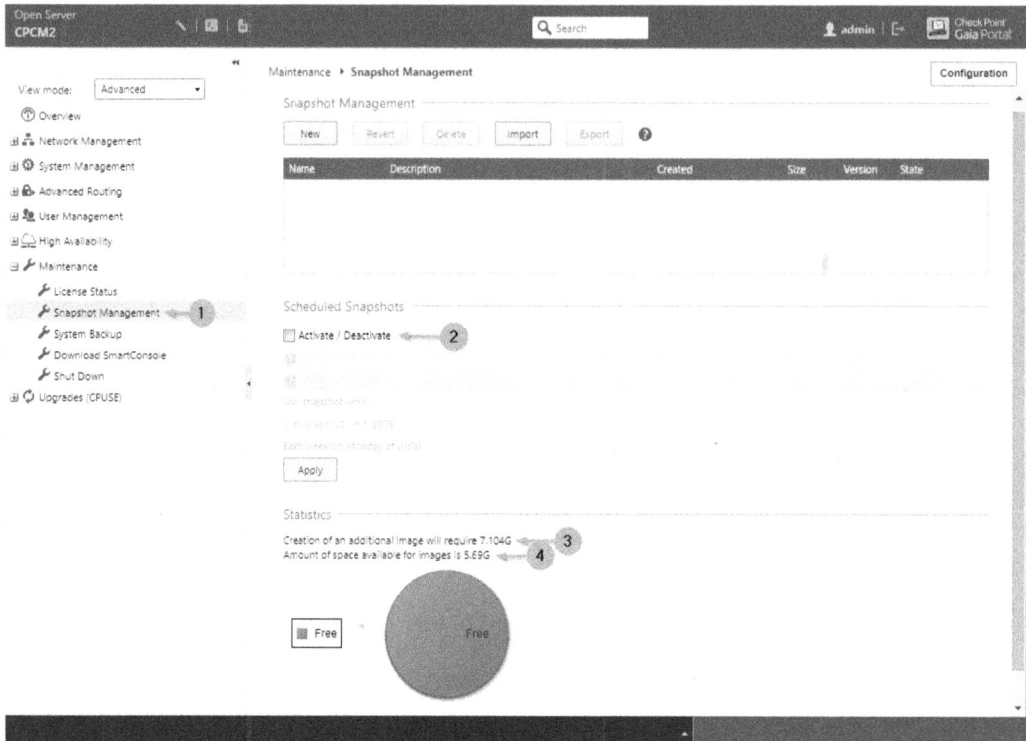

Figure 6.13 – Snapshot Management

There is an option to create scheduled snapshots and define a remote FTP or SCP repository for those, but this is only available via the CLI. Once it is configured in the CLI, it can be activated or deactivated from WebUI [2].

In the **Statistics** section of **Snapshot Management**, you can see the required space [3] as well as space available for the creation of a snapshot [4].

With the snapshots section of the chapter completed, let's move on to the server migration tools.

Server migration tools

These tools are located in $FWDIR/scripts/ and invoked with migrate_server export or migrate_server import commands with options and are used on licensed management components of Check Point infrastructure such as management, log, and SmartEvent servers.

This tool produces a portable version of Check Point configuration that can be imported on a different appliance or a platform as well as used for migration to a newer version of the product.

The `migrate_server` tool has a built-in verification mechanism that, when invoked, will provide confirmation that the process is expected to succeed or if there are issues that should be addressed.

When the `migrate_server` tool is executed, it will attempt to download the latest version of itself from the Check Point repositories before running. If your management server does not have internet connectivity, the process will result in an error.

In this case, you may have to download the migration tools package and perform its installation offline. At `https://supportcenter.checkpoint.com`, search for and open the `sk135172` SecureKnowledge article. Scroll down to *To manually install the latest version of the Check Point Upgrade Tools Package:* section [1] and click on the desired target version's download link [2], as shown in the following screenshot:

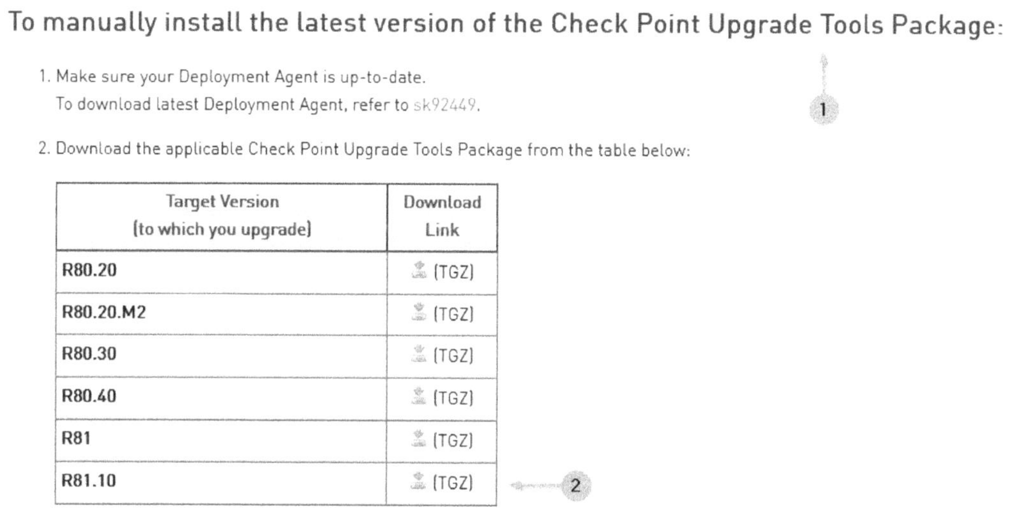

To manually install the latest version of the Check Point Upgrade Tools Package:

1. Make sure your Deployment Agent is up-to-date.
   To download latest Deployment Agent, refer to sk92449.

2. Download the applicable Check Point Upgrade Tools Package from the table below:

Target Version (to which you upgrade)	Download Link
R80.20	(TGZ)
R80.20.M2	(TGZ)
R80.30	(TGZ)
R80.40	(TGZ)
R81	(TGZ)
R81.10	(TGZ)

Figure 6.14 – Migration tools download for offline export and import operations

You will have to perform offline installation of the tool before proceeding. A detailed description of the offline package installation process is covered later in this chapter in the *Using CPUSE* section.

To execute the tool on appliances not connected to the internet, or if errors are encountered, the `-skip_upgrade_tools_check` option should be invoked.

This is an example of the tool being executed in verification (`verify`) mode, with the target version of `-v R81.10` on a server not connected to the internet:

```
CPSMS> expert
Enter expert password:

Warning! All configurations should be done through clish
You are in expert mode now.

[Expert@CPSMS:0]# cd $FWDIR/scripts/
[Expert@CPSMS:0]# ./migrate_server verify -v R81.10 -skip_
upgrade_tools_check
```

Once verification has completed successfully, execute the same command without the `verify` option and by appending the path and the destination filename. Your best destination option is the `/var/log/` directory, and the filename should be specified without a `.tgz` extension for export operation. The code is illustrated here:

```
[Expert@CPSMS:0]# ./migrate_server export -v R81.10 -skip_
upgrade_tools_check /var/log/cpsmexport_093021
```

However, for an import operation, this should be done with a `.tgz` extension, as illustrated here:

```
[Expert@CPSMS:0]# ./migrate_server import -v R81.10 -skip_
upgrade_tools_check /var/log/cpsmexport_093021.tgz
```

If the source and/or destination servers are connected to the internet, omit `-skip_upgrade_tools_check` from the command.

The `migrate_server` tool generates HTML reports for export and import processes as well as an archive containing multiple logs generated during either export or import. Should difficulties arise during the migration process, these logs must be provided to the **Technical Assistance Center** (**TAC**) for assistance in troubleshooting. The location of both HTML reports and the logs archive is displayed during and at the completion of the export and import operations.

Of all the backup and recovery methods described here, the migration tool is the only one that allows for the export and import of either unindexed log files using -l (lowercase *L*) or indexed log files using -x options. Both require a significant amount of space.

This concludes the server migration tools section.

Saving and loading the configuration

I have to remind you that while you are looking at the output of the show configuration and are seeing a list of commands, this is really the output of the database query. While individual commands could easily be copy-pasted, modified, and executed, do not attempt to do this using the entirety of a Gaia configuration.

There is a number of reasons that necessitate saving and loading a Gaia configuration, such as migration to new appliances, replacement of the existing hardware due to component failure (very rare, but it does occasionally happen), as well as offline preparation of a configuration for deployment on new appliances. This last one is common when working with clusters.

In our lab, the Gaia configuration is pretty simple, but in production environments, you may easily encounter a configuration in excess of 1,000 lines describing complex dynamic routing, hundreds of interfaces, and a variety of other settings.

Saving the configuration to a file

Save the configuration of the appliance to a file by executing save configuration <filename>—that is, save configuration cpcm2_gaia_config_092621.

You cannot specify a path; the output will be saved in your user's home directory—/home/secadmin in our case.

At this point, you may change your shell to **Bash** in Expert mode and download the resultant file using SCP or SFTP.

If you do not want to or cannot change shells to use SCP or SFTP for the file exchange, use this method as a workaround:

1. Save the configuration file with a .tgz file extension, as follows:

   ```
   save configuration <filename>.tgz
   ```

2. Enter Expert mode and move the file to the `/var/log/CPbackup/backups` directory. Note that we do not actually create an archive but simply add a `.tgz:` extension, as illustrated in the following screenshot:

```
CPCM1> save configuration cpcm1_gaia_config_092621.tgz
CPCM1> expert
Enter expert password:


Warning! All configurations should be done through clish
You are in expert mode now.

[Expert@CPCM1:0]# ls
cpcm1_gaia_config_092621.tgz
[Expert@CPCM1:0]# mv cpcm1_gaia_config_092621.tgz /var/log/CPbackup/backups/
[Expert@CPCM1:0]# ls /var/log/CPbackup/backups/ |grep .tgz
cpcm1_gaia_config_092621.tgz
[Expert@CPCM1:0]# 
```

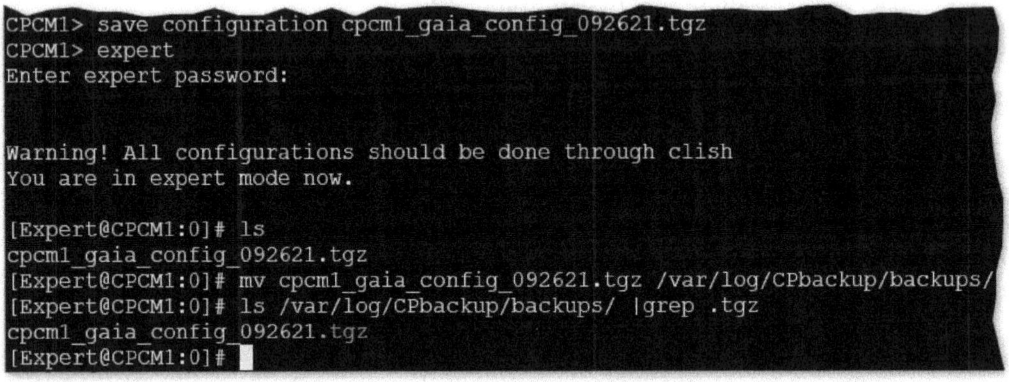

Figure 6.15 – WebUI file retrieval preparation

3. Once this is done, you may log in to the WebUI of the appliance and click the lock icon to acquire read/write permissions [1]. In the **Maintenance** section of the navigation tree [2], click on **System Backup** [3] and you should see our file listed in the **Backup** section [4]. Click it once to select and click **Export** [5], as highlighted in the following screenshot:

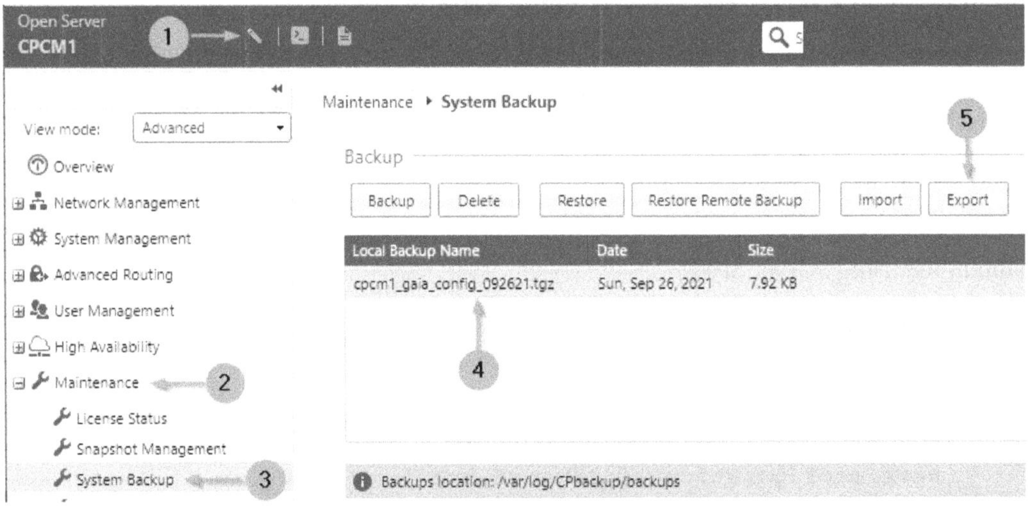

Figure 6.16 – File retrieval using WebUI

4. When presented with the **Export Backup** confirmation popup, click **OK**. Your file will be downloaded to the default browser's download location.

5. Remove the `.tgz` extension from the file to avoid accidental processing by your archival application, confirming your choice in the **Windows Rename** popup. You may now open the file using Notepad++ (or any other text editor of your choice that will preserve Unix **End of Line** (**EOL**)) for editing.

Loading the configuration

> **Note**
> This is not a part of the lab. Do not execute this; simply read this subsection.

Follow these steps to load the configuration:

1. Log on to the appliance.
2. Enter Expert mode.
3. Change the shell to Bash by executing `chsh -s /bin/bash`.
4. Upload the configuration file using the SCP/SFTP client into your user's home directory.
5. Execute an `ls` command to confirm the presence of the uploaded file in your home directory.
6. Change the shell to CLISH by executing `chsh -s /etc/cli.sh`.
7. Execute the following commands, where `<filename>` could be copy/pasted from the output of the `ls` command:

    ```
    set clienv on-failure continue
    load configuration <filename>
    set clienv on-failure stop
    save config
    ```

The `set clienv on-failure continue` command allows the configuration file to be read repeatedly until all data in it is processed. Without it, you will encounter errors on *out-of-sequence* lines, lines with multiple spaces or tabs between parameters, and so on. During the loading process, you will see a currently processed line number out of the total (*Processing line # out of #*) indicator. Depending on the size of the configuration file, you may be able to see a few iterations of the incrementing counter.

That's it for saving and loading the configuration.

Offline configuration editing and comparison

There are some situations where it makes sense to perform editing and/or compare the configuration of Gaia offline. One of the reasons for this is to ensure consistency of the organization-wide configuration parameters, such as **Domain Name Server** (**DNS**), **Network Time Protocol** (**NTP**), roles, and users. The other is a consistent definition of the clusters' topologies and routing settings.

You may, for instance, save the configuration of a single cluster member, download it, open it in the text editor, and perform a search and replace for the IPs of the interfaces and hostnames, replace host entries, save the file under a new cluster member's name, upload it to the new cluster member, and load the configuration.

Another case is the migration of a Gaia configuration to a new appliance with a different topology. For example, if you are moving to a more powerful appliance to handle the increased throughput using 10, 40, or 100 **Gigabit Ethernet** (**GbE**) interface cards, you must re-assign the interface names in both the Gaia configuration and the topology of the gateways and clusters.

In order to ensure a smooth transition to the new appliances with a later version of Gaia, you may use a **Compare** plugin or function of the text editor (Notepad++, in our case) to merge version-specific Gaia settings of the target appliance with the configuration segments of a source.

To do so, open Notepad++ on your LabHost PC, then click on **Plugins** [1] and **Plugins Admin…** [2], as illustrated in the following screenshot:

Figure 6.17 – Notepad++ plugins administration

Once the **Plugins Admin** window is opened, scroll down [1] until you see **Compare** [2] and check the checkbox. Click **Install** [3] and click **Yes** in the Notepad++ restart notification [4], as illustrated in the following screenshot:

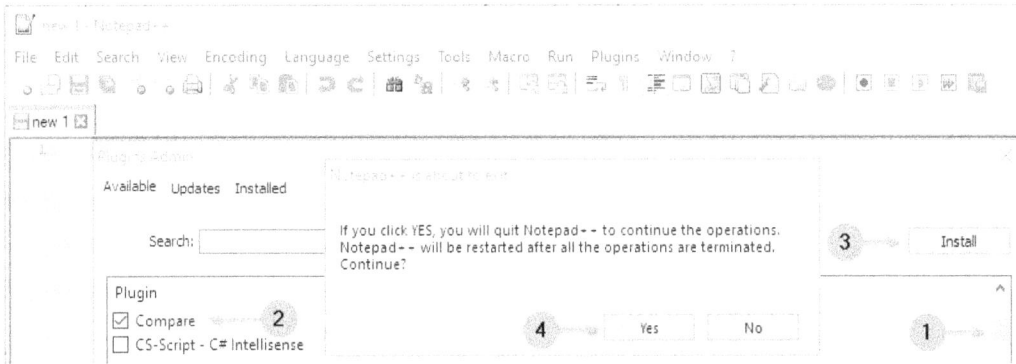

Figure 6.18 – Notepad++ Compare plugin installation

If you perform a `save configuration <filename>` command on both cluster members in our lab from the SmartConsole VM, download the resultant files to it, and then copy them to the LabHost PC, you can then perform the **Compare** action between the files to see the differences.

To do so, perform the following steps:

1. Open both files in the Notepad++ text editor.

2. Select the Notepad++ tab of the first file [1].

3. Click on **Plugins** [2], **Compare** [3], and **Set as First to Compare** [4], as illustrated in the following screenshot:

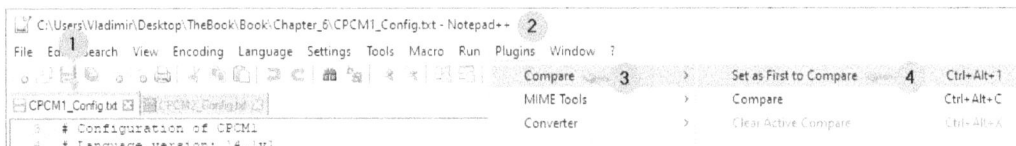

Figure 6.19 – Notepad++: Set as First to Compare

4. Click on the second file's tab in the Notepad++ text editor [1].

5. Click on **Plugins** [2], **Compare** [3], and **Compare** [4], as illustrated in the following screenshot:

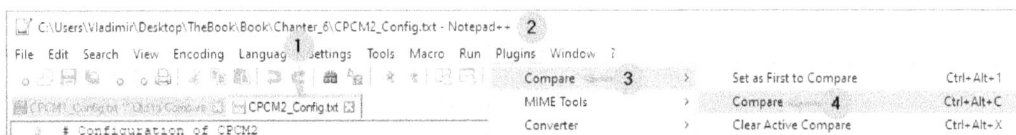

Figure 6.20 – Notepad++: choosing a second file to compare

6. You will then be presented with a split-screen view containing both files with differences highlighted. Scroll down to the section of the files containing `set interfaces` commands to note the differences. At this point, you should only see the differences in the IP addresses of the interfaces. For demo purposes, the following is the compare view of the configured cluster member `CPCM1` and `CPCM2` (with only the FTW completed on CPCM2):

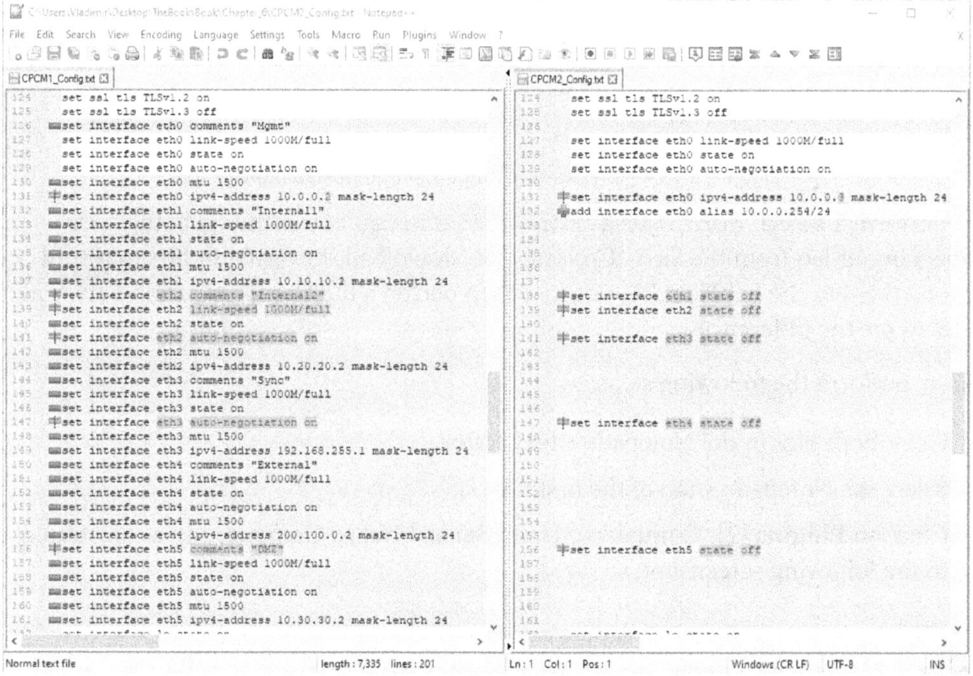

Figure 6.21 – Notepad++: comparing the configuration of cluster members

Note the difference in interface states as well. You now can copy-paste settings from source to destination, edit them as necessary, and save them under a different filename for subsequent `load configuration` operations.

We have seen how to use offline editing and comparison to simplify the Gaia configuration.

Using CPUSE

CPUSE refers to the Deployment Agent daemon's interfaces and policies. It allows for seamless and safe update and upgrade processes including installation of tools, hotfixes, and minor and major version upgrades for Gaia and Check Point components. The safety of upgrades is assured by the automatic creation of the snapshots that could be used to recover appliances to their previous good state.

Updates and upgrades performed either from Gaia WebUI, CLISH installer commands, or SmartConsole all rely on CPUSE for execution.

CPUSE in WebUI

In most instances, your gateways and management servers are allowed outbound connections to the following Check Point and Akamai domains:

- .updates.checkpoint.com
- .updates.g01.checkpoint.com
- .gwevents.checkpoint.com
- .gwevents.us.checkpoint.com
- .deploy.static.akamaitechnologies.com

When these conditions are met, the Deployment Agent itself is updated in the background to the latest version automatically. Once this happens, it will query Check Point repositories for the latest updates and upgrades for the system on which it is running. The results will be presented to you in the **Upgrades (CPUSE)** [1] | **Status and Actions** [2] section of the navigation tree of the WebUI, as illustrated in the following screenshot:

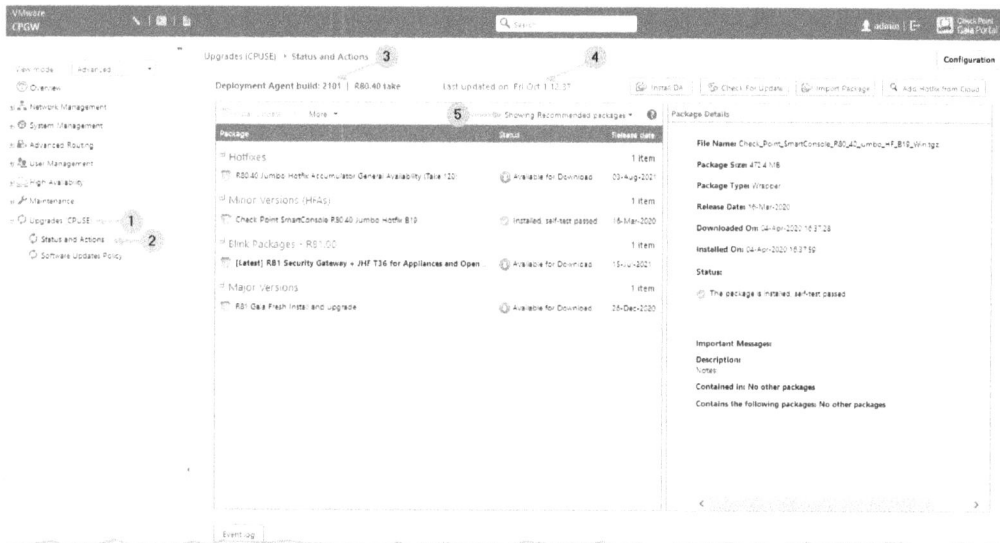

Figure 6.22 – WebUI Upgrades (CPUSE) | Status and Actions view

There, you can see the current Deployment Agent build as version [3], the last time the available updates were queried [4], and the actual updates filtered, by default, with the **Showing Recommended packages** [5] option from the drop-down menu.

Right-click on the newly available package, which will present you with the following options:

- **Download**
- **Verifier**
- **Install Update**

And then, right-click on the already installed package to see the following options:

- **Export Package**
- **Uninstall…**
- **Save Install Log**

These options appear as shown in the following screenshot:

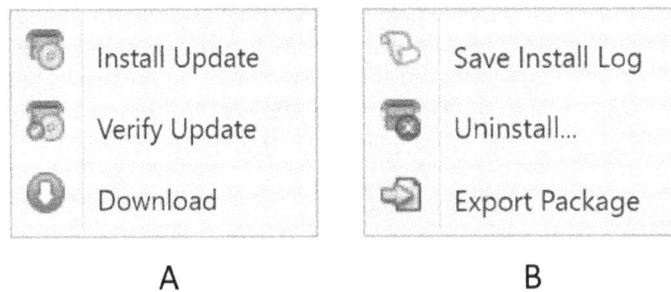

Install Update	Save Install Log
Verify Update	Uninstall...
Download	Export Package
A	B

Figure 6.23 – CPUSE right-click options (A – new package; B – installed package)

For uninstalled packages, an additional option, **Delete From Disk**, is available.

The normal installation procedure is to snapshot, back up, and save the Gaia configuration, all to a network repository, then download new update packages, execute **Verifier** to confirm that there are no compatibility or capacity issues present, and, if all is well, click **Install Update**.

Positive results of the verification process for hotfixes are one or both of these options:

- Installation is allowed.
- Upgrade is allowed.

Or, for minor or major version upgrades, one or both of these options:

- **Clean Install**: Installation is allowed.
- **Upgrade**: Upgrade is allowed.

Note that **Clean Install** does require rerunning the FTW and either manual re-configuration or `load configuration` for Gaia, and, in the case of management servers, `migrate_server import`. For gateways and non-primary management servers, you will need a SIC reinitialization and policy installation to return your appliances to a functioning state.

CPUSE has a built-in rollback capability that, if triggered by issues undetected by **Verifier**, will return your appliance to a pre-attempted update or upgrade state.

When you choose to install an update, you will be notified if a reboot is required and have the ability to defer it by unchecking the **Reboot if required** checkbox.

CPUSE in the CLI

CPUSE operations described in the preceding subsection could be accomplished via the CLI (I would not recommend this unless there is a compelling reason to do so, simply because it is not very convenient), using the sequence of commands and actions mentioned in this section. Follow these next steps:

1. In the CLISH prompt of the appliance you are working on, run the following commands:

    ```
    # Step 1
    show installer packages recommended
    # Note the Display name of the package you are interested
    in.
    # Step 2
    show installer package
    # [press spacebar and then press the Tab key]
    # Note the Num(ber) corresponding to the Display name of
    the package from step 1.
    # Step 3
    installer download 2
    # Where 2 is the number noted in step 2
    # Step 4
    # To see that your package is listed:
    show installer packages downloaded
    # Step 5
    installer verify
    # [press spacebar and then press the Tab key]
    ```

```
# Note the Num(ber) corresponding to the Display name of
the package you are installing.
# Step 6
```

installer verify 1

```
# Where 1 is the number noted in step 5.
# Wait for confirmation that installation is permitted -
it will look similar to this:
```

**Info: Initiating verify of Check_Point_R81_JUMBO_HF_MAIN_
Bundle_T36_FULL.tgz...**

**Interactive mode is enabled. Press CTRL + C to exit (this
will not stop the operation)**

Result: Package is available for installation

```
Step 7
```

installer install

```
# [press spacebar and then press the Tab key]
# Note the Num(ber) corresponding to the Display name of
the package
Step 8
```

installer install 1

```
# Where 1 is the number noted in step 7
```

Confirm that you want to continue and how to handle the reboot, as follows:

The machine will automatically reboot after install.

**Do you want to continue? ([y]es / [n]o / [s]uppress
reboot)　y**

**Info: Initiating install of Check_Point_R81_JUMBO_HF_
MAIN_Bundle_T36_FULL.tgz...**

**Interactive mode is enabled. Press CTRL + C to exit (this
will not stop the operation)**

Extracting Bundle:　　　　　5%

The installation process will iterate through various stages, as shown here:

Validating Install:　　　　　7%

```
-
```

Installing:　　　　　10%

```
-
```

And shortly after confirmation of, successful installation, the appliance will be rebooted if this option was chosen in `Step 8`, as illustrated here:

```
Result: Package R81 Jumbo Hotfix Accumulator General
Availability (Take 36) was installed successfully.
CPSMS>
Broadcast message from secadmin@CPSMS (Fri Oct  1
15:38:45 2021):

The system is going down for reboot NOW!
```

2. If this is a management server, give it a few minutes for the orderly startup of services and confirm its readiness by switching to Expert mode and executing the `$FWDIR/scripts/cpm_status.sh` command, as follows:

```
login as: secadmin
Pre-authentication banner message from server:
| This system is for authorized use only.
End of banner message from server
secadmin@10.0.0.10's password:
Last login: Fri Oct  1 14:54:39 2021 from 192.168.7.107
CPSMS> expert
Enter expert password:
Warning! All configurations should be done through clish
You are in expert mode now.
[Expert@CPSMS:0]# $FWDIR/scripts/cpm_status.sh
Check Point Security Management Server is running and
ready
[Expert@CPSMS:0]#
```

If the command returns `During initialization`, repeat it at approximately 30-second intervals until you see `Check Point Security Management Server is running and ready`. At this point, you may connect to the server using SmartConsole and commence administration.

Note that the reason for all display name and number lookups is that numbers may differ between available packages and the downloaded. In the CLI, these commands' output looks something like this:

```
** ****************************************************************** **
**                          Hotfixes                                  **
** ****************************************************************** **
Num Display name                                                  Type
1   R81 Jumbo Hotfix Accumulator General Availability (Take 36)  Hotfix
```

Figure 6.24 – CPUSE CLI package number and display name

Next, let's learn how to use CPUSE in offline mode.

CPUSE in offline mode

In highly secure environments, as well as some staging environments, certain components of your Check Point infrastructure may not be allowed internet connectivity. In these cases, you must perform offline updates.

To do so, you will have to update the Deployment Agent first, as follows:

1. On the Check Point portal, look up `sk92449`.
2. Scroll down to the **Download the latest build of Deployment Agent...** section [1], read the notes to assure that there are no impediments to its installation in your case [2] and, unless otherwise instructed by the TAC, download the **General Availability** package by clicking its **Download Link** icon [3], as illustrated in the following screenshot:

(3) Download the latest build of Deployment Agent, What's New and Installation instructions 1

Build Number	Release Date	Status	Download Link	What's New
2101	16 August 2021	General Availability	(TGZ) 3	
2113	29 September 2021	Gradual Deployment	(TGZ)	

Notes: 2

- Latest build is usually *gradually* released to all customers. Therefore, not all machines might receive the latest build at the same time.
- This package should not be used for upgrading the Deployment Agent for the R80.20SP version for Maestro & chassis products.
- Check Point always recommends to upgrade the Deployment Agent to the latest available build.

Figure 6.25 – CPUSE offline Deployment Agent download

3. In the Gaia WebUI, select the **Updates (CPUSE) | Status and Actions** option, and
    click on **Install DA** [1], as illustrated in the following screenshot:

Figure 6.26 – CPUSE offline update options

4. Click **Browse…** [1], navigate to the downloaded **Deployment Agent (DA)** package
    and click to select it. Click **Install** [2] to update the Deployment Agent, as illustrated
    in the following screenshot:

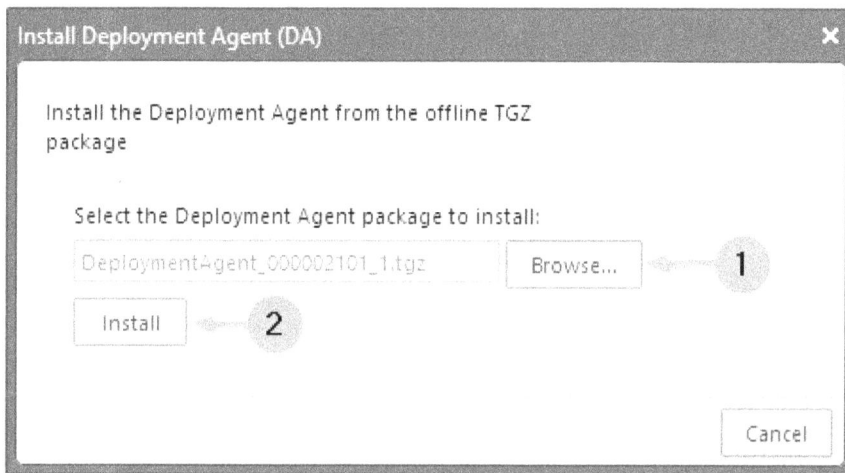

Figure 6.27 – CPUSE Install Deployment Agent update

You will see the Deployment Agent update progress popup for a few seconds
followed by confirmation of a successful update.

5. Locate the hotfix or a package you are interested in and download its **CPUSE
    Offline package** [1]. The offline package must be a `.tar` archive, as illustrated in
    the following screenshot:

Figure 6.28 – CPUSE offline package download

6. Click the **Import Package** button (*Figure 6.26* tag [2]), click on **Browse...** [1] to the downloaded `.tar` archive, to select the package, and click **Import** [2], as illustrated in the following screenshot:

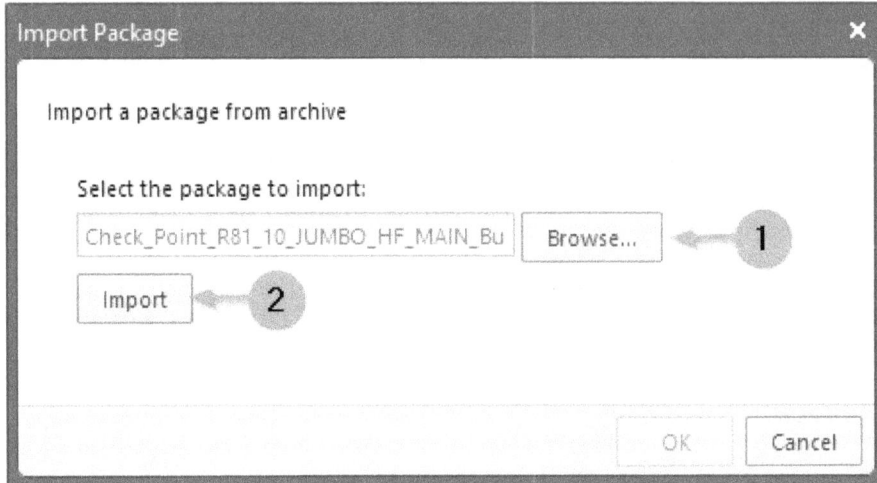

Figure 6.29 – CPUSE offline: Import Package

7. When you click the **Import** button in the previous step, the offline package is uploaded to the appliance and then the CPUSE import process performs verification of compatibility with your appliance and version. If there are no issues, the package is added as downloaded to the appliance's repository and you can perform verification and installation using the methods described in the *CPUSE in WebUI* and *CPUSE in the CLI* subsections.

In some instances, when CPUSE offline packages are not available to download on the Check Point portal (such as due to obsolescence or if those are custom hotfixes), you may export them from other components of your Check Point infrastructure where they are already downloaded or installed.

In the CPUSE WebUI, right-click the package and click **Export**. Import the resultant `.tar` archive to the destination appliance.

This concludes the CPUSE section of this chapter—we have learned how to perform the maintenance tasks necessary to keep your gateways and management server up to date.

Summary

In this chapter, we were introduced to CLISH and Expert mode. We have completed the configuration of the second cluster member candidate using the CLI, and we have covered multiple backup and recovery options. Additionally, we have learned about CPUSE, a utility for updates and upgrades of Check Point appliances.

In the next chapter, we will install the SmartConsole application and familiarize ourselves with its components, functionality, and navigation.

7

SmartConsole – Familiarization and Navigation

In this chapter, we will learn about the SmartConsole application, your primary interface for the daily administration of Check Point infrastructure components. While it may seem complex on the surface, this application, as you start working with it, will reveal its UI logic to be intuitive and accommodating. By learning to identify its various elements, you will develop a sense of how a variety of tasks could be accomplished in a number of different ways.

This chapter will follow the structure of the interface from the outside in, describing its integral components, and following the expanded views and panes, contextual action menus, search and filtering capabilities, as well as many other elements.

Learning how to take advantage of SmartConsole's capabilities will not simply make you proficient with the product but will make your experience of working with it a joy.

In this chapter, we are going to cover the following main topic:

- Introduction to the SmartConsole application and Demo Mode
- SmartConsole components, capabilities, and navigation

Technical requirements

To run SmartConsole, you will need a Windows PC or a virtual machine with the following:

- Intel Pentium Processor E2140, or a 2 GHz equivalent processor
- 4 GB RAM
- 2 GB HDD free space
- 1024 x 768 minimum display resolution (1920 x 1080 preferred and a lot more realistic)

Introduction to the SmartConsole application and Demo Mode

The SmartConsole application has a feature-rich graphical interface where most of the administration magic happens. This is where objects and policies are created, logs and events are viewed and analyzed, reports are generated, and components of your Check Point infrastructure are monitored.

When developing SmartConsole, Check Point conducted a study among security administrators of market-leading products in its category to determine how much time and how many steps are required to execute the most common security management tasks. Their developers then worked on systematically reducing complexity and streamlining these operations so that you can perform them in a fraction of the time. The resultant product is a management interface like no other and, while it was considered a gold standard in enterprise firewall administration in its earlier iterations, the R80+ versions have taken it to a whole new level.

To accommodate different administration styles and navigation tendencies (that is, some admins instinctively look for the action items on top of the screen, some on the left, right, or in the right-click contextual menus), there is generally more than one way to accomplish most administrative tasks.

Even if you do not have sufficient computing resources for the entire virtual lab, you likely have enough to install the SmartConsole application. As long as you have internet connectivity, you can use it with the free unlimited Check Point Cloud Demo mode to get your bearings in terms of the navigation and operation of the product.

Installing the SmartConsole application

While it is not yet required to install the SmartConsole application in our lab (as we will get to it a bit later), if you want to follow along and explore its interface now, you may do so by installing it on your LabHost PC or, if you are not building the lab, on any Windows physical or virtual PC with internet access.

> **Note**
>
> If you are installing it on your LabHost PC, we already have the installation file pre-downloaded in the `C:\CPBook\LabShare\Software` directory.

The following are the steps to install the SmartConsole application:

1. Doble-click the `Check_Point_R81.10_T##_SmartConsole.exe` file (where `##` is the version of the application).

2. In the **Windows User Account Control** popup, click **Yes** to proceed.

3. In the **Welcome to SmartConsole** window, check the checkbox agreeing to the **End User License Agreement (EULA)** [1]. Unless warranted, leave the **Install Directory** value as its default [2] and click the **Install** button, which has now turned green [3].

Figure 7.1 – SmartConsole EULA and installation directory

4. Monitor the installation process (this may take a few minutes).

5. At the **Thank you for installing SmartConsole** window, leave the **Launch SmartConsole** box checked and click **Finish**.

6. If we take a closer look at the SmartConsole login options, we can see multiple authentication methods [1] as well as the choice of login to either your self-provisioned management server or the Check Point SaaS version of SMS, **Cloud Management Service** [2], and the ability to log on in **Read Only** mode [3].

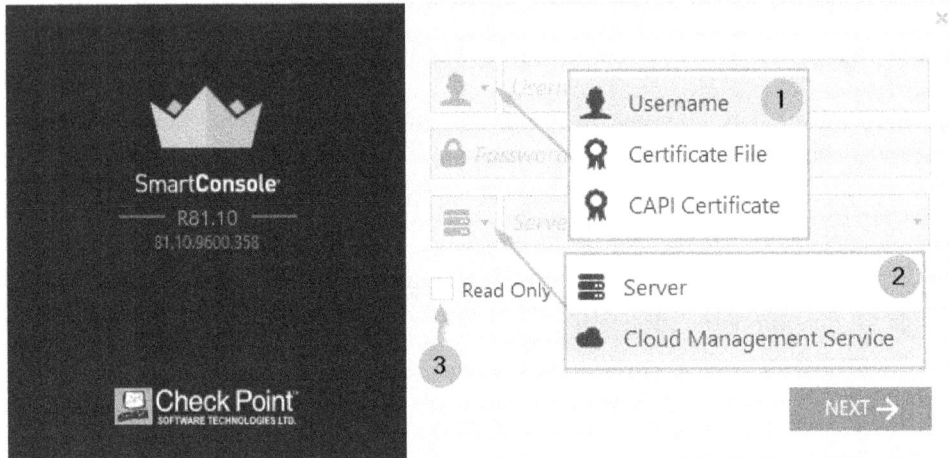

Figure 7.2 – SmartConsole login options

Initializing Demo Mode

We have successfully installed the SmartConsole application. Next, let's take a look at how to initialize Demo Mode:

1. At the SmartConsole login prompt, check the **Demo Mode** box and then click **Next**.

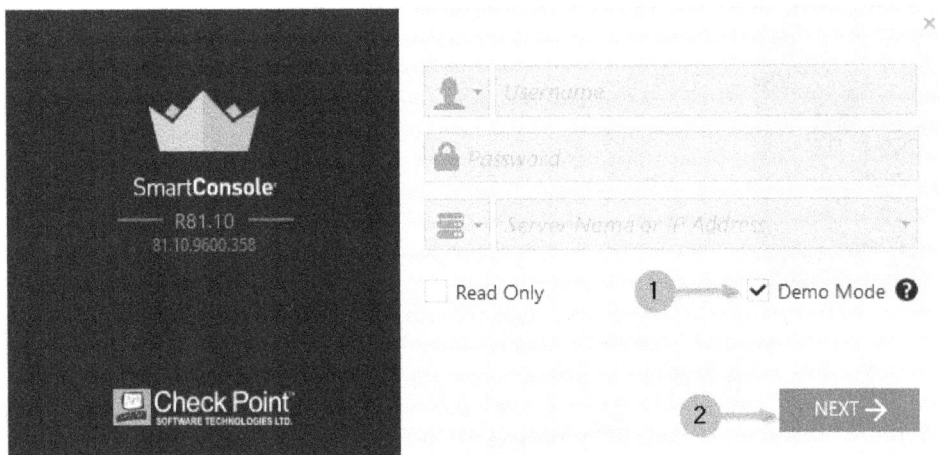

Figure 7.3 – SmartConsole – choosing Demo Mode at the login prompt

2. In the next window, leave the **Start a new demo** radio button selected [1] and either leave the product experience improvement box checked or unchecked [2], and then click **LOGIN** [3].

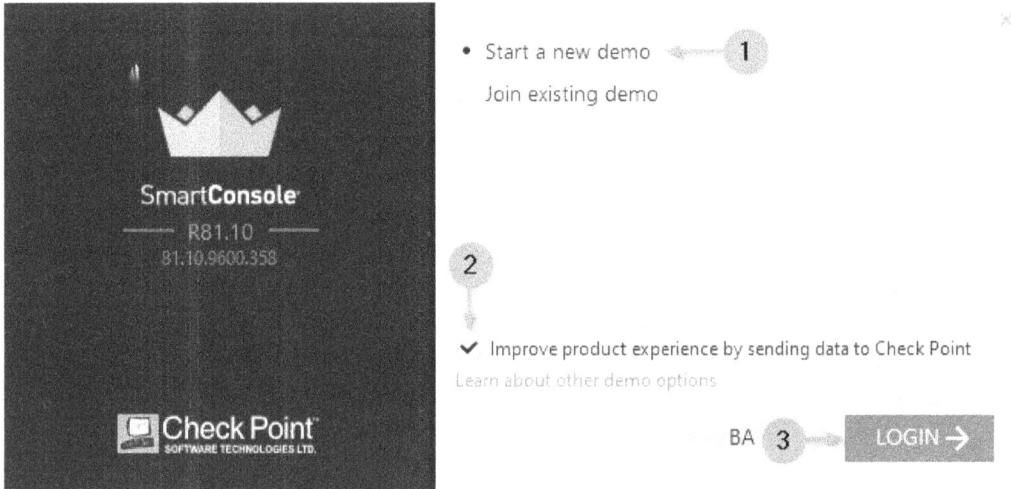

Figure 7.4 – SmartConsole Demo Mode login

3. At this point, you are presented with the screen containing instructions on how to access this demo environment if you exit the application [1], Demo ID [2], and the Demo Server availability deadline [3]. Click **CONTINUE** [4].

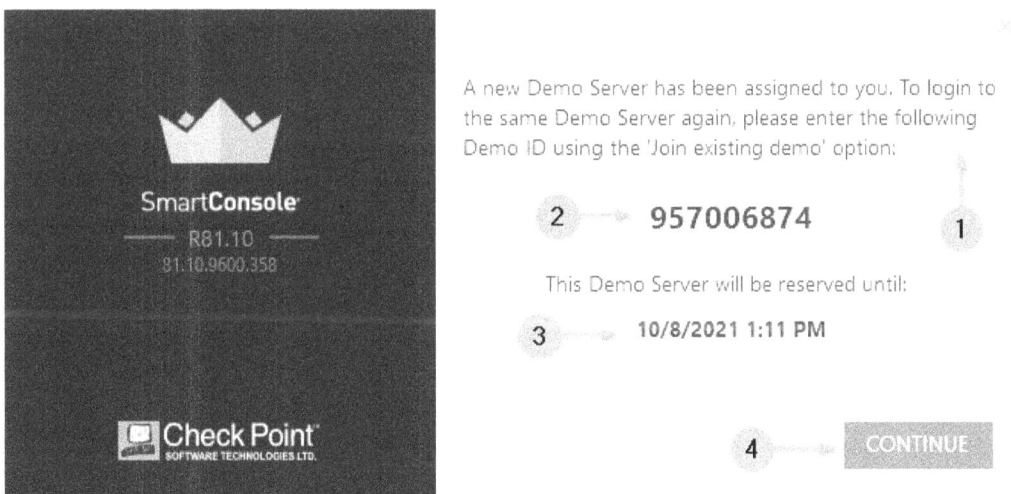

Figure 7.5 – Demo ID and availability

4. You are presented with the **First Connection to server Demo Server** screen. Note the small lightbulb icon [1] that, when moused over, manifests a small help popup explaining how the action could be accomplished. The **Fingerprint:** [2] section shows the unique server's identifier that, in production environments, should be compared with the corresponding value on the server you are connecting to via either WebUI or CLI. Click **PROCEED** [3].

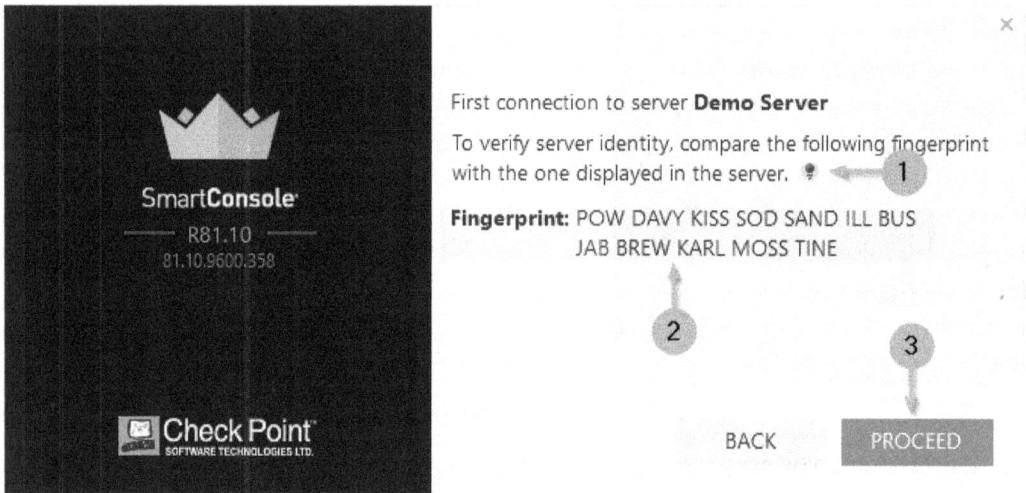

Figure 7.6 – One-time management server fingerprint verification

5. When SmartConsole opens for the first time, you are greeted with the **What's New In SmartConsole** popup containing brief descriptions of new or updated features in each major and minor version to date [1]. Each announcement [2] includes relevant links to secure knowledge articles [3], release notes, administration guides, references, and, occasionally, videos or webinars. The question mark icon in the top-right corner of the **What's New** popup [4], when clicked on systems connected to the internet, will open an online SmartConsole administration guide.

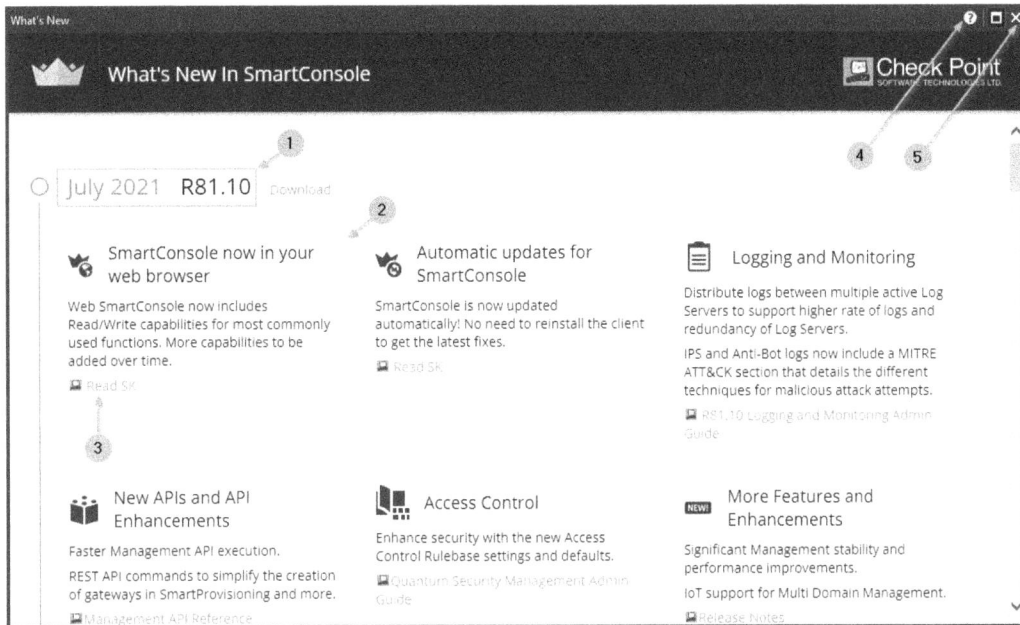

Figure 7.7 – SmartConsole What's New popup

> **Note**
>
> The What's New in SmartConsole button loads the page located at
> `https://sc1.checkpoint.com/documents/SmartConsole/`
> `WhatsNew.html` so it is dependent on PC it is running on having
> connectivity to the internet (or at least to Check Point online services).

You may safely close it to see the unobstructed application interface [5].

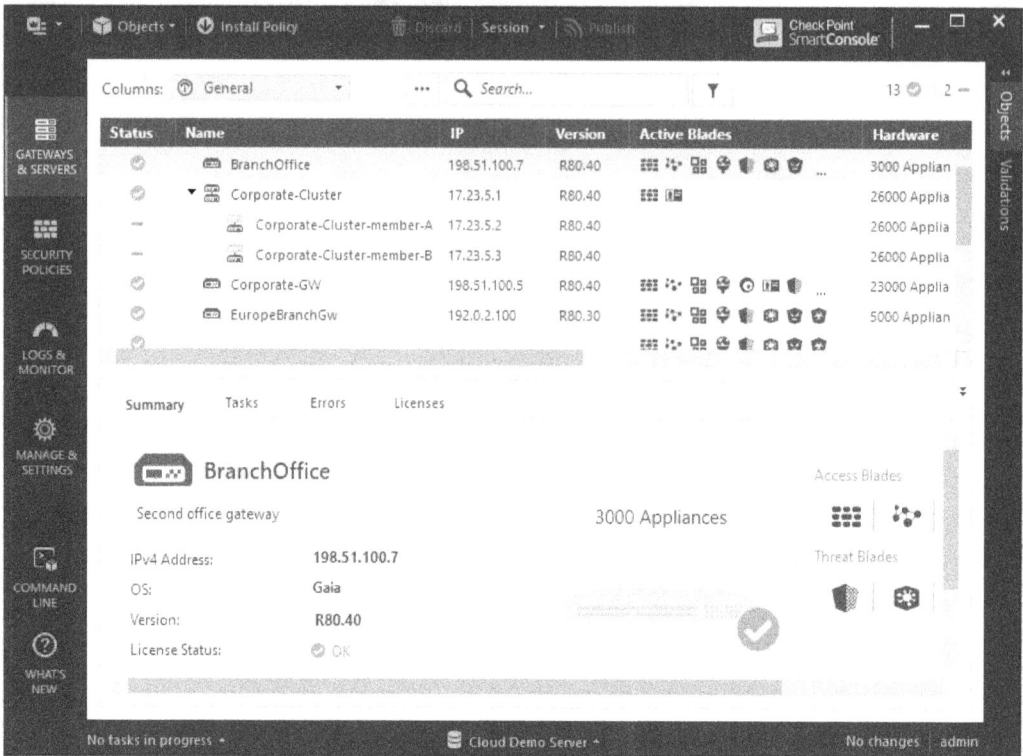

Figure 7.8 – SmartConsole main view – GATEWAYS & SERVERS

As you can see, this is a feature-rich interface and, depending on the version of the book you are reading (paper or digital), and the device you are reading it on, it may be difficult to discern all of the components (these should really be viewed on at least a 13" 1920 x 1080 screen to avoid excessive scrolling.) Let's break it down into smaller segments that we can discuss in further detail.

Demo Mode comes with some limitations, of which the following are relevant to this book:

- We cannot install a security policy.
- We cannot install a database.
- We cannot perform updates of Check Point blades and contracts.
- We cannot use high-availability management.

- We cannot access Gaia Portal (Web UI) features.
- There is no external communication from the security management server (no Active Directory Server, LDAP Server, or RADIUS Server connectivity).
- The creation of Gateway Cluster and VSX Gateway/Cluster objects is not possible.
- Access to the CLI using SSH is denied.

You can see the specific information about Demo Mode (also referred to as Cloud Demo) in Check Point `sk103431`. Moving on to the next section, we will see that even with the limitations listed above, there is ample opportunity to acquaint ourselves with the interface and its features using Demo Mode.

SmartConsole components, capabilities, and navigation

SmartConsole, in a single security domain environment, consists of the following:

1. A global toolbar
2. A session management toolbar
3. **Objects**, **Validations**, and **Sessions** panes (right vertical tabs)
4. Logged-in administrator's pending changes or publish status
5. Management server(s) status and actions
6. A task information area
7. A What's new popup recall and management CLI
8. A navigation toolbar (left vertical tabs)
9. A navigation content area (populated according to the navigation toolbar selection)

All these elements can be seen in the following screenshot:

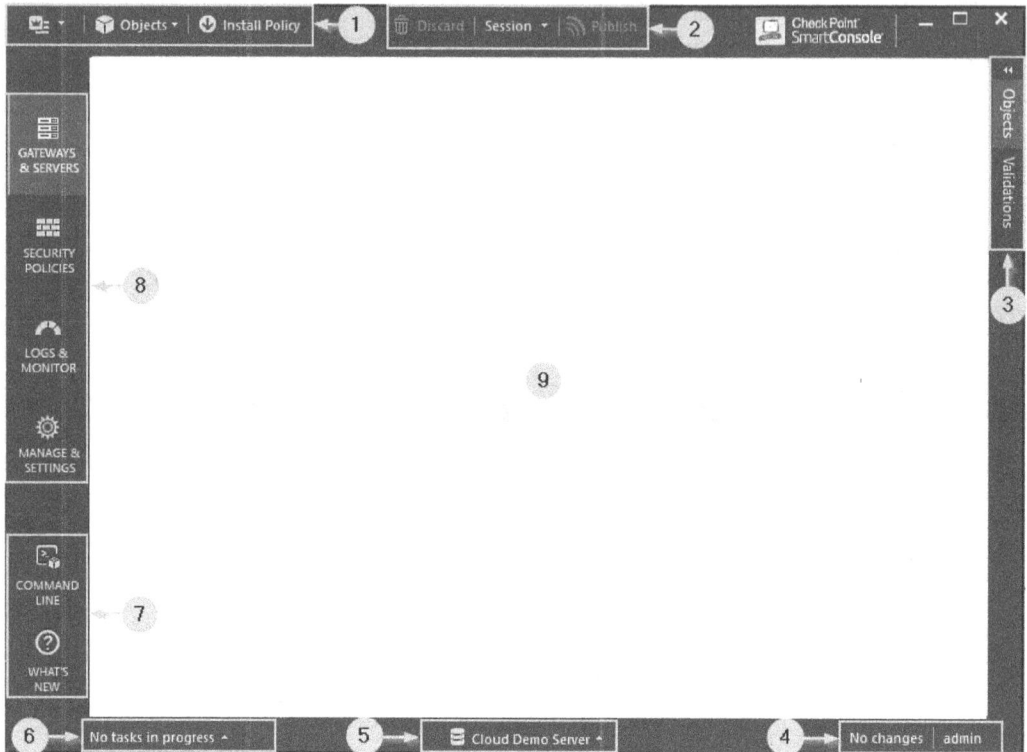

Figure 7.9 – SmartConsole main elements

Let's take a look at the functionality that each of these elements is responsible for.

Global toolbar

The global toolbar is comprised of three segments: the main SmartConsole menu [A], **Objects** [B], and **Install Policy** [C].

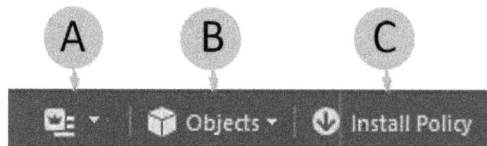

Figure 7.10 – Global toolbar

Let's discuss each segment in detail:

- When we click on the main SmartConsole menu [A] in the preceding screenshot, note that some of the elements shown here point to items that could be accessed elsewhere in the UI, and some are items unique to this menu.

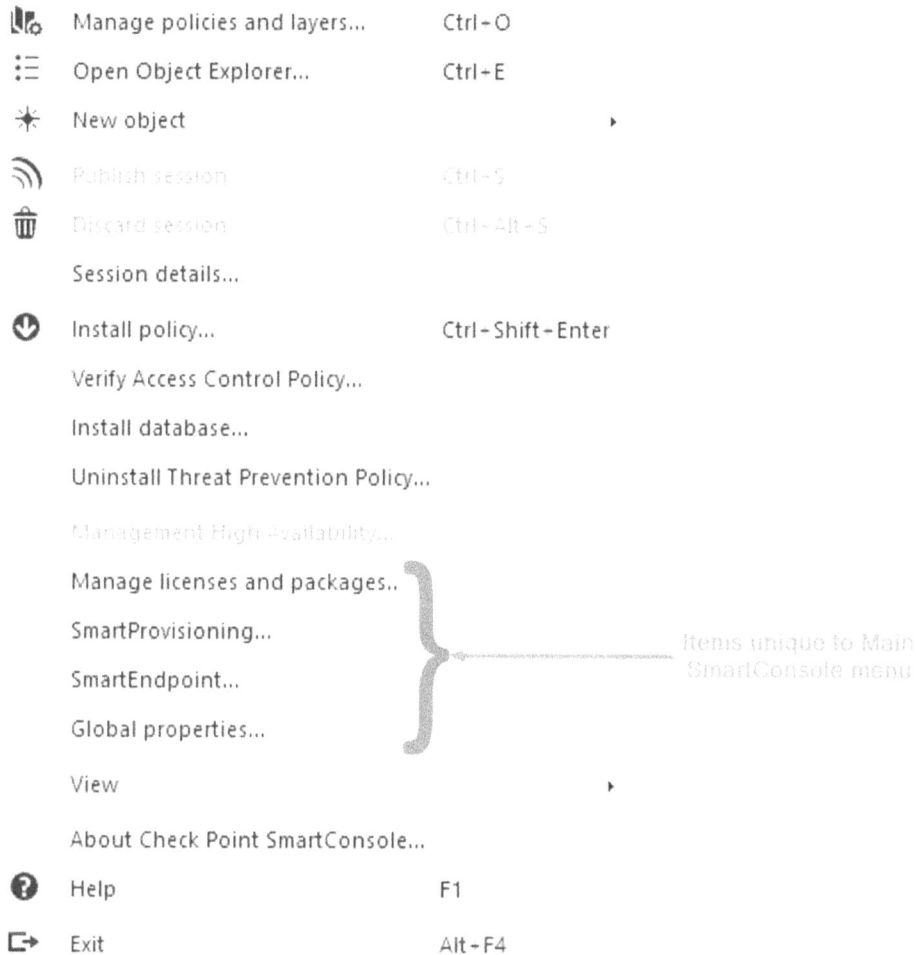

📊	Manage policies and layers...	Ctrl-O
☰	Open Object Explorer...	Ctrl+E
✳	New object	▸
🔉	Publish session	Ctrl-S
🗑	Discard session	Ctrl-Alt-S
	Session details...	
🕐	Install policy...	Ctrl-Shift-Enter
	Verify Access Control Policy...	
	Install database...	
	Uninstall Threat Prevention Policy...	
	Management High Availability...	
	Manage licenses and packages..	
	SmartProvisioning...	
	SmartEndpoint...	
	Global properties...	
	View	▸
	About Check Point SmartConsole...	
❓	Help	F1
⤷	Exit	Alt-F4

Items unique to Main SmartConsole menu

Figure 7.11 – Global toolbar main menu

The **SmartProvisioning**, **SmartEndpoint**, and **Manage licenses and packages** options are intended to open different management products that are integrated with SmartConsole. Of those, only **Manage licenses and packages** will be of interest to us later.

- **Objects** [B] in *Figure 7.10* is used to create new objects using cascading menus:

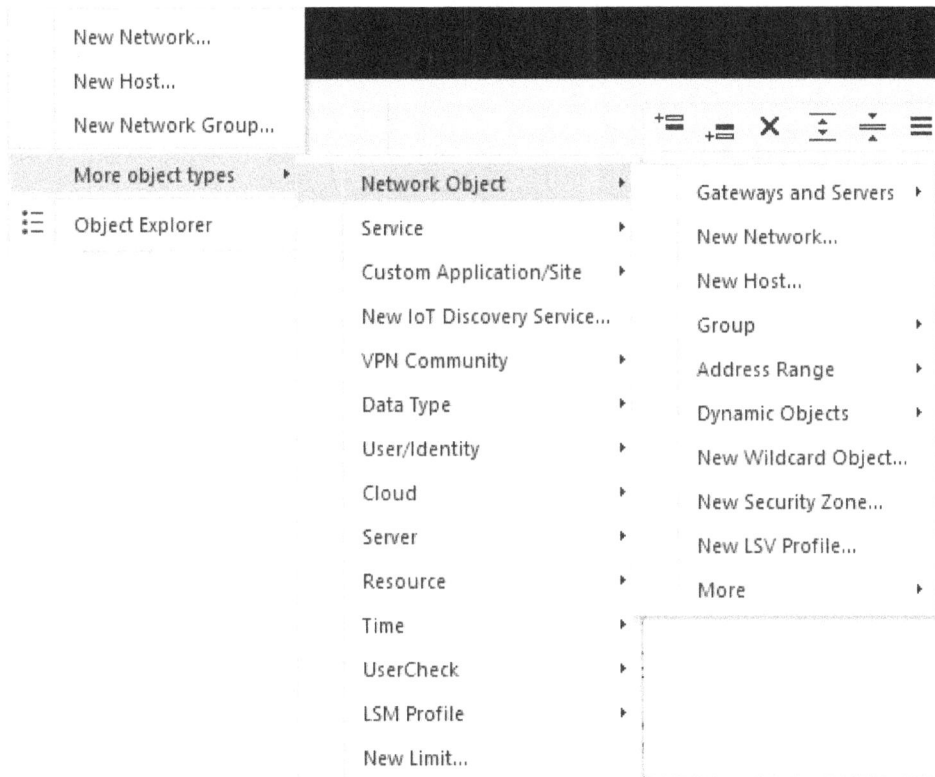

Figure 7.12 – Global toolbar, objects, cascading menu

Alternatively, it is used to call **Object Explorer**, a subsystem for navigating object categories and their branches, searching, selecting, cloning, editing, and creating new objects.

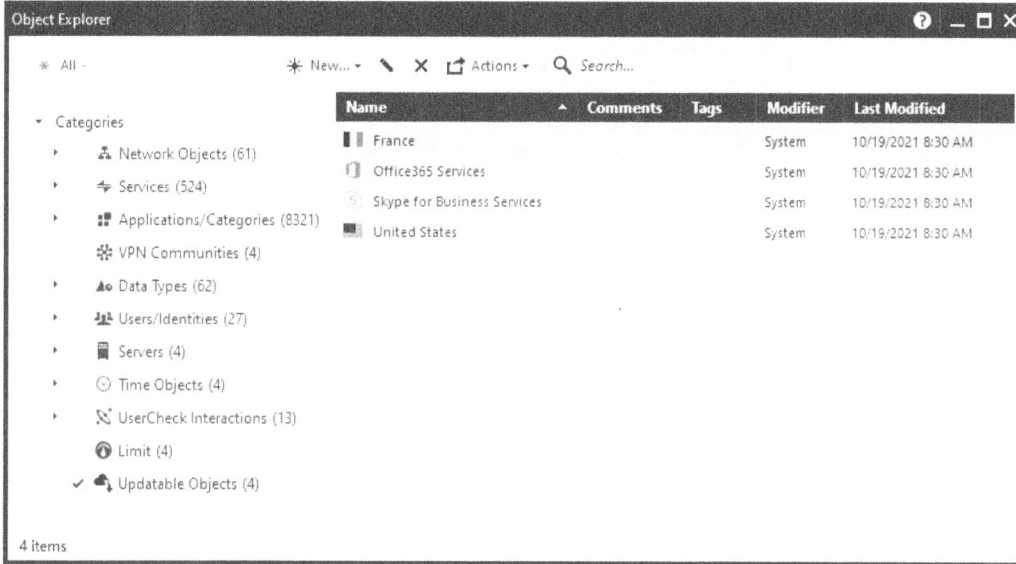

Figure 7.13 – Object Explorer

- **Install Policy** [C] in *Figure 7.10* is the global policy package installation menu that, when invoked, prompts you to select the policy you are interested in installing.

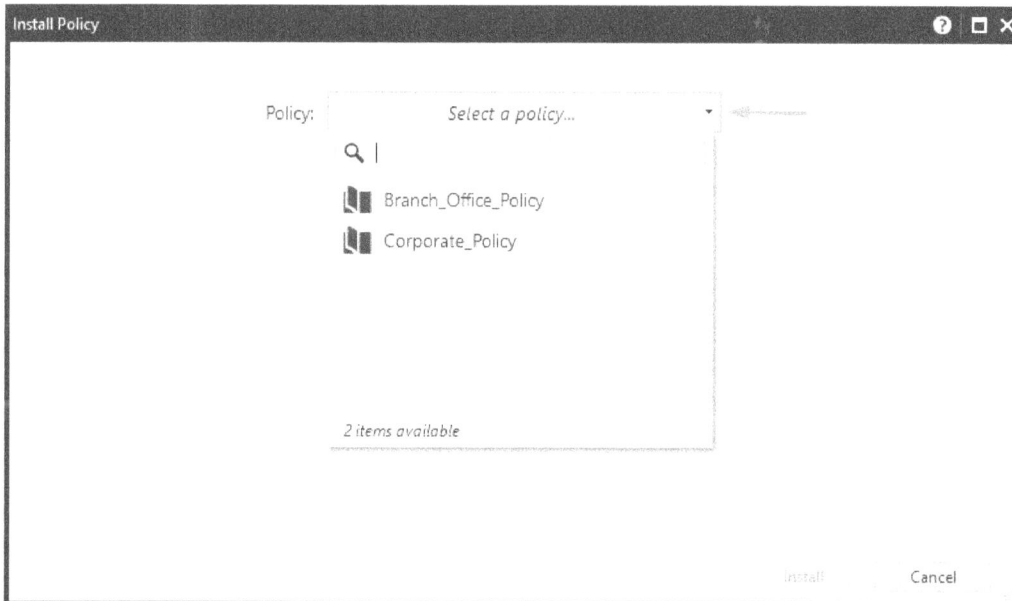

Figure 7.14 – Global toolbar – Install Policy

Session management toolbar

The session management toolbar is comprised of three components that can change their availability based on conditions. That is, **Discard** [A] and **Publish** [C] are grayed out until changes are made during your management session. **Session** [B] can be changed by an administrator to assign a name and a description that will aid other administrators in identifying the purpose of this session while it is active and, later, in audit logs. The number in the yellow circle indicates the number of uncommitted changes in your session.

Figure 7.15 – Session management toolbar

The session management toolbar can be aligned to the left, center, or right by using *Alt + L, Alt + C,* or *Alt + R* combo key presses. This may come in handy when using an RDP session with a jump host.

Objects bar and the Validations and Session panes

Before we talk about each of these components, we must perform one out-of-order operation.

On the left side of your SmartConsole application, click **MANAGE & SETTINGS** – the left vertical tab [1]. To the right of it, in the view tree, click **Preferences** [2], scroll to the bottom [3], and check the **Enable Session pane – Review all changes before you publish** [4] option. You should now see one more item titled **Session** at the bottom of your right vertical tabs [5]. Also note the double-arrow symbols [6] that, when clicked, either expand or collapse the view panes to accommodate smaller screens. Click on **Objects** [7].

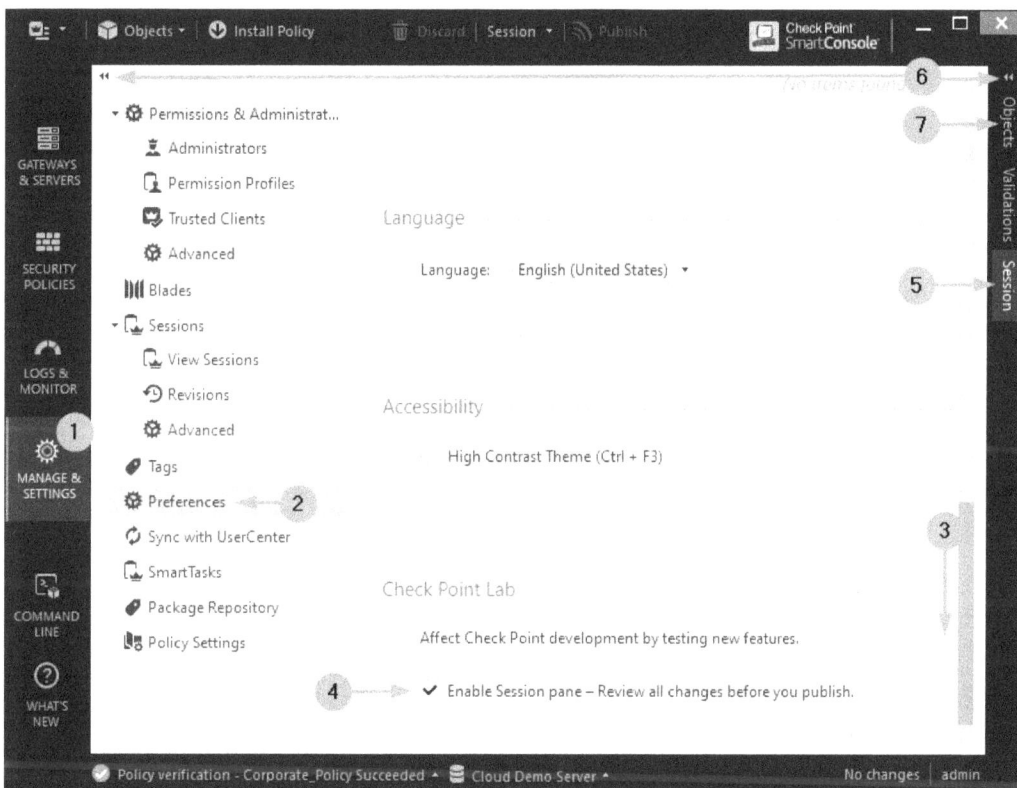

Figure 7.16 – Objects bar and the Validations and Session panes

Now, with all three right tabs shown, we can go over each:

Figure 7.17 – Objects bar and the optional Validations and Session panes

Let's discuss each element in detail:

- The **Objects** bar [A] from *Figure 7.17* presents us with the compact version of the Object Explorer. We can search for the objects [1], launch the explorer from here as well [2], drill down through object categories by double-clicking on them [3], or proceed with the creation of the objects through cascading menus by clicking **New** [4].

Figure 7.18 – Objects bar

- The **Validations** pane [B] from *Figure 7.17* can be accessed at any time to see the reasons for the failure to commit changes. As shown below, when an invalid configuration is introduced and the **Publish** action is attempted [1], we are presented with the session validation error [2] and, if our validations pane is collapsed, we can click on **Show pane** [3] to expand it. The number in the yellow disk [4] in the **Validations** tab acts as a violations counter. We can see the reason for the validation errors in the open **Validations** pane [5].

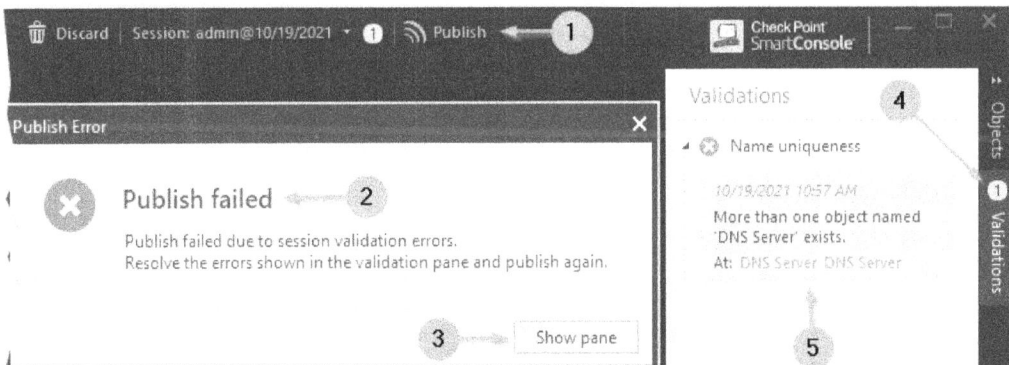

Figure 7.19 – Validations pane

- The **Session** pane [C] from *Figure 7.17* allows us to broadly track changes introduced (but not yet committed) by the administrator. It logs object creation and editing events, but does not provide details for each action. It does, however, provide double-click access to the objects, rules, and settings from the **Session Activity** [1] view. The following screenshot is representative of what the creation of the objects such as the **Time** tag [2], the **Maintenance** time window [3]. The **Rule 14.3** being created [4] and edited [5] to use these objects, as well as the change in settings for the password length requirements sequence [6], looks like:

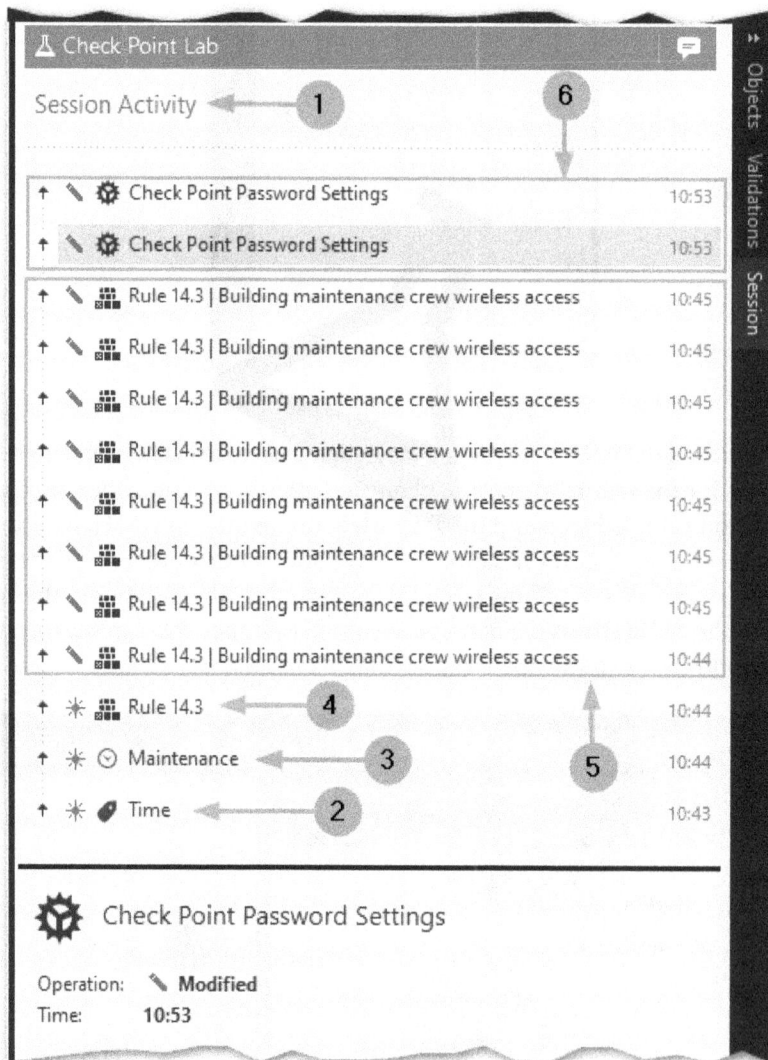

Figure 7.20 – Session pane, right vertical tab

> **Note**
>
> You can see the session activity only until it is published. Once the changes are committed, the session activity becomes blank until further changes are introduced by you.

Logged-in administrator's pending changes or publish status

The logged-in administrator's pending changes or publish status indicates the administrator logged in in the current instance of SmartConsole [1], as well as a number [2] and a list of other currently logged-in administrators [3]. In addition to that, it also lists either the number of pending (saved but uncommitted) changes or states in which the pending changes were published [4]. The following is a depiction of what two administrators, **admin** and **Saul**, are seeing throughout the session where **admin** has introduced two changes and committed them using the publish function:

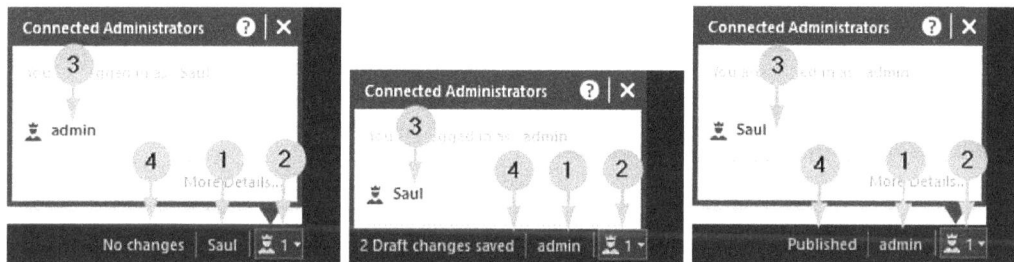

Figure 7.21 – Logged-in administrators pending changes or publish status

Management server(s) status and actions

The management server(s) status (and actions) functionality is contingent on the type of management environment you are connected to.

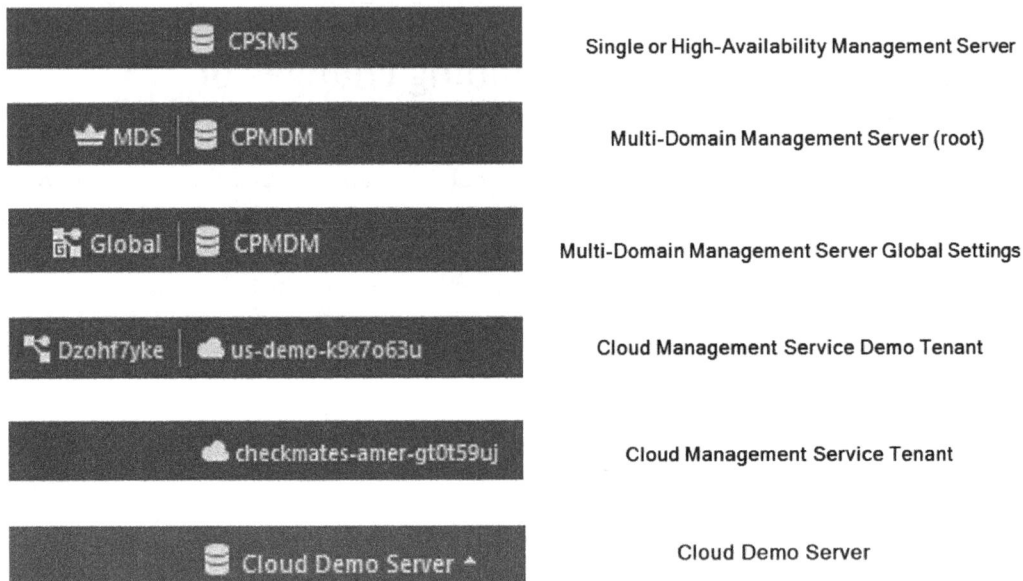

🗄 CPSMS	Single or High-Availability Management Server
👑 MDS 🗄 CPMDM	Multi-Domain Management Server (root)
🖧 Global 🗄 CPMDM	Multi-Domain Management Server Global Settings
🖧 Dzohf7yke ☁ us-demo-k9x7o63u	Cloud Management Service Demo Tenant
☁ checkmates-amer-gt0t59uj	Cloud Management Service Tenant
🗄 Cloud Demo Server ▲	Cloud Demo Server

Figure 7.22 – Management server (or environment) status

In management high availability, which includes a secondary management server, clicking on the stacked cylinder with the server's name or IP address next to it [1] and a warning indicator, if any are present [2], will produce a **High Availability Status** popup [3] containing each server's status, error messages (if present) [4], and also display the **Actions** buttons [5]. The available actions are specific for each primary and secondary management server.

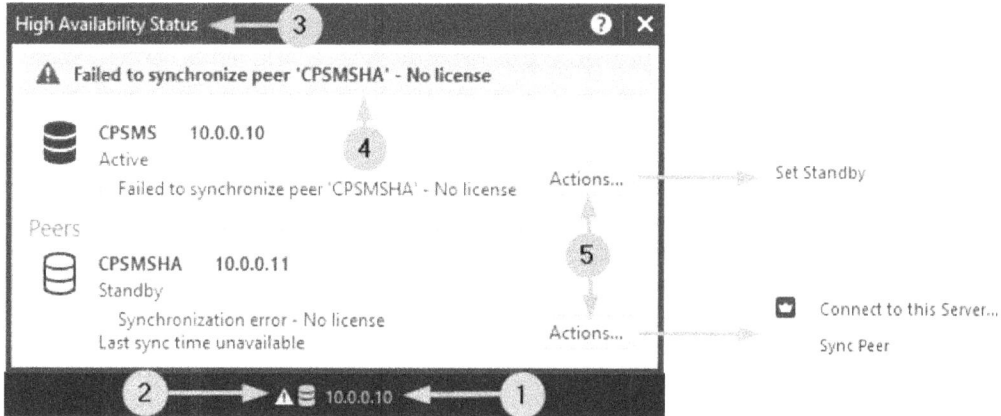

Figure 7.23 – Management server status (Management High Availability)

In the case of **Cloud Demo Server**, which we are using throughout this chapter, there is an additional menu accessible by clicking it [1] that presents us with additional options, such as a direct link to the Check Point community portal [2]. The **Experience concurrent administrators** menu [3] simply lists some of the additional administrators available in Cloud Demo Server, but, when clicked, opens up the SmartConsole login prompt with the last server's IP and last admin's username pre-filled, so it is really of limited utility presently (a bug, reported to CP). **Demo Server information** [4] and **Copy Demo ID to clipboard** [5] are actually useful as you can copy the server's IP and hostname, as well as a session ID, save those in a text file, and use them to either access the same session if you have logged out or closed an application, or to log in as the other administrators in order to experience concurrent administration. The final important option in that menu is **Extend Demo expiration time** [6].

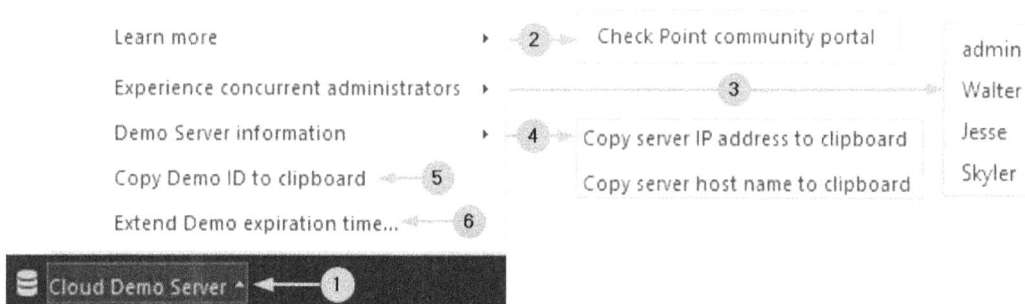

Figure 7.24 – Cloud Demo Server status and actions

You should take advantage of the **Extend Demo expiration time** option when you are planning to spend more than an hour working or experimenting. Set a reminder to extend it further, if necessary, so as not to lose all the changes you have made so far. When this option is clicked, you are presented with the countdown timer [1] and can set the desired extension interval for your session [2].

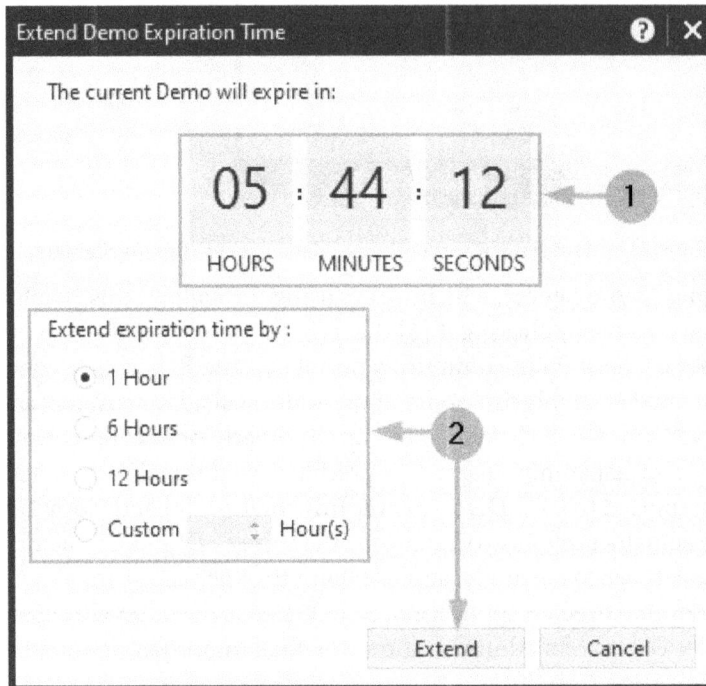

Figure 7.25 – Cloud Demo session expiration and extension

Task information area

The task information area, in its default non-expanded state, displays either the status of the currently running task with the progress bar or the status of the last completed task. These include either automated background actions, such as periodic subscription feed updates for IPS, AV, AB, application control, and URL filtering, as well as the actions you have performed manually, such as policy or database installations and management high availability manual synchronization.

When clicked, the task information area [1] expands in a **Recent Tasks** pop-out window that contains a chronological record of completed and currently running tasks. All tasks are displayed by default, but you can sort them according to success or failure [2]. Manually executed tasks can be filtered by the current session's administrator [3]. By default, all tasks executed by all administrators are shown [4]. When moused over, the truncated status description [5] produces a complete description popup. The **Details** option [6] in each, when clicked, opens another window containing extended task progress or results information. If you would like to keep an eye on the tasks in progress, which could be handy when multitasking but would like to ascertain the successful conclusion of the process, pin [7] the **Recent Tasks** window for it to remain open when navigating away from it.

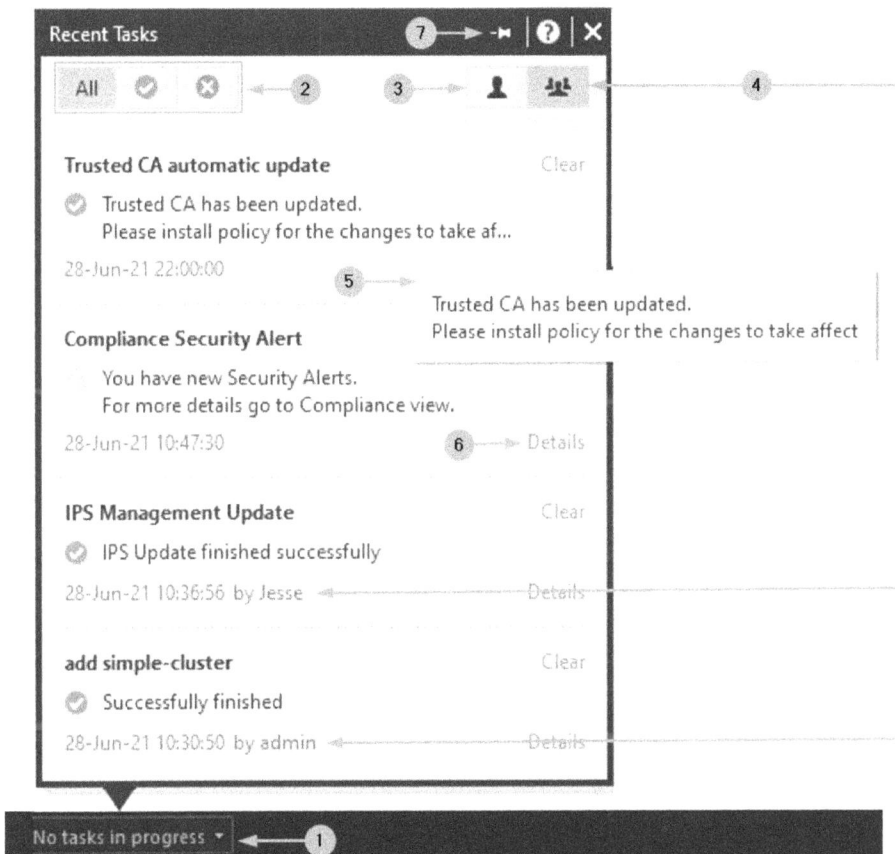

Figure 7.26 – Task information area, Recent Tasks

The WHAT'S NEW popup recall and management script CLI and API

The **WHAT'S NEW** popup recall [A] and management script CLI and API [B] are only grouped incidentally since both are located in the bottom-left section of the SmartConsole application:

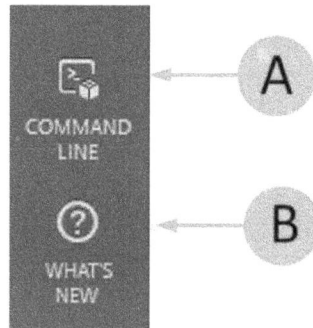

Figure 7.27 – COMMAND LINE (management CLI) and WHAT'S NEW

The **COMMAND LINE** icon, when clicked, opens up the window to the management CLI (not a shell, just an API running on your management server) within SmartConsole. Unlike all other management CLI access methods, this one does not require additional authentication since it relies on that of the existing session. You can enter commands manually, copy and paste, or open space-separated files [1]. The icon labeled [2] provides access to the online references for the management API.

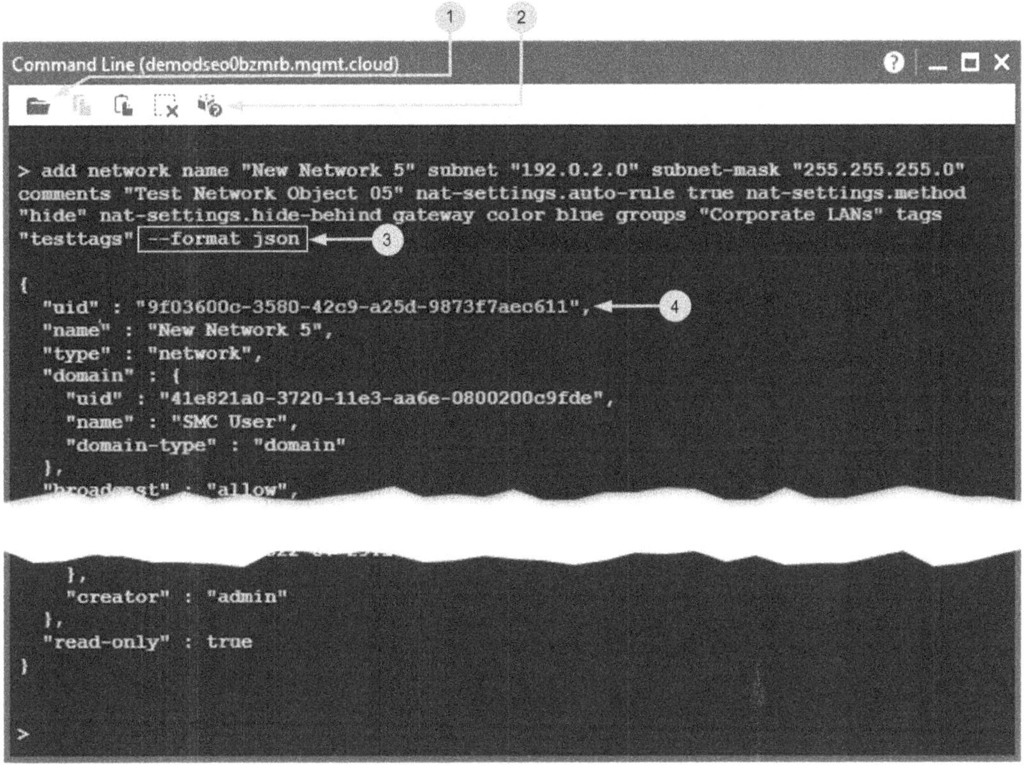

Figure 7.28 – SmartConsole, Management CLI

In the preceding example, the add network name "New Network 5" subnet "192.0.5.0" subnet-mask "255.255.255.0" comments "Test Network Object 05" nat-settings.auto-rule true nat-settings.method "hide" nat-settings.hide-behind gateway color blue groups "Corporate LANs" tags "testtags" --format json command is used to do the following:

- Create a network object with a description.
- Configure its **Network Address Translation** (**NAT**) settings.
- Set its UI object's color to blue.
- Add it to the group.
- Tag it.

`--format json` [3] appended to the end of the command provides detailed object information that, among other things, contains its **Unique Object Identifier** (**uid**) that could be used for absolute references to the object even if it is renamed at a later date. Without it, a successfully executed command would simply return you to the prompt while the commands containing errors would return a verbose explanation for the failure.

It is extremely convenient for moderate bulk operations such as the addition of a large number of objects from space-separated files.

For instance, we can create a file containing the following lines:

```
add host name host_test1 ip-address 10.0.0.111
add host name host_test2 ip-address 10.0.0.112
add host name host_test3 ip-address 10.0.0.113
```

Then we can open it from the **Command Line** window [1]. We will see a sequential line processing and a return to a > prompt [2] and then, in the **Session** tab's [3] **Session Activity** window, we can see the objects being created [4], as shown here:

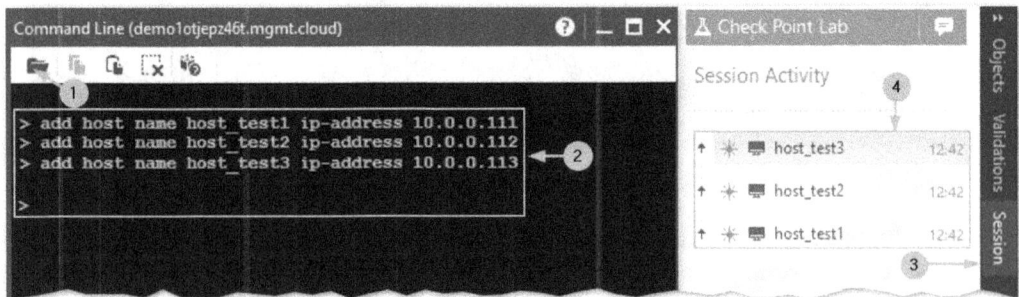

Figure 7.29 – SmartConsole, Management CLI, bulk object creation

Note that in order to commit the changes introduced via SmartConsole's built-in CLI, you must do so by either clicking on the **Publish** icon in the session management toolbar or by using the *Ctrl + S* combination keypress.

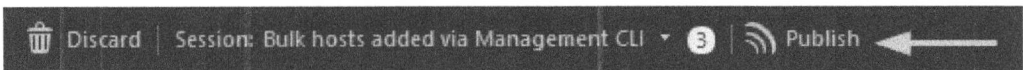

Figure 7.30 – SmartConsole CLI, Publish (commit) changes

Navigation toolbar

With the exception of **Multi-Domain Management** (**MDM**) servers (which are not covered in this book), the navigation toolbar consists of the following four tabs:

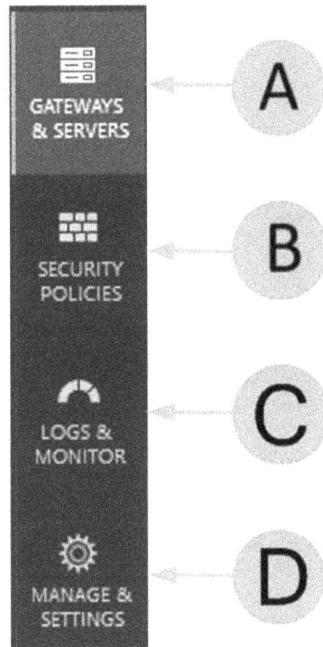

Figure 7.31 – SmartConsole navigation toolbar (left tabs)

These are **GATEWAYS & SERVERS** [A], **SECURITY POLICIES** [B], **LOGS & MONITOR** [C], and **MANAGE & SETTINGS** [D]. Each, when selected (clicked), opens up a different view in the central pane of the SmartConsole application. Let's go over each to learn about them. In the interests of saving on space, the icon representing each navigation tab is shown above the content of the central pane in each subsection below.

GATEWAYS & SERVERS [A]

This is the view where all management servers and gateways of your Check Point environment are shown, new ones can be created, and a variety of actions can be performed on each.

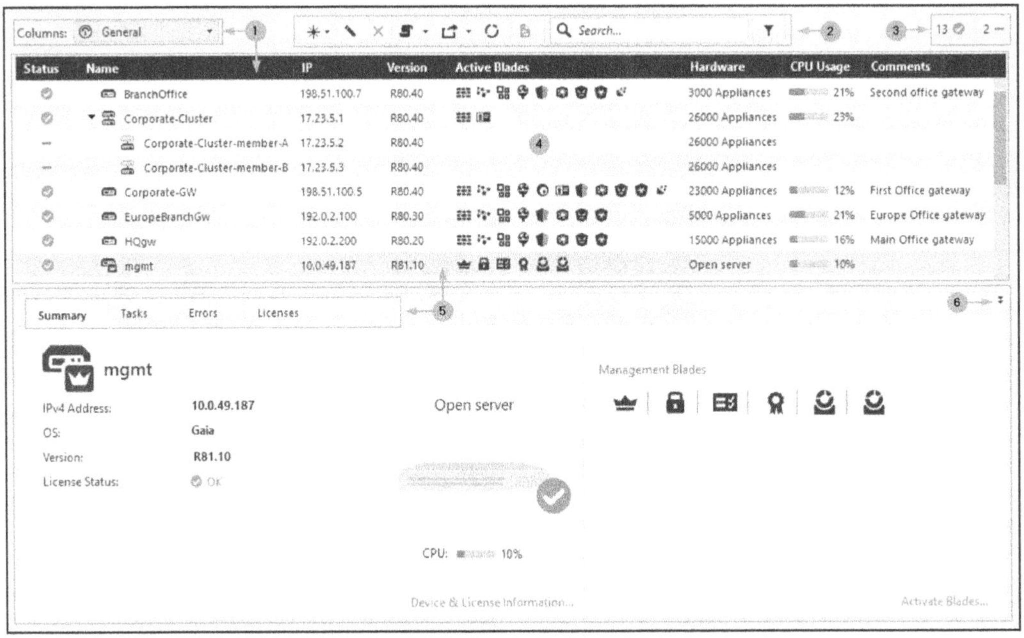

Figure 7.32 – SmartConsole – GATEWAYS & SERVERS

It consists of the following elements, which can be seen in the preceding screenshot.

The Columns menu [1]

This is a drop-down menu where each selection, in addition to common data, such as status, name, and IP address, shows different choice-specific information. You can further refine the resulting view by right-clicking on the **Columns:** header bar and checking (or unchecking) individual headers.

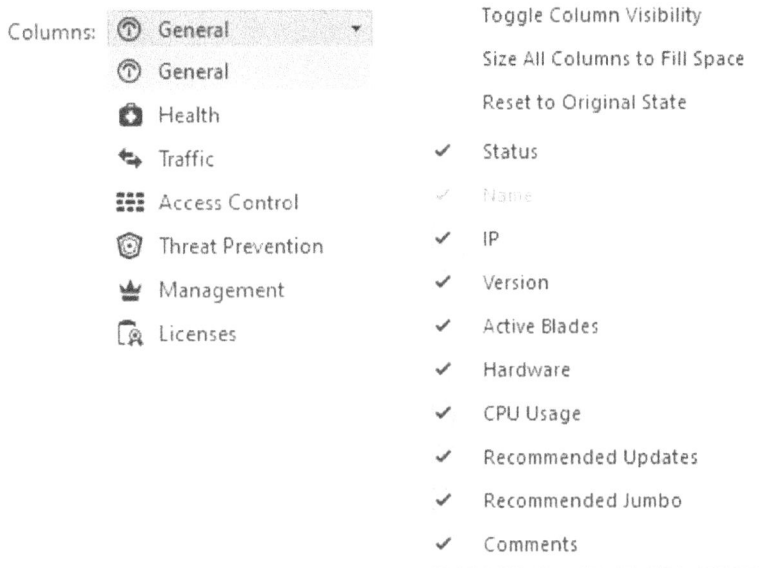

Figure 7.33 – SmartConsole – GATEWAYS & SERVERS | Columns menu

The action menu [2]

When your SmartConsole window is expanded wide enough, the action menu looks like this:

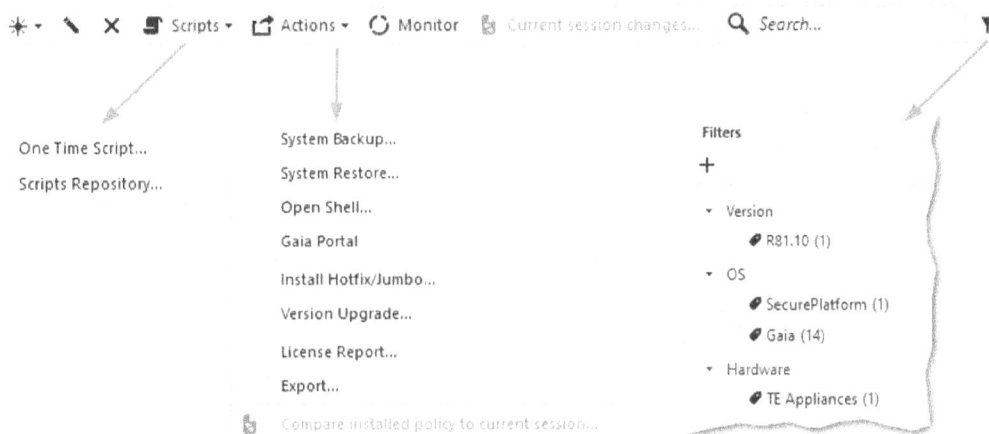

Figure 7.34 – SmarConsole – GATEWAYS & SERVERS | Actions menu, search and filter bar

The different elements are discussed as follows:

- The icons/buttons, from left to right, allow you to create, edit, and delete Check Point objects, execute scripts, and perform certain actions on selected objects.

- From the **Actions** menu, we can perform system backups and restores, features and security updates, and version upgrades. In the background, these actions utilize the same backup and update mechanisms that we discussed earlier in this book's Gaia chapters.

> **Note**
>
> Note that system backup, system restore, updates, and upgrades from SmartConsole are only possible for the gateways or clusters, not management servers.

The **License Report**, **Export**, and **Compare installed policy to current session** actions are global. The **License Report** action will generate a report covering all your Check Point appliances, and the **Export** action will generate a CSV version of the current view with chosen columns. The **Compare installed policy to current session** action will generate a change report depicting changes in access control and threat prevention rules as well as changes in objects.

- When you begin to type in the search window, it changes color to focus your attention on the fact that the objects shown in the view are dynamically filtered. To empty the search, click **x**. When switching between navigation tabs, the search is cleared automatically.

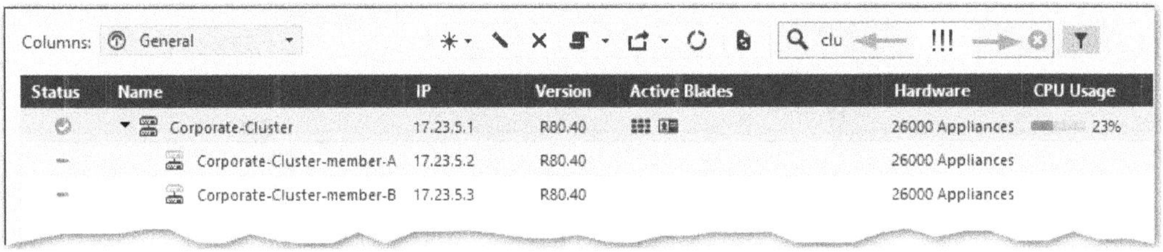

Status	Name	IP	Version	Active Blades	Hardware	CPU Usage
⟳	▾ Corporate-Cluster	17.23.5.1	R80.40	▦ ▯▮	26000 Appliances	▬▬ 23%
--	Corporate-Cluster-member-A	17.23.5.2	R80.40		26000 Appliances	
--	Corporate-Cluster-member-B	17.23.5.3	R80.40		26000 Appliances	

Figure 7.35 – SmartConsole – GATEWAYS & SERVERS | Dynamic search

- Filter, the last icon/button in the **Actions** toolbar, when clicked, opens another pane inside and on the right of the gateways and servers view. In it, you can check the box corresponding to one of the system tags [1] to narrow down the scope of the appliances in your view. To clear the filter, simply uncheck the same box. If you click **x** [2] above the selected filter, (when moused over, the **Remove filter** popup is shown), and the entire filter type disappears from the **Filters** pane. To add it again, click on + [3], check the empty box in the filter search popup [4], and then click **OK** [5].

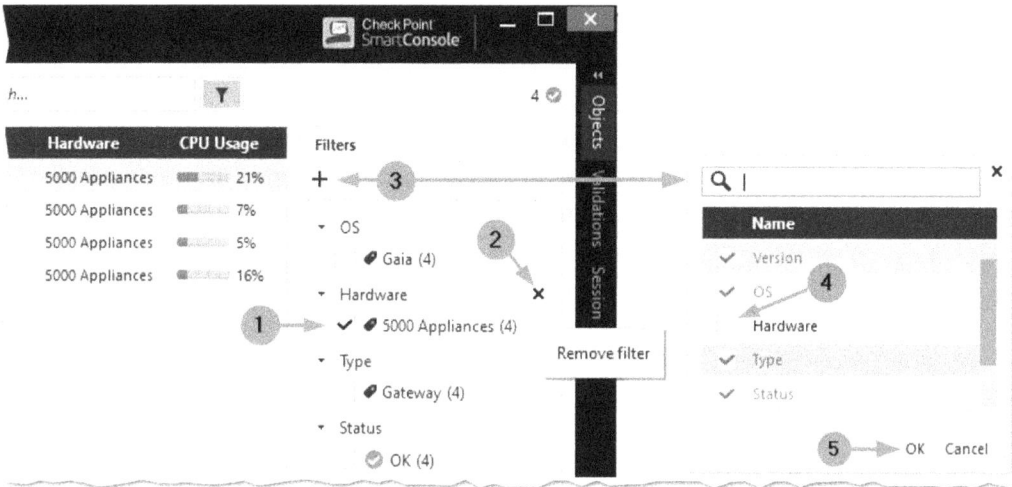

Figure 7.36 – SmartConsole – GATEWAYS & SERVERS | Filters

Object status and counters [3]

Located in the top-right section of the **GATEWAYS & SERVERS** view, this is simply there to give you an at-a-glance view of the number of objects and their states. If your organization uses hundreds of gateways, those with the problem states may not appear in the main view until scrolled down. Thus, if any unexpected indicators are present, we are alerted and can filter or sort the view by status.

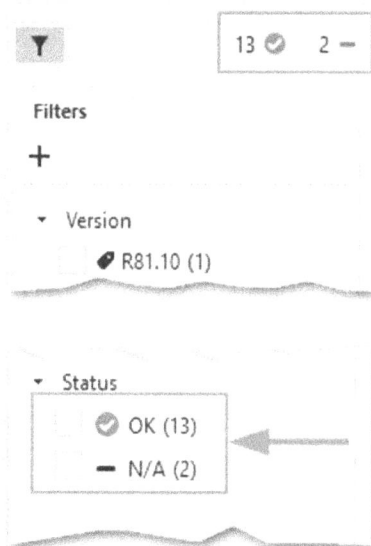

Figure 7.37 – SmartConsole – GATEWAYS & SERVERS | Status and counters

List of appliances [4]

Located in the navigation view (pane) top section is the list of all the Check Point appliances defined in your infrastructure. We already know that the rest of the information in there is dependent on the option selected in the **Column:** drop-down menu. When an appliance is selected, either by means of single-clicking or by using the arrow up/down keys, the selected line is highlighted in a light-blue color.

Right-clicking on any of the appliances in the list presents us with the menu containing options already described in *The action menu* section, as well as a few additional items.

Figure 7.38 – SmartConsole – GATEWAYS & SERVERS | Right-clicking the object menu

The **Where Used** option allows us to track down the dependencies of the other objects on the selected appliance, determine which policies and rules this object is being used in [1], and replace it [2] in groups and rules, with another one [3], as shown in the following screenshot:

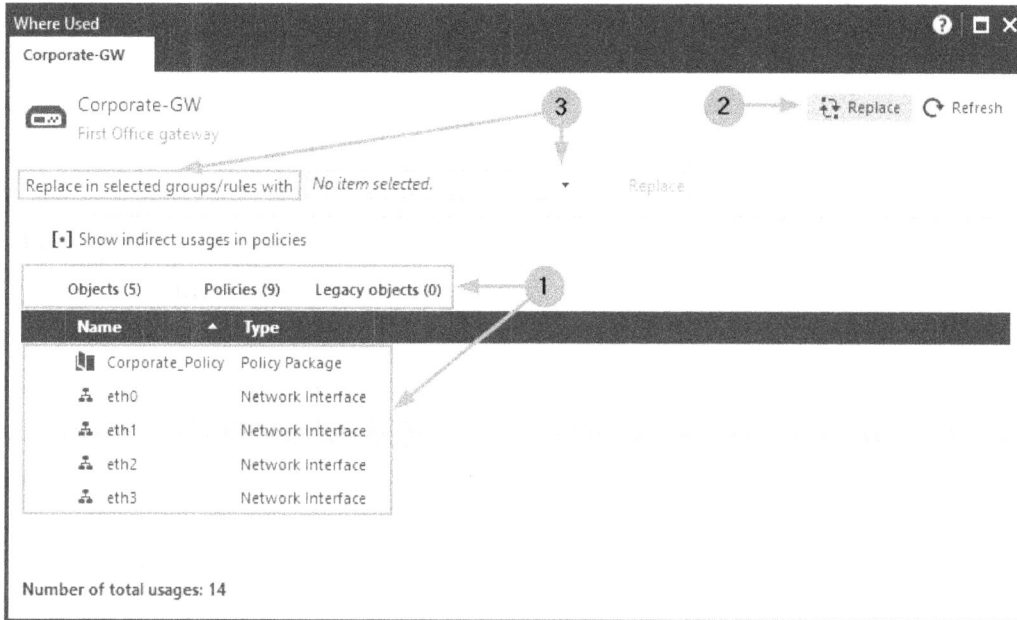

Figure 7.39 – SmartConsole – GATEWAYS & SERVERS | Where Used

The **Copy To Clipboard** option only copies the name of the gateway and its description in text format, while **Copy As Image** copies the screenshot of the selected line. These options could be useful when creating documentation.

Additionally, I would like to mention that, when clicked, the column headers in this view are used to sort the content of the list by either alphabetical, ascending, descending, or status order. It is handy when trying to find appliances on a particular version, in a particular state, or to see where possible utilization issues may be present.

Appliance-specific tabs [5]

By default, when an appliance is selected, you will see four tabs: **Summary, Tasks, Errors,** and **Licenses:**

- The **Summary** tab provides at-a-glance data regarding the appliance, its state, the presence of warnings or errors that could be accessed by clicking on corresponding links [1], as well as the list of active blades (features) and the link to the appliance object's properties where those could be activated or deactivated [2].

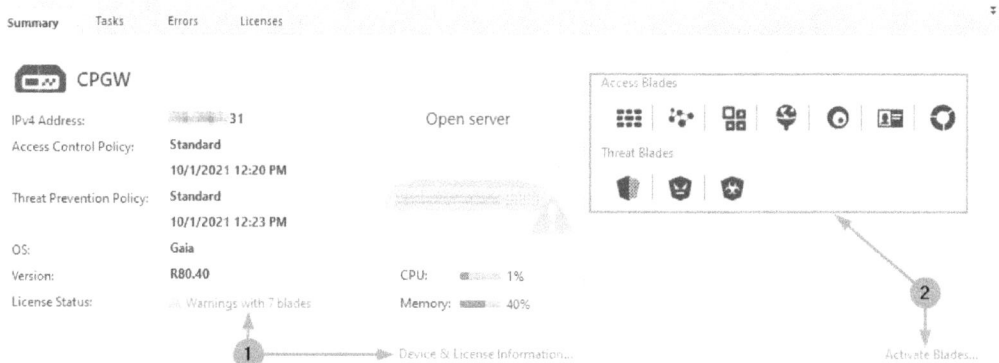

Figure 7.40 – SmartConsole – GATEWAYS & SERVERS | Appliance's Summary tab

- The **Tasks** tab contains the list of tasks executed by scripts from the SmartConsole's **Actions/Scripts** options, as well as the results, time, and the name of the administrator who has run the script. Since we are not permitted to execute scripts in Demo Mode, I am substituting this example from one of my lab environments for illustration.

In the following screenshot, each task, when double-clicked [1], opens a **Task Details** window. In it, you can scroll through the tasks by clicking *up/down* arrows [2]. You can copy the content of a task by clicking the stacked page icon [3] to view much more detailed information and paste it into a text editor for review. Note that the copied text uses a comma as the end-of-line character. To make it more readable when using Notepad++, perform the following steps:

A. Paste the copied text from the clipboard into the new file.

B. Press *Ctrl + H*.

C. In the **Find What:** field, enter , (a comma).

D. In the **Replace with:** field, enter \n.

E. In **Select Search Mode**, choose **Extended (\n, \r, \t, 0, x…)**.

F. Click **Replace All**.

And you are presented with properly formatted output. If you simply want to see the abbreviated results within SmartConsole, in the **Task Details** window, click on **Show Results**, and the results will be presented in the **Task Results** window [4].

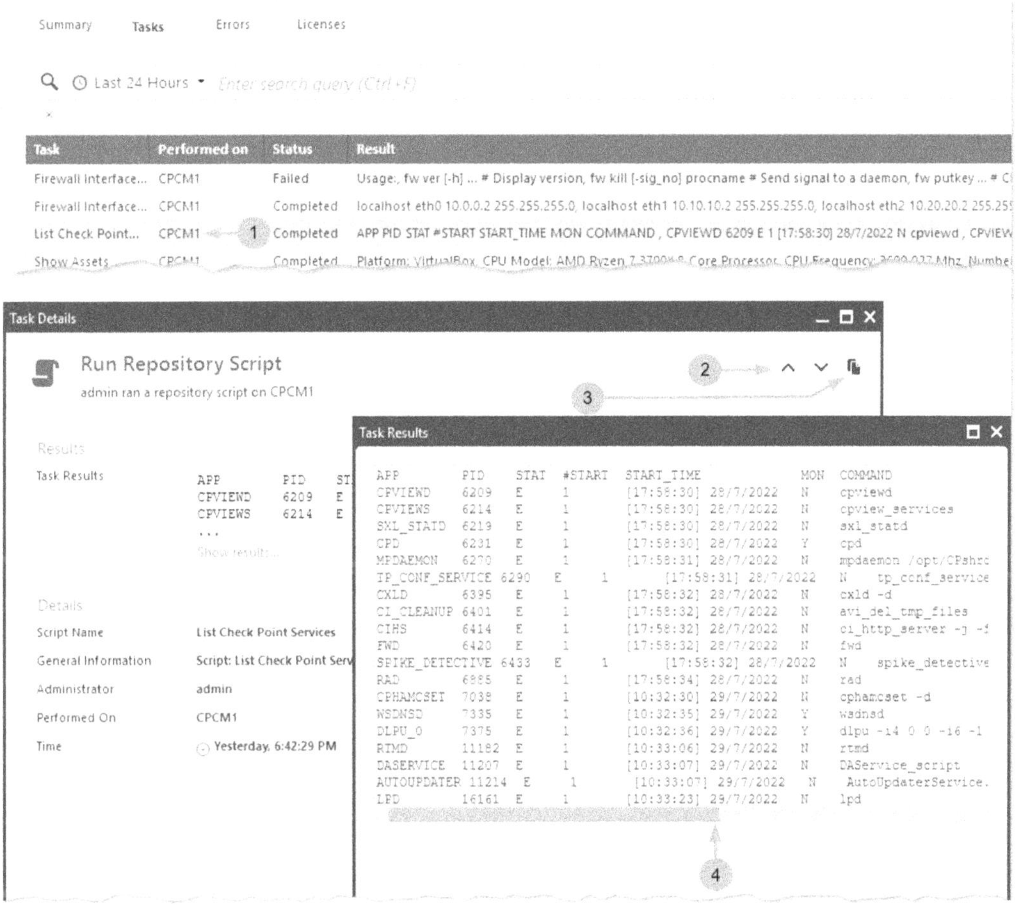

Figure 7.41 – SmartConsole – GATEWAYS & SERVERS | Appliance's Tasks tab

- The **Errors** tab will display the blade (function) – specific warnings, such as the pending license expiration of the selected appliance:

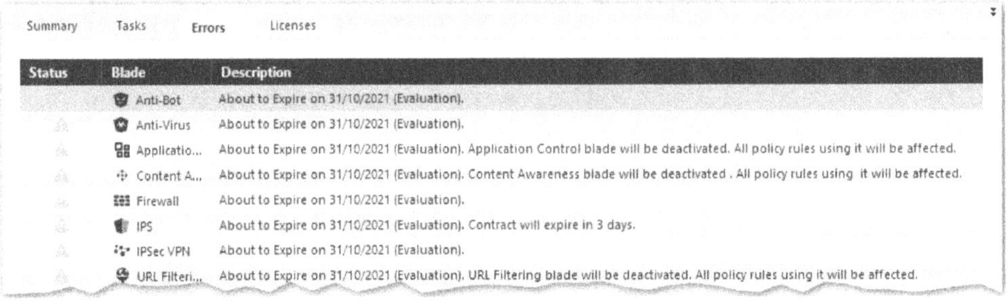

Figure 7.42 – SmartConsole – GATEWAYS & SERVERS | Appliance's Errors tab

A typical start of an administrative troubleshooting workflow may begin like this:

A. Open SmartConsole.

B. Note the number of the appliance with warnings or errors in the right-top at-a-glance counters and status indicator, sort the appliances by clicking on the **Status** column header to get problematic units listed on top, click one to select, and note the warnings or error information in the **Summary** tab.

C. From there, either open the **Device & License information** monitoring utility to investigate or click on the **Errors** tab to see a list of the issues encountered to determine the seriousness and priority of the warnings displayed there.

- The **Licenses** tab, when clicked, performs the pull operation from the appliance to get the up-to-date licenses from it. Once licenses are retrieved, for each, a checkbox, followed by **IP Address**, **Expiration Date**, **Certificate Key** (**CK**), and the product's **stock-keeping unit** (**SKU**), are displayed.

If the checkbox next to a particular license is selected, we can click the **Remove** button to, for instance, get rid of the expired, but still present, licenses. Clicking the **Add** button allows us to do just that, using either a licensing string or a license file. In the top-right corner of the tabs pane is the pane show/hide toggle switch that looks like stacked arrows.

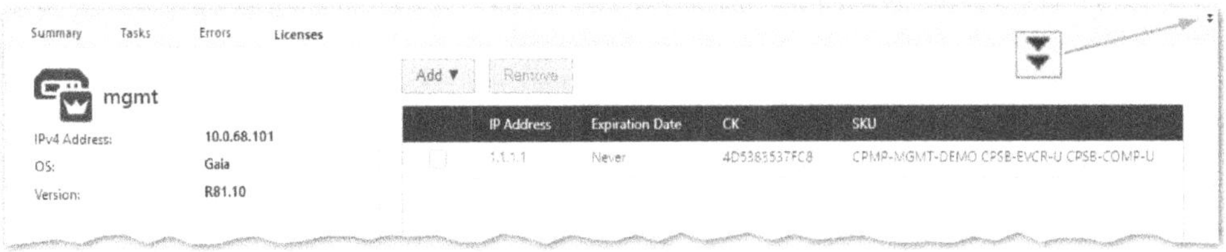

Figure 7.43 – SmartConsole – GATEWAYS & SERVERS | Appliance's Licenses tab

SECURITY POLICIES [B]

This is the view where most of the day-to-day administration is done, the components of security policies are defined, and rules are created. On top, tabs corresponding to individual policy packages are displayed. We may have multiple policy packages opened by clicking + and choosing from those available [1]. In a view tree [2], we can choose the policy aspect we are interested in working on. Depending on our choice, the tools [3] and central pane [4] are populated with relevant content. In **Shared Policies** [5], **Inspection Settings** is always present, and **Mobile Access** and **Data Loss Prevention** (**DLP**) could be added. When **Inspection Settings** is clicked, this generates a popup window where profiles for the deep packet, protocol parsing, and **VoIP** inspection settings can be created and assigned to specific gateways, and where the exceptions for the inspection settings are configured. The pane show/hide toggle icon is located in the top-left corner [6].

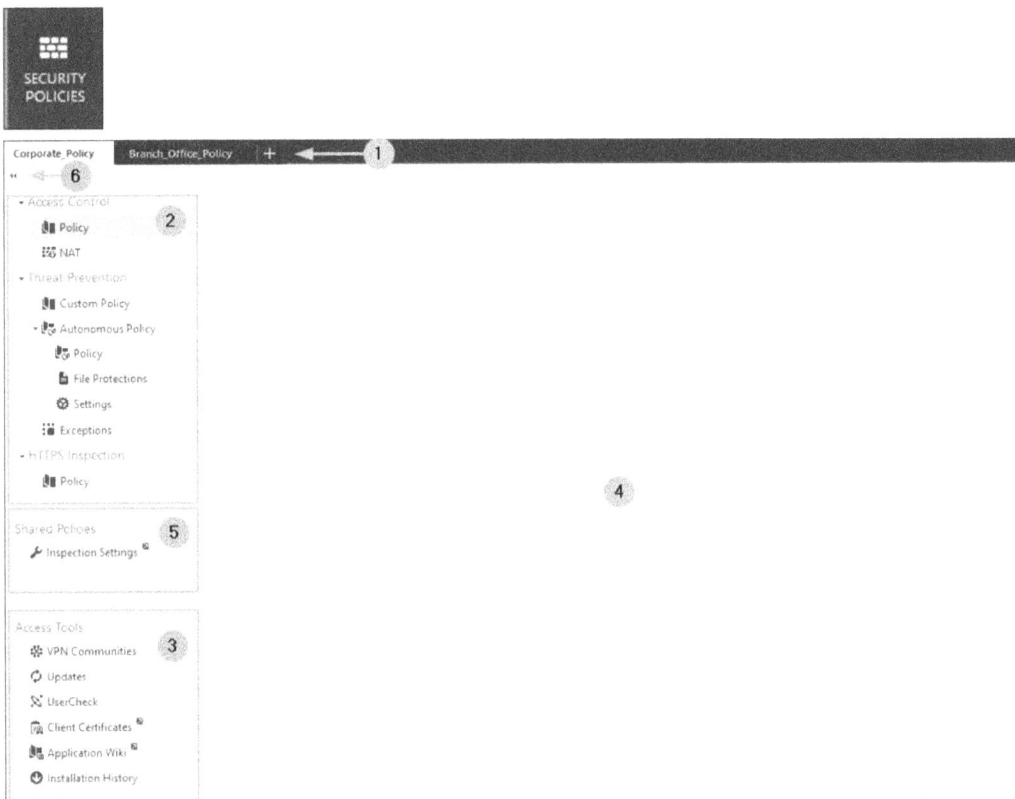

Figure 7.44 – SmartConsole – SECURITY POLICIES

Access Control policies can comprise the following features (blades):

- **Firewall**

- **Application Control and URL Filtering**

- **IPsec VPN**

- **Mobile Access**

- **Identity Awareness**

- **Content Awareness**

The access control policy view consists of the following components:

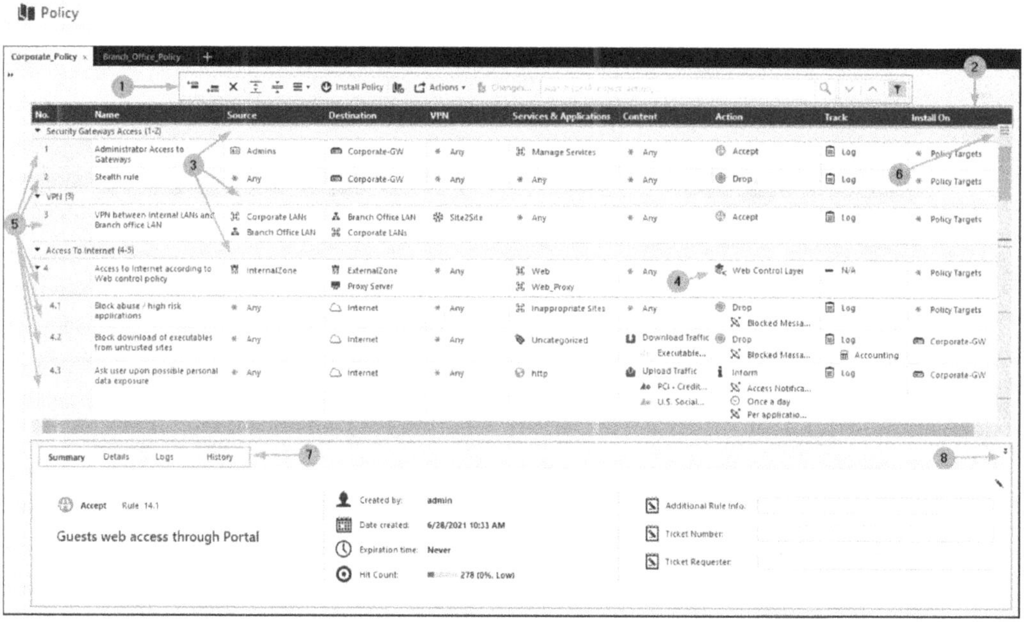

Figure 7.45 – SmartConsole – SECURITY POLICIES | Access Control, Policy

Let's discuss each of these components in detail in the following sections.

Action menu [1]

An elaborate view of the action menu is shown in the following screenshot:

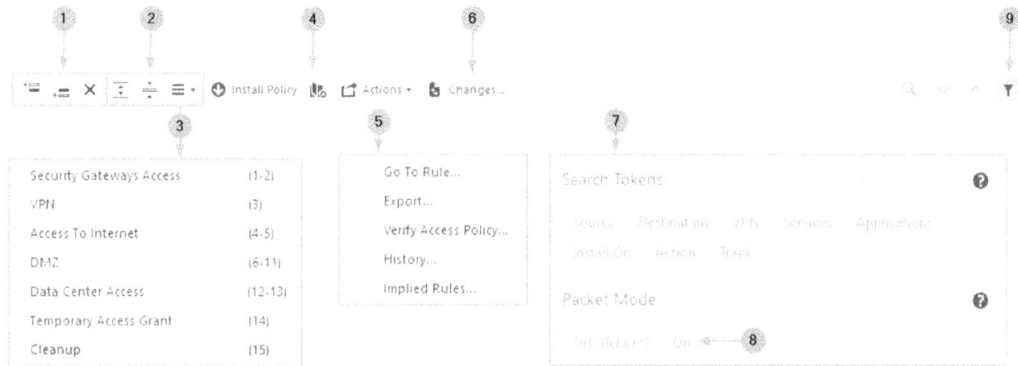

Figure 7.46 – SmartConsole – SECURITY POLICIES | Access Control, Policy, and Actions menu

In the preceding screenshot, the first three icons [1] allow us to create new rules above or below the one currently selected or to delete selected rule(s) if multiple rules are selected by clicking while pressing the *Ctrl* button on the keyboard.

Three section navigation icons [2] allow us to either collapse or expand all the named policy sections (described just a bit later) or to navigate to a particular section. **Install Policy** is used to initiate the current policy installation process.

Edit policy [4] opens up a window where policy types, layers, and functions (blades) for the currently selected policy package are defined and the policy package installation targets are selected.

Actions [5], when clicked, expands to a drop-down menu where we can choose to **Go To Rule** (either by means of a number or UID), **Export**, which produces the CSV file reflecting a visible policy structure and objects, and **Verify Access Policy**, which analyzes the policy following modifications for inconsistencies. **History**, when clicked, produces a pop-up window where the sessions with the number of changes made to the policy attributed to individual administrators are chronologically presented. From this window, each revision could be opened in a read-only mode in a separate SmartConsole instance. Additionally, the audit logs for each revision are available, detailing every action. The **Implied Rules** option deserves a separate mention, and we will discuss this further in the book.

Changes [6] produces a detailed, printable report for unpublished (uncommitted) changes, listing the creation of, and changes to, objects, rules, and specific properties.

The search box [7], when clicked, produces a window containing shortcuts to **Search Tokens** (syntax abbreviations, such as `src:`, `dst:`, `action:`, and so on) and a **Packet Mode** [8] **Off/On** selector. When **Packet Mode** is **On**, your search query will show all the rules that the data packet will match even if your search parameter(s) are not explicitly defined in the rule. When **Packet Mode** is **Off**, only the rules containing your search parameter(s) are returned.

Packet mode On

Packet mode Off

Figure 7.47 – SmartConsole – SECURITY POLICIES | Policy, Actions menu, Packet Mode Search

The filter icon [9] allows us to toggle between viewing only those rules that match our query (default behavior) or to see them in the context of the entire policy. When you perform the search, the results are highlighted in orange.

Columns [2]

The visible columns selector is available by right-clicking and is different in each view. In access control policies, it is comprised of the common fields, such as **Hits**, **Name**, **Source**, **Destination**, **VPN**, **Services & Applications**, **Action**, **Time**, **Track**, **Install On**, and **Description,** as well as **Content** (available when a **Content Awareness** blade is enabled on at least one of your gateways). Generally, it is a good idea to have all of the checkboxes selected unless you are pressed for screen space. You can re-order the column appearance by clicking + holding and dragging the header of the column to the desired location. The first column, **No.**, is mandatory and cannot be disabled or removed from view.

Section titles [3]

Section titles are used to help organize policies by grouping relevant rules together. These could be created anywhere in our policies by right-clicking on a rule number in a desired location of your policy and choosing one of the **New Section Title | Above/ Below** options. Once created, it will include all the rules below until the next section title is encountered in the policy. Section titles are not only descriptive, but also allow you to collapse or expand the view of the rules contained in each section either individually, by clicking a small triangle on the left side of the section title, or in bulk, by using section title options in the actions menu [2]. The three horizontal lines (*burger*) icon, either in the action menu or on the right edge of each section tile [3], are used to jump to the section you are interested in seeing, making it easier to work with long policies.

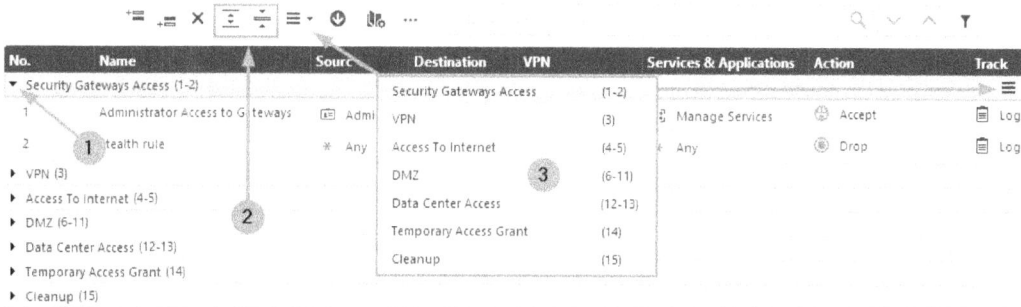

Figure 7.48 – SmartConsole – SECURITY POLICIES | Policy, Action menu, Section titles

Inline layers [4]

We will address the security policy layers in a later chapter, but for now, it is sufficient to know that one of the options in organizing a security policy **Rule Base** (a collection of the individual rules) is to use **inline layers**. Inline layers are recognized by the presence of the expansion icon [1] to the left of the rule number as well as the layer icon [2] in the **Action** field of the rule. When the expansion icon is clicked, the inline layer expands with the offset rules numbered with decimals after the parent rule's number.

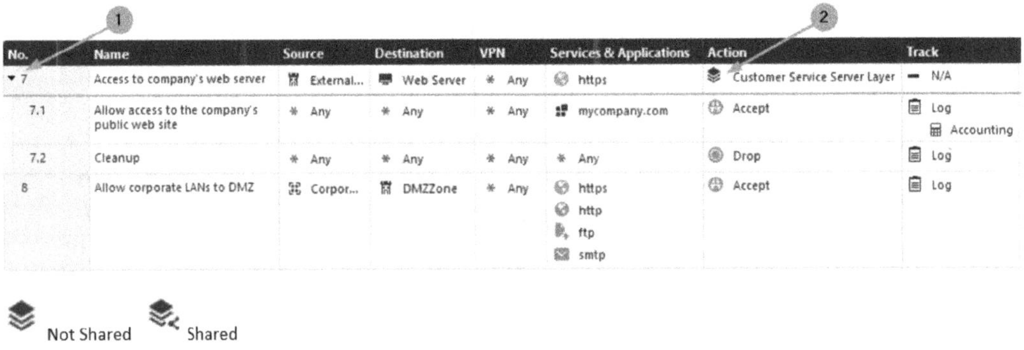

No.	Name	Source	Destination	VPN	Services & Applications	Action	Track
▼ 7	Access to company's web server	External...	Web Server	* Any	https	Customer Service Server Layer	— N/A
7.1	Allow access to the company's public web site	* Any	* Any	* Any	mycompany.com	Accept	Log / Accounting
7.2	Cleanup	* Any	* Any	* Any	* Any	Drop	Log
8	Allow corporate LANs to DMZ	Corpor...	DMZZone	* Any	https / http / ftp / smtp	Accept	Log

Not Shared   Shared

Figure 7.49 – SmartConsole – SECURITY POLICIES | Policy, Action menu, inline layers

Two icons in the preceding screenshot are depicting layers that could be identified as either not shared or shared.

Rules [5]

We'll talk more about rules in the next chapter's *Making rules* section, but there are a few things about rules I would like to point out presently.

The rule action menu is accessed by right-clicking on the rule number in the **No.** column. Besides obvious choices, such as **New Rule**, **Delete**, **Cut**, **Copy**, and **Paste**, there are a few items of interest in it. The **Disable** [1] option does just that, allowing you to keep a rule in a policy in case it needs to be reactivated. The **Rule Expiration** [2] option calls the **New Time** menu, where the time interval and recurrence of this rule's state are defined. It is handy for use with contractors, consultants, and auditors. **Copy Rule UID** and **Copy as Image** [3] should be used for documentation (UID specifically may be important for scripting or command-line operations). **Hit Count** [4], when enabled in the columns selector, will display either absolute or relative rule utilization values within a predefined timeframe. **Show Logs** [5] lets us see either the per-rule logged activity or the rule modification history logs described later in the *Rule-specific tabs* section.

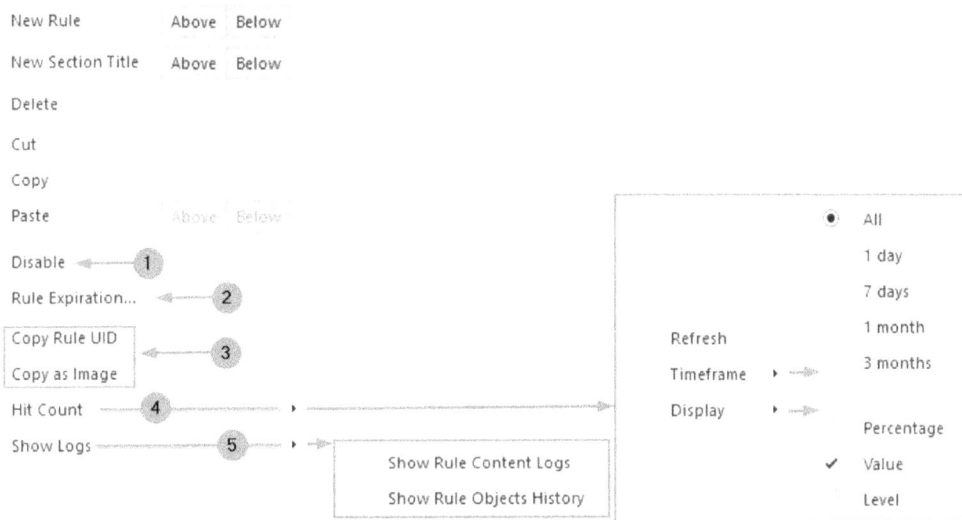

Figure 7.50 – SmartConsole – SECURITY POLICIES | Policy, Rules menu

The rule number, as shown in the policy, is only relevant to this session. If rules are created or deleted above the referenced rule, their number in the **No.** column will change. Do not refer to rules according to their numbers in documentation; define short and unique names reflecting the rule's purpose, fill out the description fields and include `rule UID ()`, as shown in the following screenshot:

872405ec-e9f7-4b6b-9af9-e495338536f3

Figure 7.51 – SmartConsole – SECURITY POLICIES | Policy, Rule references in the documentation

The search for the rule in the policy by UID is presently available via shortcut only. Use the *Ctrl + G* key combo to reveal the **Go to Rule** pop-up window, paste the UID in it, and press **OK** to locate the referenced rule.

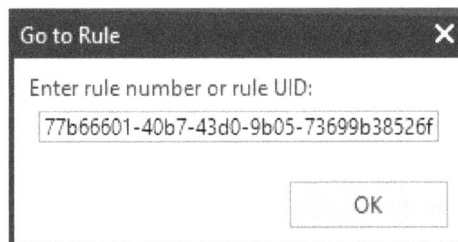

Figure 7.52 – SmartConsole – SECURITY POLICIES | Policy, searching rules according to UID

Indicators [6]

The three horizontal gray bars icon is only visible once your policy is long enough to warrant a vertical scroll bar's appearance. When clicked, you are presented with the indicators [1] selector. For each selected category, colored horizontal lines appear in the scrollbar [2] with spaces between the bars indicating the relative sizes of the sections and locations of **Edited**, **Locked**, **Section**, **Selection**, and **Search result** in the policy. You can get to the desired location in the policy by clicking on one of these colored bars.

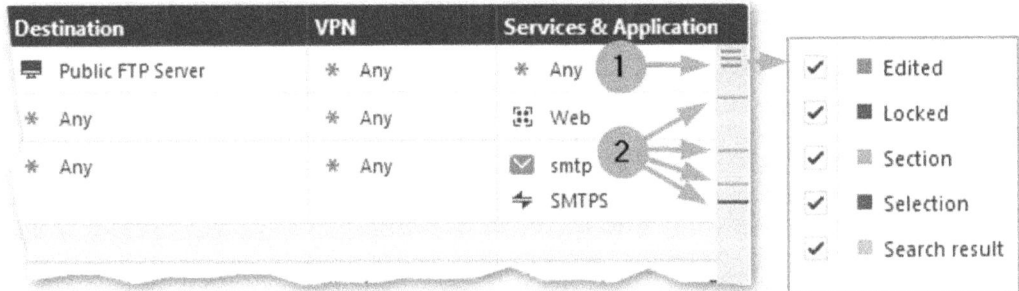

Figure 7.53 – SmartConsole – SECURITY POLICIES | Policy, Indicators

Rule-specific tabs [7]

By default, only four tabs are available: **Summary**, **Details**, **Logs**, and **History**:

- The **Summary** tab identifies the currently selected rule [1], along with information relevant to its creation, the author, the time of expiration (if defined) [2], and three custom field names and strings [3]:

> **Note**
> While you can edit the custom field names [3], these are global values and, once edited, they will be applied to all the rules in all policy packages. The string values are unique in each rule, if filled out.

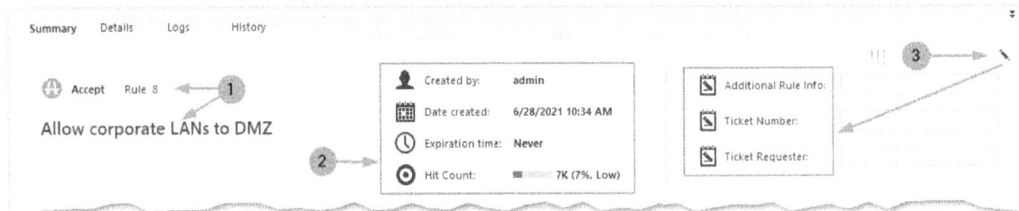

Figure 7.54 – SmartConsole – SECURITY POLICIES | Policy, Per-rule tabs, Summary

- The **Details** tab provides an expanded view of the selected rule. It allows the expansion of the groups to see their content [4], the IP addresses associated with objects [5], and the ports associated with services [6]. The tab-specific options menu [7] can be used to choose visible columns.

Figure 7.55 – SmartConsole – SECURITY POLICIES | Policy, Per-rule tabs, Details

- The **Logs** tab displays the rule-specific logs (denoted as **Current Rule** [x] in the search query field) [8] for **Last 24 Hours** by default [9]. You can use either manual or automatic refresh of the logs in this tab [10]. The logs options menu [11] is located to the right of the search field and we'll cover this in the *LOGS & MONITOR* section of this chapter.

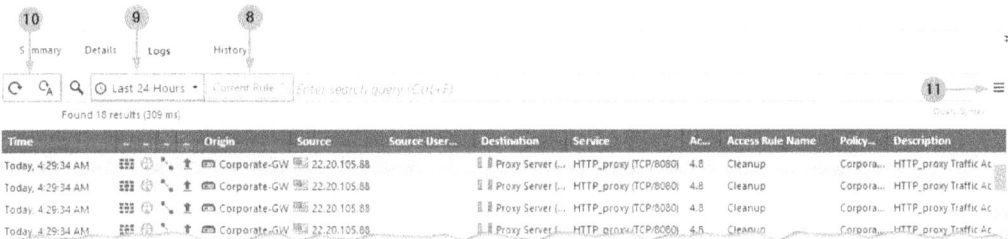

Figure 7.56 – SmartConsole – SECURITY POLICIES | Policy, Per-rule tabs, Logs

- The **History** tab displays audit logs, containing information about rule creation and subsequent modifications, filtered by the current rule for **All Time** [12]. It has the same logs options menu as the **Logs** tab [13]. The number of query results and duration are displayed on the left under the time interval selector [14].

Figure 7.57 – SmartConsole – SECURITY POLICIES | Policy, Per-rule tabs, History

Additional policies

Network Address Translation (**NAT**) is a subject that deserves a section of its own in the book and will be addressed later in *Chapter 10, Working with Network Address Translation*.

Another policy type that could be enabled or found in the access control section is the **Quality of Service** (**QoS**), but it is not covered in this book.

A **global** HTTPS inspection policy is created automatically with an access control policy, but is only enforced on the gateways with HTTPS inspection functionality enabled and if those gateways are defined as the policy installation targets.

LOGS & MONITOR [C]

Information in the **LOGS & MONITOR** section is useful not only for Check Point administration, but can provide invaluable data to SOC analysts, risk managers, and corporate boards, when SmartEvent is implemented, and additionally, when the **Compliance** blade is licensed and implemented, to auditors and compliance specialists.

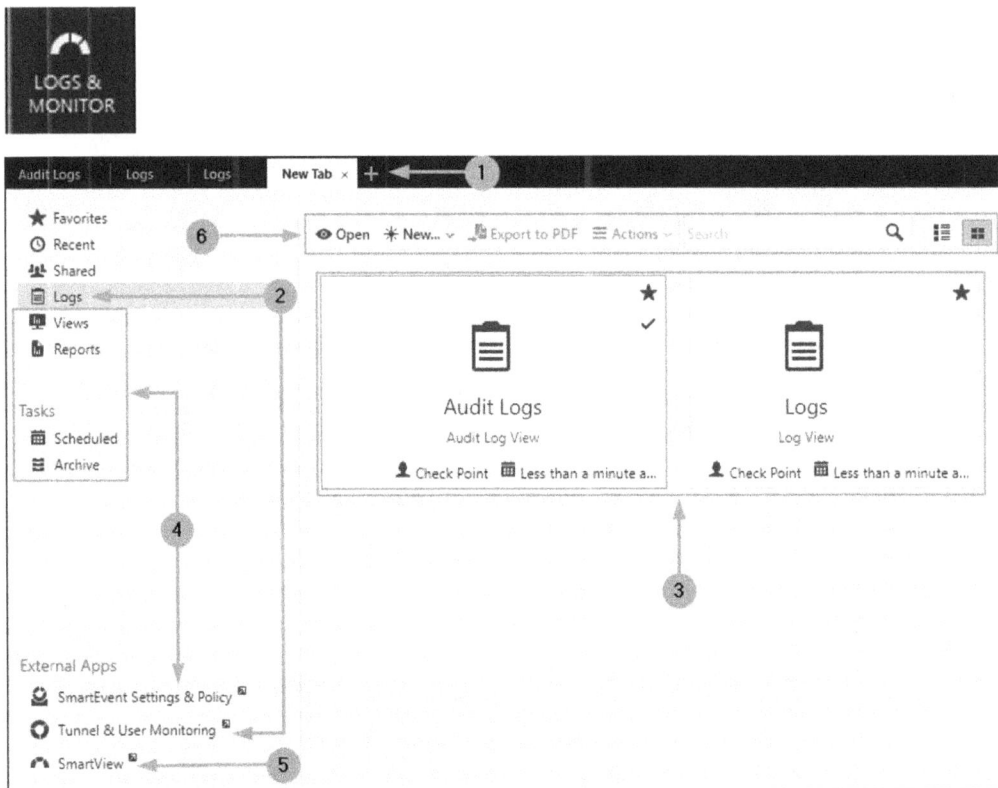

Figure 7.58 – SmartConsole – LOGS & MONITOR | Logs (No SmartEvent)

The **LOGS & MONITOR** view allows administrators to open multiple tabs. Each time a new tab is opened using the + sign [1], you are presented with the view shown in the preceding screenshot.

With a basic management license, only **Logs** and **Tunnel & User Monitoring** [2] in the **External Apps** section are available. When **Logs** is selected in the navigation pane, **Audit Logs** and general logs can be opened from the central pane [3]. **Views**, **Reports**, and **Shared**, as well as **SmartEvent Settings & Policy** under **External Apps** [4], are only accessible when the **SmartEvent** blade is enabled and licensed. **SmartView** [5] in **External Apps** is only visible when **SmartEvent** is enabled, but can be accessed without it using `https/<ip_or name_of_the_management_server>/smartview`. This allows access to the **LOGS & MONITOR** view from the web browser.

Two types of logs, **Audit Logs** and general **Logs** [3], can be accessed from the central pane. **Views**, **Reports**, and the ability to share them, when customized, among multiple administrators, as well as **Scheduled** tasks, are only available with the SmartEvent functionality enabled and a license applied. In the **External Apps** section [4], in addition to **Tunnel & User Monitoring** that is always available, **SmartEvent Settings & Policy** and **SmartView** options are shown when SmartEvent is enabled. This last one is likely an omission by Check Point, as SmartView is accessible directly without SmartEvent. The actions menu [6] is largely irrelevant without SmartEvent. The list and squares icons on the right side of the actions menu are there for selecting either a thumbnail or a table view.

Let's look at both log options (**Audit Logs** and **Logs**) to see how those could be utilized. Many of the actions and options shown apply to both types of logs, but will be described in the context of **Logs**.

Audit Logs

Audit Logs contains data related to changes in the configuration and other administrative actions. It is similar to that of a Rule History tab but covers all actions in your environment.

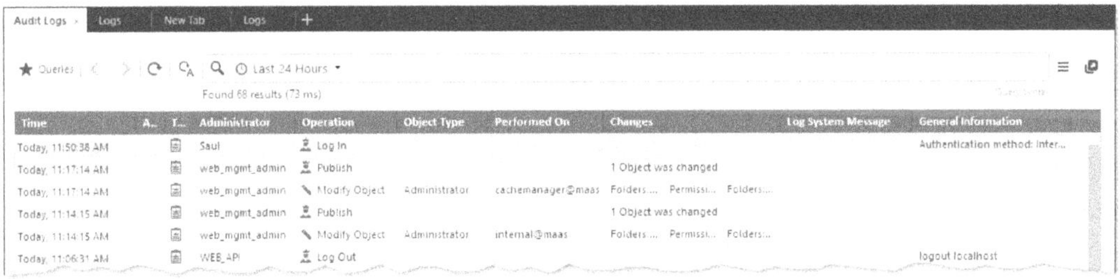

Figure 7.59 – SmartConsole – LOGS & MONITOR | Logs, Audit log

Logs

Logs, on the other hand, track all the actions that your gateways are performing for which tracking is enabled in the rules or elsewhere in the configuration settings. Looking at the typical **Logs** screen, we can discern the following items of interest:

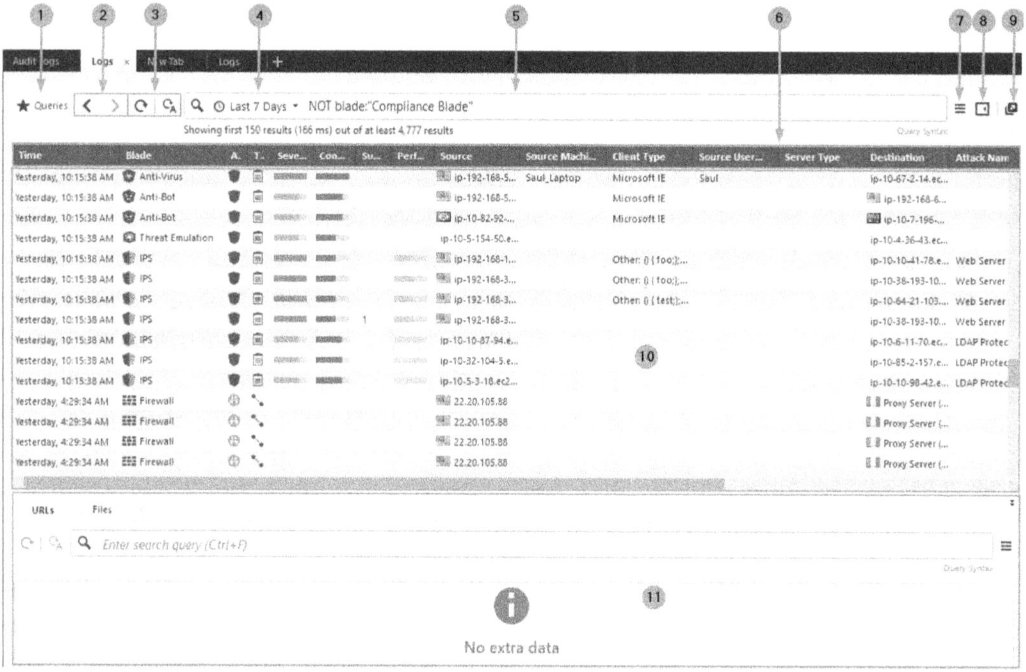

Figure 7.60 – SmartConsole – LOGS & MONITOR | New Tab | Logs, Logs

Let's discuss each component in detail:

- **Queries [1]: Queries**, when clicked, allows us to use several predefined search parameters as well as create and organize our own favorites.

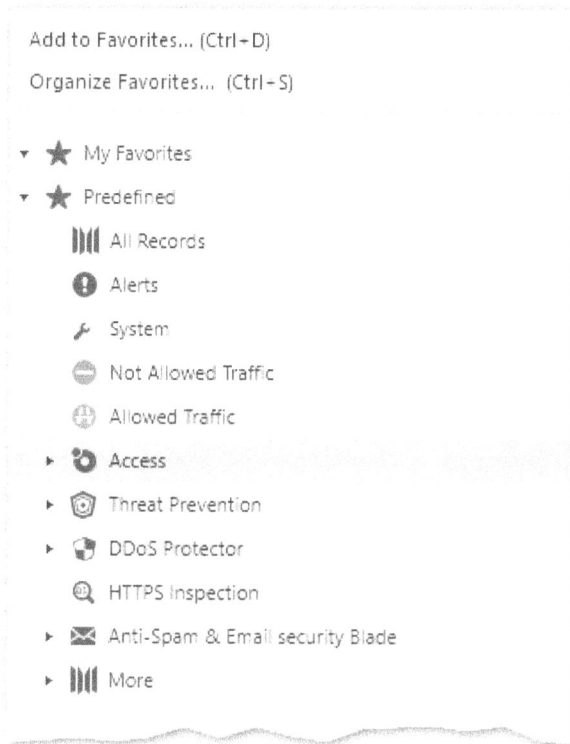

Add to Favorites... (Ctrl+D)

Organize Favorites... (Ctrl+S)

▾ ★ My Favorites

▾ ★ Predefined

    ▥ All Records

    ❶ Alerts

    ⚲ System

    ◉ Not Allowed Traffic

    ◉ Allowed Traffic

  ▸ ◉ Access

  ▸ ◉ Threat Prevention

  ▸ ◉ DDoS Protector

    ◉ HTTPS Inspection

  ▸ ✉ Anti-Spam & Email security Blade

  ▸ ▥ More

Figure 7.61 – SmartConsole – LOGS & MONITOR | Logs, Logs, Queries

- **Search history navigation arrows [2]:** The search history navigation arrows allow us to move between results of sequential queries; in other words, as you are adding more parameters to a complex query, you may want to execute it a few times as it is being constructed, so that if minor modification of the final step is required, you can *step back* to change it.

- **Manual and automatic refresh options [3]:** Manual and automatic refresh options enable logs to either remain still while you are working on them, which is useful when working with past events, or to update dynamically, which is more suitable for real-time troubleshooting.

- **Time interval setting [4]:** The time interval setting has a number of predefined options to choose from as well as the **Custom Time** option, where you define the start and end dates and times of your query.

- **Search query field [5]**: The search query field is automatically populated if you select the predefined query or where you can enter it manually. The search could be performed on a single word, a "*composite string*", using single ? or multiple characters and * wildcards (the * wildcard can be used in the IP address, if the specific address is not known, to cover the entire network). You can combine **Field Keywords** (column headers typed using lowercase characters or their aliases) with values using the `<field name>:<value>` format and join those using the Boolean operators AND, OR, and NOT. This search query is a representative example:

  ```
  blade:(Threat) AND (severity:(High OR Critical)) AND (NOT
  source:1.01.0.53)
  ```

When you click in the search query window, the **Add Search Filter** menu with predefined fields appears below. You can click on any of the available options *to only append the field to the end of the query* irrespective of where, inside the body of the query, your cursor is located.

One of the options shown in the **Add Search Filter** menu is **Other Fields**, which contains a very long list of available options.

If you want to narrow the scope of the search query in a single tab and ensure its persistence when working on the rest of the query, the **Constant Filter** option is available. To enable it, right-click in the search query field and, in the pop-up window, type the partial query and name it using a short description. When **OK** is clicked, your query is represented as a distinct rectangular shaded icon with the option to delete it in its top-right corner. You can repeat this process to add multiple constant filters.

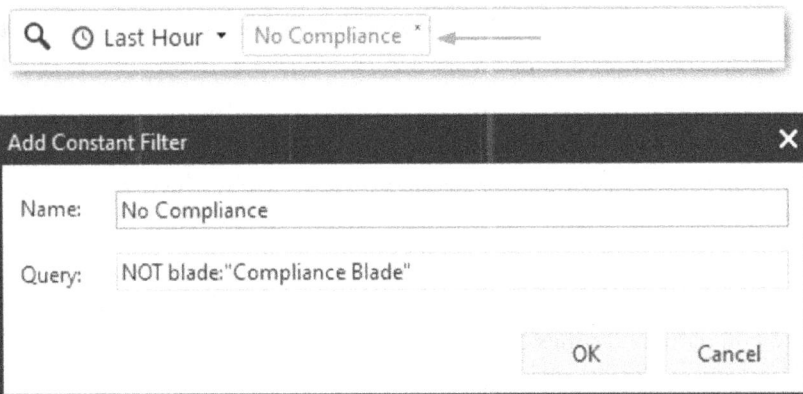

Figure 7.62 – SmartConsole – LOGS & MONITOR | Logs, Logs, Queries, Constant Filter

You may do the same with any existing query by selecting it and using the right-click menu to add that query to **Constant Filter**.

- **Columns headers [6]**: Columns headers reflect the current columns profile and, when right-clicked, present you with the options to hide the chosen column, choose a different profile, edit the current one, or create a custom profile. It is set by default to **Automatic Profile Selection** and I recommend leaving it in place until you are fluent with the product.

- **Logs options [7]**: Logs options, when clicked, allows access to a cascading menu shown in the following screenshot, allowing us to open an archived log file, change the view in **Results Pane** to a more compact **Table** format, disable host-to-IP address resolution (enabled by default), hide the identities of users (when working with support), or spawn the new tab with the same query or a new window outside of the parent SmartConsole application (useful when working on policy and looking at the results side by side).

The **Custom Commands** option opens a pop-up window where you can define external tools (such as the Windows `tracert` command), which could be executed from within **Logs Results**, passing the value of the field you are mousing over as a parameter. A **Create an API logs query** option is also available for the developmentally inclined.

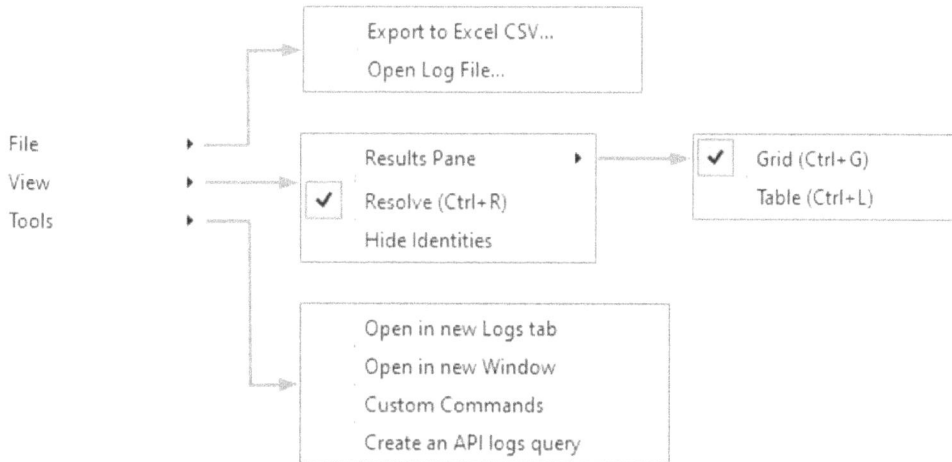

Figure 7.63 – SmartConsole – LOGS & MONITOR | Logs, Logs, Options

- **Filters pane [8]**: The filters pane, when toggled, opens **Tops** and the **Log Servers** pane, shown in the following screenshot. The **Log Servers** tab is available only if multiple management and/or log servers are present. **Tops** is a *global* statistical output of the top sources, destinations, services, and so on, for the defined time interval. The **Log Servers** tab lists all the log servers you are querying. In cases where log servers are deployed in different geographies and only serve local gateways, we can uncheck them for the duration of the session. If this is done, the **Origin** column of logs will contain only those log servers that remain selected.

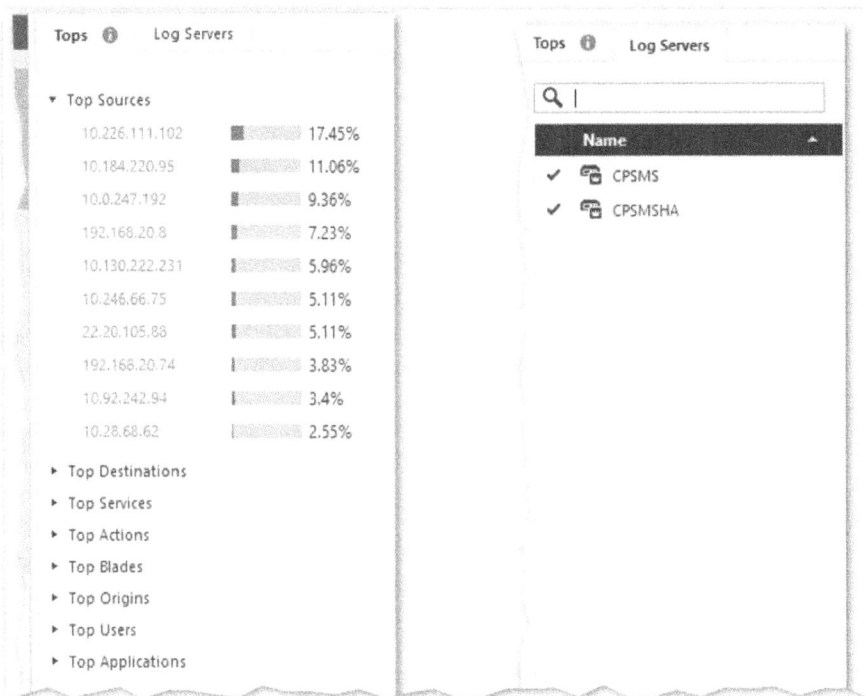

Figure 7.64 – SmartConsole – LOGS & MONITOR | Logs, Logs, Filters Pane

- **Undock [9]**: Undock does the same thing as the **Open in new Window** option for use either in a widescreen or a multi-monitor setup, or to share a specific logs window in a conference setting.

- **Results pane [10]**: The results pane is where logs are displayed. When you have located the log entry of interest, you can double-click it to open a **Log Details** window, a compact representation of all relevant data with links allowing rapid navigation jumps to the rules and profiles, as well as the creation of exemptions and external references, such as vulnerability and remediation data, when applicable. Note that the **Log Details** window always contains all relevant information irrespective of which **Columns Profile** is selected.

The tree icons in the top-right corner [a] allow us to move up and down between filtered logs while remaining in the **Log Details** window and to copy a text version of the displayed log to the clipboard. Additional tabs [b] are present when relevant. The **Matched Rules** tab will list all the rules in all layers of the policy that processed the traffic that generated this log entry.

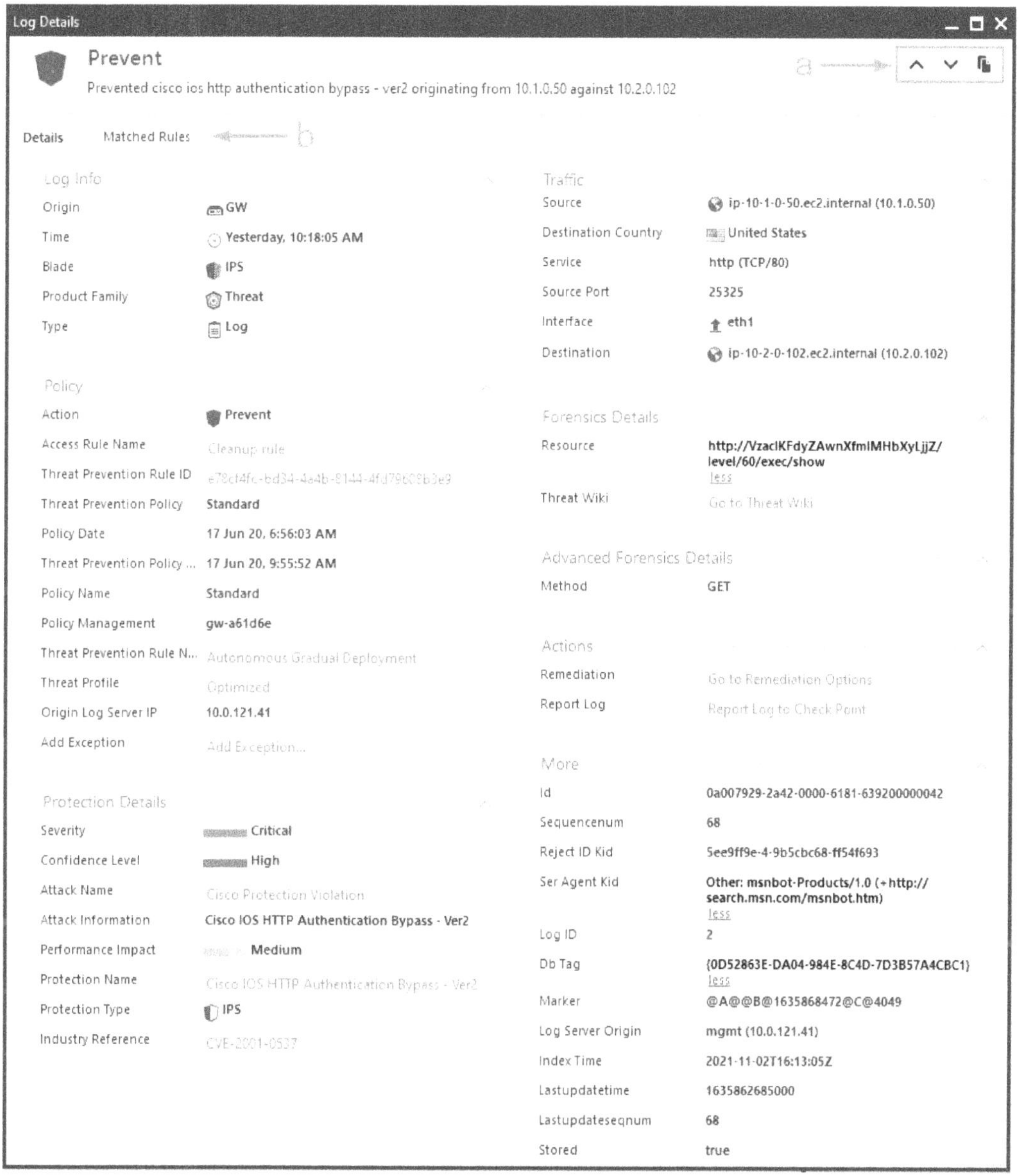

Figure 7.65 – SmartConsole – LOGS & MONITOR | Logs, Logs, Log Details

Within the results pane, the right-click action menu is **Blade**- (feature) and field-specific. For instance, as shown in the following screenshot, right-clicking on the **URL Filtering** icon in the log's **Blade** field [1] gives us the ability to **Add to Filter** [2] only **URL Filtering**-specific parameters. Right-clicking on the object in the **Source** field of the log [3], followed by **Add to Filter** [4], shows the IP address of the host. The additional option of **Create Host** [5] is now available and the **Actions** option [6] now lists the tools that could be invoked using the IP address as a parameter.

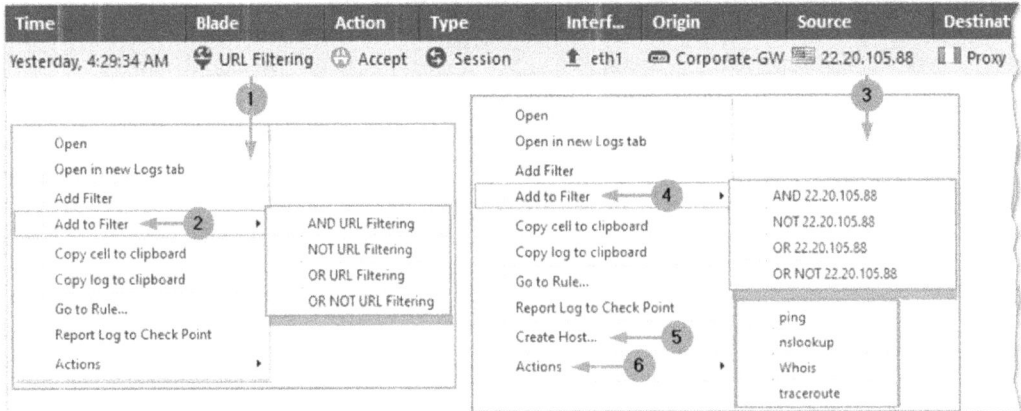

Figure 7.66 – SmartConsole – LOGS & MONITOR | Logs, Logs, Actions – Example 1

Right-clicking on the **Blade** field in the **IPS** log [7] presents us with the options of **Open Protection** [8], **Add Exception** [9], and **Go To Advisory** [10].

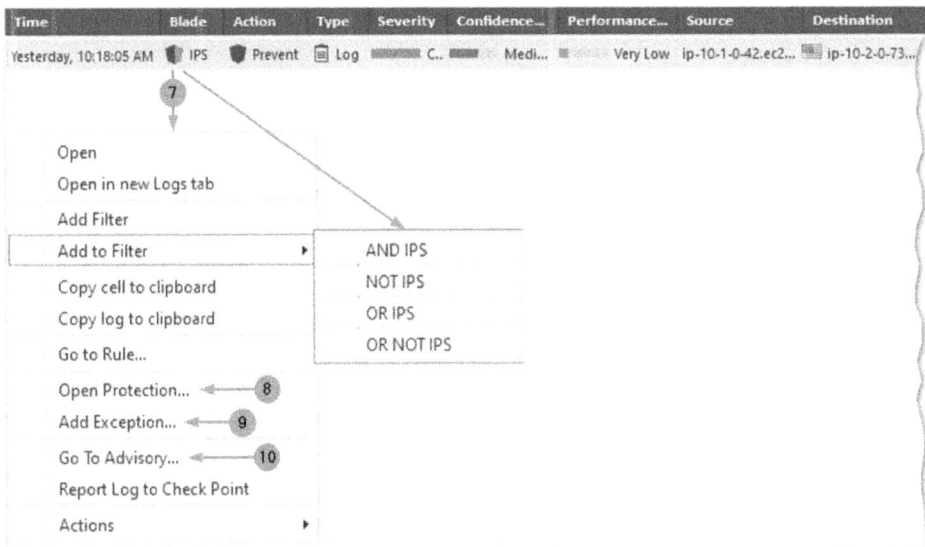

Figure 7.67 – SmartConsole – LOGS & MONITOR | Logs, Logs, Actions – Example 2

- **Log-specific tabs section** [11]: Log-specific tabs may contain additional information for each log entry, such as **Connections**, **URLs**, and **Files**, if sufficient depth of tracking is configured in the rule that produced the log.

MANAGE & SETTINGS [D]

This is, essentially, a collection of administrative tools and settings.

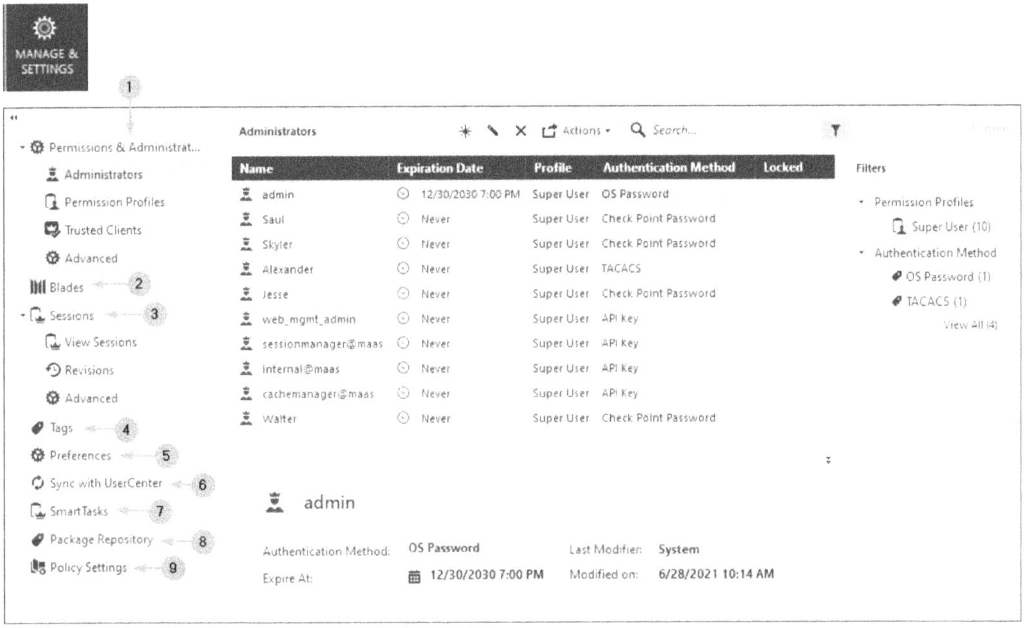

Figure 7.68 – SmartConsole – MANAGE & SETTINGS

This view contains a number of items in its navigation tree and we will go over the numbered items in the preceding screenshot to get a general idea of what they represent:

- **Permissions & Administrators** [1] is comprised of:

  - **Administrators**, where accounts are created or edited, their permission profiles assigned, and the authentication methods are defined.

  - **Permission Profiles**, containing predefined profiles and where additional profiles with granular administrative permissions can be created.

  - **Trusted Clients**, where Check Point administration sources could be restricted to IPs and networks.

- **Advanced**, where the defaults can be configured for authentication methods, account expiration, minimum password length and handling, SmartConsole idle session timeout, and login restrictions (the number of attempts before being locked out and its duration).

- **Blades** [2], where general product settings, as well as those of individual features, can be adjusted. In addition to **General**, which contains **Global Properties** and **Inspection Settings**, settings for the following blades can be adjusted from within this section of SmartConsole: **Application Control & URL Filtering**, **Content Awareness**, **Threat Prevention**, **Compliance**, and **Management API**.

  Data Loss Prevention, **Mobile Access**, **Anti-Spam & Mail**, and **HTTPS Inspection** settings can be configured in SmartDashboard, a legacy interface that will be launched automatically.

- **Sessions** [3], containing:

  - **View Sessions**, which, when selected, presents us with the list of administrative sessions and their state, showing the number of objects locked for editing, the number of pending changes, and the date/time the owner of the session has logged in and out. From this view, with the target session selected [1], and using the **Actions** menu [2] (also accessible by right-clicking), we may **Publish & Disconnect** [3], **Discard & Disconnect** [4], **Disconnect**, **Take over** [6], or review the **Changes** report [7] if our permissions profile allows it.

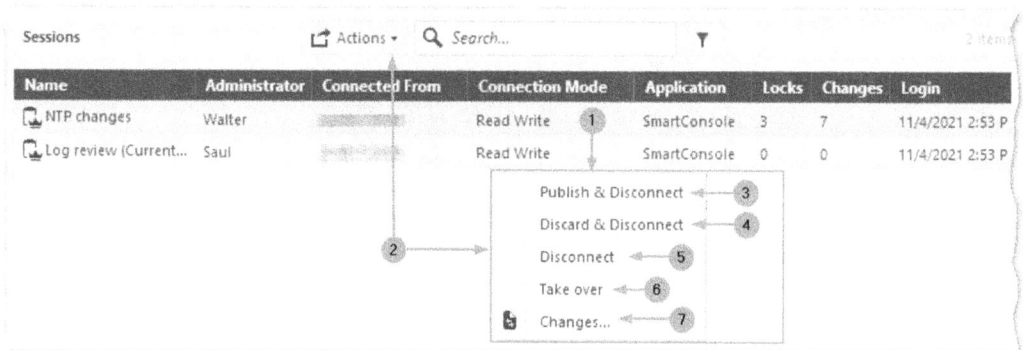

Figure 7.69 – SmartConsole – MANAGE & SETTINGS | Sessions, View Sessions

- **Revisions,** which allows us to rapidly roll back to a previous known good state, discarding all the changes introduced after the selected revision. To determine the scope of the changes we'll be discarding, select the target revision [1] and then, from its **Actions** menu [2], choose **Changes** [3] and the **Compare selected with current session** [4] option to generate a change report. Since this report covers only the policies and common objects, check **Audit Logs** [5] to verify that no system configuration changes apart from those in the report have taken place, or that it is safe to discard them.

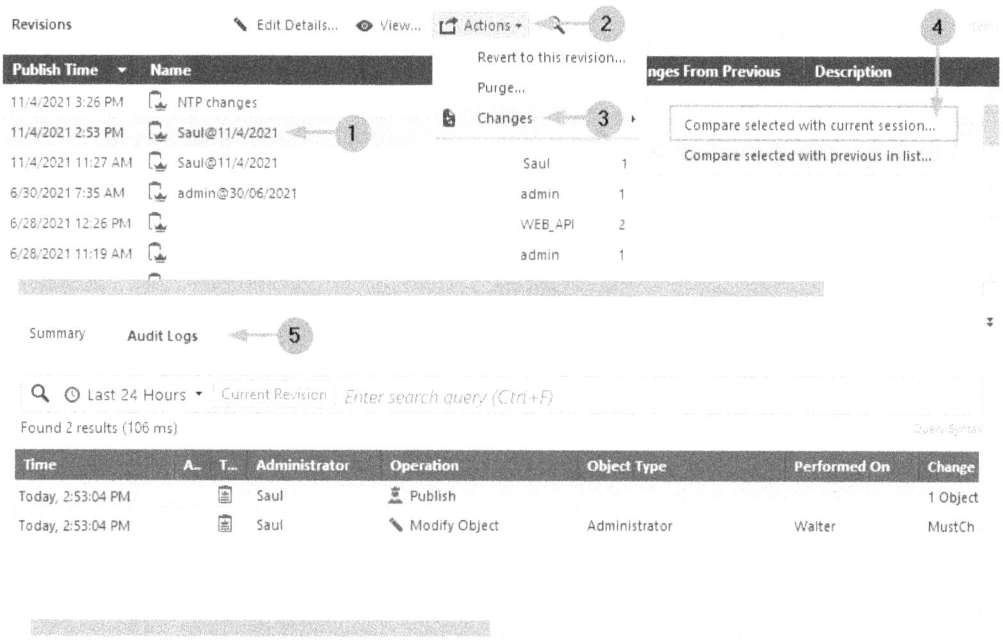

Figure 7.70 – SmartConsole – MANAGE & SETTINGS | Sessions, View Sessions

- **Advanced,** where the session settings allow each administrator to have a single SmartConsole session in this environment or to allow multiple concurrent sessions. We may also configure the format of automatically generated session names and can enforce the mandatory creation of session descriptions.

- **Tags** [4], used for centralized search tags review creation, and editing. Although tags can be created by any administrator with the rights to modify objects, this is the place where we can determine their creators and the most recent editors.

- **Preferences** [5], used for the following purposes:

  - Choose IP address versions (IPv4 and/or IPv6), the subnet display format (mask length or subnet mask), and network object creation based on the route refresh interval.

  - Specify the SmartConsole custom login message and logo.

  - Reset dismissed warnings, and enable debugging of the application.

  - Add SmartConsole extensions.

  - Choose the language.

  - Enable a high contrast theme.

  - Enable Check Point's lab features (where we have enabled the **Sessions** pane earlier in this chapter, in the *Objects, Validations, and Sessions Panes* section).

- **Sync with UserCenter** [6], for dynamic inventory management, proactive support, and, when relevant, a more efficient ticket resolution.

- **SmartTasks** [7], where trigger- and action-based automation could be configured to either run custom scripts or to send custom web requests based on four predefined triggers.

- **Package Repository** [8], where hotfixes and upgrade packages could be uploaded for subsequent centralized installation on gateways and clusters.

- **Policy Settings** [9], where the rule base cell settings for empty source, destination, and services and applications are defined. By default, when a new rule is created, these cells contain the value **Any** along with the **Drop** as default action. Here, you can replace **Any** with **None**, if so desired. We can also define the behavior of the system when **None** is encountered in the rule base, such as generating warnings, preventing publication, or doing nothing. Using **None** instead of **Any** in a highly sensitive environment ensures that accidental removal of the last object from the rule will not cause overly permissive access by changing the empty field to **Any**.

This concludes the overview of the SmartConsole features and navigation. There is certainly more to it than what can be addressed in a single chapter and, for that reason, I highly recommend downloading Check Point's offline version of the HTML-based SmartConsole R81.10 Help package from `https://supportcontent.checkpoint.com/documentation_download?ID=114606`.

Summary

In this chapter, we have covered the components, capabilities, navigation, and variety of administration options available in a SmartConsole. It should be easy for you to recognize the components of views, invoke and take advantage of context-specific action menus, searches, filters, and logs. You are now aware of the flexibility this interface is offering and are better equipped to achieve your administrative tasks using it. For those only beginning to work with the product, Cloud Demo (or Demo Mode) provides immediate access to a safe modeling environment that I highly encourage you to take advantage of. As SmartConsole itself is self-updatable, I recommend spending some time perusing the **What's New** section to quickly get up to speed with the latest advances in its continuous evolution.

In the next chapter, we will take a look at policy packages, layers, and layouts, as well as some examples of how those could be used to better address the security requirements of different environments.

8
Introduction to Policies, Layers, and Rules

In this chapter, we will learn about policy packages, blades used in Access Control policies, and their use in layers. We will look at the possible policy organization methods, rules' structure and capabilities, and their placement based on the packet flows and use of acceleration technology.

In this chapter, we are going to cover the following main topics:

- Access Control policies, layers, and rules
- Packet flows and acceleration
- Best practices for Access Control rules
- Application Control and URL Filtering layer structure
- Logs, tracking depth, and oddities

Access Control policies, layers, and rules

In the previous chapters, we learned how to navigate SmartConsole. Now, let's take a look at Access Control policies, layers, and rules.

Policies

Policies comprise layers containing rules. The basic Access Control policy is, itself, a single layer.

If a policy consists of more than one layer, you may think of the top layer as a collection of coarse filters and subsequent layers as finer filters.

Policies can be organized into policy (or ordered) layers and/or inline layers.

Access Control policies comprise layers with select features. Using **Menu | Manage policies and layers…** [1], policies [2] can be created [3] or, through the **Actions** menu [4], cloned [5], and the resultant clones renamed:

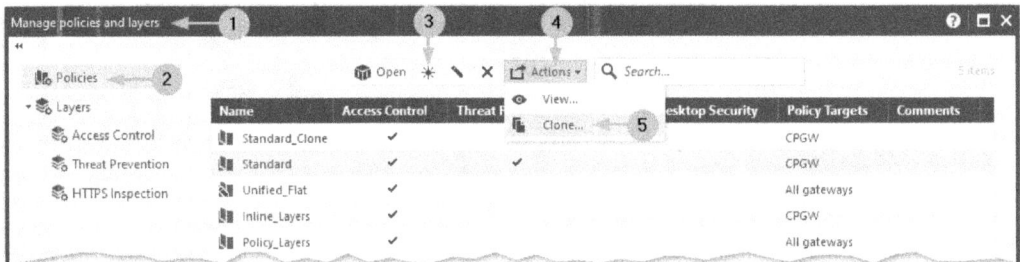

Figure 8.1 – Policy and layer management, creating or cloning policies

Cloning is convenient if the redesign of the security policy is required. Using cloned policies allows for a fast fallback by reinstalling the original policy, in case of unforeseen complications.

> **Note**
> While policy changes can be rolled back by loading the original policy, any changes made to the objects since the clone was made will persist.

The default Access Control policy, called **Standard** [1], is created automatically and contains a single layer with a firewall [2] feature enabled:

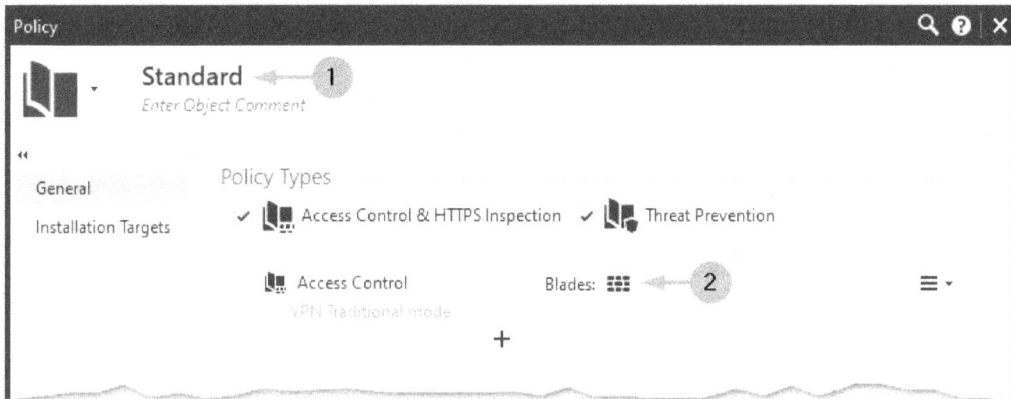

Figure 8.2 – Default Standard policy

Multiple policies and a common object database

When deciding on the policies that will be used in your organization, for example, a single policy (applicable to all gateways), or a few policies (with separate policies for data centers, headquarters, and branches), or a unique policy for each location, keep in mind that, unless you are working with multi-domain management, you are relying on a common object database. Changes to objects that are published and installed on one of the **installation targets** (discussed later in this chapter) will remain pending for all others until policies are installed on them. In cases where policies are routinely installed on some of the gateways and only seldomly on others, you may find a backlog of changes made by different administrators in multiple sessions [1] since the last policy installation date [2] on specific gateways, as shown here:

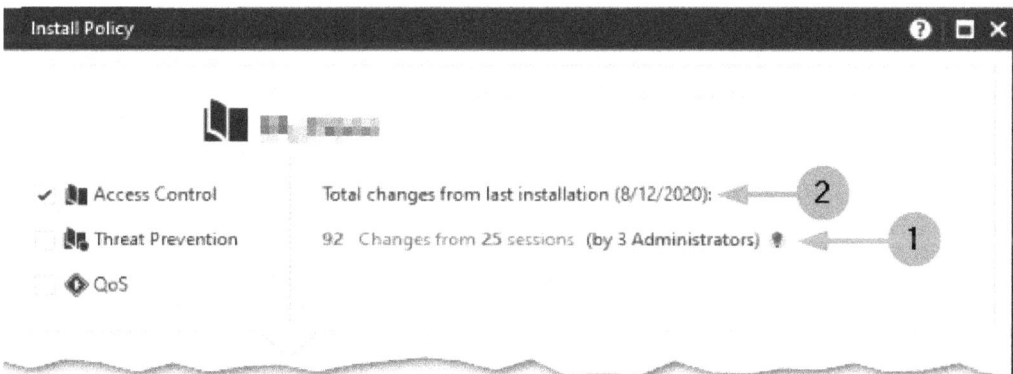

Figure 8.3 – Changes lag in multi-policy implementations

To avoid this situation, either try to schedule periodic policy installations across your infrastructure or install all policies whenever changes are made to any of the objects. Otherwise, you may have to review **Audit Logs** for these changes before installing a lagging policy.

Layers

Layers can be used in two different modes: policy (ordered) layers and inline layers. Layers can be shared and reused in multiple rules or policies.

Ordered layers

Ordered layers are executed sequentially. Traffic accepted by any rule in the top (or a higher) layer is then matched against the rules of the next layer. Additional ordered layers are created by using the + button in the **Access Control** section of the policy editor [1] and choosing **New Layer…** [2] in the layer browser:

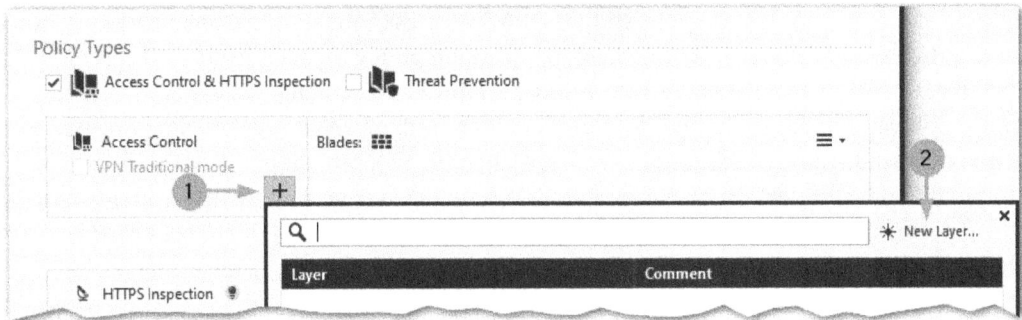

Figure 8.4 – Creating a new policy layer

Once **Layer Editor** is open [1], name the new layer according to its intended application [2] and choose the blades (functions) that it will be responsible for [3]:

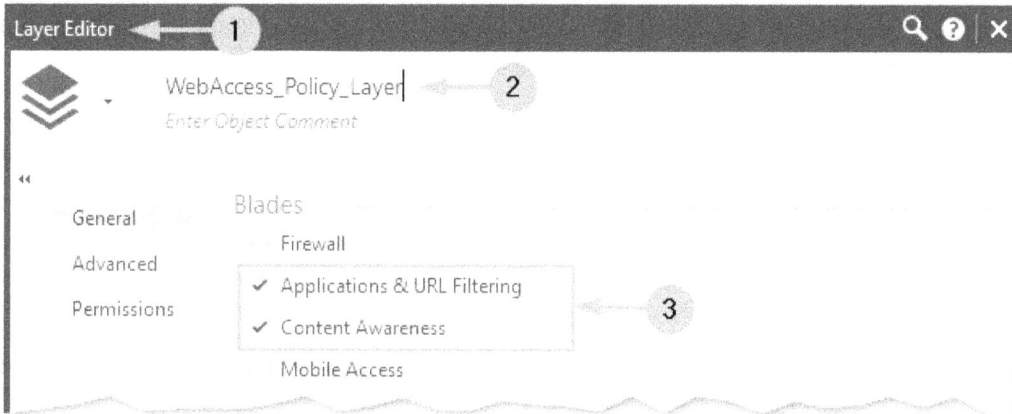

Figure 8.5 – Defining the new layer's functions in Layer Editor

In the **Policy** editor [1], we see the second layer containing the selected features [2]:

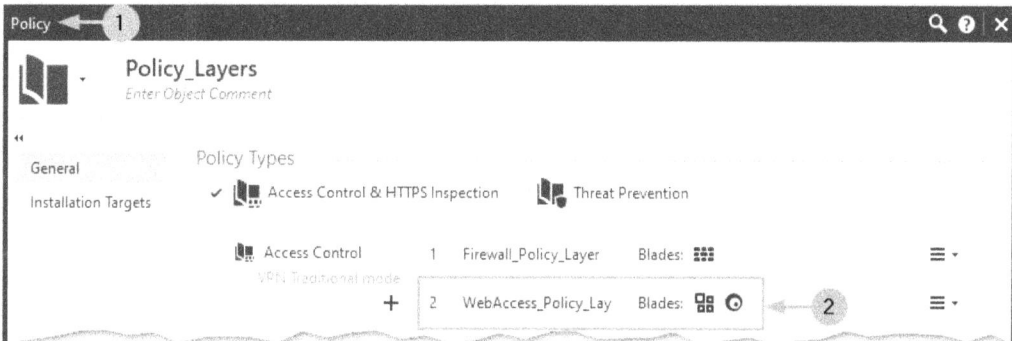

Figure 8.6 – Ordered layers view in the Policy editor

When the **Policy** change depicted in *Figure 8.6* is confirmed by clicking the **OK** button (not shown), the resultant policy structure is visible in the view tree:

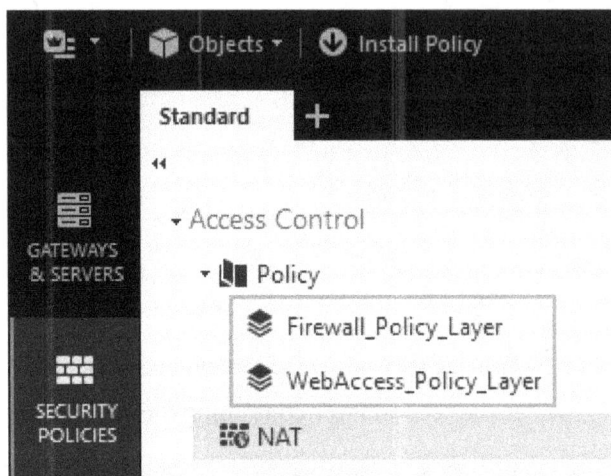

Figure 8.7 – Ordered layers view in SECURITY POLICIES

By default, when the first additional ordered layer is added to the policy, the top layer is automatically named **Network**. I suggest renaming it to reflect the layer's purpose and enabled blades.

Inline layers

Inline layers are executed on traffic matching a rule that contains a layer in its **Action** column. When creating a new inline layer, it is best to set the new section title in which it'll be contained [1]. Then, in a rule that should contain more specific filters, click the drop-down arrow in the **Action** column [2], click on **Inline Layer** [3], and **New Layer…** [4]:

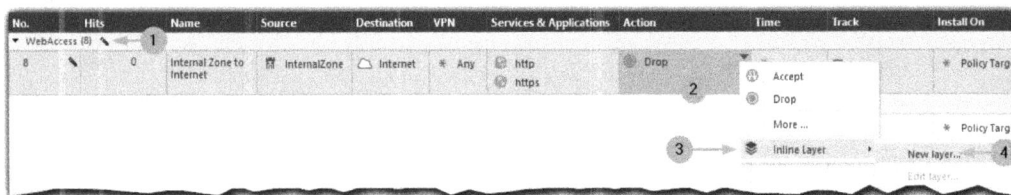

Figure 8.8 – Creating an inline layer

You'll be greeted with the same Layer Editor as shown in *Figure 8.5*. Name the layer appropriately and select the blades you need.

The typical use of additional layers, either policy (ordered) or inline, is to add **Application Control (APCL)/URL Filtering (URLF)**, as well as **Content Awareness** capabilities.

Note that layers with APCL/URLF and Content Awareness have inherent firewall functionality, and there is no need to explicitly select the **Firewall** option when these layers are created. This allows the use of sources, destinations, and services in rules contained in those layers.

Dedicated layer administration

Layers can have individual permissions assigned to them, allowing for delegation of administration. This is accomplished by navigating to **MANAGE & SETTINGS** [1], expanding **Permissions & Administrators** [2], selecting **Permissions Profiles** [3], and clicking on the *create new* icon [4]. Then, in a pop-up window, name a new profile to reflect its purpose [5], click **Access Control** [6], and select **Edit layers by the selected profiles in a layer editor** [7]:

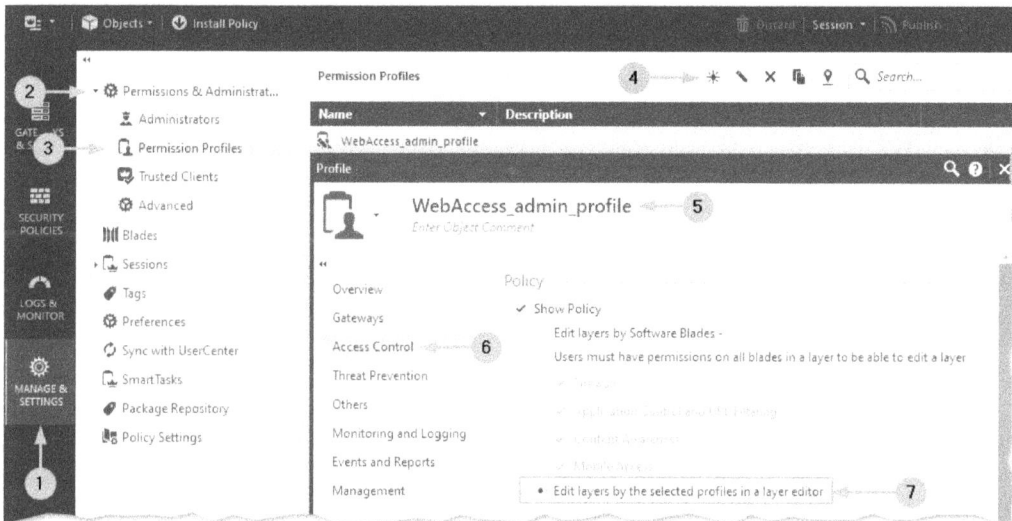

Figure 8.9 – Creating layer-specific permission profiles

Once this is done, complete the profile configuration by clicking **OK**.

If policy layers are used, as in the **TestPolicy** screenshot shown here [1], choose the layer in question [2], right-click it, and click **Edit Policy…** [3]:

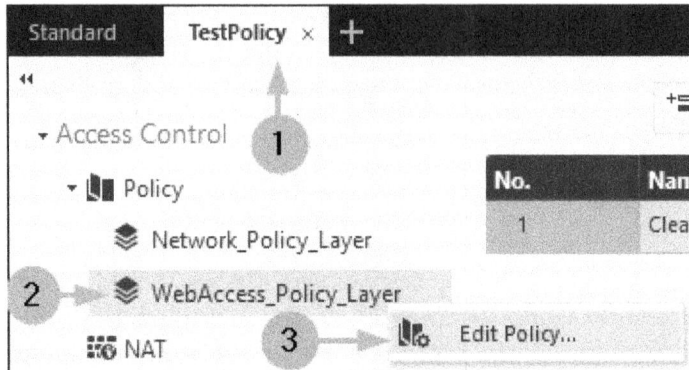

Figure 8.10 – Editing policy layer properties

In a selected policy's pop-up window [1], click on the options menu of the ordered layer [2] and click **Edit Layer…** [3]:

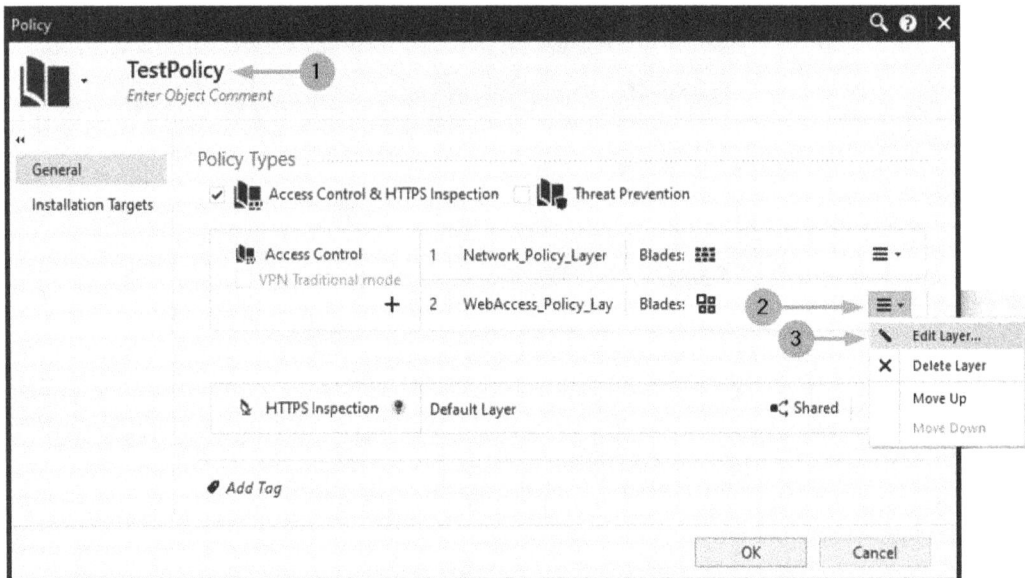

Figure 8.11 – Invoking the layer editor for the policy layer

For inline layer(s), in the **Action** column of the unified policy [1], right-click the *layers* icon [2], click on **Inline Layer** [3], and then on **Edit layer...** [4]:

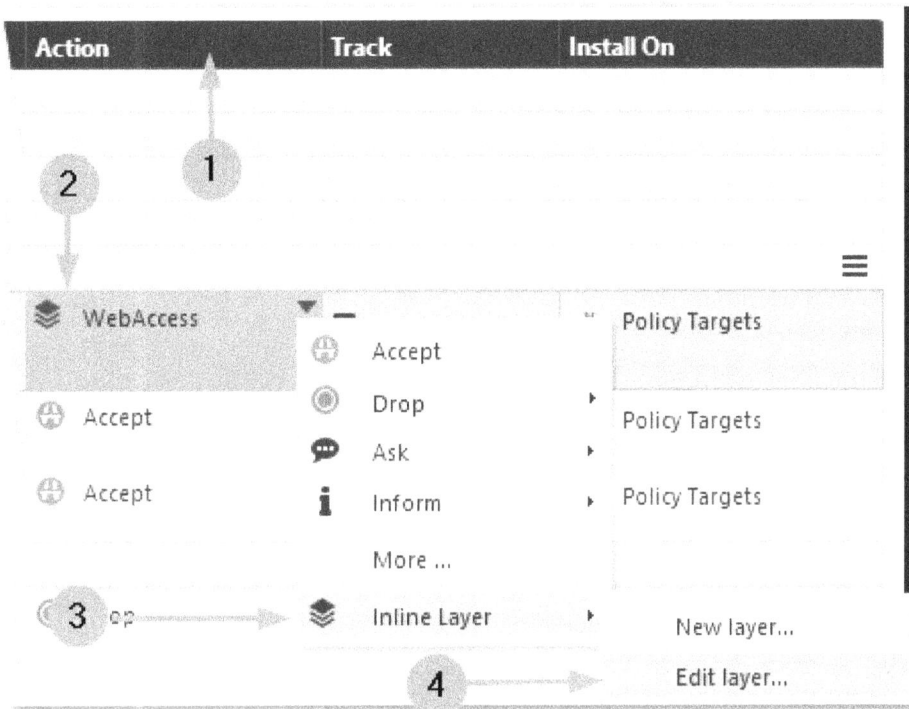

Figure 8.12 – Invoking the layer editor for Inline Layer

In either the ordered layer or inline layer cases, an identical **Layer Editor** window is opened displaying the selected layer's name [1]. Click on **Permissions** on the left [2] and, in the bottom portion of the screen under **Select additional profiles that will be able to edit this layer**, click + [3], and select the profile we created earlier in *Figure 8.9* [4]. Click **OK** in this and the parent window to confirm the change:

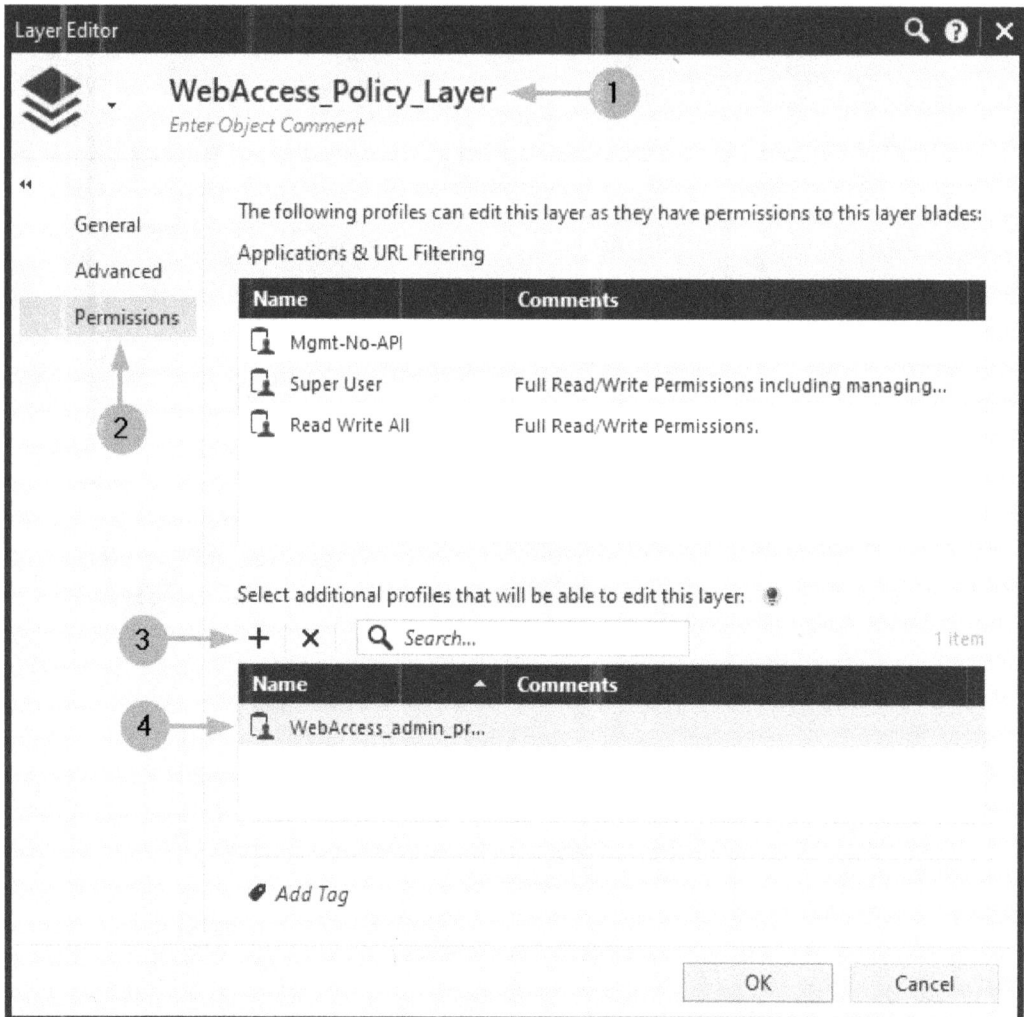

Figure 8.13 – Assigning dedicated layer administrators

Administrators with selected profiles will then be able to create or modify rules, have limited object type editing capabilities, and publish the changes for this layer.

> **Note**
>
> Starting with version R81.20, additional permission profiles allowing administrators to submit changes for review, acceptance, and publishing by others are available.

Attention! Layers cannot be installed individually, but only as part of the overall Access Control policy. When a permission profile is assigned to an administrator for a layer allowing policy installation, all published changes by all other administrators outside of that layer will take effect as soon as the policy is installed.

Access Control policy structure preferences

Ordered layers, when used without inline layers, are a less efficient way to organize policies. They are most frequently encountered when the current Check Point environment is upgraded from earlier versions, where APCL/URLF was a function of a separate policy. It makes sense to maintain those if the resultant APCL/URLF layer contains a large number of rules. You are also limited to the use of two ordered layers if the policy targets are gateways running versions R77.30 or earlier. In this case, the first ordered layer can only be Firewall/VPN, and the second, APCL/URLF. The same applies to the SMB gateways running R77.20, the last version of R77, for that category of appliances.

Inline layers are more suitable for new environments, or where a dedicated ordered APCL/URLF layer contains a small number of rules that could easily be migrated to inline layers. An added advantage of inline layers is more granular administration, as individual inline layers containing the same blades may have their own permission assigned to a dedicated administrator or a group.

In large companies, this may allow country-, office-, or department-specific administration of inline layers by local admins while the overarching common security policy section is controlled from headquarters. Another advantage of inline layers is consolidated logging; as parent rules of inline layers do not have tracking options, log entries are created when traffic is processed by the nested rules. In ordered layers, rules in each layer are typically logged, so more logs are generated, impacting log storage and search efficiency.

Ordered and inline layers can be combined into a single policy. One example is the top ordered layer containing common firewall rules for gateways at all locations with inline firewall layers specific to particular sites. The second ordered layer may contain multiple inline layers, each configured with specific sites' APCL/URLF rules.

While you have the capacity to create complex policies, I strongly suggest keeping them as simple as possible.

Shared layers

Layers can be designated as **shared**. This is accomplished in **Layer Editor** [1] by checking the **Multiple policies and rules can use this layer** checkbox [2] under **Sharing**:

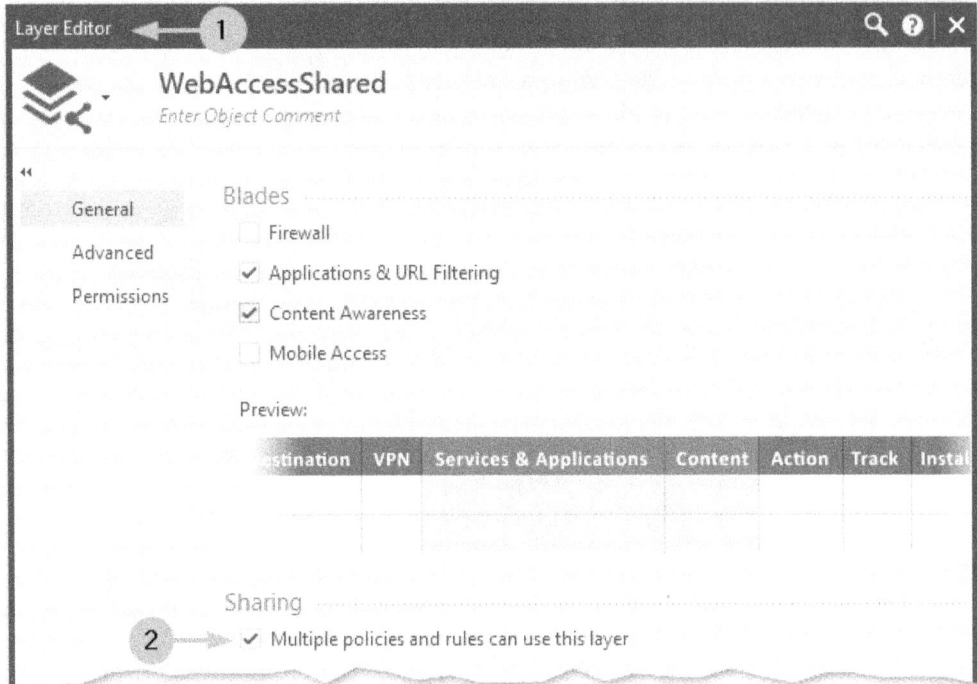

Figure 8.14 – Shared layers

This allows reuse of the same layer in the following:

- **Multiple policies**: For instance, having a common APCL/URLF and Content Awareness layer applied to different sites in your organization, each governed by an individual policy

- **Rules in the same policy**: Such as applying common URLF capabilities to different departments

A policy containing shared layers can be installed on a specific target gateway or group without immediately affecting the rest of the gateways in your infrastructure. That said, see this caveat: If the changes are made in a shared layer, but the policies containing it are installed only on select gateways, it is not immediately apparent which gateways are enforcing the latest updated policy. In this case, you can use the **Install Policy** action to see if any of the changes are lagging for a specific policy. This is shown on the policy installation confirmation screen, and you can safely cancel out of it.

Rules

Let's look at the three types of rules used in the Access Control policies: **explicit**, **implicit**, and **implied**.

Explicit rules

Rules that we can define and see in the policies and layers are called **explicit rules**.

In addition to explicit rules, there are normally invisible rules called implicit and implied rules.

Implicit rules

Implicit rules are the invisible last rules (also known as **Implicit Cleanup Action**) in layers. Implicit rules are there to determine how the traffic is treated if it did not match any of the explicit rules in a layer. The behavior of implicit rules for individual layers is defined in **Layer Editor** [1] for a select layer, under **Advanced** [2], then the **Implicit Cleanup Action** [3] properties:

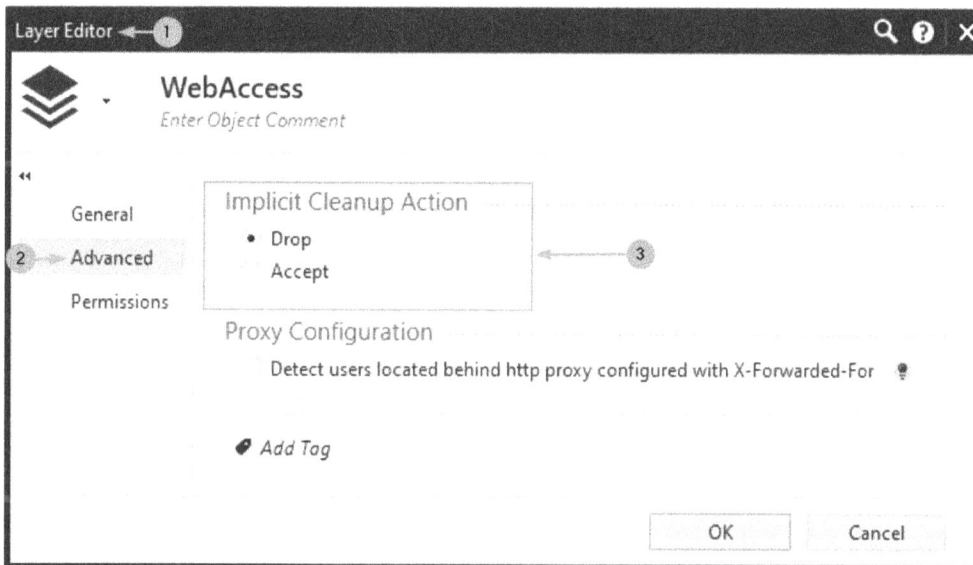

Figure 8.15 – Implicit rules (Implicit Cleanup Action)

> **Important Note**
> When constructing your layers, always create an explicit cleanup rule with the action identical to that of **Implicit Cleanup Action** and configure relevant tracking settings.

Implied rules

Implied rules are created in the Rule Base as a section of Global Properties. They are not editable but can be selectively enabled or disabled in **Menu | Global Properties | FireWall**. These rules are preconfigured to allow connectivity for a variety of services used by gateways, such as connectivity to AAA servers, and log transfers to management servers, for example:

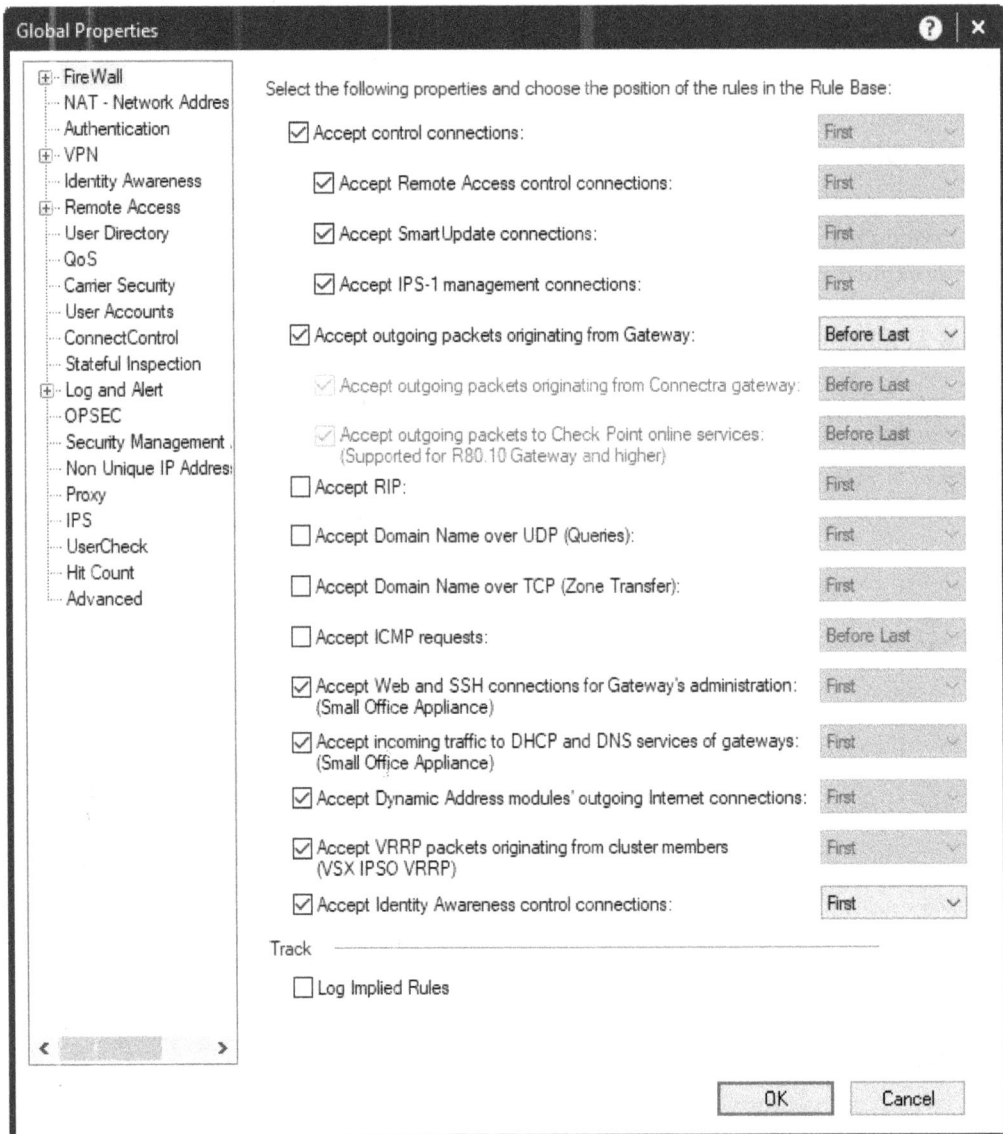

Figure 8.16 – Implied rules configuration

Or, as actual rules, in the properties of the Access Control policy via **Actions | Implied Rules**:

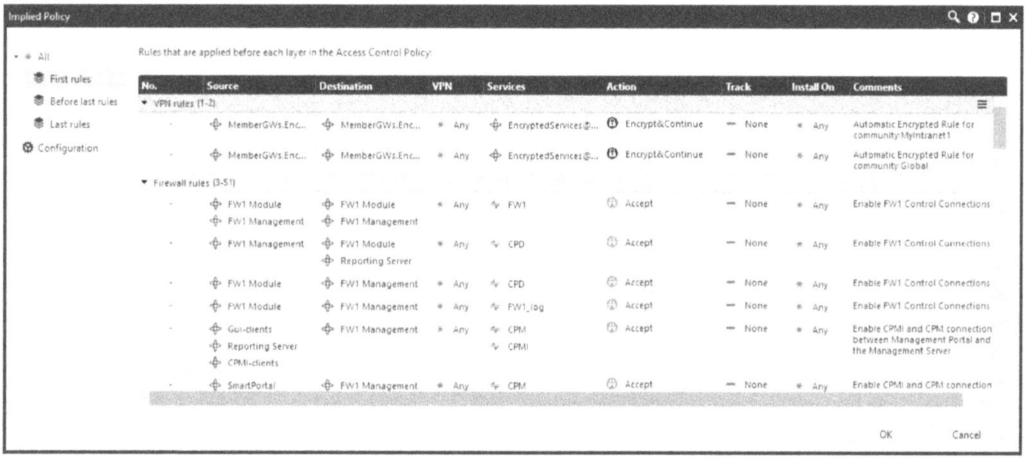

Figure 8.17 – Implied rules view

Important Note

Although it looks like these are the individual policy's implied rules settings (accessible via the policy's **Actions | Implied Policy | Configuration**), these are global parameters and will override the settings found in **Global Toolbar | Menu | Global Properties**. Not all the implied rules are enabled by default. You may disable those that are not required in your environment.

Rule processing order

The overall order of rule processing looks like this:

1. Implied first (applied before explicit rules)

2. Explicit rules (all except last)

3. Implied before last (applied before the last explicit rule)

4. Explicit last (typically, it is *any source to any destination, drop, and log*)

5. Implied last (applied after the last explicit rule, typically empty, as explicit last rules drop all packets)

6. Implicit cleanup action (by default, it is *any source to any destination, drop, and do not track*)

By default, implied and implicit rules are not logged to avoid excessive telemetry. It is possible to enable logging for all implied rules all at once (globally, for all your policies), when it is useful for the duration of troubleshooting sessions, or if you have a specific requirement to log.

Important Note

If you are enabling logging of the implied rules for troubleshooting, set a reminder to disable logging when your troubleshooting session is complete, publish changes, and install policies.

I should also point out that Check Point's own documentation occasionally swaps implied and implicit rule names for no apparent reason. The definitions used here are based on the list of rules displayed by using **Show Implied Rules** in the policies' **Action** menu.

The combination of implied, implicit, and explicit rules is referred to as a Rule Base.

Access Control Rule Base components

Let's go over the components of the Access Control Rule Base by looking at the column headers and corresponding content or functionality available in each field:

Column	Description
No.	Rule numbers. For inline layers, each level is separated by a period (for example, 3.7 is the seventh rule in rule number 3 of the parent layer. Number values can change if additional rules are introduced above the current one.
Hits	The number of times connections match the rule.
Name	An arbitrary name assigned by admins. I suggest keeping those uniform and reflecting their purpose (for example, `Internal networks to Internet Accept`).
Source	Static, dynamic, or composite network objects that define where the traffic originated from or is destined for.
Destination	
VPN	VPN communities to which this rule applies.
Application & Sites	Services, Categories, and Applications & Sites. If the Firewall-only blade is enabled, only the Services options are available and are matched on the port numbers only. To match on a protocol signature, APCL/URLF must be enabled in the layer. Mobile applications if the Mobile Access blade is enabled.

Column	Description
Content	Bi- or uni-directional control of the file types or data types (for example, you can block or accept the download of executable files or allow the upload and download of PDFs). The data type allows for the same actions to be enforced on the content of the files that are being transferred, such as credit card numbers or medical records. Note that this capability is limited to a small number of services and is not intended as a comprehensive DLP (Data Loss Prevention) solution.
Action	Either the Inline Layer object or a response of the firewall when the rule is matched. Could be either a simple action, such as Accept, Drop, or Reject, or with interactive UserCheck messages, such as Ask or Inform. UserCheck should be used in APCL/URLF rules only if relying on seeing these messages in a browser. Could be used in other rules if client PCs have the UserCheck client application installed. Avoid using Reject unless warranted. Bandwidth limit per rule for uploads, downloads, or both. Custom UserCheck and Limit objects could be created. Enable Identity Captive Portal to redirect HTTP(S) traffic to an authentication portal for user authentication before further connections from the same source are processed.
Time	A single or a recurring time frame when the rule is active.
Track	Logging depth (Log, Detailed Log, or Extended Log). Alerts (pop-up, Simple Network Management Protocol (SNMP), email, or user-defined scripts). Types (per connections, per sessions, or firewall sessions). Accounting (to track data transfers in connections, upload and download bytes, and browsing time).
Install On	Gateway(s) on which the current rule is enforced (if defined) or a default value of Policy Installation Targets (specified in the policy's properties).
Comment	Could be used for a longer description of the purposes of the rule or additional notations.

Table 8.1 – Columns of the Access Control Rule Base

Note that not all these columns are available by default. The **Content** column becomes visible only after the **Content Awareness** blade is selected in the layer's properties, and several others could be made visible by right-clicking on the column header and checking the empty boxes.

Now, just because all of these fields and options are available, it does not necessarily mean that all of them should be used in a single layer. Read on to learn about the packet flows and the reasons for using multiple layers.

Packet flows and acceleration

Given today's threat landscape and an ever-increasing emphasis on security, you would think that every packet traversing, entering, or leaving our network must be inspected with prejudice before being allowed to pass. This, clearly, is not the case, as doing that would simply use up unnecessary CPU cycles and negatively impact performance, while not improving security in the least.

Check Point utilizes the following proprietary technologies to achieve the best possible traffic processing speed for its firewalls:

- **CoreXL** is the load balancing mechanism that assigns CPU cores for either acceleration and packet routing duties, **Secure Network Distributor** (**SND**), or for deep packet inspection, **FireWall Worker** (**fwk**).

- **SecureXL** is the traffic routing and acceleration mechanism that increases throughput and connection rate using either software or dedicated acceleration cards (SecureXL instances) and chooses the fwk processes that will handle deep packet inspection for a given connection (Dynamic Dispatcher).

- **Multi-Queue** is the mechanism that designates multiple CPU cores to process high-volume traffic arriving at the same network interface for acceleration and packet routing.

- **HyperFlow** is the mechanism allocating multiple cores assigned for deep packet inspection (fwk processes) to individual sessions carrying a massive amount of traffic. This feature is currently in the Early Availability stage and, at the time of writing, it is not known when it'll be released for General Availability.

All these technologies/processes are interacting with each other in real time to dynamically allocate processing resources for traffic flow optimization and to ensure that maximum advantage is taken of the available physical or virtual hardware.

From a security policy organization perspective and rules configuration, we are primarily interested in the SecureXL traffic acceleration mechanism (the term SecureXL is interchangeably applied to a device, driver, acceleration driver, or acceleration layer).

Let's take a look at what it actually does: as connections are being established, state-related information from packets is used to populate and maintain dynamic state tables. For RPC- and UDP-based traffic, virtual sessions are created to track connectionless protocols.

Throughout this process, eligible connections are offloaded to SecureXL connection tables to speed up the processing of subsequent packets and improve throughput. Accept, drop (if enabled), and **network address translation** (**NAT**) templates are created to improve session (connection) rates:

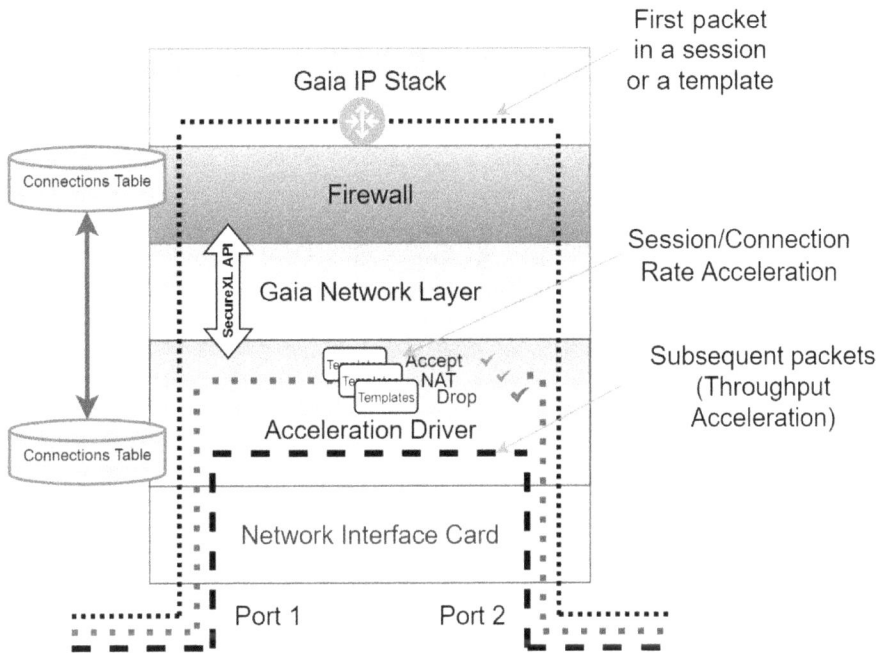

Figure 8.18 – SecureXL process outline

Some or all of the subsequent packets of particular established connections may be handled by the SecureXL acceleration driver's optimized inspection operations.

Accept templates improve the rate at which new connections for the existing sessions are established. They contain source and destination IP addresses, destination port numbers, as well as ingress and egress interfaces. Wildcards are used for the source port numbers. New connections for the existing sessions may be established faster if an Accept template is present.

NAT templates allow accelerated connections to avoid taking a *long* trip through the firewall layer to be subjected to the NAT policy.

Drop templates are disabled by default, as those are based on preconfigured connection rates. If you enable them during a normal firewall operation state, you may run into an unfortunate situation where they will be engaged by the normal spike in your traffic patterns. They are, however, a valuable tool to be used when under **denial-of-service** (**DOS**) attacks.

To determine whether part or all of a session or connection can be accelerated, we must look at the packet flow:

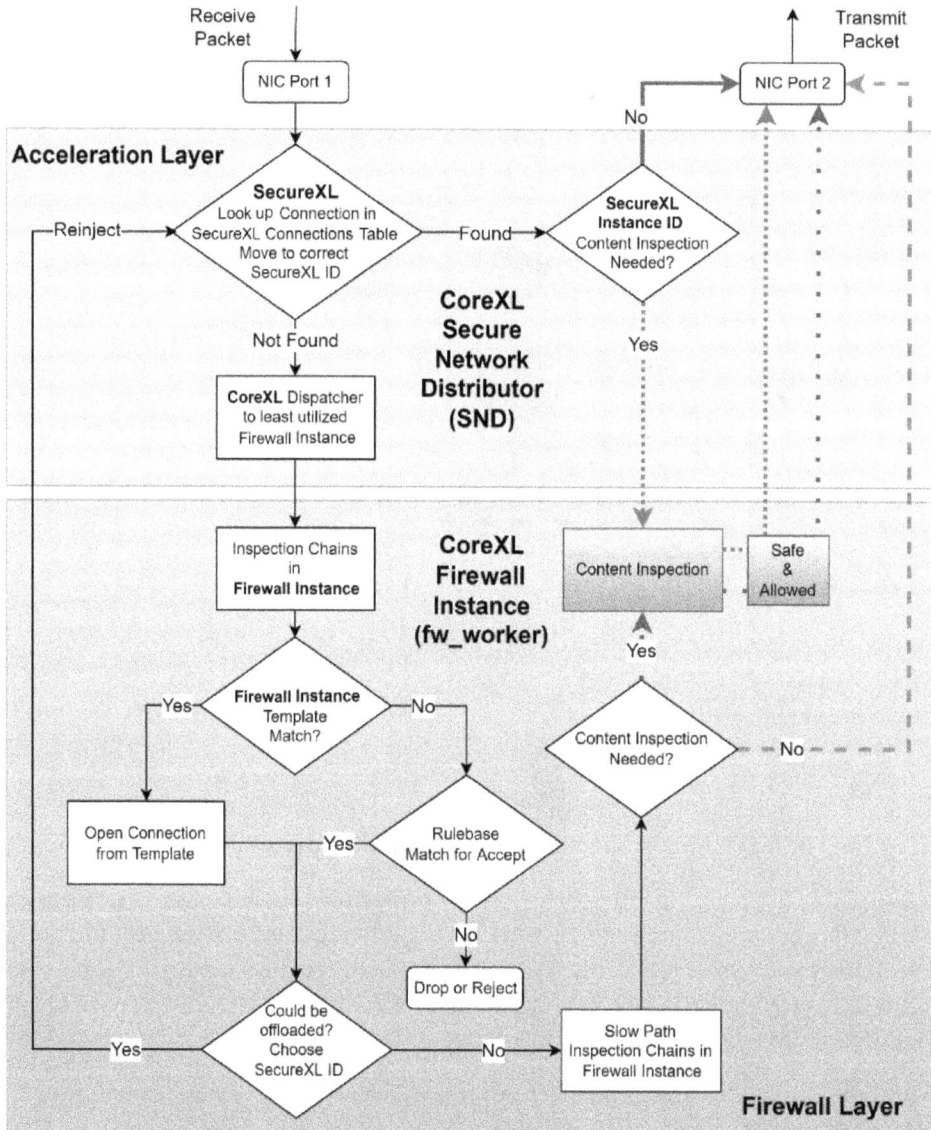

Figure 8.19 – Packet flow general description (simplified)

Note that this decision process takes place in both directions; that is, if the preceding diagram is applied to the first packet of a three-way TCP handshake (SYN), then, for a response (SYN-ACK), mentally swap the NIC Port 1 and NIC Port 2 in place to visualize that. For an eligible connection, SYN-ACK is returned via a fully accelerated path.

> **Note**
>
> In the latest versions of the R81.X releases, the number of available paths has increased to nine and may continue growing as Check Point continuously works on ways to handle more traffic in the acceleration layer. For our purposes, it is immaterial.

Let's expand a bit on the inspection chains and content inspection modules shown in *Figure 8.19*, to see how many hoops the packet may have to jump through if so warranted.

Inspection chains

Inspection chains refer to the sequences of inspection modules the traffic passing through the gateway is subjected to. There are two inspection chains, **in chain** [1] and **out chain** [2], allowing for specific actions to be taken on traffic as it enters or exits the gateway. The number of inspection chain modules will vary depending on the release version and the blades activated on the gateway. If you would like to see those, and to see which ones are part of the acceleration layer [3] or firewall layer [4], use the `fw ctl chain` command in `Expert` mode:

```
[Expert@CPCM1:0]# fw ctl chain
in chain (21):
     0: -7fffffff (0000000000000000) (00000000) SecureXL stateless check (sxl_state_check)
     1: -7ffffffe (0000000000000000) (00000000) SecureXL VPN before decryption (vpn_in_before_decrypt)
     2: -7ffffffd (0000000000000000) (00000000) SecureXL VPN after decryption (vpn_in_after_decrypt)
     3:        6 (0000000000000000) (00000000) SecureXL lookup (sxl_lookup)
     4:        7 (0000000000000000) (00000000) SecureXL QOS inbound (sxl_qos_inbound)
     5:        8 (0000000000000000) (00000000) SecureXL inbound (sxl_inbound)
     6:        9 (0000000000000000) (00000000) SecureXL medium path streaming (sxl_medium_path_streaming)
     7:       10 (0000000000000000) (00000000) SecureXL inline path streaming (sxl_inline_path_streaming)
     8:       11 (0000000000000000) (00000000) SecureXL Routing (sxl_routing)
     9: -7f800000 (ffffffff91b68310) (ffffffff) IP Options Strip (in) (ipopt_strip)
    10: - 1ffffff8 (ffffffff91b790b0) (00000001) Stateless verifications (in) (asm)
    11: - 1ffffff7 (ffffffff91b068a0) (00000001) fw multik misc proto forwarding
    12:        0 (ffffffff92849a10) (00000001) fw VM inbound  (fw)
    13:        2 (ffffffff91b6c3f0) (00000001) fw SCV inbound (scv)
    14:        5 (ffffffff918a93d0) (00000003) fw offload inbound (offload_in)
    15:       20 (ffffffff9284d0d0) (00000001) fw post VM inbound  (post_vm)
    16:   100000 (ffffffff9280a3f0) (00000001) fw accounting inbound (acct)
    17: 7f730000 (ffffffff91ac1370) (00000001) passive streaming (in) (pass_str)
    18: 7f750000 (ffffffff92633f70) (00000001) TCP streaming (in) (cpas)
    19: 7f800000 (ffffffff91b682a0) (ffffffff) IP Options Restore (in) (ipopt_res)
    20: 7fb00000 (ffffffff91ea0090) (00000001) Cluster Late Correction (ccl_in)
out chain (15):
     0: -7f800000 (ffffffff91b68310) (ffffffff) IP Options Strip (out) (ipopt_strip)
     1: - 1fffff0 (ffffffff92631040) (00000001) TCP streaming (out) (cpas)
     2: - 1ffff50 (ffffffff91ac1370) (00000001) passive streaming (out) (pass_str)
     3: - 1f00000 (ffffffff91b790b0) (00000001) Stateless verifications (out) (asm)
     4:        0 (ffffffff92849a10) (00000001) fw VM outbound (fw)
     5:       10 (ffffffff9284d0d0) (00000001) fw post VM outbound  (post_vm)
     6: 7f000000 (ffffffff9280a3f0) (00000001) fw accounting outbound (acct)
     7: 7f700000 (ffffffff92631530) (00000001) TCP streaming post VM (cpas)
     8: 7f800000 (ffffffff91b682a0) (ffffffff) IP Options Restore (out) (ipopt_res)
     9: 7f850000 (ffffffff91e9fb70) (00000001) Cluster Local Correction (ccl_out)
    10:  7f900000 (0000000000000000) (00000000) SecureXL outbound (sxl_outbound)
    11:  7fa00000 (0000000000000000) (00000000) SecureXL QOS outbound (sxl_qos_outbound)
    12:  7fb00000 (0000000000000000) (00000000) SecureXL VPN before encryption (vpn_in_before_encrypt)
    13:  7fc00000 (0000000000000000) (00000000) SecureXL VPN after encryption (vpn_in_after_encrypt)
    14:  7fd00000 (0000000000000000) (00000000) SecureXL Deliver (sxl_deliver)
[Expert@CPCM1:0]#
```

Figure 8.20 – Inspection chain modules

From this, we can infer that packets can take multiple paths, with accelerated traffic being the shortest, and the traffic relying on firewall instance slow chains and content inspection being the longest.

Content inspection

Content inspection is a complex process, but to glance at what is happening under the hood, let's look at the following figure:

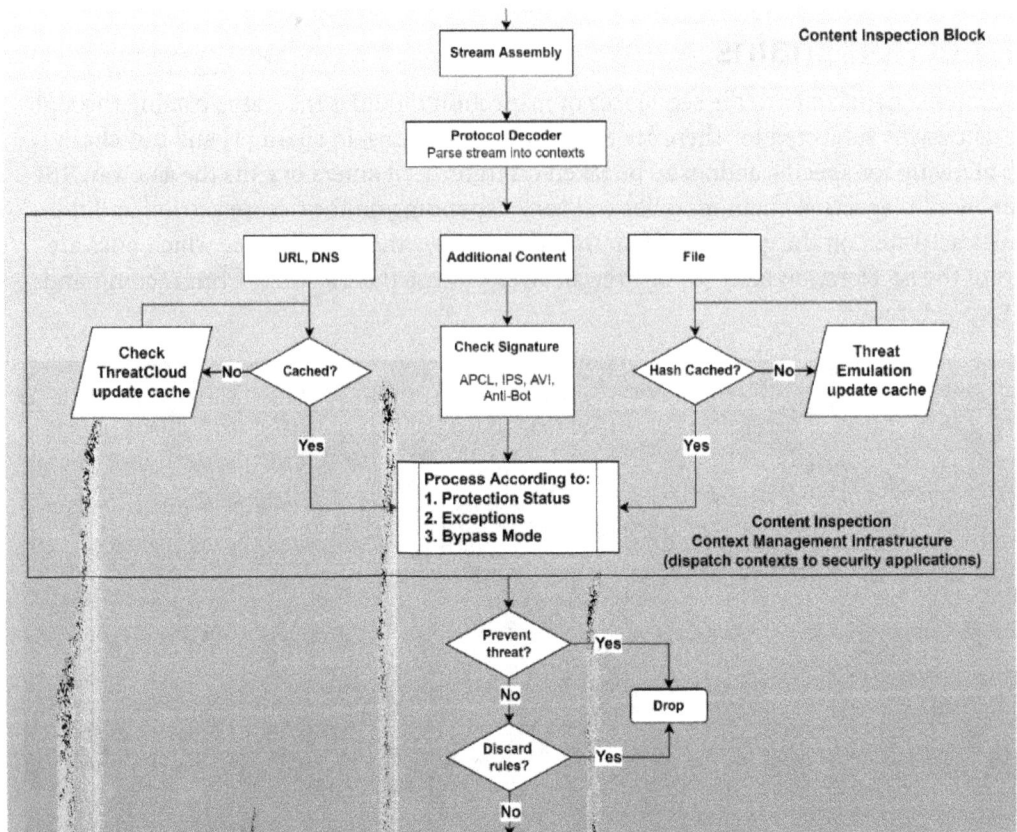

Figure 8.21 – Content inspection, simplified

The previous diagram is a close replica of a segment from the original, created by Moti Sagey and later refined and expanded by Valeri Loukine from Check Point. I highly recommend taking time to read the *Security Gateway Packet Flow and Acceleration with Diagrams* article by Valeri, as it is exceptionally clear and simple, and can be found here:

```
https://community.checkpoint.com/t5/General-Topics/Security-
Gateway-Packet-Flow-and-Acceleration-with-Diagrams/td-p/40244
```

Now, seeing what is involved in getting a packet from one interface to another, the reason for the acceleration of eligible packets should become evident. It stands to reason then, that if we can exempt some of the traffic from content inspection, such as that being explicitly trusted, requiring least latency, or where deep packet inspection will not provide additional benefits, our policies should be constructed with it in mind. For ultra-low latency applications where Access Control is still important, such as high-frequency trading or market data delivery, Check Point, in collaboration with NVIDIA, released Quantum Lightspeed appliances capable of 3 microseconds latency.

Best practices for Access Control rules

Now, with what we have learned in the previous section, let's combine Check Point's own best practices for Access Control rules as printed in their user guide, with a few additional suggestions:

- When a new policy is created, a single explicit cleanup rule is automatically included. Change its **Track** settings to **Log**.

- On top of the policy, create a rule allowing **https** and **ssh_version_2** access to the gateways and cluster members from the IPs of your Check Point administrators' PCs. This rule, together with the next, the **stealth rule**, will limit the exposure of your gateways if Gaia's **System Management | Host Access | Allowed Hosts** contains default settings allowing connectivity from any IP address.

- The second rule from the top should be created, named the **stealth rule**, and configured to deny direct access to the gateways from **Any** source.

- Create section titles above these three rules describing their purpose.

- Create additional session titles describing the structure of your policy, so that the rest of the rules will be created under the corresponding sections.

- Create **Firewall/Network** rules to explicitly accept safe traffic. If inline layers are used, add an explicit cleanup rule to drop everything else for each such layer.

- Create an ordered layer relying on content inspection after the **Firewall/Network** ordered layer. Alternatively, put rules that examine Access Roles, applications, **Data Type**, or **Mobile Access** in an inline layer as part of the **Firewall/Network** rules. In the parent rule of the inline layer, define the source and destination only.

- Share ordered layers and inline layers when practical.

- If your environment contains gateways with version R77.X, use a two ordered layers structure with **Firewall/Network** in the first, and APCL/URLF in the second. The policy applied to R77.X cannot contain a **Mobile Access** blade or **Content Awareness**.

- In layers relying on content inspection, place rules with objects defined in the **Content** field closer to the bottom. Rules using **File Types** objects should be higher than those containing data types.

From the packet flow diagram (*Figure 8.19*), we know that if content inspection is required, it will always prevent the full acceleration of traffic.

Looking at possible scenarios with or without content inspection:

- Firewall/IPSec VPN blades only are enabled in a layer, with threat prevention disabled (can be fully accelerated).

- Firewall/IPSec VPN blades only are enabled in a layer, with threat prevention enabled. Select sources, destinations, and services are using empty threat prevention profiles (can be fully accelerated).

- Firewall/IPSec VPN and any other blades are enabled in a layer, with threat prevention enabled. Traffic is not exempt from threat prevention (content inspection is engaged).

- Firewall/IPSec VPN and any of the other blades are enabled in a layer, with threat prevention disabled (content inspection is engaged).

Combining this with best practices, we may derive the following:

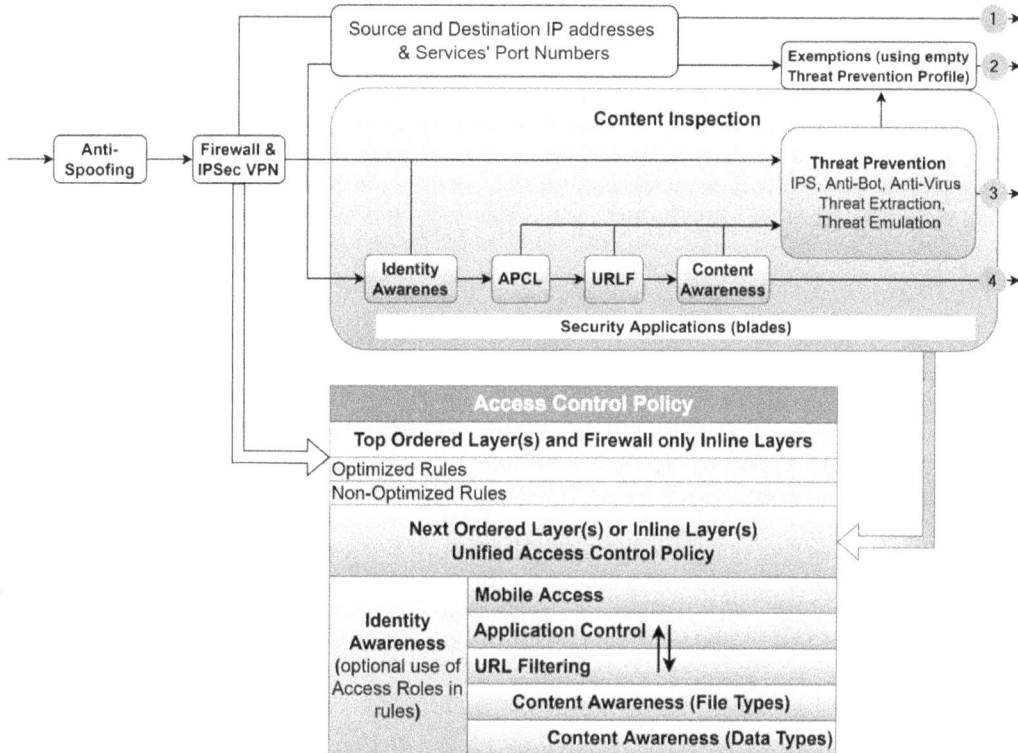

Figure 8.22 – Policy organization based on best practices

To see when rules are considered **non-optimized**, see *sk32578* (`https://supportcenter.checkpoint.com/supportcenter/portal?eventSubmit_doGoviewsolutiondetails=&solutionid=sk32578`) *SecureXL Mechanism | (3) Factors that adversely affect SecureXL performance.* As an example, rules with **RPC, DCOM,** or **DCE-RPC** services will disable accept templates' creation for all rules located below. These rules, therefore should be placed as close to the end of the **Firewall/Network** layer as possible.

Threat prevention exemptions

> **Note**
> Threat prevention is outside of the scope of this book, but since it is relevant to the structure of the policy and placement of the rules, see this brief subsection for references.

Threat prevention is enabled or disabled on a policy level, not per layer. It consists of the rules allowing the use of sources, destinations, services, and profiles. Custom profiles could be created, and individual security blades enabled or disabled. Create a threat prevention profile, name it `Empty` [1] and disable all the blades in it.

Use this profile with specific sources [2], destinations [3], and services [4] to exempt specific traffic from content inspection:

No.	Name	Protected Scope	Source	Destination	Services	Action	Track
▶ 1	Exemptions from Content Inspection	＊ Any	🖥 Host_A	🖥 Host_B	⚡ Service_X	📋 Empty	─ None

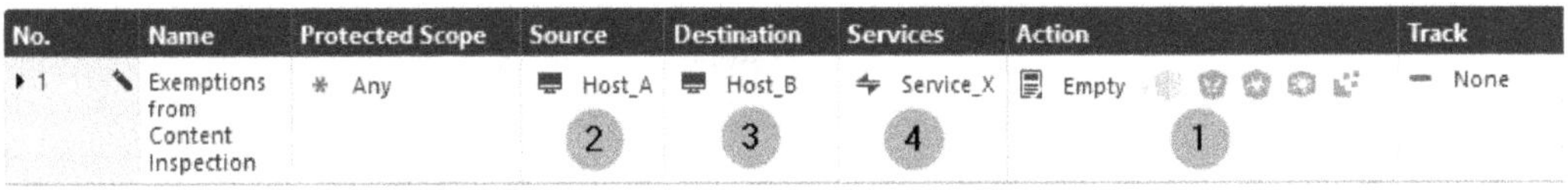

Figure 8.23 – The use of an empty threat prevention profile for content inspection exemptions

Figure 8.24 describes a basic Access Control policy structure with firewall/IPSec VPN blades enabled in the top (ordered layer) and a single inline layer with APCL/URLF enabled blades:

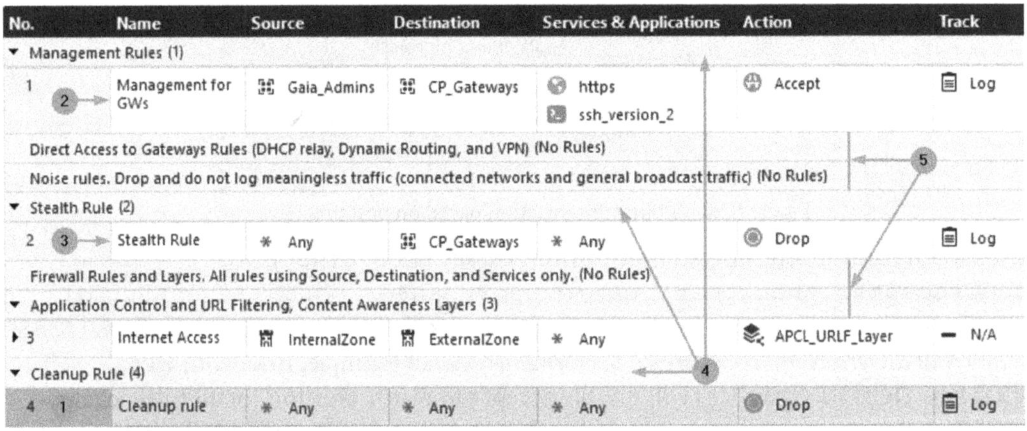

No.	Name	Source	Destination	Services & Applications	Action	Track
▼ Management Rules (1)						
1	Management for GWs	⅜ Gaia_Admins	⅜ CP_Gateways	☺ https 📱 ssh_version_2	☺ Accept	📋 Log
Direct Access to Gateways Rules (DHCP relay, Dynamic Routing, and VPN) (No Rules)						
Noise rules. Drop and do not log meaningless traffic (connected networks and general broadcast traffic) (No Rules)						
▼ Stealth Rule (2)						
2	Stealth Rule	＊ Any	⅜ CP_Gateways	＊ Any	◉ Drop	📋 Log
Firewall Rules and Layers. All rules using Source, Destination, and Services only. (No Rules)						
▼ Application Control and URL Filtering, Content Awareness Layers (3)						
▶ 3	Internet Access	🏠 InternalZone	🏠 ExternalZone	＊ Any	📚 APCL_URLF_Layer	─ N/A
▼ Cleanup Rule (4)						
4	Cleanup rule	＊ Any	＊ Any	＊ Any	◉ Drop	📋 Log

Figure 8.24 – Creating an Access Control Rule Base according to best practices

We will go through functional policy creation in *Chapter 11*, *Building Your First Policy*, but for now, let's see how the traffic is matched to the rules in your policies.

Column-based matching

In mature organizations with exceptionally granular Access Control policies, it is common to see very long rule bases with hundreds, or even thousands, of rules. As packets making it through the firewall layer require a Rule Base lookup, the length of the Rule Base exerts a toll on performance.

To speed up this process, Check Point uses column-based matching.

Rule-matching in policies works in a columnar plus top-down order. For instance, in a firewall-only enabled policy, the first packet is progressively matched against three fields in this order:

1. The **Destination** column

2. The **Source** column

3. The **Services & Applications** column

The Rule Base is collapsed to a possible matched rules array. Then, top-down matching takes place until the first matching rule is encountered and enforced.

We can use **Packet Mode** search (described in the previous chapter) to demonstrate the column-based matching.

> **Note**
>
> Packet Mode search works best in either flat (single-layer) policies or policies with inline layers. In cases of ordered layers, Packet Mode search filters only the rules affecting traffic in the currently selected layer.

Figures 8.25 and *8.26* illustrate how traffic matches the rule allowing alerts and notifications from the **Check Point Security Management Server** (**CPSMS**) (10.0.0.10) to reach a **MailServer** (10.30.30.6) using the SMTP protocol (port 25):

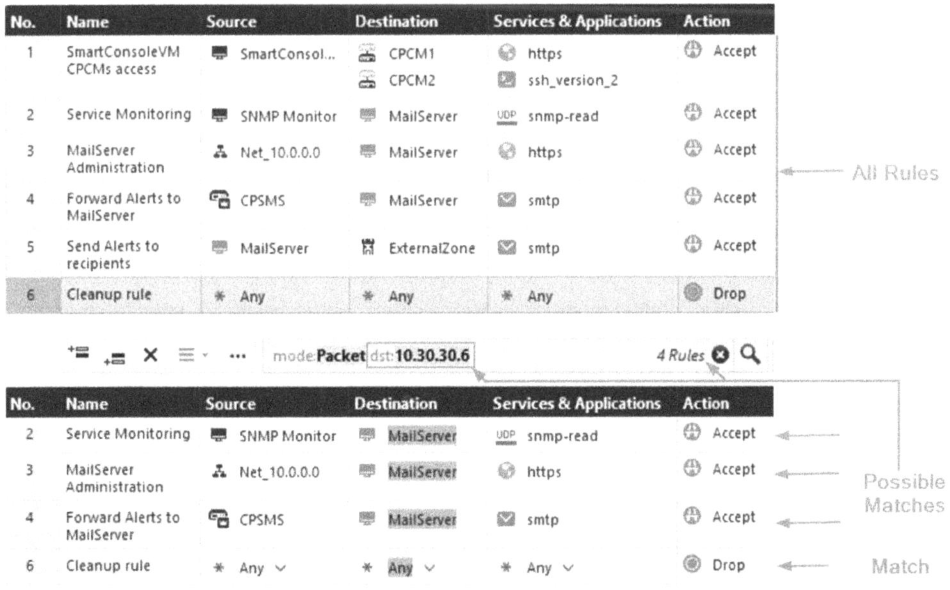

Figure 8.25 – Column-based rule matching, part 1

Continuing the process, we see the following:

Figure 8.26 – Column-based rule matching, part 2

If additional blades are enabled in the layer, and the possible matching rules contain access roles, categories, application, and/or content parameters, the first packet is insufficient for policy enforcement decisions.

> **Note**
> Policies containing layers with blades other than Firewall/IPSec VPN are referred to as unified Access Control *policies*. The term should really be unified Access Control *layers*.

A unified policy classifies the content of subsequent packets until enough is known for traffic identification. As in the previous example, the more that is known about the connection, the more rules are eliminated from the matched rules array.

Examples of classification objects are access roles for source and destination fields, protocols, applications, services, file and content (types), and the direction of transfers:

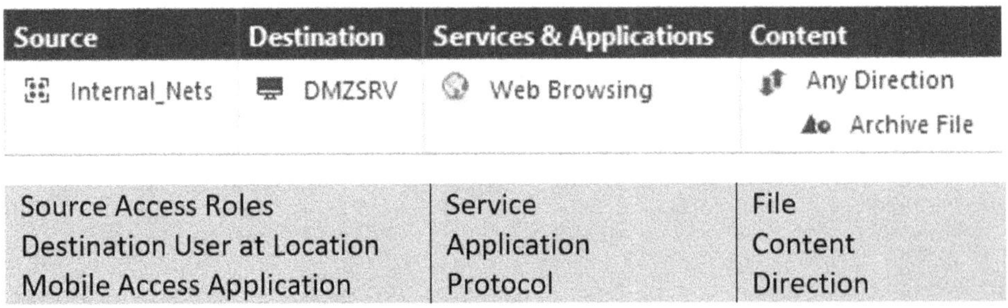

Figure 8.27 – Classification objects

Consider an attempt to download a `.zip` file from Google Drive, looking at the Rule Base here:

No.	Source	Destination	Services & Applications	Content	Action	
1	InternalZone	Internet	http https	Any Direction Archive File	Drop Blocked Message...	
2	InternalZone	Internet	Critical Risk High Risk	Any	Drop Blocked Message...	Match Possible
3	InternalZone	Internet	http https	Any	Accept	
4	Any	Any	Any	Any	Drop	

Figure 8.28 – APCL/URLF plus Content Awareness rules column-based matching

The first packet of the connection (SYN) will be matched on all the rules but will be allowed to proceed by rule number 3.

The CoreXL Dynamic Dispatcher forwards SYN to the **Context Management Infrastructure** (**CMI**). The unified policy initiates the classification of the packet. Classifier Apps execute on the packet, to create Classifier Objects to match each column in the policy.

After completion of the three-way TCP handshake, the client in the Internal Zone sends an HTTP GET request to download the file.

The HTTP header of the client request contains the host, `lh3.googleusercontent.com`, which URLF determines not to be a critical or high-risk category (as it is considered a medium-risk category) and is allowed to proceed by rule number 3. Rule number 2 is eliminated from the possible matches array.

No.	Source	Destination	Services & Applications	Content	Action	
1	InternalZone	Internet	http https	Any Direction Archive File	Drop Blocked Message...	Match Possible
2	InternalZone	Internet	Critical Risk High Risk	Any	Drop Blocked Message...	No Match
3	InternalZone	Internet	http https	Any	Accept	Match Possible
4	Any	Any	Any	Any	Drop	

Figure 8.29 – APCL/URLF plus Content Awareness rules column-based matching

Protocol streaming and parsing are used to extract the file from the HTTP body in the server's response. The pattern matcher determines the file type is the archive.

The result is returned to the classifier for matching against the remaining rules, and the first match is found to be rule number 1:

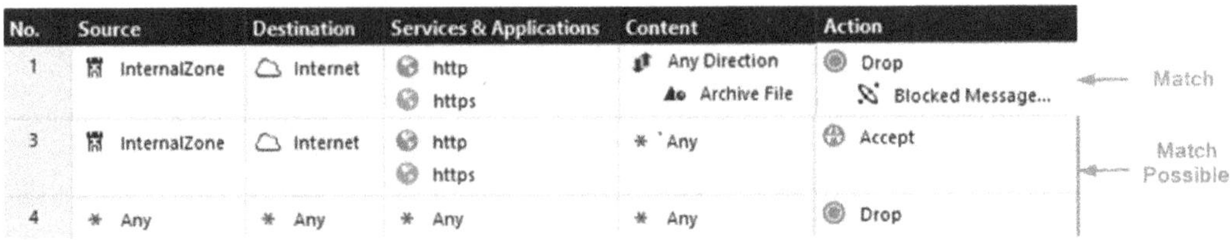

No.	Source	Destination	Services & Applications	Content	Action	
1	InternalZone	Internet	http https	Any Direction Archive File	Drop Blocked Message...	← Match
3	InternalZone	Internet	http https	Any	Accept	← Match Possible
4	Any	Any	Any	Any	Drop	

Figure 8.30 – APCL/URLF plus Content Awareness rules column-based matching

Kudos to Bob Bent of Check Point for his posts on the subject.

APCL/URLF layer structure

At the time of writing, Check Point's own user guide for versions R81 and up contained this incorrect statement:

"5. Create an Application Control Ordered Layer after the Firewall/Network Ordered Layer. Add rules to explicitly drop unwanted or unsafe traffic. Add an explicit cleanup rule at the bottom of the Ordered Layer to accept everything else.

Alternatively, put Application Control rules in an Inline Layer as part of the Firewall/Network rules. In the parent rule of the Inline Layer, define the Source and Destination"

The screenshot from the official documentation is as follows:

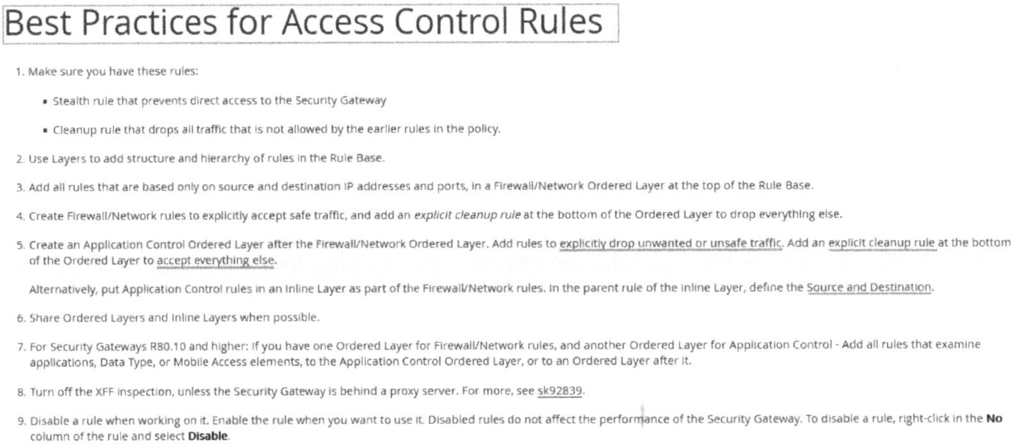

Best Practices for Access Control Rules

1. Make sure you have these rules:
   - Stealth rule that prevents direct access to the Security Gateway
   - Cleanup rule that drops all traffic that is not allowed by the earlier rules in the policy.

2. Use Layers to add structure and hierarchy of rules in the Rule Base.

3. Add all rules that are based only on source and destination IP addresses and ports, in a Firewall/Network Ordered Layer at the top of the Rule Base.

4. Create Firewall/Network rules to explicitly accept safe traffic, and add an *explicit cleanup rule* at the bottom of the Ordered Layer to drop everything else.

5. Create an Application Control Ordered Layer after the Firewall/Network Ordered Layer. Add rules to explicitly drop unwanted or unsafe traffic. Add an explicit cleanup rule at the bottom of the Ordered Layer to accept everything else.

   Alternatively, put Application Control rules in an Inline Layer as part of the Firewall/Network rules. In the parent rule of the Inline Layer, define the Source and Destination.

6. Share Ordered Layers and Inline Layers when possible.

7. For Security Gateways R80.10 and higher: if you have one Ordered Layer for Firewall/Network rules, and another Ordered Layer for Application Control - Add all rules that examine applications, Data Type, or Mobile Access elements, to the Application Control Ordered Layer, or to an Ordered Layer after it.

8. Turn off the XFF inspection, unless the Security Gateway is behind a proxy server. For more, see sk92839.

9. Disable a rule when working on it. Enable the rule when you want to use it. Disabled rules do not affect the performance of the Security Gateway. To disable a rule, right-click in the **No** column of the rule and select **Disable**.

Figure 8.31 – Check Point's Best Practices for Access Control Rules error

If you were to follow this recommendation, then traffic to any IP and port on the internet would be allowed, unless explicit rules are present in the APCL/URLF layer to drop it.

Instead, I suggest using the following approach:

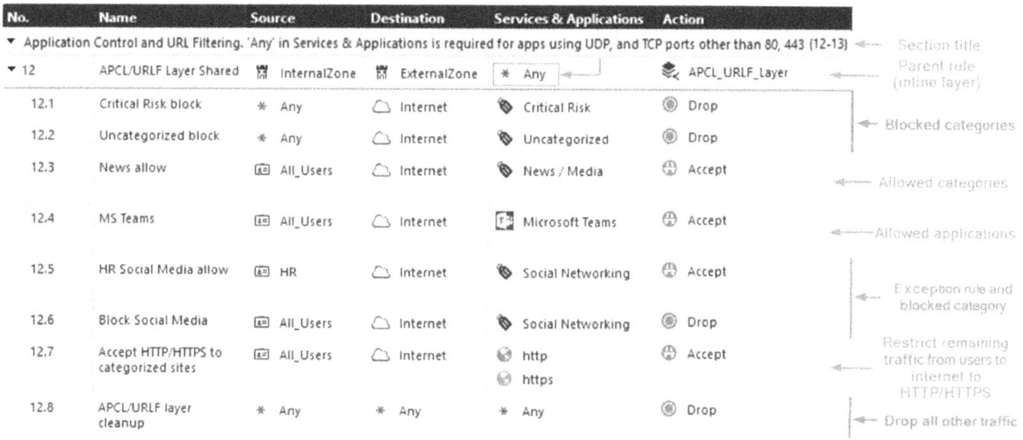
No. Name Source Destination Services & Applications Action
Application Control and URL Filtering. 'Any' in Services & Applications is required for apps using UDP, and TCP ports other than 80, 443 (12-13)
12 APCL/URLF Layer Shared InternalZone ExternalZone Any APCL_URLF_Layer
12.1 Critical Risk block Any Internet Critical Risk Drop
12.2 Uncategorized block Any Internet Uncategorized Drop
12.3 News allow All_Users Internet News / Media Accept
12.4 MS Teams All_Users Internet Microsoft Teams Accept
12.5 HR Social Media allow HR Internet Social Networking Accept
12.6 Block Social Media All_Users Internet Social Networking Drop
12.7 Accept HTTP/HTTPS to categorized sites All_Users Internet http https Accept
12.8 APCL/URLF layer cleanup Any Any Any Drop
Section title
Parent rule (inline layer)
Blocked categories
Allowed categories
Allowed applications
Exception rule and blocked category
Restrict remaining traffic from users to internet to HTTP/HTTPS
Drop all other traffic

Figure 8.32 – Correct structure for the APCL layer

Let's go over the preceding figure:

- At the top, the section title describes the inline layer.
- Rule 12 (the parent rule of the inline **APCL_URLF_Layer**). While this example is using **InternalZone** as a source, you may use a network or a group of networks instead. The use of **ExternalZone** for the destination is, in my opinion, the best way to define *internet* in the **Firewall/Network** policy layer, as the actual **Internet** object is only available in layers with the **Application Control & URL Filtering** blade enabled.

The use of **Any** in the **Services & Applications** field of rule 12 is explained by the following: many applications, especially those that incorporate voice or video streaming, such as **Cisco Webex Teams** [1] in the following example, using UDP and additional TCP ports other than 80 and 443 [2]:

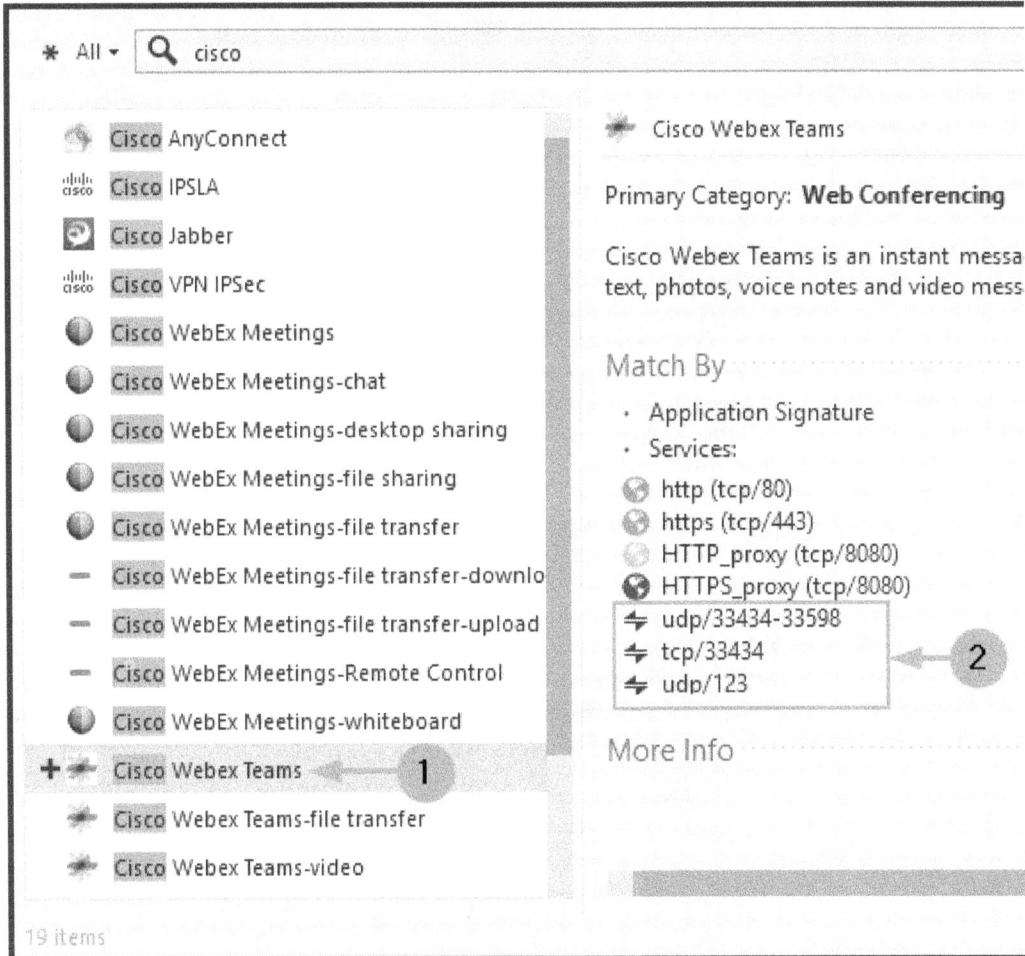

Figure 8.33 – Applications using UDP and TCP ports other than 80 and 443

The APCL/URLF blade's properties as shown in the following screenshot [1] state that we are matching web applications on HTTP/HTTPS [2], but this is not the same as allowing all relevant traffic for those applications:

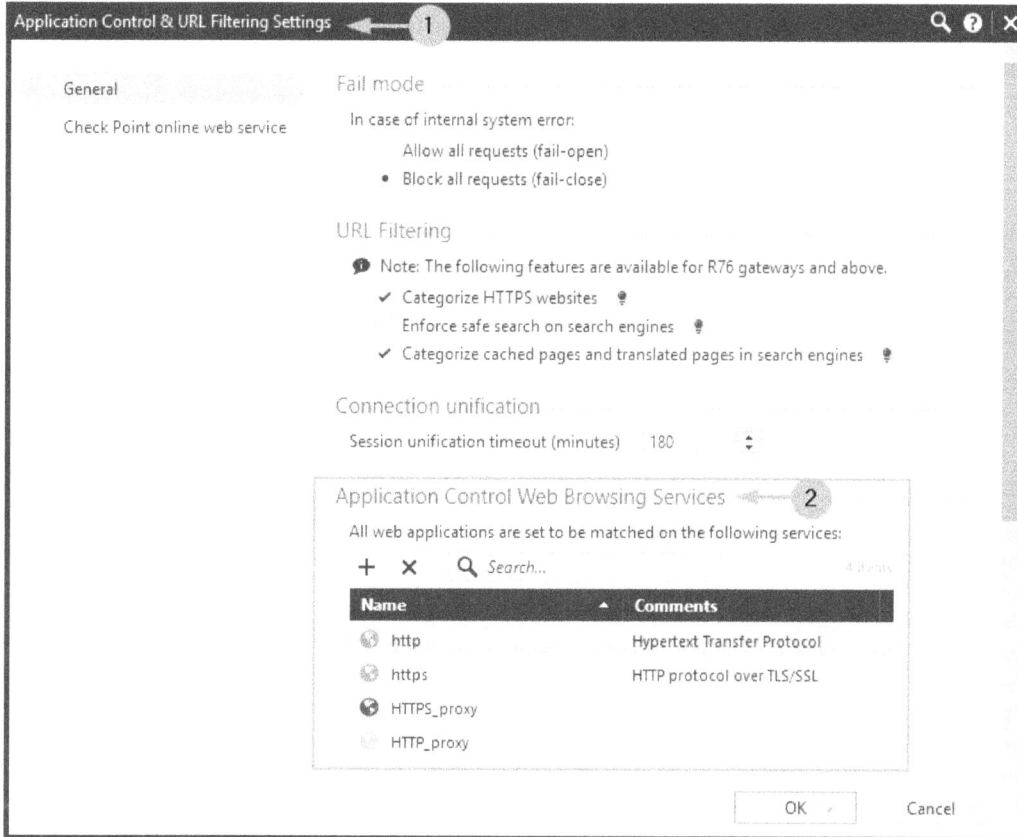

Figure 8.34 – Web application match-on services

Imagine that we have rule-tracking set for the **Application Session and Connections**. We should see all the relevant connections for the session, HTTP(S), as well as other protocols. In the rules depicted in *Figure 8.32*:

- Rules **12.1** and **12.2** (blocked category) are there to prevent all web access to the **Critical Risk** and **Uncategorized** categories of sites.

- Rule **12.3** (allowed category) allows access to **News/Media** categories of sites to **All_Users**. This restricts access based on a combination of networks and users' AD group membership. The use of access roles is possible when the **Identity Awareness** blade is also enabled in the layer. We'll talk about identity awareness in a later chapter. You may use IP ranges, network objects, or groups instead.

- Rule **12.4** (allowed application) allows access for **MS Teams** to **All_Users**.

- Rule **12.5** in tandem with rule **12.6** creates an exemption for access to the **Social Networking** categories for Human Resources from all the rest **All_Users** prohibition.

- Rule **12.7** allows access to the rest of the URLs (or web applications) not explicitly denied by previous rules. This is achieved by limiting access to the relevant protocols only.

- Rule **12.8** Is the **APCL/URLF layer cleanup** rule, dropping the rest of the attempts and logging them.

To generalize this example of the inline layer for APCL/URLF, we can simplify it as follows:

1. 12.1 and 12.2 Blocked Categories, including uncategorized; (Block)

2. 12.3 Allowed Categories; (Accept)

3. 12.4 Allowed Applications; (Accept)

4. 12.5 Blocked Categories *exemptions*; (Accept)

5. 12.6 Blocked Categories; (Block)

6. 12.7 Allowed Web Browsing to the rest; (Accept)

7. 12.8 Block all other traffic; (Drop)

If even finer control is needed, you can use additional (nested) inline layers.

Actions and user interactions (UserCheck)

APCL/URLF rules allow Check Point administrators to craft interactive rules. By combining **Drop** with **Blocked Message** [1], you inform users that an attempt to access a site, category, or web application is violating the company security policy [2]. Instead of seeing the `This Site can't be reached (ERR_CONNECTION_RESET)` message that a simple **Drop** would result in and logging it as such [3], the action is logged as **Block** [4]:

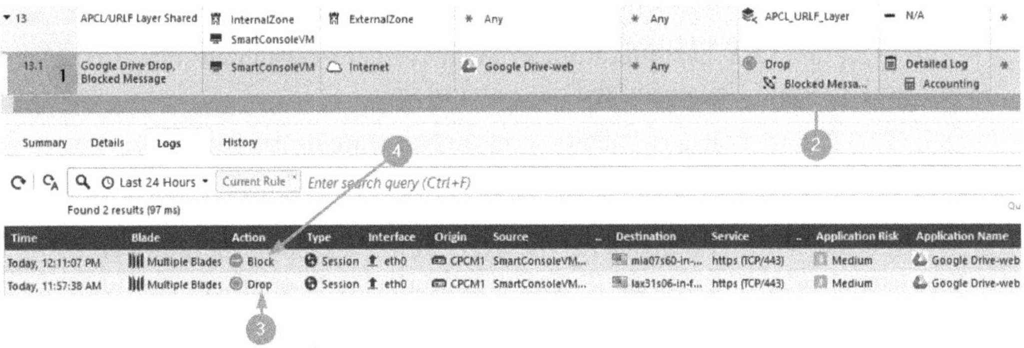

Time	Blade	Action	Type	Interface	Origin	Source		Destination	Service		Application Risk	Application Name
Today, 12:11:07 PM	Multiple Blades ⊘ Block	⊕ Session ↥ eth0	CPCM1	SmartConsoleVM...	mia07s60-in-...	https (TCP/443)	Medium	Google Drive-web
Today, 11:57:38 AM	Multiple Blades ⊘ Drop	⊕ Session ↥ eth0	CPCM1	SmartConsoleVM...	lax31s06-in-f...	https (TCP/443)	Medium	Google Drive-web

Figure 8.35 – UserCheck in APCL/URLF rules

Users are presented with an informative message that does the following:

1. Displays a reason for denial of access
2. Allows submission of a reclassification request (typically takes 24 hours)
3. Provides a **Reference** number for the incident handling

The preceding informative messages can be seen in the following screenshot:

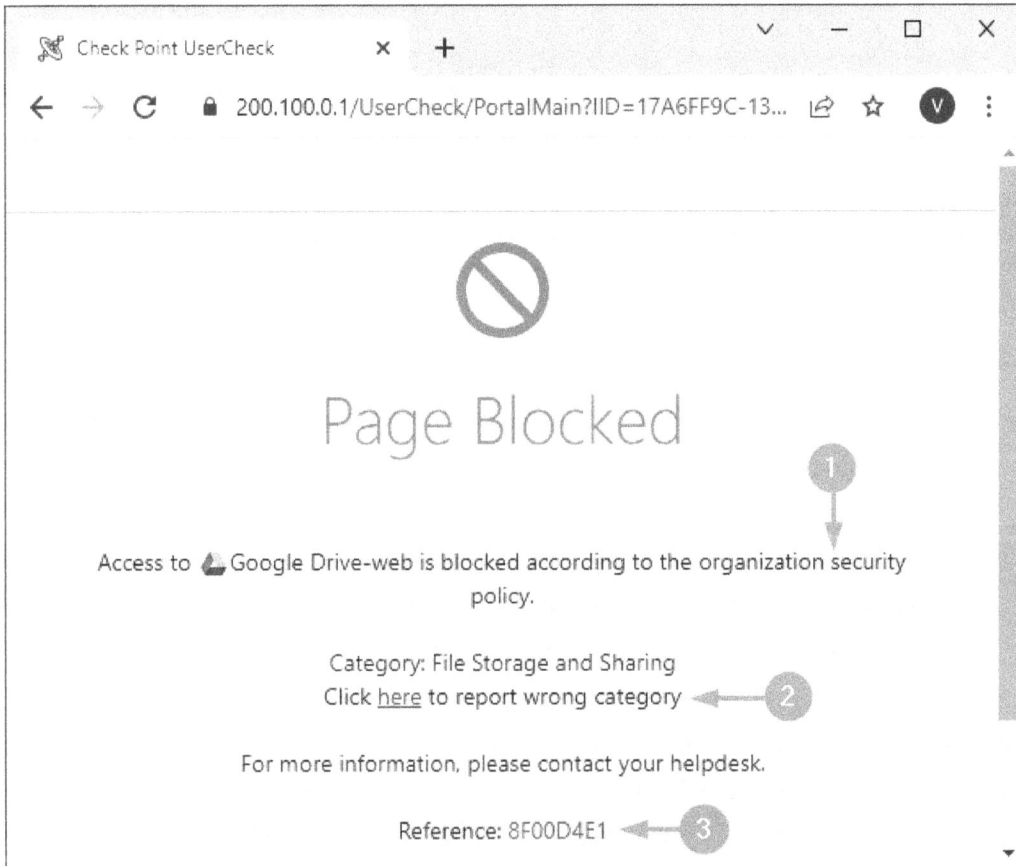

Figure 8.36 – UserCheck Page Blocked according to security policy

A reference number [1] can be used in tickets submitted to Check Point administrators for review or corrective actions, to easily retrieve relevant logs [2]:

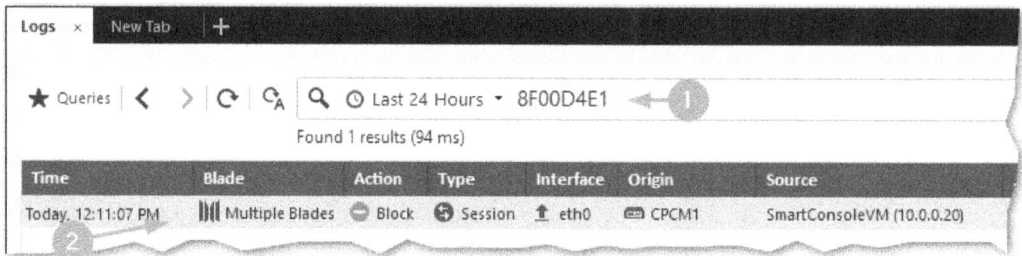

Figure 8.37 – Log retrieval by incident ID reference number

Additional interactions are possible, such as informing users before they are successfully granted access, checkbox consent to abide by the company policy, and redirecting to an external portal (authenticated via a generated pre-shared secret). For the **Ask/Inform** [1] actions, you can define **UserCheck frequency**, confirmation conditions (Per rule/category/application/site/data type), and transfer speed limit [2].

Enable Identity Captive Portal [3] is only relevant when the following applies:

- Identity awareness is enabled on the gateway with **Browser-based Authentication**.

- Identity awareness is used in the layer.

The following screenshot illustrates various **Action Settings** available to us:

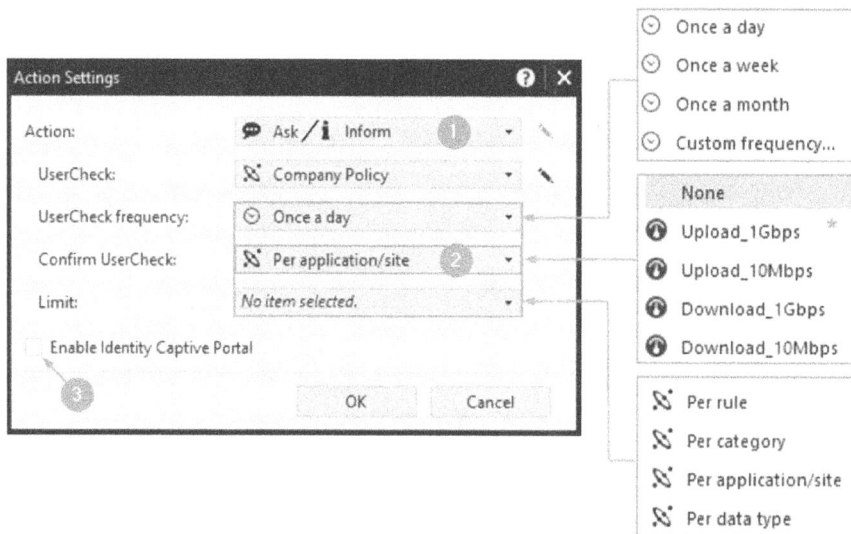

Figure 8.38 – Action Settings

For the **Accept** action, only **Limit** can be selected.

> **Note**
>
> Besides the four pre-defined **Limit** values, additional values and transfer directions can be created using **Objects** | **New** | **More** | **Limits**. Limits can be bi- or uni-directional. Limits are applied per rule, not per individual connection.

Administrators can modify existing, or create new, UserCheck actions. I strongly suggest keeping all original items in place, creating clones for each, and using those for customization and in rules.

Content Awareness

Content Awareness is not a **data loss prevention** (**DLP**) solution. It will certainly improve your overall security posture if you do not have a dedicated solution, but it is not intended to replace one.

In a nutshell, Content Awareness lets you prevent the upload, download, or transfer in any direction of either all or certain types of files, and files containing certain types of data. It works over a limited number of services. It is enabled by default to work with HTTP, HTTPS, HTTP(S) Proxy, SMTP, and FTP.

Additional options are **CheckPointExchangeAgent** and **Squid_NTLM** (I am uncertain as to the reason for this last one).

You can create custom data types, but the file types are limited to those provided by Check Point. At the time of writing, there are 62 different file and data types that can be used in the Content Awareness rules.

To use this feature in your rules, the Content Awareness blade should be enabled on the gateway(s) in your policy's Installation Targets and selected in the properties of the layer.

When creating rules for Content Awareness, I recommend explicitly specifying services on which they are enforced, as shown in Rule 14.3 [1] in the following screenshot. Rule 14.4 [2] does not have services specified and you are permitted to install it, but, looking at that rule, we may infer that Content Awareness is enforced on any service:

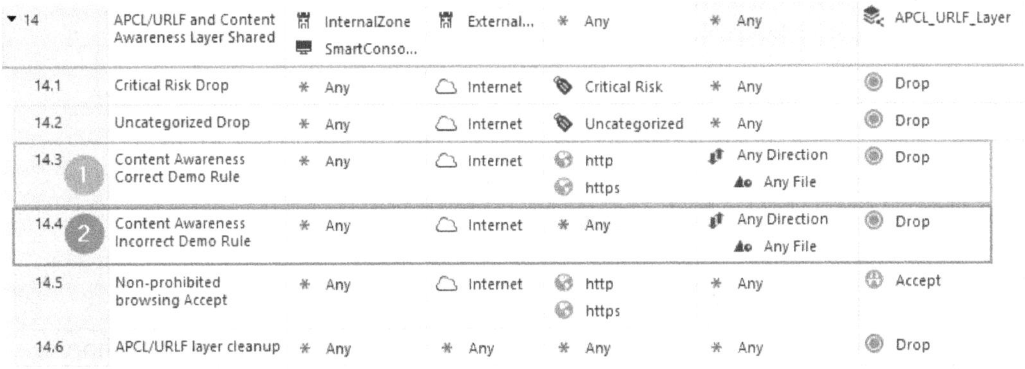
14
APCL/URLF and Content Awareness Layer Shared
InternalZone
SmartConso...
External...
Any
Any
APCL_URLF_Layer
14.1
Critical Risk Drop
Any
Internet
Critical Risk
Any
Drop
14.2
Uncategorized Drop
Any
Internet
Uncategorized
Any
Drop
14.3
1
Content Awareness Correct Demo Rule
Any
Internet
http
https
Any Direction
Any File
Drop
14.4
2
Content Awareness Incorrect Demo Rule
Any
Internet
Any
Any Direction
Any File
Drop
14.5
Non-prohibited browsing Accept
Any
Internet
http
https
Any
Accept
14.6
APCL/URLF layer cleanup
Any
Any
Any
Any
Drop

Figure 8.39 – Correct and incorrect rule formats for Content Awareness

This assumption would be incorrect, due to the limitations of the Content Awareness blade.

There is a reason I've chosen to write about Content Awareness after the *Actions and User Interactions* section.

> **Note**
>
> Do not use **UserCheck** in the **Action** field of the Content Awareness rules.

For instance, if the rule is configured to **Drop** with **Blocked Message** [1] and you attempt to transfer a file, there will not be a browser-based notification, and in the rule logs you will see the **Redirect** as **Action**:

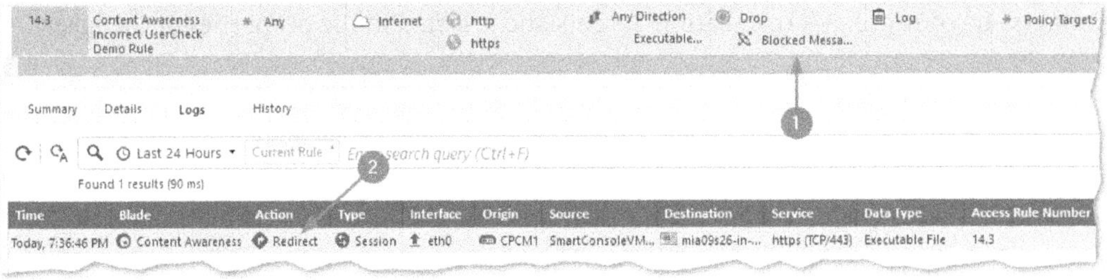
14.3
Content Awareness Incorrect UserCheck Demo Rule
Any
Internet
http
https
Any Direction
Executable...
Drop
Blocked Messa...
Log
Policy Targets
Summary
Details
Logs
History
1
Last 24 Hours
Current Rule
search query (Ctrl+F)
2
Found 1 results (90 ms)
Time
Blade
Action
Type
Interface
Origin
Source
Destination
Service
Data Type
Access Rule Number
Today, 7:36:46 PM
Content Awareness
Redirect
Session
eth0
CPCM1
SmartConsoleVM...
mia09s26-in-...
https (TCP/443)
Executable File
14.3

Figure 8.40 – Content Awareness rule with UserCheck, Redirect Action

The good news is that file transfer was prevented:

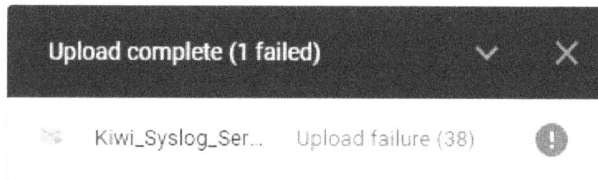

Figure 8.41 – File transfer prevented with the Content Awareness blade, no UserCheck

The not-so-good news is that you do not have a log entry to prove it to your auditors. **Redirect** is actually an indicator that Check Point cannot display the message in the browser window and is forwarding it to **UserCheck Client**. UserCheck Client is available for download from SmartConsole, but I do not recommend using it. This is an old Windows-specific installer. To keep the behavior of your rules and corresponding logs consistent, simply use the **Drop** or **Accept** actions for Content Awareness.

You may encounter **Redirect** in logs for application-specific rules, but it will not be perceived as a data loss event.

Now that we've seen how the rules are constructed and the available actions, let's take a look at how they are tracked and examine a few common cases that may be difficult to interpret without additional explanation.

Logs, tracking depth, and oddities

Tracking for Access Control rules should be configured on a per-rule basis. Either right-click in the **Track** field of the rule or hover over it with your mouse and click on the drop-down arrow in its top-right corner [1]. The small drop-down box gives you the ability to set some of the options using a one-click operation.

Clicking on **More** [2] opens a more extensive **Track Settings** dialog box where, in addition to the same options as above, you can choose the logging level (for rules in layers with APCL/URLF enabled) [3], set **Alert** [4], choose the **Accounting** option [5], and select whether **Log Generation** is set **per Connection** [6] or **per Session** [7]:

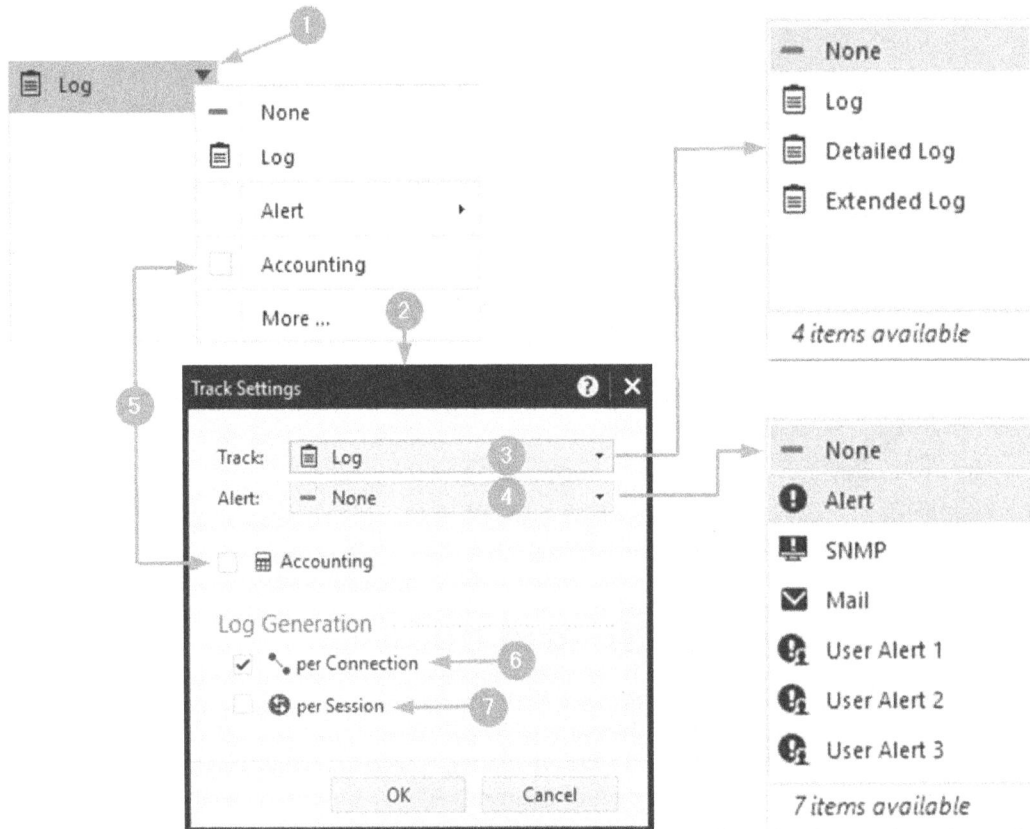

Figure 8.42 – Track Settings configuration

Note that the **Detailed Log** and **Extended Log** options are available only in rules located in layers with either APCL/URLF, Content Awareness, or Mobile Access blades enabled.

When either **Detailed Log** or **Extended Log** are selected [8], **Track Settings** acquires an additional option, **Enable Firewall sessions** [9]:

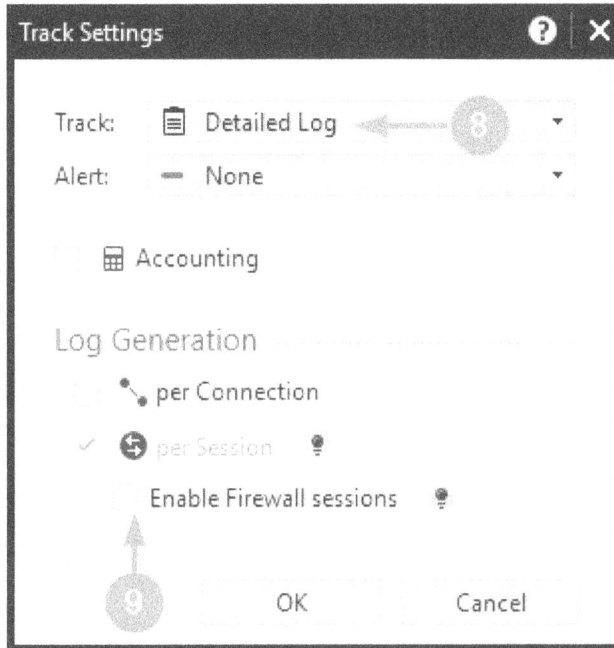

Figure 8.43 – Track settings for rules in APCL/URLF layers

This is what each of the **Track Settings** options is for:

- **None**: No logging is taking place. Used for *noise* filtering rules.

- **Log**: The basic logging level is enabled by default in all new **Firewall/Network** layer rules. This is the only option available in **Firewall/Network** only layer rules besides **None**. If used in the APCL/URLF layers with an application specified in the rule, that application will be logged. If used in a layer with **Content Awareness** and file/data type specified, that file/datatype will be logged.

- The **Detailed Log** and **Extended Log** options become available if either one of these blades is enabled in the layer: **Application Control & URL Filtering**, **Content Awareness**, or **Mobile Access**.

- **Detailed Logging**: Expands basic **Log** in APCL/URLF layers to include identified sites and/or applications, even if they are not specified in the rule. As a best practice, use it for cleanup rules.

- **Extended Logging**: Expands **Detailed Logging** with raw URLs and filenames in APCL/URLF layers.

- **Accounting**: Available for all logging options. Provides statistics relevant to connection durations, the amount of transferred data, and the egress interface. Incremented every 10 minutes for existing connections. Necessary for relevant reports in SmartEvent.

- **Per Connection**: Used to log individual connections in sessions. The default option for all firewall rules.

- **Per Session**: The default option for all APLC/URLF rules. All sites/applications that are identified by that blade are logged at the start of a session. This log is not updated for the next 3 hours (default setting) or until the session is terminated.

- **Per Firewall Session**: Becomes available in APCL/URLF layers when either the **Detailed Log** or **Extended Log** options are set in the **Track** field. Enable for all non-application specific APCL/URLF rules with the **Accept** action (that is, when simple service(s) are specified in **Services & Applications**).

- **Alerts**: Can only be enabled when at least one tracking option is chosen. By itself, simply generates **Log** with the type **Alert**:

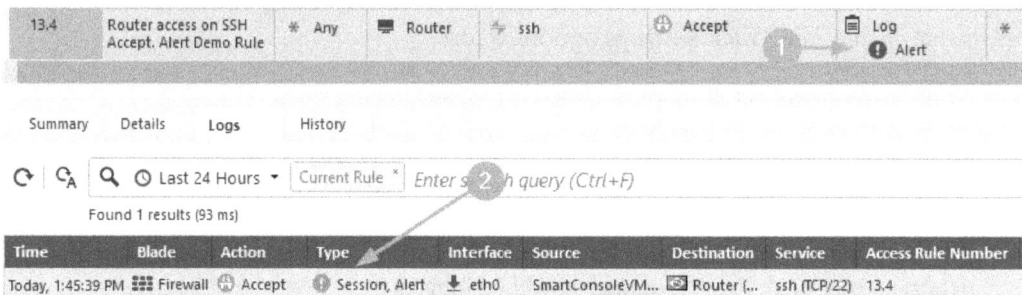

Figure 8.44 – Log type alert

This is useful when trying to draw attention to specific traffic in logs (filtered by **Type**) in **Logs**.

The other options for alerts are self-explanatory and could be configured in **Menu | Global Properties | Log and Alert | Alerts**.

Based on my experience (and the recommendations by Timothy Hall who presented on this subject at the 2022 Check Point Experience conference, and was kind enough to discuss this subject with me), these are the preferred settings for logging:

Tracking Settings Recommendations					
Blade(s)	Track	Accounting	Per Connection	Per Session	Enable Firewall Sessions
Firewall Only	Log	Optional	Yes	No	N/A
Content Awareness	Log	No	No	Yes	Yes
APCL/URLF with or without Content Awareness	Detailed Log	Yes	No	Yes	Yes

Figure 8.45 – Recommended log generation settings

> **Note**
>
> These are my personal recommendations, and your situation may require different settings.

If, for instance, you need to see NAT data in your APCL/URLF rules for troubleshooting, you may add a rule above the one you are working on – specify a narrow combination of source, destination, and services, and in its **Track** field, select **Detailed Log** with **Sessions** and **Connections**.

Oddities – CPEarlyDrop and insufficient data passed

There are a few cases that tend to cause headaches for administrators that I'd like to cover at this point. Both of those are shown in the general logs in the following screenshot, filtered by destination IP. Since logs are presented in reverse chronological order, the first one is on the bottom [1] and the second one is on top [2].

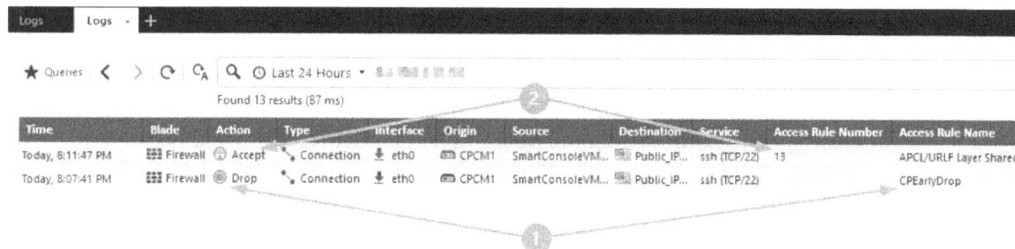

Figure 8.46 – Logging edge cases

Let's examine these cases to see how they are manifested, and their causes.

CPEarlyDrop

When we attempt to initiate an SSH connection from sources allowed by the rules in the Application Control layer [1], we expect it to fail on the **APCL/URLF layer cleanup** rule 13.5 [2]. There is nothing in the rule's logs to indicate that has occurred [3].

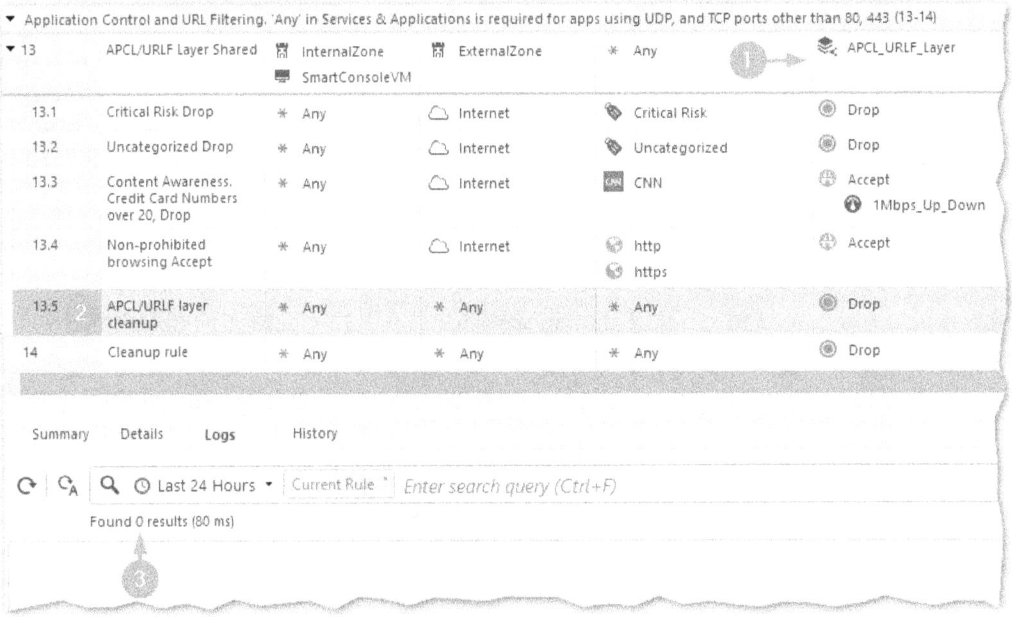

Figure 8.47 – CPEarlyDrop – no logs in the presumed matching rule

Instead, we must go to the general logs by way of **LOGS & MONITOR | New Tab | Log View** and hunt it down, (as shown in the first scenario depicted in *Figure 8.46*).

When we open the log card for the event, we see the following note [1]: **Early Drop: blocking the connection before final rule match. To learn more see sk111643 http://supportcontent.checkpoint.com/solutions?id=sk111643**, and **Access Rule Name** is **CPEarlyDrop** in **APCL_URLF_Layer** [2]. Clearly, there is no such rule between rules 13.1 and 13.5, so what is going on?

This is Check Point's policy execution optimization mechanism at work. It is allowed to drop the attempted connection as soon as it determines that there is nothing in your policy that will allow it to succeed. This is great for security and performance but is mildly irritating for administrators.

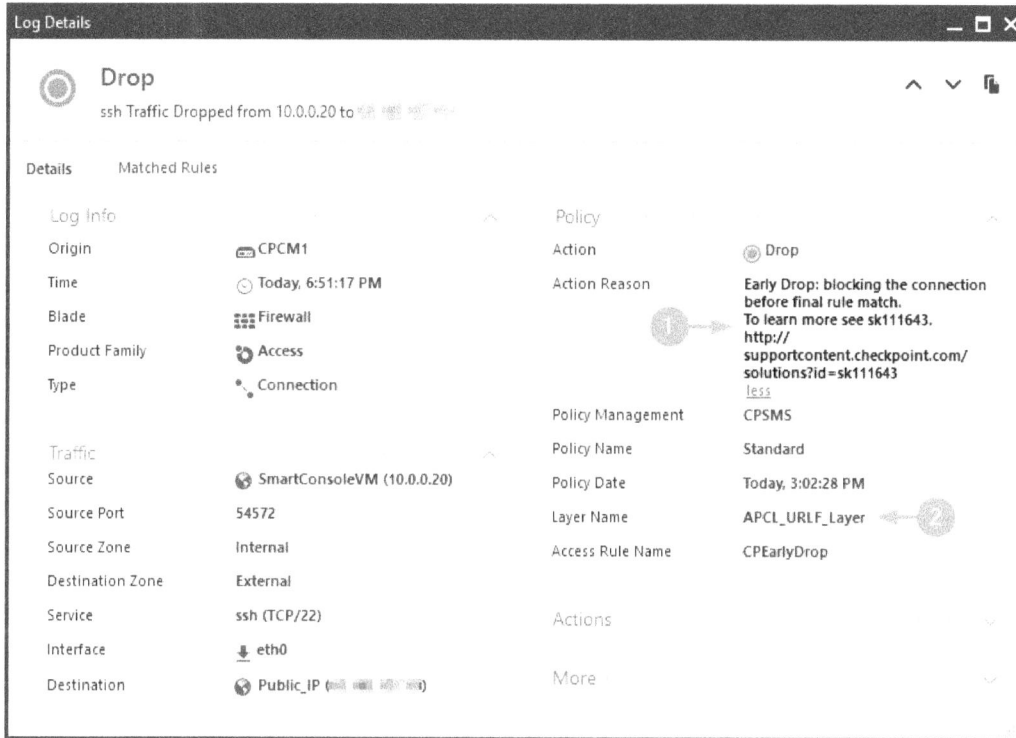

Figure 8.48 – CPEarlyDrop log card

Insufficient data passed

Let's take a look at the second case. In the following policy, we have a rule allowing an SSH connection to a specific host with a public IP [1]. When we attempt to connect to it, the connection fails and there is nothing in the logs for the corresponding rule [2]:

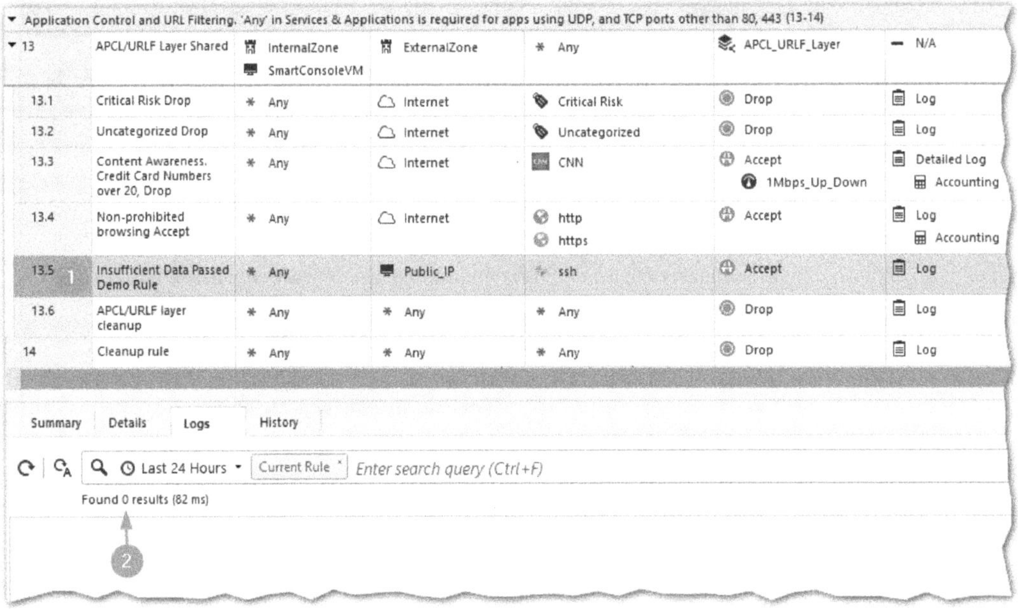

Figure 8.49 – Insufficient Data Passed, no logs in the expected location

On the odd chance that we've missed something, we can check both cleanup rules, one for the APCL/URLF layer and a general one below it (rules 13.6 and 14), but there is nothing there either. Did we just lose the packet? Not really. This situation is possible if the destination is not responding. In my case, it is a random public IP that is not listening on port 22 [1]. We are presented with the reason: **Connection terminated before the Security Gateway was able to make a decision: Insufficient data passed. To learn more see sk113479** [2]. Remember, that this is the Application Control layer. For TCP connections, a service is identified only after the server's response is received. The first packet is allowed to pass and is attributed to the parent rule, which is rule 13 [3] in our case. The connection is, in fact, only an *attempted connection* at this point.

Log Details   — ☐ ✕

Accept
ssh Traffic Accepted from 10.0.0.20 to ▨ ▨ ▨ ▨   ▨▨▨   Non-responding host

Details   Matched Rules

Log Info

		NAT	
Origin	CPCM1	Xlate (NAT) Source IP	SmartConsoleVM (200.100.0.20)
Time	Today, 8:11:47 PM	Xlate (NAT) Source Port	27981
Blade	Firewall	Xlate (NAT) Destination Po..	0
Product Family	Access	NAT Rule Number	6
Type	Connection	NAT Additional Rule Num...	0

Traffic

		Actions	
Source	SmartConsoleVM (10.0.0.20)	Report Log	Report Log to Check Point
Source Port	55010		
Source Zone	Internal	**More**	
Destination Zone	External	Id	6eb31b9b-5e41-caf5-6212-e6b1000000..
Service	ssh (TCP/22)		more
Interface	eth0	Marker	@A@@B@1645374669@C@40009
Destination	Public_IP (▨ ▨ ▨ ▨)	Log Server Origin	CPSMS (10.0.0.10)
		Id Generated By Indexer	false
		First	false

Policy

Action	Accept	Sequencenum	2
Reason	Connection terminated before the Security Gateway was able to make a decision: Insufficient data passed. To learn more see sk113479.	Security Outzone	ExternalZone
		Nat Rule Uid	66fe2601-ce1d-429a-ae44-9cd503a64836
	less		less
Policy Management	CPSMS	Db Tag	{50DC5597-4353-1B43-84E9-112953142 0D8}
Policy Name	Standard		less
Policy Date	Today, 3:02:28 PM	Logid	0
Layer Name	Standard Network	Description	ssh Traffic Accepted from 10.0.0.20 to ▨ ▨ ▨
Access Rule Name	APCL/URLF Layer Shared		less
Access Rule Number	13		

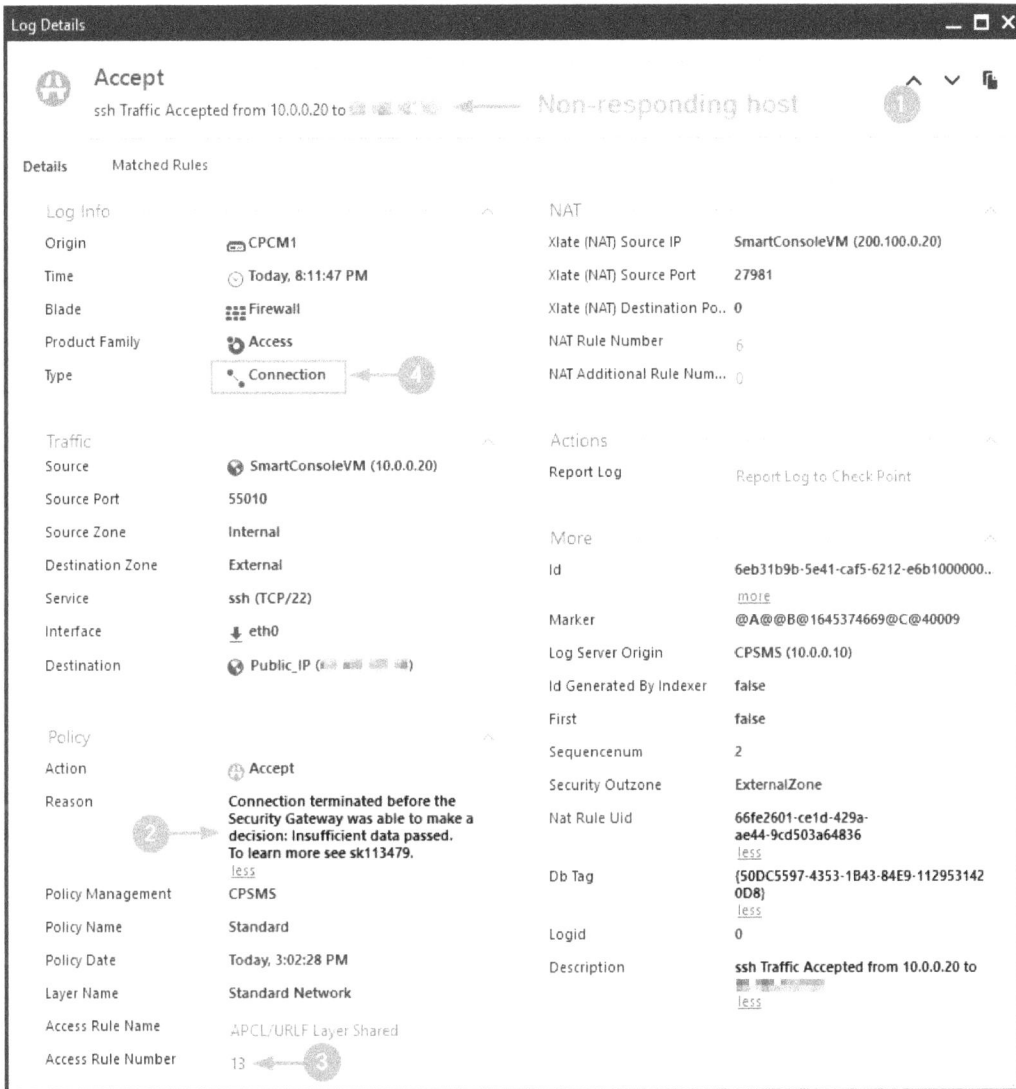

Figure 8.50 – Insufficient data passed, log for connection to a non-responding host

If you know without a shadow of a doubt that there is a host listening for a connection on the correct service port, there are a few possible explanations:

- You are experiencing asymmetric routing. In this case, the server's responses are being sent via a different route and are not returning to the gateway's interface from which this connection has originated.

- You are prohibited from connecting to that host on that port, by either an external or host-based firewall on the receiving end.

We've seen that the logs in Check Point contain a wealth of information. We have also learned that while **Rule Logs** is a convenient option, when troubleshooting, the use of **Log View** is required. I would also suggest systematic and periodic log reviews; filtering out expected logs and looking at the dropped traffic generated on internal and DMZ hosts, you can easily spot anomalies. In these cases, reach out to the system's owners to determine the reasons for this traffic. Most often, if it is dropped by the firewall and no one is complaining, it shouldn't be there.

Summary

In this chapter, we have covered policies, layers, and rules, and learned about a variety of ways policies could be structured. We have been introduced to a number of performance optimization technologies, such as CoreXL and SecureXL. We've discovered how the properties of layers and the placement of rules impact traffic acceleration and affect latency. We've learned about column-based matching and how it works with firewall/network rules, as well as rules in layers with a content inspection. We have also learned about the best approaches to the Application Control layer structure, Content Awareness, and track settings for logs based on active blades in the layers. We also discussed and explained edge cases for missing rule logs.

The next chapter is a mix of theory and practice. We will cover secure internal communication, internal certificate authority, and create a cluster object. Additionally, we will learn about object types and create some of the objects that will be needed for our first security policy.

9
Working with Objects – ICA, SIC, Managed, Static, and Variable Objects

In this chapter, we will learn how to create, modify, and use objects. We will cover the topics of **Internal Certificate Authority** (ICA) and **Secure Internal Communication** (**SIC**) and how these factor into the creation of other Check Point managed objects. We will create our first high-availability cluster as well as most of the rest of the objects representing components of our lab. We will also address the need for variable objects and learn about some of those.

In this chapter, we are going to cover the following main topics:

- Working with objects
- Object categories
- Introduction to Internal Certificate Authority and Secure Internal Communication
- Gateways and servers

- Creating networks and Host objects
- Variable objects

Working with objects

There are numerous ways in which objects can be created or modified within SmartConsole. To recap some of the ways learned in lessons from *Chapter 7, SmartConsole – Familiarization and Navigation,* and to show a few more object-specific options, let's take a look at how it is done:

1. We can create objects by going to the **Objects** menu [1], using the **New...** option in the right **Objects** tab [2], opening **Object Explorer** from either of these two menus [3], and using the **New** option in it or the context-specific **New** option [4], as shown in the following screenshot for the **GATEWAYS & SERVERS** view:

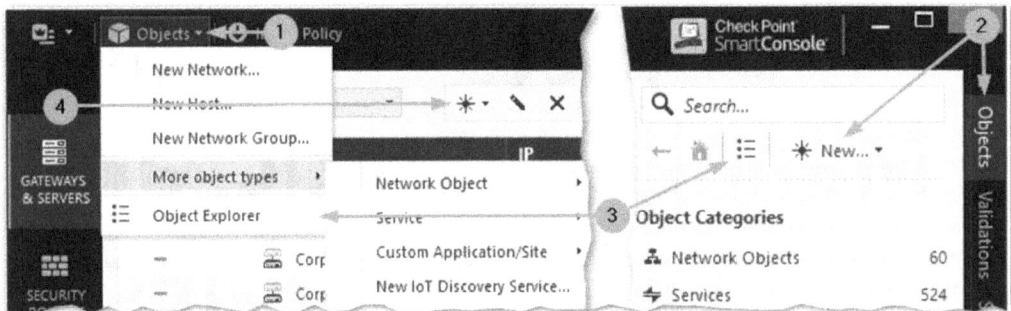

Figure 9.1 – Creating new objects options

2. To insert objects in the rules, we may click the + sign in any of the cells of our rules [1] to access the **Object Browser** [2]. To add existing objects to the cell, we can use a freeform search [3] and then, in the filtered list, click on the + sign to the left of the desired object(s) so that it turns into a checkmark [4]. We can also use the **New** option [5] within **Object Browser** to create additional objects:

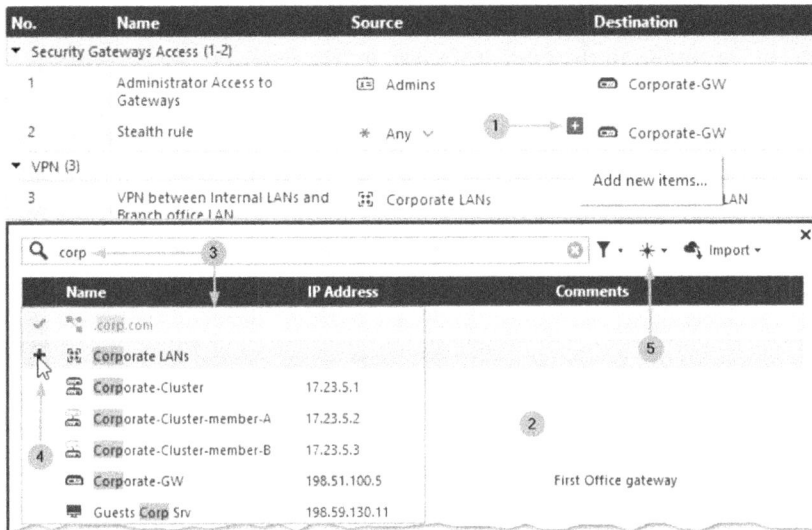

Figure 9.2 – Selecting objects for inclusion using Object Browser

3. We can select the objects in the existing rules by left-clicking, then right-clicking
    to call the context menu, where we can choose from the **Edit Object…** [1], **Group
    Selected Objects…** [2] (we are prompted with the new group name request),
    Clone… [3] (will use same properties but require a unique name), **Remove** [4]
    (that is, the selected object from the selected cell in the rule – it does not delete an
    object from database), and **Where Used…** [5] (with or without the **Replace** action)
    options. We may also check the **Negate Cell** option [6], but it is a cell-wide property,
    so if multiple objects in this cell are present, the entire effective content of that cell
    will be inverted:

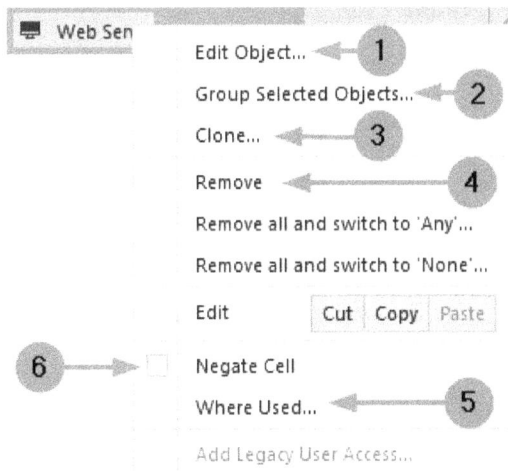

Figure 9.3 – Object context menu in the rules

4. We can also select **Delete** [1] to delete objects from the database in the UI through either a context menu in the **Object Explorer** [2] or the right-hand **Objects** tab [3]:

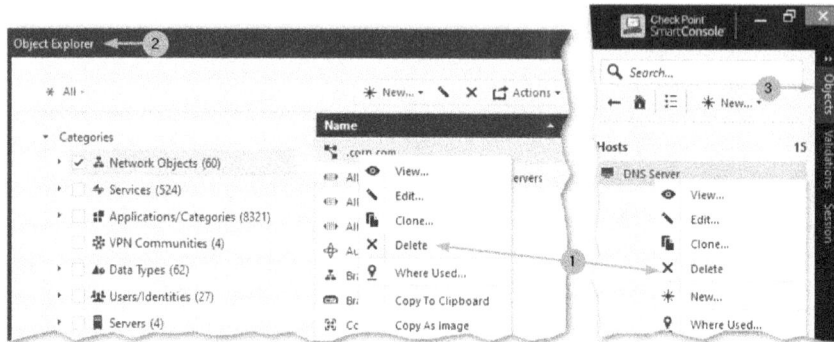

Figure 9.4 – Object context menu in the Object Explorer and right-hand Objects tab

Now that we have covered a variety of ways to work with objects in SmartConsole, let's get back to our lab and start creating objects that we'll use later in our policy.

Object categories

In SmartConsole, click the **Objects** tab [1] on the right side of the interface to see the general object categories available [2]. To dive deeper and see all the available object types within those categories, click on the **Object Explorer** icon [3] and then on the expansion triangle to the right of the **New...** button [4]. When you do, you'll see an expanded list of main categories:

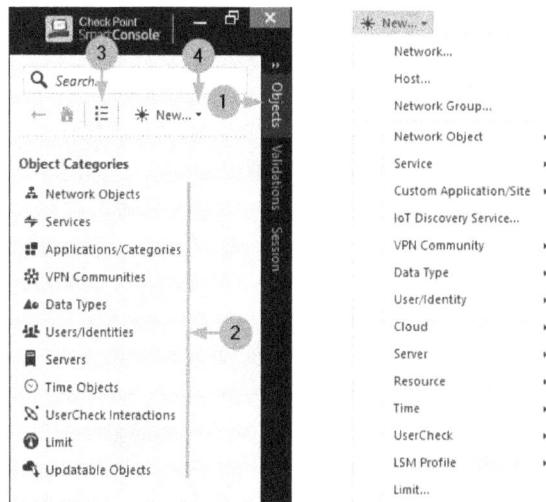

Figure 9.5 – Object Categories

There are too many predefined object types to list but to give you an idea, here is an abbreviated list of object types within these categories:

Object Categories	Object Types
Network Objects	Check Point gateways and servers, interoperable devices (VPN peer gateways), hosts, networks, IP ranges, zones, domains, dynamic objects, and logical servers (load balancers)
Services	Services and service groups
Applications/Categories	Applications, categories, and mobile applications
IoT Discovery Service	Third-party services for IoT visibility
VPN Communities	Site-to-site or remote access VPN communities and their topologies, such as star (hub and spoke) or mesh
Data Types	File attributes and data types such as US social security numbers, credit card numbers, and medical record numbers
Servers	AAA servers (RADIUS and TACACS) or groups, trusted and subordinate certificate authorities, Syslog, and SIEMs
Users/Identities	Locally defined users and groups or those managed by external identity services or identity providers
Cloud	Either explicit objects in any number of different cloud service providers' environments, such as AWS, Cisco ACI/ISE, Google Cloud Platform, Microsoft Azure, Kubernetes, Nuage, OpenStack, VMware NSX/NSX-T/vCenter, Generic Data Center, and IoT Cloud Adapter or the results of a data center query to all data centers, specific or a select group
Resource	URI, URI for QoS, SMTP, FTP, TCP, CIFS, and MMS
Time	Time (date and time ranges) and time groups
UserCheck	Ask, Inform, and Drop
LSM Profile	A number of large-scale management options
Limit	Upload, download, or bidirectional bandwidth rates
Updatable Objects	Lists of the IP addresses and the FQDNs maintained by various service providers

Table 9.1 – Object Categories

In addition to the specific categories listed here, there are two broad categories that objects fall into: **static** and **variable**.

Static and variable object categories

Note that the term **variable object** is not used by Check Point, but I will shortly explain what this refers to.

Static objects are either those whose properties are defined by Check Point administrators or those that are imported from external sources (data centers) and remain unchanged for the lifespan of the object or unless manually edited. Hosts, networks, and IP ranges are examples of static objects. Changes in the properties of these objects require the installation of a policy to take effect.

Variable objects, on the other hand, are placeholders – they could be thought of as groups with dynamic membership determined outside of the objects database. Changes in the properties of these objects do not require policy installation. We will cover variable objects later in this chapter.

Now that we've seen the object categories and types of objects in each category, let's get back to our lab and proceed with the creation of the objects for the left side of the environment described in our lab topology diagram.

> **Returning to the Lab**
>
> Verify that the following VMs, **CPSMS**, **SmartConsole**, **CPCM1**, **CPCM2**, **Router**, and **ADDCDNS**, are running in VirtualBox.
>
> Install the SmartConsole application in **SmartConsole VM** from the \\ LabShare mapped drive, open the SmartConsole application, and log on to CPSMS using the IP address 10.0.0.10, username admin, and password CPL@b8110.
>
> In our lab environment, the only object present in **GATEWAYS & SERVERS** when we are logged in via SmartConsole is **CPSMS**, our management server.

We should now create a cluster object **CPCXL**, comprised of two member gateways, **CPCM1** and **CPCM2**, as well as other objects used in our lab. Before we proceed with these tasks, let's introduce two additional concepts, SIC and ICA.

Introduction to Internal Certificate Authority and Secure Internal Communication

Both **Internal Certificate Authority** (**IAC**) and **Secure Internal Communication** (**SIC**) are related to the use of certificates. Let's take a look at them in the order of dependency.

Internal Certificate Authority

When the first management server is installed (and designated as the primary during Gaia First Time configuration Wizard), an ICA is created. ICA is responsible for the issuance, renewal, and revocation of certificates to all components of your Check Point infrastructure for SIC. ICA is additionally used to generate VPN certificates to gateways and clusters used for authentication between the same VPN community members, as well as user certificates for either internally managed users or users managed on LDAP servers.

Most operations of ICA are performed in SmartConsole objects' properties using one-click operations, including certificates for internally managed users. Additionally, the `cpconfig` CLISH utility and a dedicated ICA Management Tool on the management server (for advanced operations) are available.

For users managed on LDAP servers, use the ICA Management Tool. It is capable of bulk operations based on files created with an LDAP query. This is warranted in larger environments that do not have either a dedicated PKI or a third-party **Certificate Authority** (**CA**).

If a dedicated PKI is present or third-party CA services are used, an additional trusted CA object(s) must be created and the external CA's own management tools used for the administration of user certificates. Additionally, gateways authenticating these users should have a certificate issued by the same CA. Gateways may have multiple VPN certificates issued by ICA as well as other trusted CAs.

The ICA Management Tool is disabled by default.

> **Note**
> The following step should be performed only if you are building a new environment or are working in a lab environment. In existing production environments, do **not** attempt to modify the internal CA without fully accounting for consequences and having a comprehensive plan tested in a lab environment.

If, at the time of creation of your first (primary) management server, its hostname does not contain a domain name, reinitialize the internal CA using cpconfig: choose the (6) Certificate Authority option and change the name of the internal CA to the host.domain format. In our case, it would look as follows:

```
CPSMS> cpconfig

This program will let you re-configure your Check Point Security Management
Server configuration.

Configuration Options:
----------------------------

(1)   Licenses and contracts

(2)   Administrator

(3)   GUI Clients

(4)   SNMP Extension

(5)   Random Pool

(6)   Certificate Authority

(7)   Certificate's Fingerprint

(8)   Automatic start of Check Point Products

(9) Exit

Enter your choice (1-9) :6

Configuring Certificate Authority...

========================================
The Internal CA is initialized with the following name: CPSMS


Do you want to change it (y/n) [n] ? y


Please enter the name of this Internal CA: cpsms.mycp.lab

Are you sure you want to change the Internal CA name (y/n) [n] ? y

Trying to contact Certificate Authority. It might take a while...

 Certificate was created successfully

cpsms.mycp.lab was successfully set to the Internal CA

Done
```

Figure 9.6 – Changing the ICA name

> **Note**
> In high-availability management, the ICA name could be different from that of the primary management server. In that case, an additional DNS A record pointing to the primary/active management server should be created.

Secure Internal Communication

SIC is a secure communication channel between Check Point infrastructure components. It assures confidentiality, integrity, and authenticity of communication between nodes.

It uses SIC certificates issued by ICA with TLS for the creation of secure channels and relies on AES 128 for encryption.

To establish initial trust, a communication initialization process relying on activation keys is used for the gateways and Check Point servers to securely request and obtain SIC certificates.

Once the communication initialization process is successfully completed, **Trust Established** is displayed in the general properties of gateways, clusters, and non-primary management server components. We may invoke an on-demand test of the SIC and be presented with one of three outcomes:

- **Communicating** – when the SIC is successfully established and subsequent communication between Check Point hosts and gateways will be secure

- **Not Communicating** – when the management server can reach the gateway, but SIC fails

- **Unknown** – when communication issues between the management server and the node in question are encountered

SIC is dependent on the **Check Point Daemon Process** (**CPD**) responsible for the communication between modules that include licensing, policy push/fetch to and from gateway, and online updates.

> **Important Note**
> SIC requires the time on the participating components to be approximately in sync. It will tolerate small deviations.

For our lab specifically, having VMs suspended, powered down, and/or without access to NTP servers causes time drift on your Check Point hosts and Windows servers. Before continuing, please set the time manually on all powered-up components in the lab to the same value as displayed on your LabHost PC.

For Check Point CPSMS, CPCM1, and CPCM2, log on to Gaia and perform the following:

```
CPXXX> show date
Date 02/02/2022
CPXXX > show time
Time 18:19:17
```

If the values you are seeing are incorrect, set them to an accurate date/time, as shown here (`CPXXX` represents any of the Check Point hostnames in our lab):

```
CPXXX > set date 2022-03-03
CPXXX > set time 10:15:00
CPXXX > show date
Date 02/03/2022
CPXXX > show time
Time 10:15:06
CPXXX > save config
CPXXX >
```

…and to set the proper date and time in your Windows VMs (SmartConsole and ADDCDNS), in a PowerShell console in each VM, perform:

```
PS C:\Users\Administrator> set-Date "03/03/2022 10:46:00 AM"
```

Now, with the time synchronized, we can move on to creating the rest of the components of our lab, starting with the CPCXL cluster, comprising two cluster member gateways. In the process of doing this, we'll also learn more about gateways and servers.

Gateways and servers

Gateways and servers, although a subset of the **Network Object** category, refer to the Check Point gateways, clusters, VSX systems, their components (virtual systems, virtual routers, and virtual switches), and Check Point Hosts (management high availability, log, SmartEvent, or SmartEvent Correlation servers). Let's create our cluster CPCXL.

Activation keys

If you recall, during the FTW process for gateways, destined to become cluster members CPCM1 and CPCM2, in **Secure Communication to Management Server** prompt [1], we specified an activation key [2]:

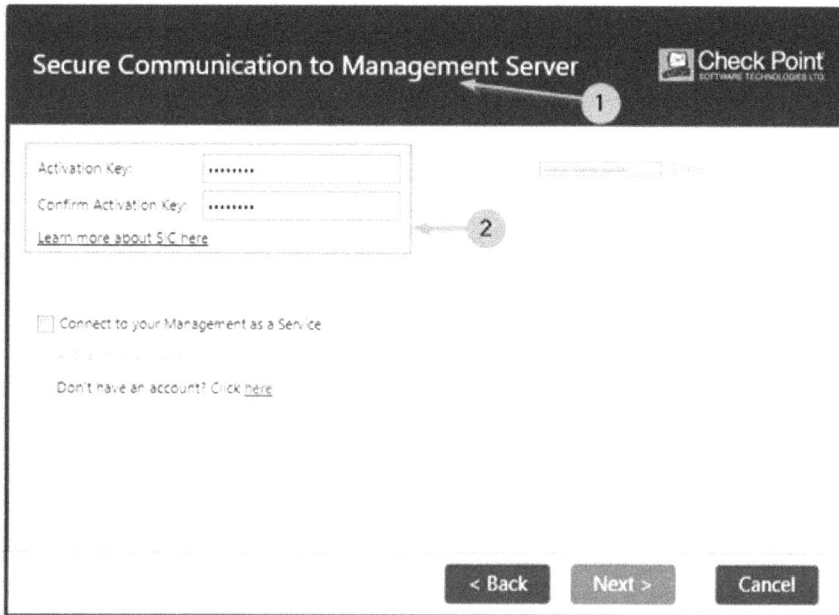

Figure 9.7 – SIC activation key configuration during Gaia FTW

We will use these keys in SmartConsole during the creation of the cluster and the addition of members gateways to it, as described in the next section.

Creating a gateway cluster

Gateways and clusters could be created in either Wizard or Classic Mode. Let's build one using Wizard Mode:

1. Open SmartConsole and select **GATEWAYS & SERVERS** [1]. Click on the new icon [2], then click on **Cluster** [3] and **Cluster...** [4]:

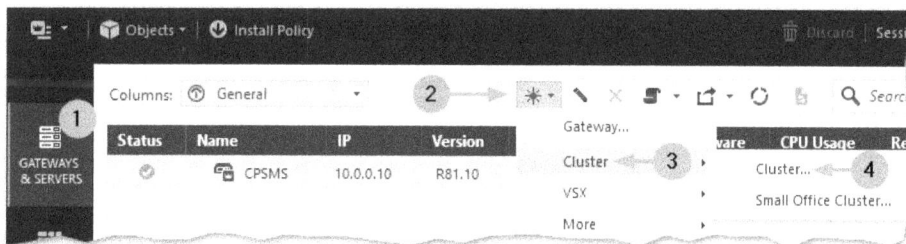

Figure 9.8 – Creating a new cluster object from the GATEWAYS & SERVERS context menu

2. In the **Check Point Security Gateway Cluster Creation** window [1], you are prompted with two ways of creating a cluster. Click on **Wizard Mode** [2]:

Figure 9.9 – Gateway Cluster Creation Wizard Mode

3. In **Cluster General Properties** [1], in the **Cluster Name** field, type CPCXL [2], and in **Cluster IPv4 Address** field, type 200.100.0.1 [3]. Verify that in the **Choose the Cluster's Solution** section, the **Check Point ClusterXL** [4] and **High Availability** [5] options are selected. Click **Next** [6]:

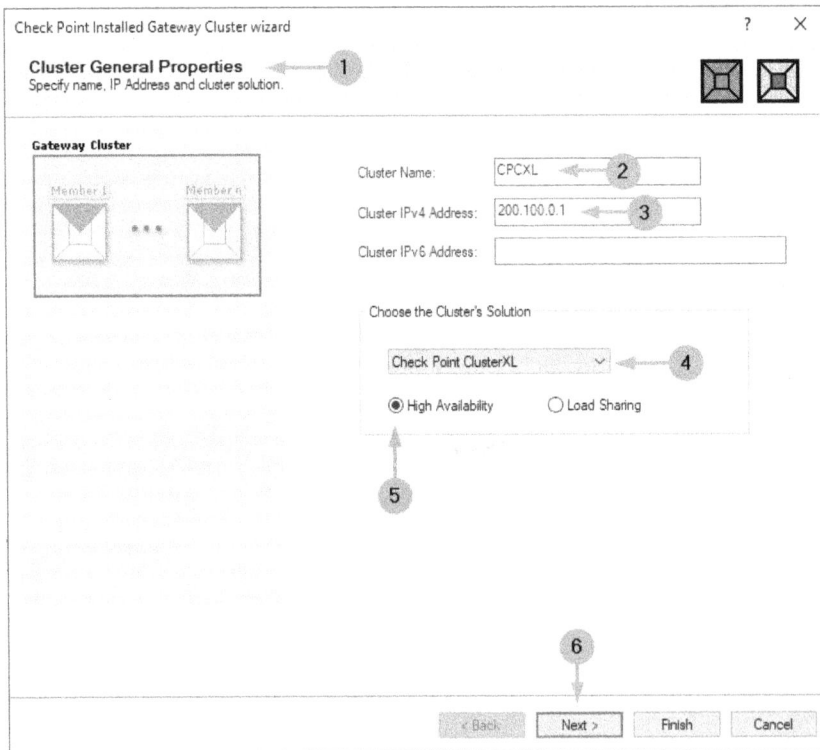

Figure 9.10 – Wizard Mode, Cluster General Properties

4. In the **Cluster members' properties** window [1], click **Add...** [2] and then **New Cluster Member...** [3]:

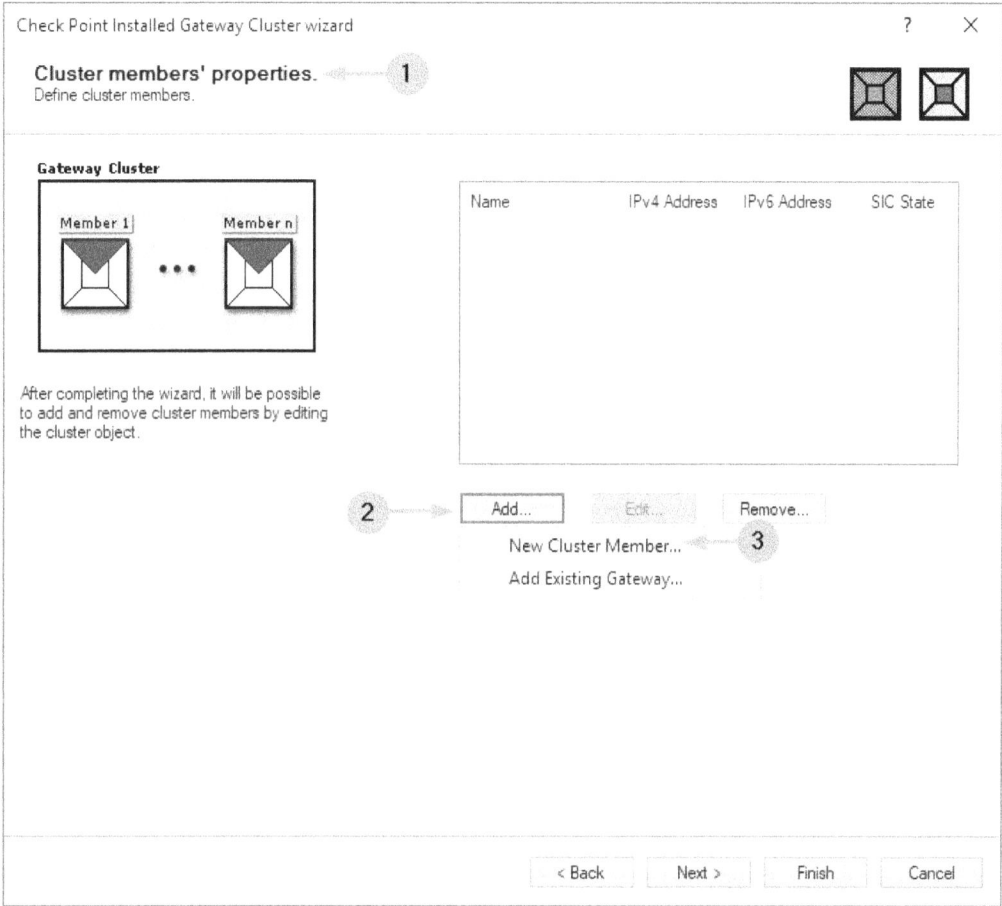

Figure 9.11 – Cluster members' properties, adding a cluster member

5. In **Cluster Member Properties** [1], set the name as CPCM1 [2] and the IP address of its management interface, which is 10.0.0.2 [3]. In the **Activation Key** and **Confirm Activation Key** fields, enter the same key as was used during the Gaia configuration for cluster members, CPL@b8110 [4]. Note the **Trust State** status is **Uninitialized** [5] and click **Initialize** [6]:

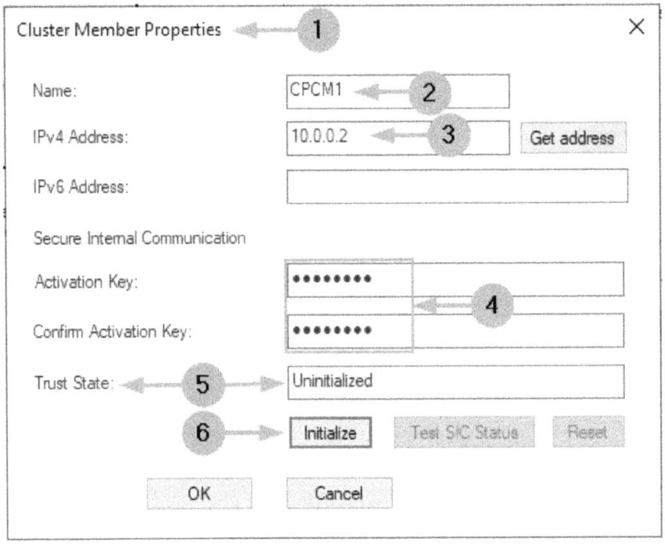

Figure 9.12 – Cluster Member Properties and Trust State initialization

6. You will see the trust state change to **Trust established** [1]. Click **OK** [2]:

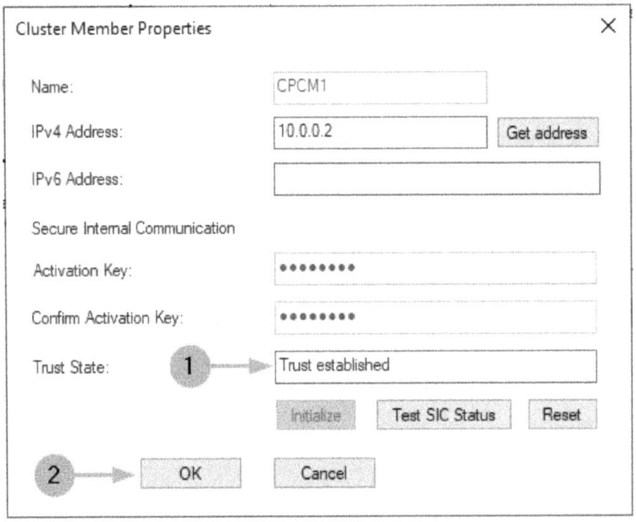

Figure 9.13 – Cluster Member Properties Trust established

7. Now, in **Cluster members' properties**, you will see the first cluster member added with its SIC state as **Trust established** [1]. Click **Add...** to repeat the process for the second cluster member [2]:

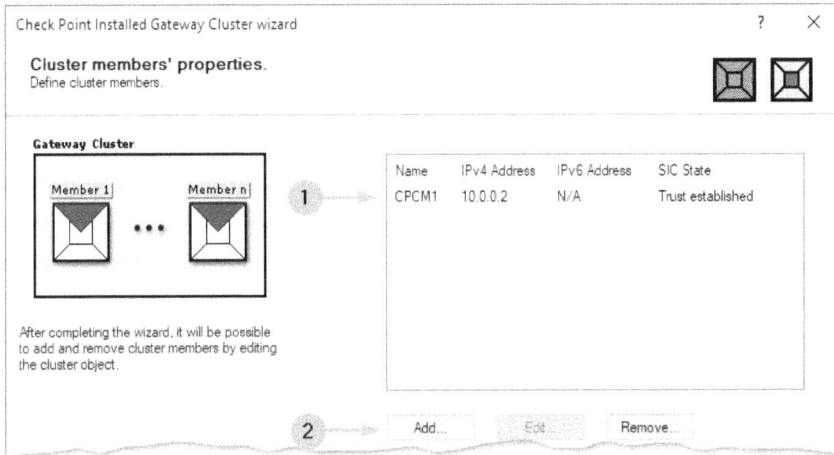

Figure 9.14 – Cluster members' properties, adding a second member

8. Repeat the process, adding the second cluster member, CPCM2, using the IP address of its management interface, 10.0.0.3. Once completed, your **Cluster members' properties** screen should have both members showing a SIC state of **Trust established** [1]. Click **Next** [2]:

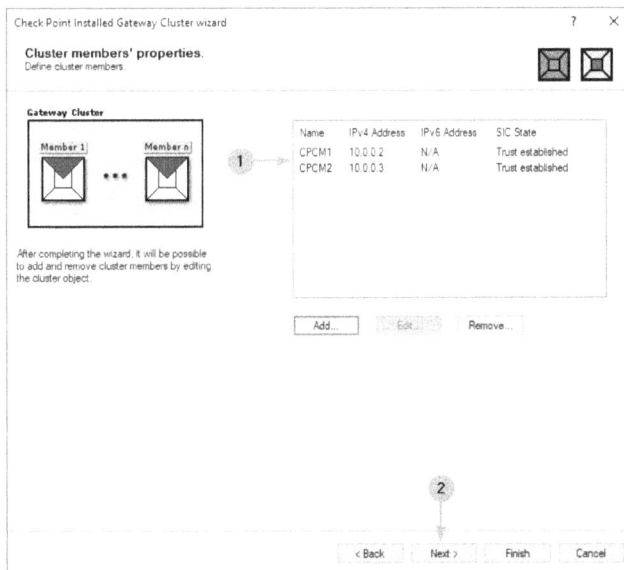

Figure 9.15 – Cluster members' properties, Trust established showing for all members

9. In the **Check Point Installed Gateway Cluster wizard** window, note the description of the next stage, the configuration of the topology of the cluster, and the roles of each connected network [1]. When done, click **Next** [2]:

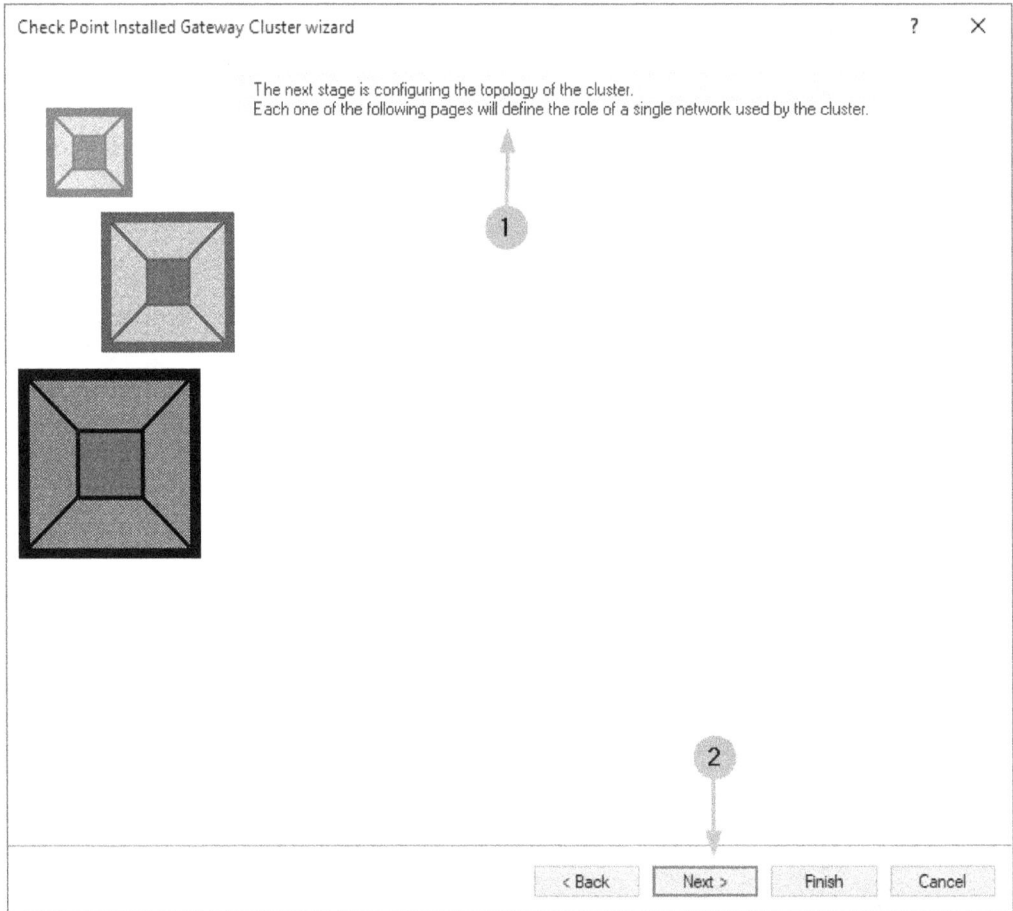

Figure 9.16 – Gateway cluster wizard, moving on to topology configuration

10. You will be presented with the **Cluster Topology** window [1], containing one of the networks in its **IPv4 Network Address** and **Net Mask** fields [2]. All the networks in our lab, except **Sync**, are represented by a cluster interface. Check the **Representing a cluster interface** option [3] and assign a virtual **Interface IPv4** address and **Net Mask** [4], as per the following screenshot. Click **Next** [5] to continue:

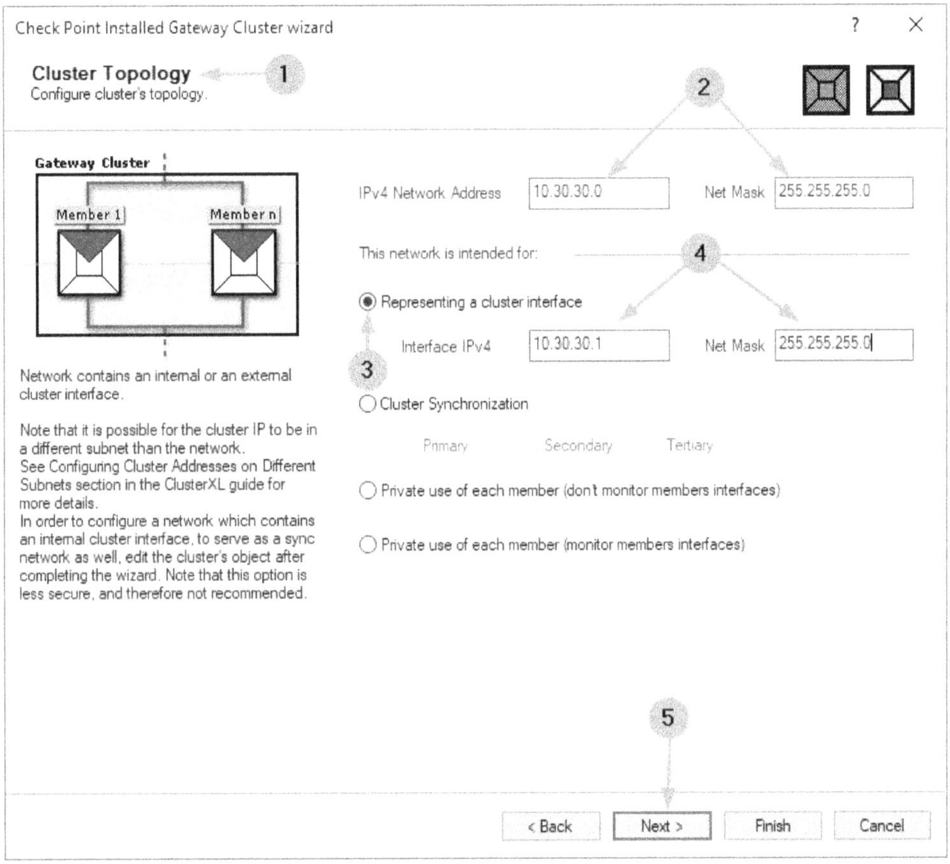

Figure 9.17 – Gateway cluster wizard, cluster topology for cluster interfaces

11. Repeat the process for the rest of the networks using the IP addresses of the cluster's virtual interfaces. When **IPv4 Network Address** representing a sync network is encountered [1], only the **Cluster Synchronization** [2] and **Primary** [3] options should be selected. This network does not have a virtual IP address assigned, as it does not represent a cluster interface:

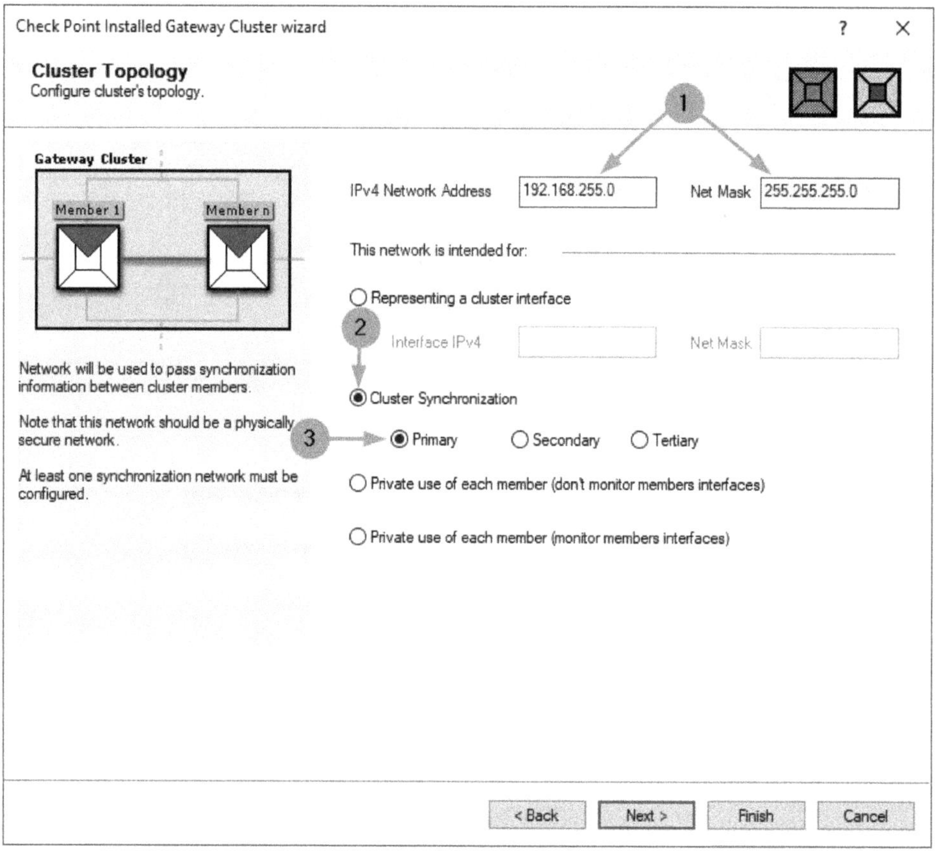

Figure 9.18 – Gateway cluster wizard, cluster topology for sync interfaces

12. Continue the process until you see **Cluster Definition Wizard Complete** [1] and click **Finish** [2]:

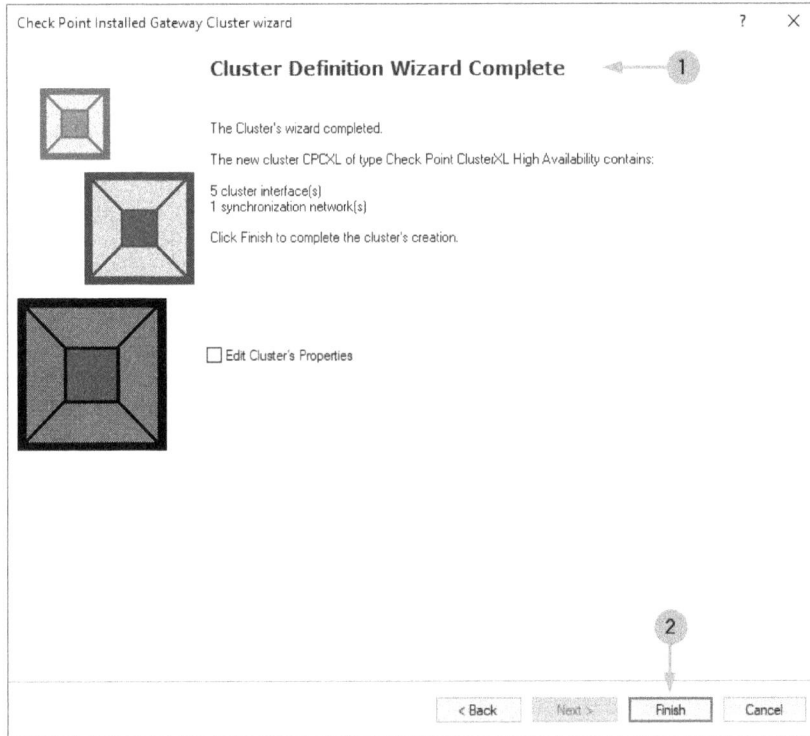

Figure 9.19 – Gateway cluster wizard completion

13. Now, in the **GATEWAYS & SERVERS** view of SmartConsole [1], we can see the **CPCXL** cluster object [2]. Note the number of changes in the **Session** field and click **Publish** to commit [3]:

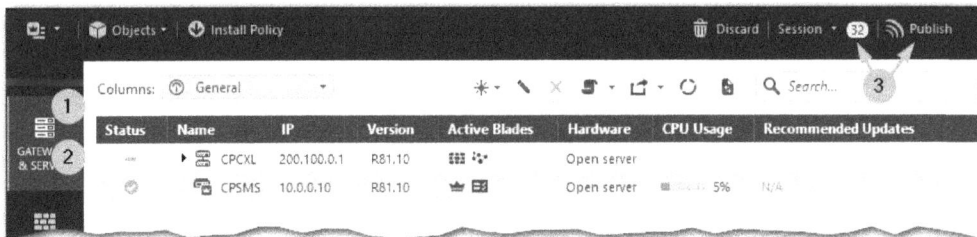

Figure 9.20 – SmartConsole GATEWAYS & SERVERS new cluster view

14. In SmartConsole, double-click on the **CPCXL** cluster object and in **Gateway Cluster Properties - CPCXL** [1], click on the **Network Management** tree option [2]. Note that the interfaces' names are not yet normalized [3], which is common for Open Server installations of Check Point. This issue is never encountered when using Check Point appliances. Also note that in **Topology** [4], the **External** network is recognized based on its IP address being outside of the RFC 1918-defined ranges [5]. To accurately enumerate the names of the interfaces, click on the drop-down arrow to the right of **Get Interfaces** [6]:

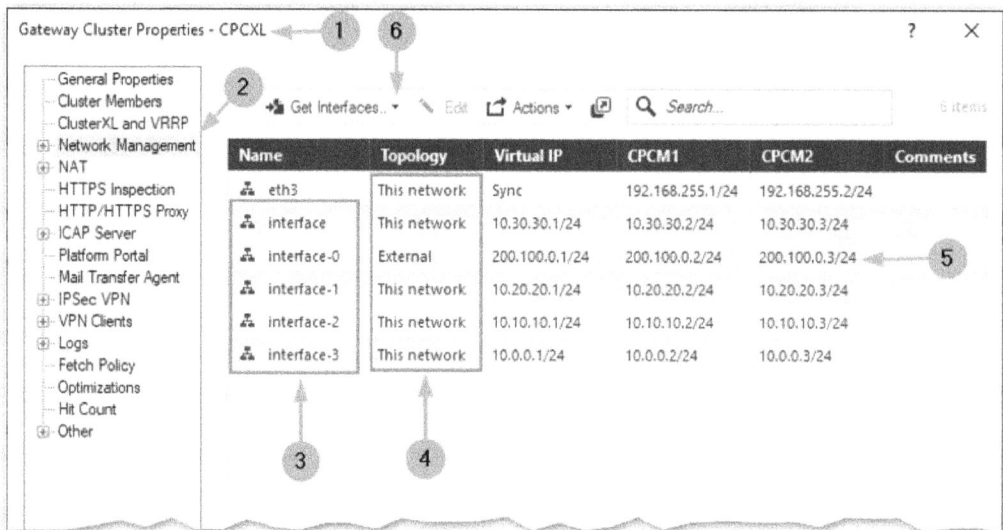

Figure 9.21 – Gateway Cluster Properties interface enumeration

15. **Important Note**: The **Get Interfaces…** drop-down list [1] contains two options: **Get Interfaces With Topology** and **Get Interfaces Without Topology**. I recommend *never* using the first option. **Get Interfaces With Topology** may wreak havoc in your environment years down the road. NIC swaps, topology reconfigurations, and anti-spoofing settings that were never properly reflected in configuration may all come into effect. The much safer option is **Get Interfaces Without Topology** [2]:

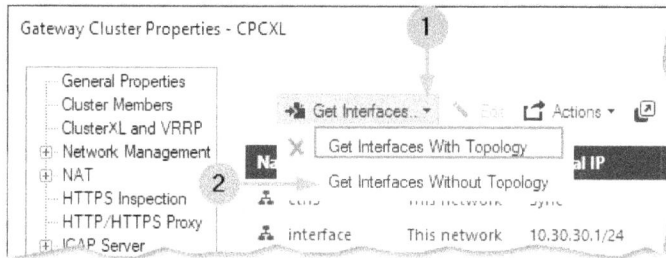

Figure 9.22 – Gateway Cluster Properties Get Interfaces Without Topology

16. Although you may still get a scary prompt [1], nothing except interface names and relevant IP addresses will be adjusted. Click **Yes** to proceed [2]:

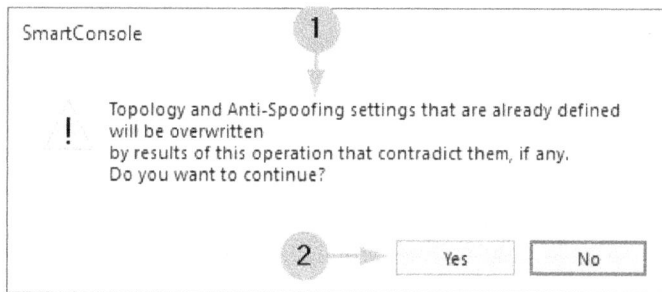

Figure 9.23 – Topology and Anti-Spoofing warning

17. Now our interfaces are properly named and sorted:

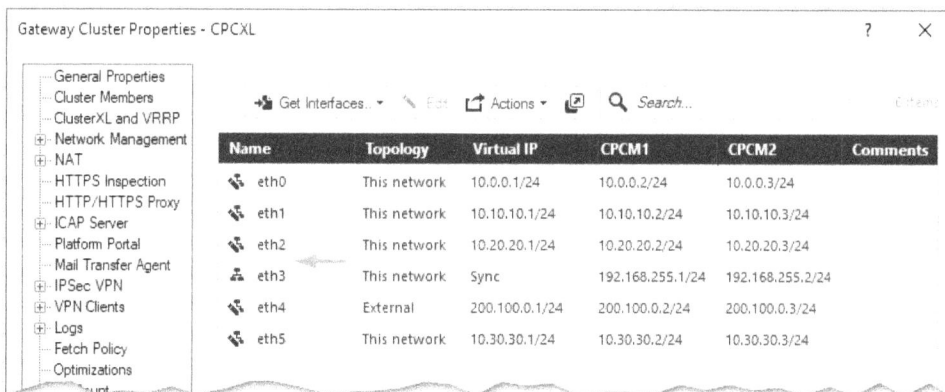

Figure 9.24 – Gateway Cluster Properties with accurately enumerated interfaces

Now, let's define the **Topology** settings for the interfaces by assigning relevant zones and entering interface comments.

18. Double-click on **eth1**. In the **Network: eth1** properties window [1], click once on **Enter Object Comment** [2] and type `Internal1`. In the **Topology** section, click on **Modify...** [3]:

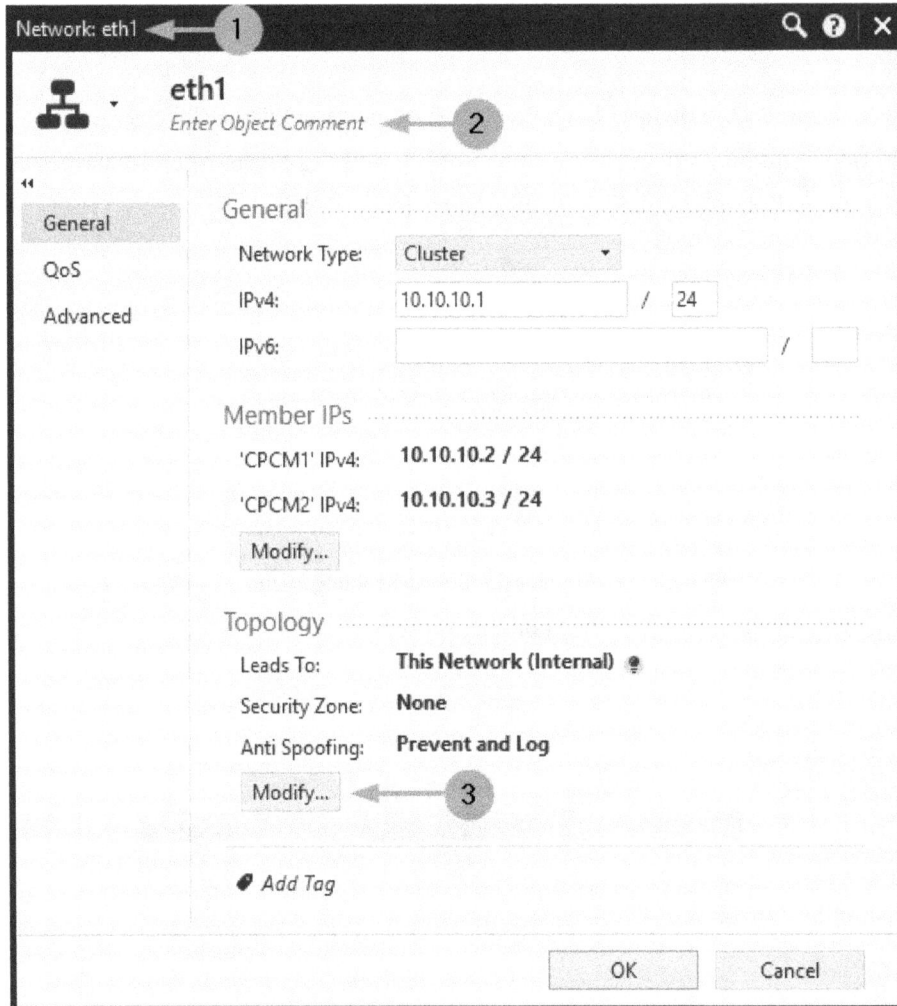

Figure 9.25 – Interface topology modification

19. In the **Topology Settings** pop-up window's [1] **Security Zone** section, check **According to topology: InternalZone** [2]. Note that you have the ability to manually specify security zones there by choosing the **User defined | Specify Security Zone** option. Click **OK** [3] here and in a parent window to commit the topology changes:

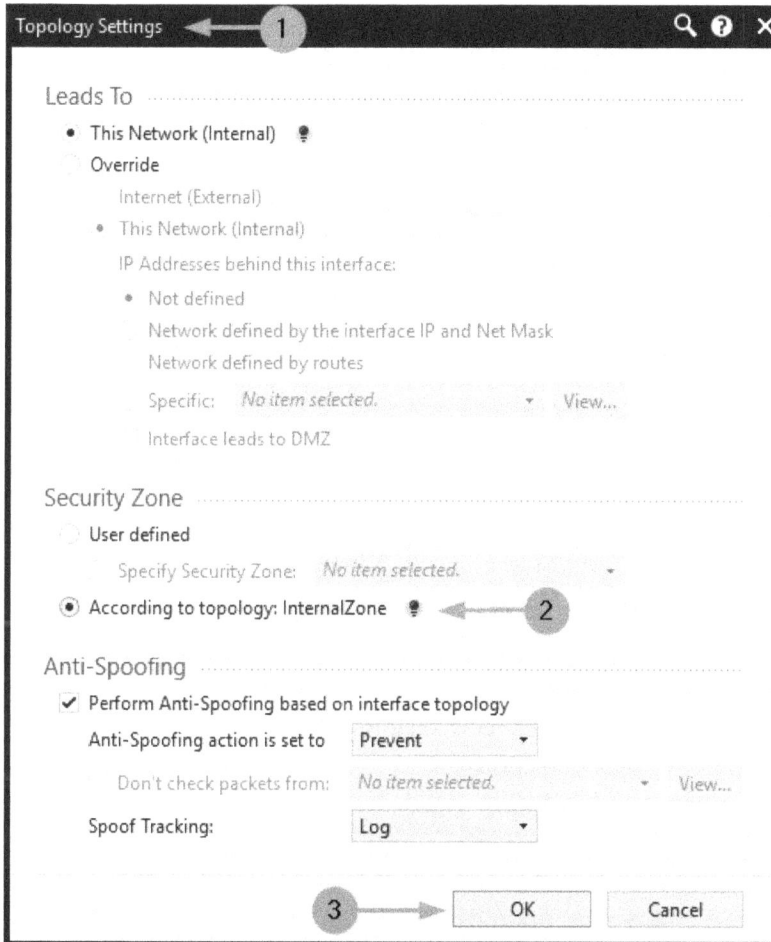

Figure 9.26 – Interface Topology Settings configuration

Repeat the same actions for the **eth2** interface using `Internal2` for the **Enter Object Comment** field and choosing **InternalZone**.

For **eth0**, set the comment to **Mgmt**.

For **eth3**, set the comment to **Sync**.

For **eth4**, set the comment to **External** and modify the **Topology Settings** page's **Security Zone** section to **According to topology: ExternalZone**.

For **eth5**, set the comment to **DMZ**, modify the **Topology Settings** page's **Leads To** section to **Override**, and select **This Network (Internal)** and **Network defined by the interface IP and Net Mask**; check the box for **Interface leads to DMZ**. Change the **Security Zone** setting to **According to topology: DMZZone**.

When all the actions described are completed, the **Network Management** screen of **Gateway Cluster Properties** should look like this:

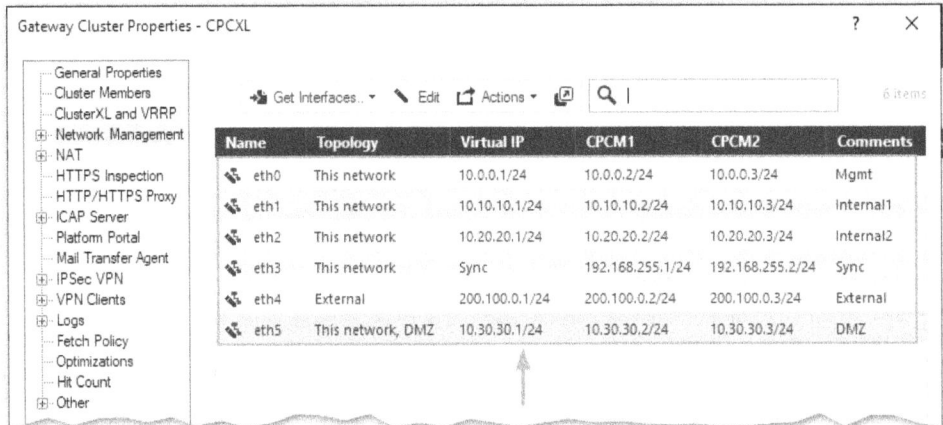

Figure 9.27 – Gateway Cluster Properties Network Management desired state

20. Click **OK** to complete the cluster's network configuration.

21. You may see a pop-up window describing the platform portal page address, as shown in the following screenshot [1]. This is material for single gateways only; as in the case of clusters, you'll be redirected to a specific active member's address. Click **OK** [2] to complete the process:

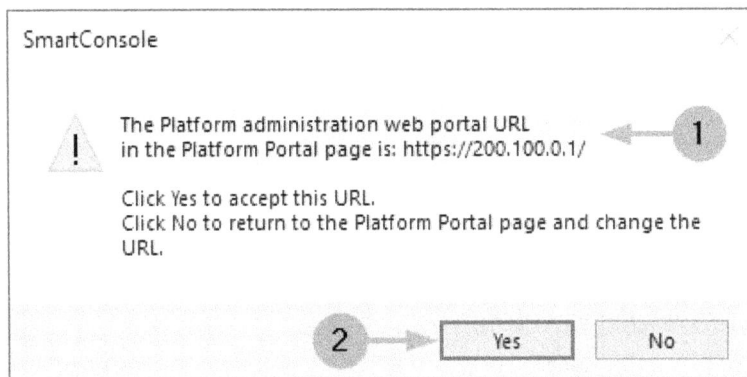

Figure 9.28 – Platform administration web portal notice and cluster applicability

22. Publish the pending changes, including a brief description of the changes in this session. Develop a habit of including short descriptions of your actions by including a short note summarizing them. If you have a ticketing system, include the ticket number that these actions are a part of [1] and click **Publish** [2]:

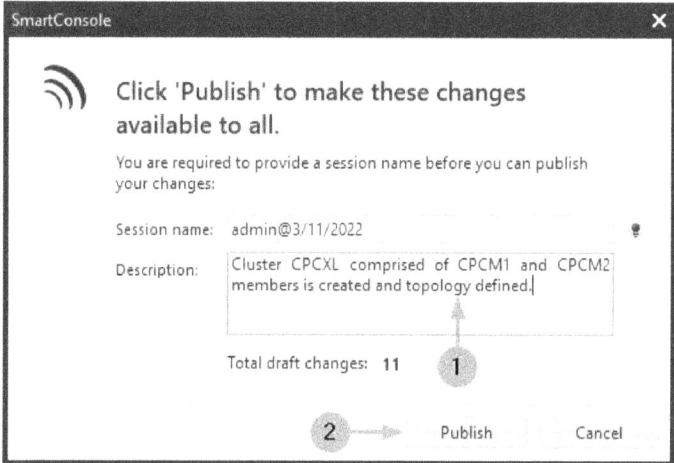

Figure 9.29 – SmartConsole publish changes and session description

23. Our cluster is now created. In the **GATEWAYS & SERVERS** view [1], you should now see that the status of the cluster and individual cluster members (visible when you click the triangle icon) [2] has changed to green [3]:

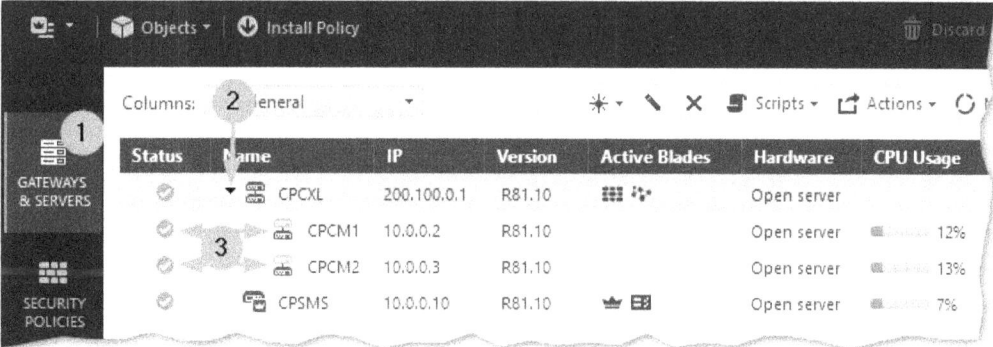

Figure 9.30 – GATEWAYS & SERVERS new cluster state

The CPCXL cluster's state is now sufficient for us to carry on with further studies. You have probably noticed that there are a lot more options in the navigation tree of the gateway cluster properties. We will be exploring some of those later in the book as we touch on the relevant subjects.

In *Chapter 12, Configuring Site-to-Site and Remote Access VPNs*, we'll use Classic Mode to create an additional gateway for our lab.

To illustrate how to create additional server objects (also referred to as a Check Point Host object), let's click on the **New** icon [1] in the **Actions** menu of the **GATEWAYS & SERVERS** view, click **More** [2], and then click **Check Point Host...** [3]:

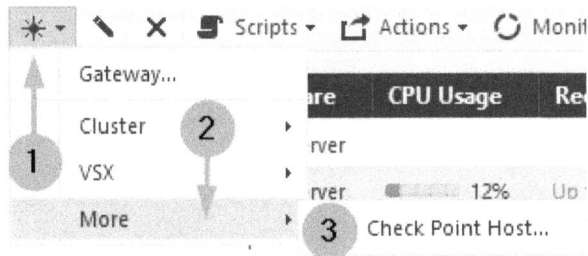

Figure 9.31 – Creating a Check Point Host

In the **Check Point Host** window, fill in **Name** [2] and **IPv4 Address** [3]. Note that there is a **Communication...** button present and **Secure Internal Communication** is set to **Uninitialized** [4]. The settings in the **Platform** section of the window are defaulted to the same version and hardware as our primary management server [5] and could be adjusted as needed. Click on the **Management** tab [6]. Here, you'll see the selection of features/blades available for the installation on the Check Point Host object. If we were implementing a redundant management server, both **Network Policy Management** and **Secondary Server** would be selected [7] as well as **Logging & Status** [8]. There should be no **Network Security** or **Custom Threat Prevention** options selected unless you are creating an all-in-one (also referred to as standalone) management and gateway object:

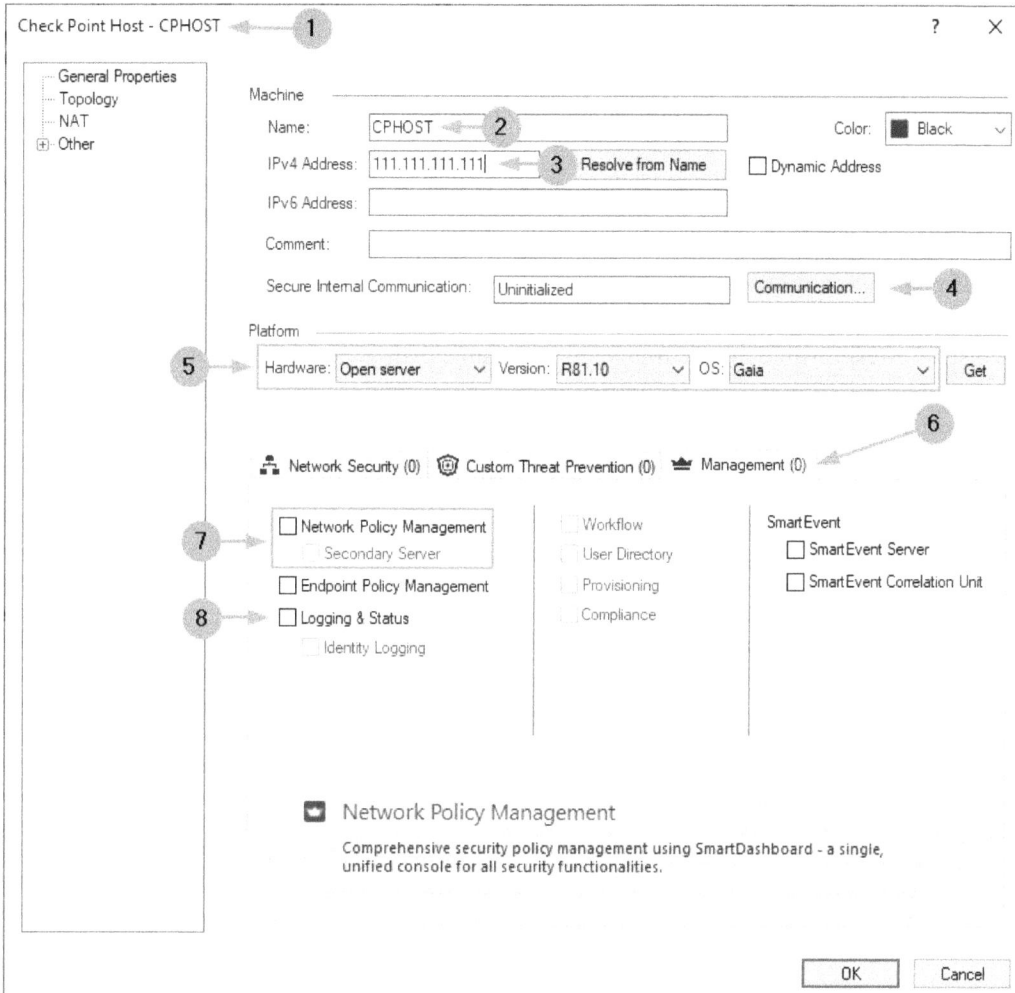

Figure 9.32 – Check Point Host object properties

When any of the blades in the **Management** tab are selected, relevant sections appear in the navigation tree on the left side of that window. The topology definition for the Check Point hosts is more of a record-keeping function since there are no anti-spoofing configuration options present.

It is now safe to cancel out of this object and discard any pending changes in SmartConsole.

Anti-Spoofing

In *Figure 9.26*, at the bottom of the **Topology Configuration** window, we could see the **Anti-Spoofing** section. Let's zoom in on it in *Figure 9.33* [1]. Spoofing is the method of falsifying the source address of packets in order to bypass access controls. The **Anti-Spoofing** settings should almost always be set as shown in the following screenshot, with **Perform Anti-Spoofing based on interface topology** checked [2], **Anti-Spoofing action is set to** set to **Prevent** [3], and **Spoof Tracking** set to **Log** [4]. If in your environment you are for some reason seeing spoofed traffic hitting your gateway, it is almost always an indication of an improperly configured topology or malicious activity. If you are a new administrator responsible for the management of an existing Check Point installation, I recommend checking the **Anti-Spoofing** settings on all gateways in your environment. Should you see option [2] disabled (unchecked), [3] set to **Detect**, or **Spoof Tracking** set to **None**, you are likely dealing with asymmetric routing in a poorly configured network. If there is a legitimate reason for ignoring some of the spoofed traffic, use the **Don't check packets from** option [5]:

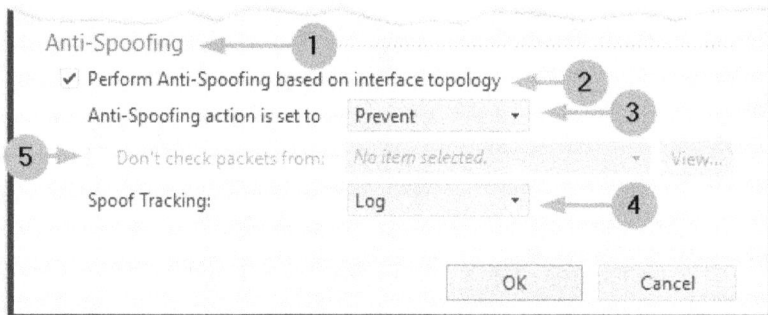

Figure 9.33 – Anti-Spoofing settings

Some cases where it is justifiable to have anti-spoofing disabled are as follows:

- Check Point Security CheckUp installations, typically performed by your reseller during **proof of concept**

- Check Point appliance deployed in a dedicated threat detection capacity

In both cases, you are mirroring traffic from multiple networks to the Check Point appliance.

Additionally, this may be warranted when packet broker appliances are in use and in other edge cases.

This covers the gateways and servers introduction and we can now move on to creating the rest of the objects representing components of our infrastructure.

Creating networks and Host objects

Now is the time to add the rest of the components of our lab infrastructure to the objects database. In *Chapter 7, SmartConsole – Familiarization and Navigation*, we saw how to do that using the management CLI. This time around, we'll be using the UI to achieve this and to see the configuration options available in these objects' properties.

Networks

New Network, **New Host**, and **New Network Group** are available as permanent shortcuts under both the **Objects** menu on the top left of SmartConsole and the **New** button in the right-hand **Objects** tab. Click on **Objects | New Network**. In the **New Network** [1] window's **Enter Object Name** field [2], enter Net_10.0.0.0. In the **Enter Object Comment** field [3], enter Mgmt Network. In the IPv4 **Network address** field [4], enter 10.0.0.0, and in the **Net mask** field [5], enter 255.255.255.0. Click on the **Add Tag** icon [6], type management, and press *Enter*. The **Groups** option [7] allows you to add this network to an existing network group(s) or to create a new group in place. **NAT** [8] will be discussed in *Chapter 10, Working with Network Address Translation* as a separate subject. Your first network object should look similar to the following screenshot. Click **OK** [9] to complete the network creation process:

Figure 9.34 – Network object

Open **Object Explorer** [1]. In its search field, enter 10.0.0.0 [2]. As you type, you will see the number of possible results shrinking until only objects containing the typed value are present. Right-click on **Net_10.0.0.0** [3] and click **Clone...** [4]:

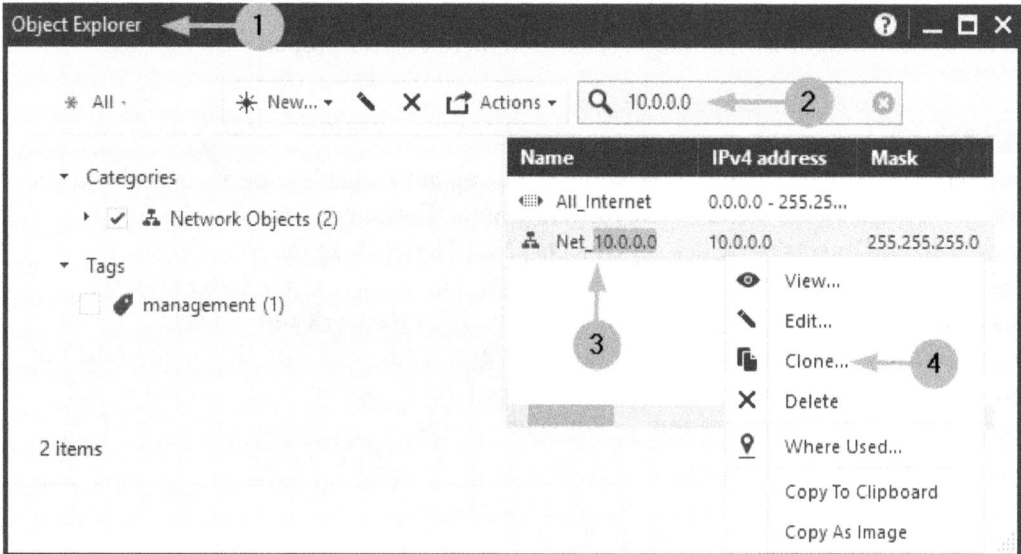

Figure 9.35 – Searching for and cloning objects

In the **New Network** window that opens after **Clone...** is clicked, the name of the cloned object is appended with _Clone, but the rest of the values are identical. Click in each field to replace the values as per the following table for each cloned network. Skip the first one, as it is already created. The **Object Comment** and **Tag** values are optional:

> **Note**
> I do not recommend using spaces or special characters in object names. This
> may cause issues with scripting and automation as these could be interpreted as
> delimiters. The use of an underscore or a dash is preferable.

Object Name	Object Comment	Network Address	Net Mask	Tag
Net_10.0.0.0	Mgmt Network	10.0.0.0	255.255.255.0	management
Net_10.10.10.0	Int1 Network	10.10.10.0	255.255.255.0	Internal1
Net_10.20.20.0	Int2 Network	10.20.20.0	255.255.255.0	Internal2
Net_10.30.30.0	DMZ Network	10.30.30.0	255.255.255.0	dmz
Net_192.168.255.0	Sync Network	192.168.255.0	255.255.255.0	sync
Net_200.100.0.0	Ext Left Network	200.100.0.0	255.255.255.0	external; left
Net_200.200.0.0	Ext Right Network	200.200.0.0	255.255.255.0	external; right
Net_172.16.16.0	Int Right Network	172.16.16.0	255.255.255.0	internal; right

Table 9.2 – Lab network objects

That's it for the networks and we can now move on to the hosts.

Hosts

Continuing in the **Object Explorer**, click **New | Host**. In the **New Host** window [1], in the **Enter Object Name** field, type SmartConsole_VM [2]. In the **Enter Object Comment** field, type SmartConsole, WebUI and SSH client [3], and in the **Machine** section's **IPv4 address** field, enter 10.0.0.20 [4]. Click **Add Tag** and enter management [5]. **Groups** [6] allows us to select or create new network groups that this host will be a member of. **Network Management** [7] is just for record keeping in case the host is multihomed (containing multiple interfaces with different IP addresses). **NAT** [8] will be discussed separately in *Chapter 10, Working with Network Address Translation*. **Advanced** [9] allows you to configure SNMP parameters for the host to allow either retrieval of the name, location, contact, and description values or setting them from the object's properties, if both read and write communities are configured. In my experience, this option is seldom used in practice. **Servers** [10] allows us to choose one or more of three available options: **Web Server**, **Mail Server**, or **DNS Server**:

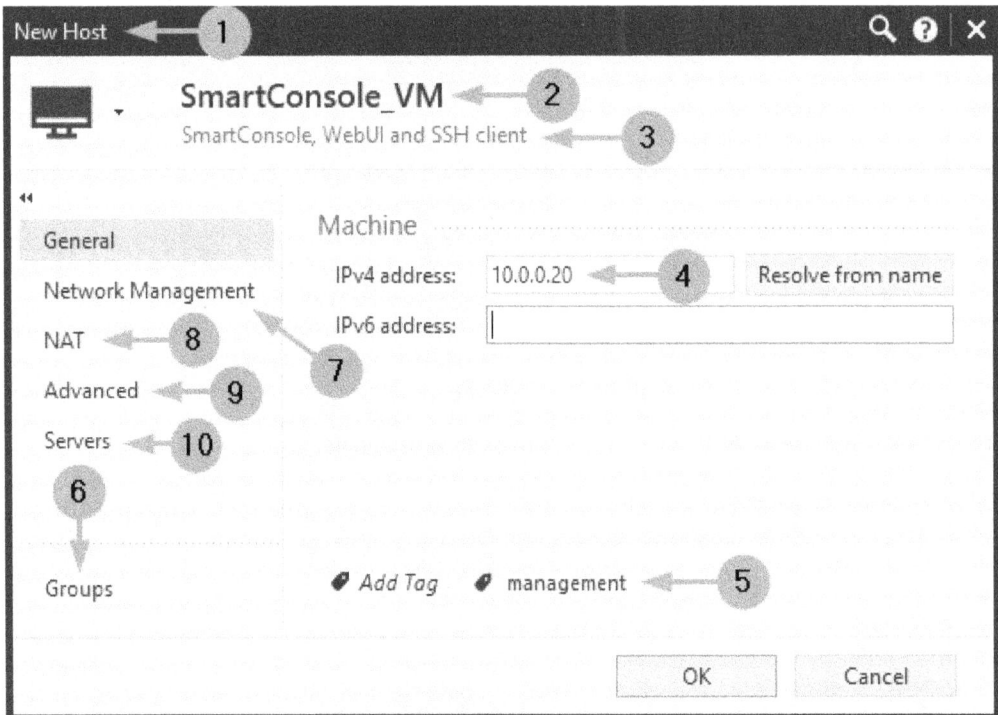

Figure 9.36 – Host object

Each selected option in **Servers** is then manifested in the object's properties navigation tree, which contains the **Configuration** and **Protections** branches [1]. There, we can define the server's **Operating System** [2], **Application Engine** [3], and **Ports Configurations** [4] properties and select the gateways protecting it [5]. The host's server configuration is relevant for threat prevention, which is not a focus of this book, but just to illustrate, when we click on **Servers** and check **Web Server**, this is the view of the **Configuration** branch:

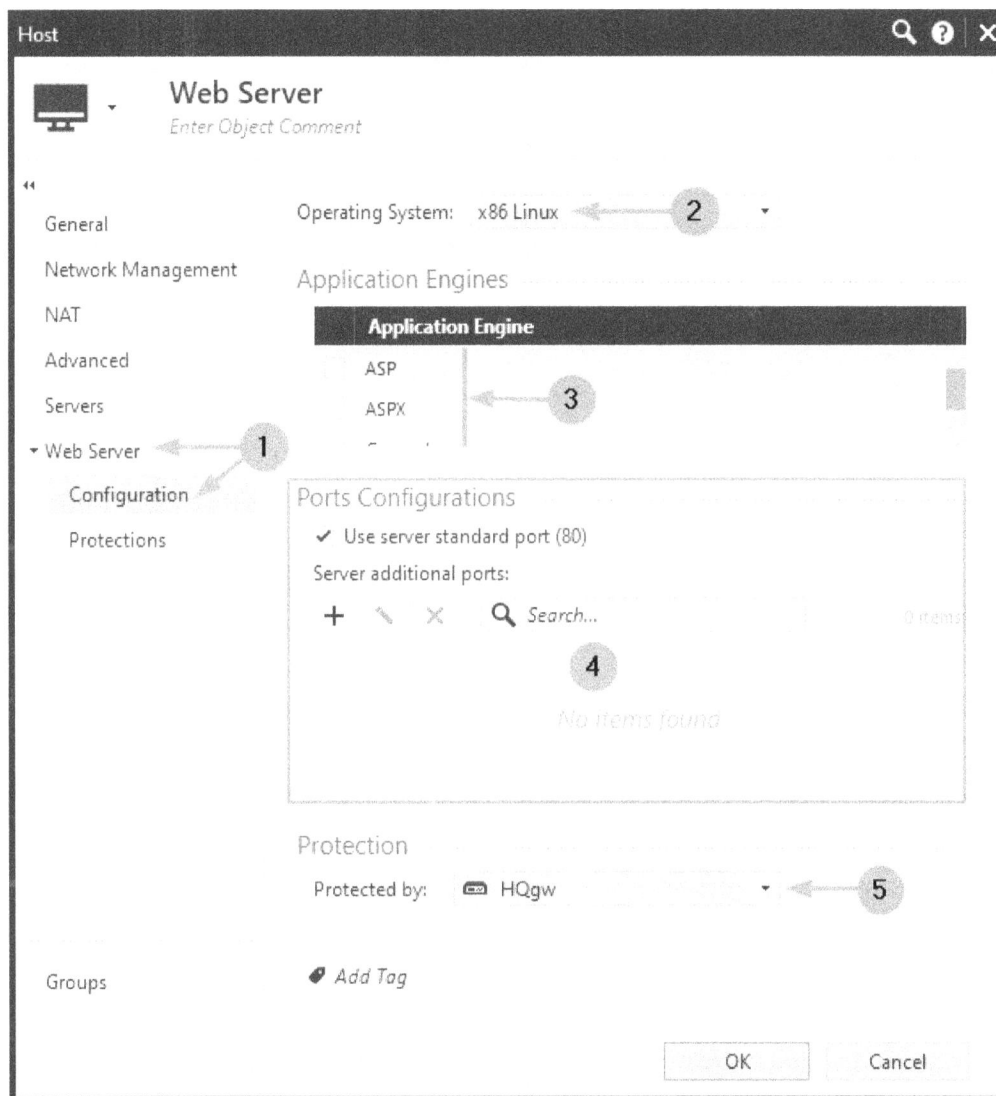

Figure 9.37 – Additional Configuration options for Host defined as a Web Server

Selecting the **Protections** branch [1] shows a warning about the dependency of these settings on the chosen threat prevention profile and points you to **Help** [2]. Object-specific help is a click away; just select the question mark icon [3]. Server-specific protection options can also be selected and adjusted here [4]:

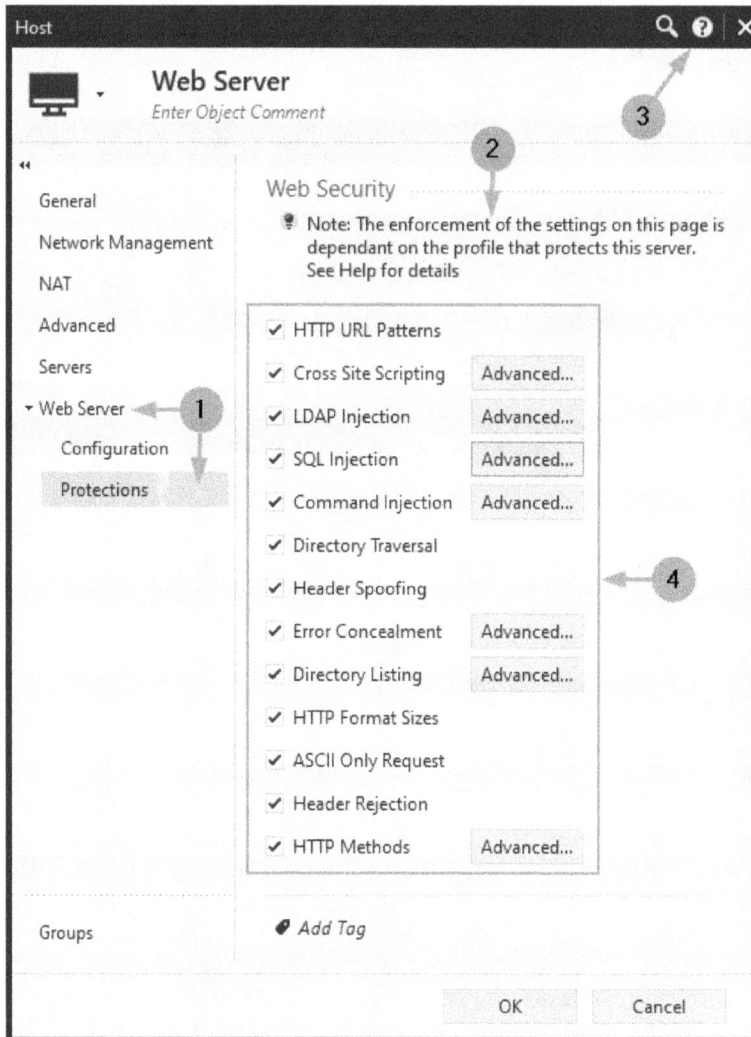

Figure 9.38 –Protections options for a Host defined as Web Server

If you have selected one of the server options for our SmartConsole_VM object, return to the server branch and uncheck all the boxes there before clicking **OK**.

Now, you may create the rest of the objects using the **New | Host** method for each, using the values from the following table. Omit `SmartConsole_VM` as it has already been created:

Object Name	Object Comment	IPv4 Address	Tag
`SmartConsole_VM`	`SmartConsole, WebUI and SSH client`	`10.0.0.20`	`management`
`LeftHost`	`Generic Windows host`	`10.10.10.10`	`internal1`
`ADDCDNS`	`AD Domain Controller, DNS server`	`10.20.20.10`	`internal2`
`DMZSRV`	`DMZ Web Server`	`10.30.30.5`	`dmz`
`RightHost`	`Generic Windows host`	`172.16.16.10`	`internal`
`RouterLeft`	`VyOS eth1`	`200.100.0.254`	`external`
`RouterRight`	`VyOS eth2`	`200.200.0.254`	`external`

Table 9.3 – Lab Host objects

You may now publish the pending changes to commit them to the database.

Also recall that in the *The WHAT'S NEW popup recall and management scripting CLI and API* section of Chapter 7, *SmartConsole – Familiarization and Navigation*, we learned how to create objects in bulk.

In subsequent chapters, we'll be creating a few more objects, such as network groups, but this is enough for us to get started. Read on to learn about some of the other types of objects that allow greater flexibility in overall security posture management.

Variable objects

Earlier in this chapter, I mentioned **variable objects**, and now is the time to expand on them a bit. Unlike static objects, which either have static IP addresses or use DHCP but are represented by static objects, such as networks or ranges in the policies, variable objects' addresses can be changed dynamically and are not confined by predefined networks or ranges. As was mentioned earlier, these could be thought of as groups with conditional membership defined outside of the Check Point objects database.

I would have used the term dynamic objects to describe this category, but that term is already reserved by one of the object types under the broader variable objects definition.

So, let's start with dynamic objects since this term has just come up.

Dynamic objects

When a new dynamic object [1] is created in **New | Network Object | Dynamic Objects | Dynamic Object**, the object itself is empty – it does not have anything in its properties except a name [2], description [3], and tags [4]. All except the name are optional:

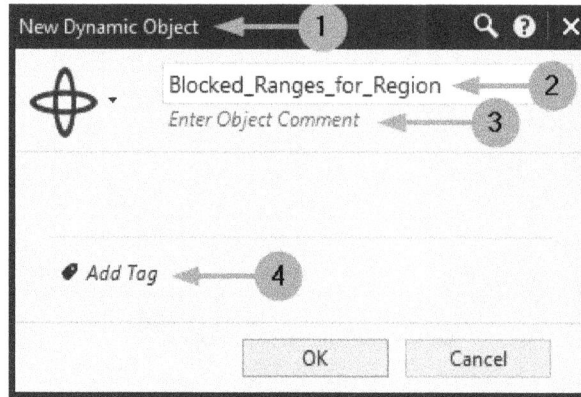

Figure 9.39 – Dynamic object properties in SmartConsole

It could now be included in a rule of the access control policy:

Name	Source	Destination	Services & Applications	Action
Dynamic Object Use Example 2	SmartConsole_VM	Blocked_Ranges_for_Region	Any	Drop

Figure 9.40 – Dynamic object in access control policy rule

This policy must then be installed on gateways. Policy installation is only required once for a new dynamic object to become active.

Then, identically named objects must be created and the content of these objects must be populated via the CLI *on each gateway* where you want to enforce the rules containing it.

Specifically for clusters, dynamic objects should be created and populated on each cluster member.

There is currently no API for dynamic objects, which is bad, but they are very easy to manipulate, which is good. It does require some scripting to fully take advantage of their capabilities.

Dynamic objects consist of a numbered list of IP ranges. Individual IPs should be presented as a range with duplicate addresses.

The same dynamic object may have a completely different list of IP addresses on each gateway or a cluster (such as a site-specific list) it is installed on. Subsequent changes to the dynamic object are done via the CLI and take effect immediately without the need to install the policy. On high-availability clusters, it becomes active when populated on an active cluster member.

Dynamic objects on gateways could be loaded from file, created as empty, created with data, and updated with changes. Individual ranges could be added or deleted to/from dynamic objects.

In Expert Mode, use `dynamic_objects -h` to see the command options.

Here is an example of the creation of a dynamic object on the gateway and the addition of the second range consisting of a single IP address:

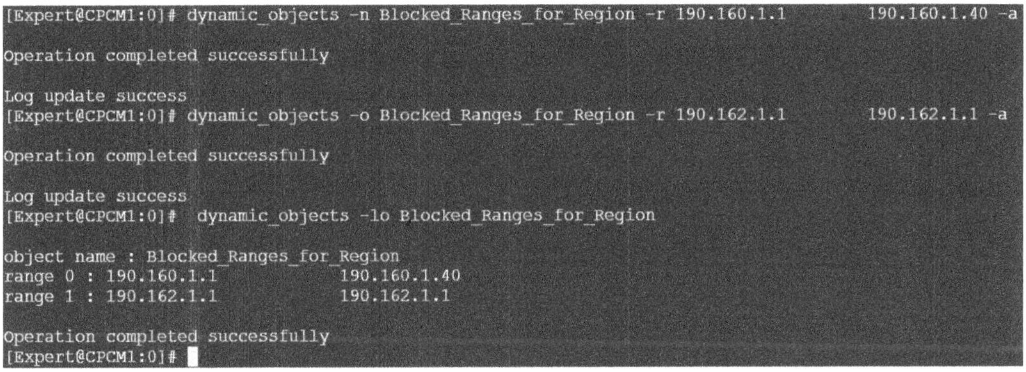

```
[Expert@CPCM1:0]# dynamic_objects -n Blocked_Ranges_for_Region -r 190.160.1.1        190.160.1.40 -a

Operation completed successfully

Log update success
[Expert@CPCM1:0]# dynamic_objects -o Blocked_Ranges_for_Region -r 190.162.1.1        190.162.1.1 -a

Operation completed successfully

Log update success
[Expert@CPCM1:0]#  dynamic_objects -lo Blocked_Ranges_for_Region

object name : Blocked_Ranges_for_Region
range 0 : 190.160.1.1         190.160.1.40
range 1 : 190.162.1.1         190.162.1.1

Operation completed successfully
[Expert@CPCM1:0]#
```

Figure 9.41 – Dynamic object CLI tools

> **Note**
>
> If you have decided to create a dynamic object as an exercise, delete it now from the CLI and SmartConsole.

Note that while the creation of a dynamic object in SmartConsole and its addition to the rule are logged in the audit logs, CLI operations with dynamic objects are logged in the logs but not indexed and, therefore, are not searchable.

The effect on traffic by the rules containing the dynamic objects, however, is indexed and easily searchable by the name of the object if tracking on the rule is enabled.

Additionally, access control policy installation on the cluster where a dynamic object is present on one member but not the other will succeed and generate a warning:

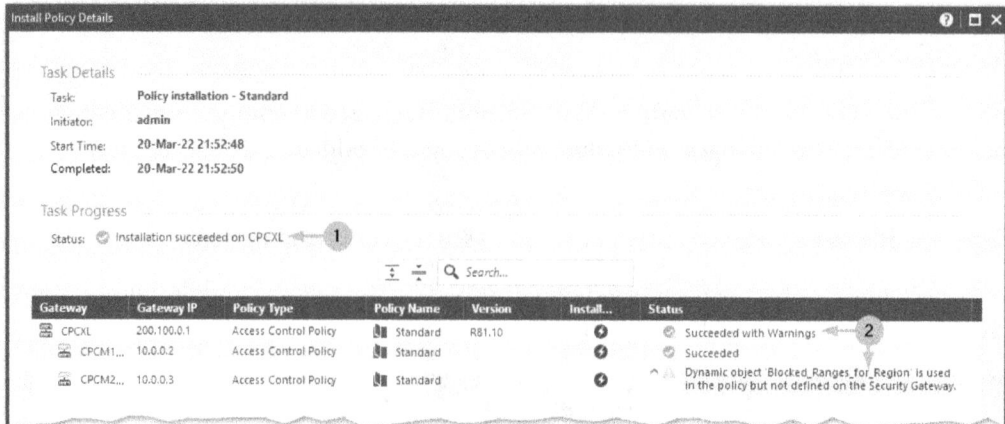

Figure 9.42 – Policy installation warning – dynamic object not defined on cluster member

But discrepancies between the contents of the dynamic objects will not be flagged, as those may be intended by the administrator.

An example of a script for creating and dynamically populating blacklists in dynamic objects can be found at `http://opendbl.net/#checkpoint.html`.

Zones (conditional)

Zones can be used in a very similar fashion to dynamic objects. This is possible if a zone is assigned to the interface with topology defined as shown in *Figure 9.43*.

For example, in interface **eth2** [1], on the **Topology Settings** page (accessible when **Modify...** [2] is clicked), **Override** [3], **This Network (Internal)** [4], and **Network defined by routes** [5] are enabled and **Security Zone** is set to either **User defined** or **According to topology** [6]:

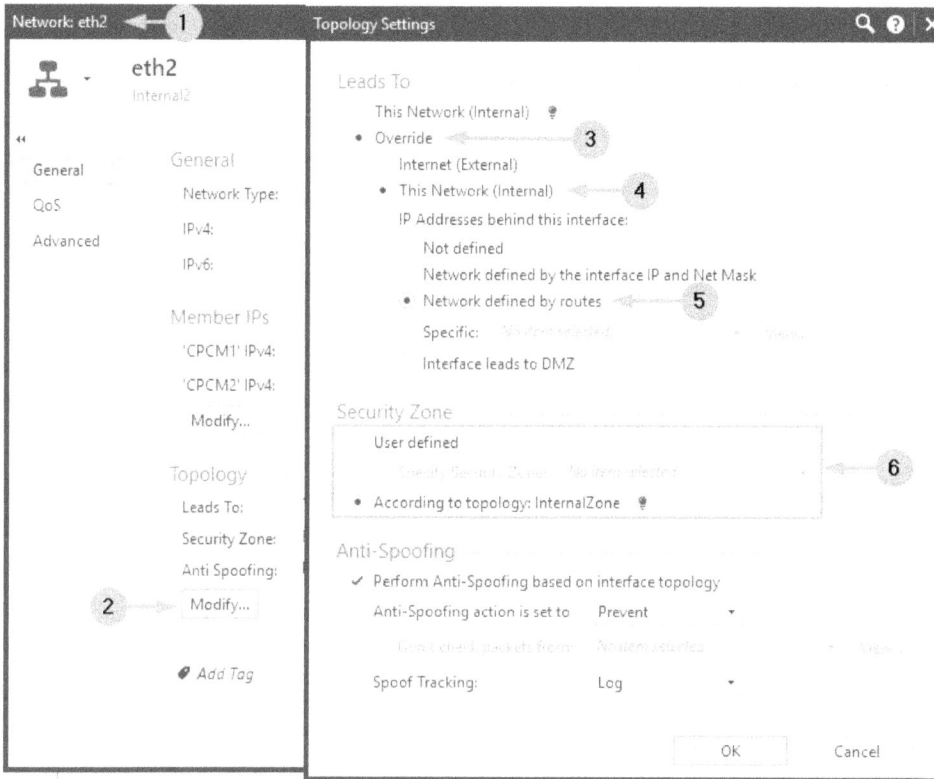

Figure 9.43 – Zones with dynamic properties based on routing

In this case, if either static routes relevant to that interface are changed or you are using dynamic routing (all configured in Gaia), rules containing this zone object will be dynamically adjusted.

Anti-spoofing, too, will account for the dynamic changes in the topology without the installation of the policy.

Domains

Domains provide a very simple and convenient way to define objects that may resolve to different IPs based on DNS records. The default option that is checked in the properties of **Domain** objects [1] is **FQDN** [2] and it should remain that way. The **Domain** object with **FQDN** is defined in the .domain.tld format or, as an example, .hackaday.com [3], *with the period preceding the rest of the string.*

For a simple domain object such as `.domain.tld`, an additional query for `.www.domain.tld` is automatically performed. Query results are cached for 60 seconds. Results returning responses of NXDOMAIN (not existing domain) will suppress the next query for 60 minutes. If you need to use services other than www, construct objects with prefixes as `.service.domain.tld`, in this example, `.api.dnsimple.com`.

> **Note**
>
> I am not sure why Check Point has elected to use this format for FQDN domain objects, as the standard FQDN format requires the period to be appended at the end. Perhaps this is just an attempt to easily recognize the properly formatted FQDN name if the length of the domain name is such that we cannot see the end of the string onscreen.

Domain objects are created by going to **New** | **More** | **Network Object** | **More** | **Domain**.

Figure 9.44 – Domain (FQDN) object

Even though the domain objects are defined, created, and modified in SmartConsole, we must use associated CLI tools *on the gateways* where the policies containing these objects are installed, and not on the management server. Normally, there is no need to resort to this unless you are troubleshooting issues with domain objects.

Use `domains_tool  -h` to see the available options:

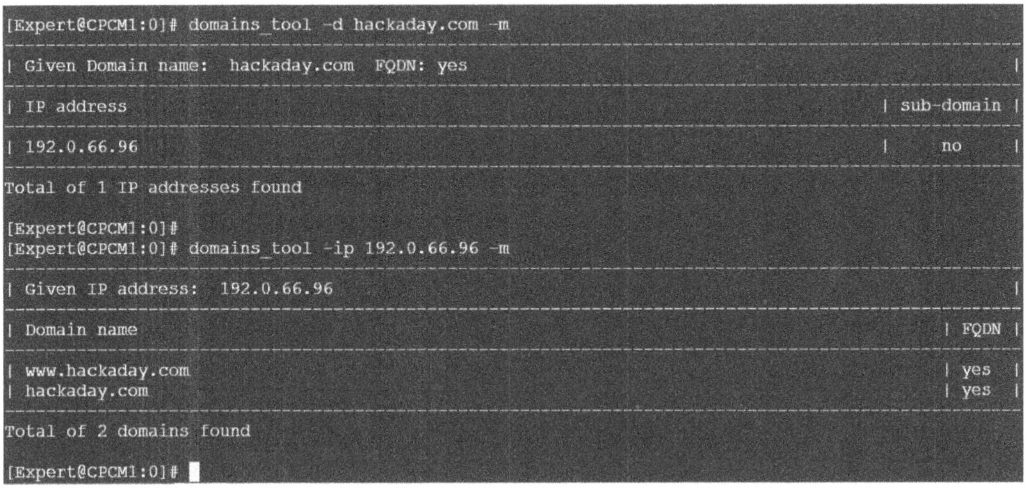

```
[Expert@CPCM1:0]# domains_tool -d hackaday.com -m

| Given Domain name:  hackaday.com  FQDN: yes                                  |

| IP address                                                    | sub-domain |

| 192.0.66.96                                                   |    no      |

Total of 1 IP addresses found

[Expert@CPCM1:0]#
[Expert@CPCM1:0]# domains_tool -ip 192.0.66.96 -m

| Given IP address:  192.0.66.96                                               |

| Domain name                                                   |  FQDN |

| www.hackaday.com                                              |  yes  |
| hackaday.com                                                  |  yes  |

Total of 2 domains found

[Expert@CPCM1:0]#
```

Figure 9.45 – Domain object CLI tools

These objects are simple to create, but we must ensure that the DNS resolution is working from the gateways and that clients are relying on these objects using the same DNS servers as the gateways.

Wildcards cannot be used with FQDN domain objects.

Refrain from using non-FQDN domain objects, as these rely on reverse DNS lookups and are known to cause more problems than benefits.

The only possible justification for non-FQDN domain objects, in my opinion, is if they are used for internally managed resources and you have a perfectly designed and implemented **IP Address Management** (**IPAM**) solution handling your DNS.

Updatable objects

The need for updatable objects manifested shortly after the mass adoption of SaaS, such as Microsoft Office 365. To allow clients to safely connect to these platforms, administrators had to download periodically updated lists of addresses and domains and incorporate those in different ways in their access control policies. To address these issues, a mechanism relying on the automated update of the relevant IP addresses and domains has been created. While limited in scope, the number of vendors and services is continually growing, and the most common ones are already available.

Because these objects are created elsewhere, they are added to the cells of the rules by clicking the + sign in the highlighted cell and, from the **Object Browser**, clicking **Import** [1], and then clicking **Updatable Objects...** [2]:

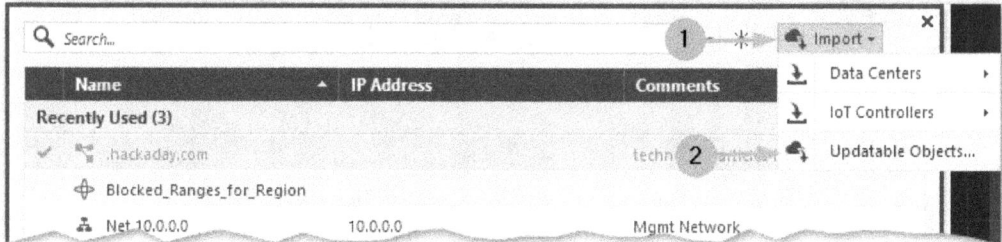

Figure 9.46 – Import of Updatable Objects from Object Browser

With the **Updatable Objects** browser window [1] open, we can see the top tier of the list of available objects:

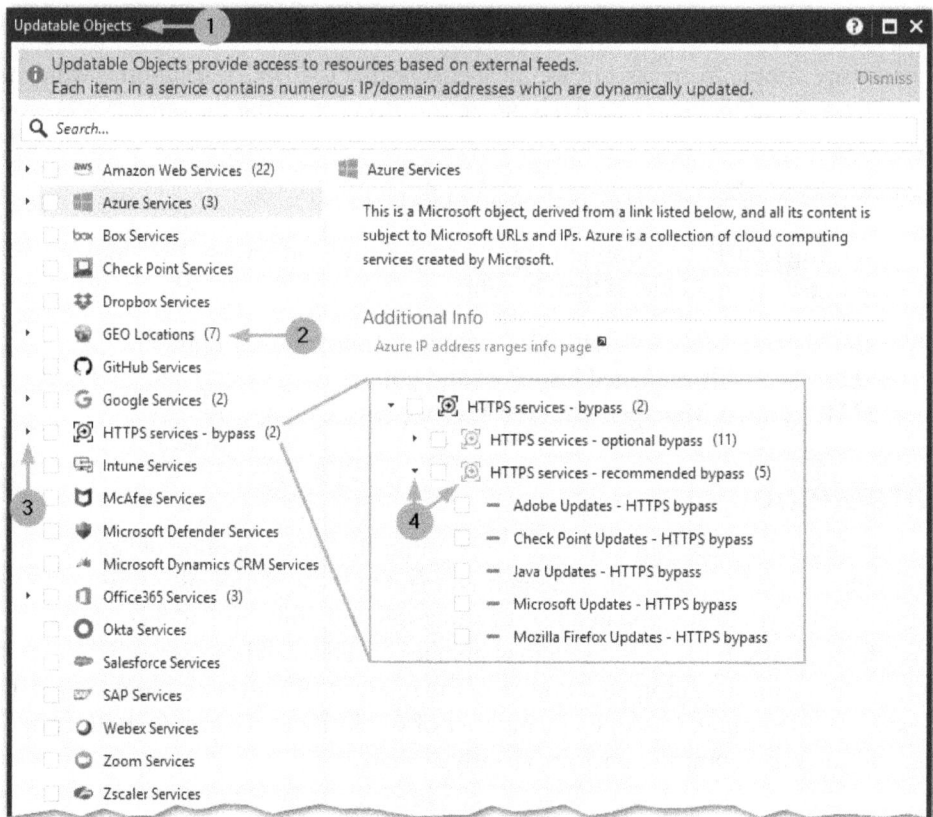

Figure 9.47 – Updatable Objects browser

Most of the items on that list you can recognize, but I would like to highlight a few things:

- **GEO Locations** [2] allows country-specific granular access control.

- The expansion triangles to the left of some of these objects [3] expand to the multiple underlying tiers of increasing granularity, where available.

- **HTTPS Services - bypass** [4] contains the nested list of **HTTPS services recommended bypass** that could be used in your HTTPS inspection policies to prevent inspection and to avoid unnecessary utilization spikes.

If, for instance, you must allow access to Microsoft Office 365 to a subset of your users residing on a specific network, the rule in your policy may look like this:

Name	Source	Destination	Services & Applications	Action
Net_10.0.0.0 to Office 365 Accept	Net 10.0.0.0	Office365 Worldwide Services	http https	Accept

Figure 9.48 – Access policy rule with updatable object

Updatable objects are simple composites, that is, they combine the functionality of domain objects and dynamic objects. The differences are that we do not have to manually provision those on the gateways, and they are updated in the background automatically without further administrative intervention.

If we are troubleshooting rules with updatable objects, we can use the `domains_tool` command with the `-uo` flag followed by the object name in quotes (the shown output is limited to 10 lines):

```
[Expert@CPCM1:0]# domains_tool -uo "Office365 Worldwide Services" | head -10

Domain tool looking for domains for 'Office365 Worldwide Services' and its children
 objects:

Domains name list for 'Exchange Online Worldwide Services':

       [1] attachments.office.net
       [2] im.outlook.office.com
       [3] eur02.admin.protection.outlook.com
       [4] autodiscover.cnxmd.onmicrosoft.com
       [5] *.outlook.com
[Expert@CPCM1:0]# []
```

Figure 9.49 – Updatable objects CLI operations using domains_tool

To see the IP addresses and ranges associated with the object, we may have to use `dynamic_objects` with the `-uo_show` undocumented flag and a keyword from the name of the object, which is `Office` in our case (shown output is limited to five lines):

```
[Expert@CPCM1:0]# dynamic_objects -uo_show | grep -A 5 "Office"
object name : CP_MS_Office365_Worldwide
range 0 : 13.107.6.152          13.107.6.153
range 1 : 13.107.6.171          13.107.6.171
range 2 : 13.107.18.10          13.107.18.11
range 3 : 13.107.18.15          13.107.18.15
range 4 : 13.107.64.0           13.107.131.255
[Expert@CPCM1:0]#
```

Figure 9.50 – Updatable objects CLI operations using dynamic_objects

Check Point gateways and management server(s) must be able to access `updates. checkpoint.com` and `dl3.checkpoint.com` for updatable objects to function properly.

Access roles

Access roles are the ultimate tool for the implementation of **zero-trust** concept in your environment. These are complex-composite objects that can be configured to utilize users, groups they belong to in external directory services, security groups from cloud-based services, Cisco Identity Services Engine TrustSec **Security Group Tags** (**SGTs**), networks, and types of clients they are connecting from, to enforce access control.

Access roles, Identity Awareness, Check Point **CloudGuard Controller**, and Identity Collector are used together to achieve perfect granularity for just-in-time access for precisely identified sources in dynamic environments.

This subject is, unfortunately, large enough to deserve its own book. To get acquainted with it, I'd recommend starting with Identity Awareness Administration Guide, a close to 300-page document that can be downloaded from `https://downloads. checkpoint.com/fileserver/SOURCE/direct/ID/103840/FILE/CP_R81_ IdentityAwareness_AdminGuide.pdf`.

Variable objects in DevOps and DevSecOps

We've now seen that there are multiple ways in which Check Point can exercise access control using objects that are fluid in nature. This allows us to partially defer security functions to trusted administrators of other services, such as Active Directory administrators moving users from one group to another, or to developers (in instances where security group memberships are assigned to the assets hosted in cloud services). This ability, coupled with automation and overlaying security posture management products, gives us the capability to address security dynamically and at scale.

That said, I recommend limiting access control for critical core services configured to use simple objects while using variable objects and associated capabilities for the rest of the operations in your environments.

Kudos to Kaspars Zibarts for his presentation on this subject at CPX 2022.

Summary

In this chapter, we have learned how to work with objects and discovered different categories of objects. We have learned about ICA and SIC. Using that, we were able to create our first Check Point cluster and configured objects for the rest of the components in our lab environment. In addition, we have learned about several types of variable objects and how to work with some of them.

In the next chapter, we'll be learning about **Network Address Translation** (**NAT**) and how it can be used by Check Point administrators to achieve a variety of goals.

10
Working with Network Address Translation

While commonly referred to as **Network Address Translation (NAT)**, this term covers both NAT and **Port Address Translation (PAT)**. In this chapter, we will learn how to use NAT (and PAT) to ensure that our traffic can reach its destinations. We will explore both automatic and manual NAT options. Additionally, we will see when the use of static or dynamic NAT is appropriate and look at a few specific cases where additional configuration in Gaia might be required to achieve the desired address translation. As you read about various configuration examples, you'll be instructed to either apply the relevant ones to our lab or not, for those that are used for demonstration and explanation. While NAT is a universal subject in networking, it is important to understand Check Point's implementation specifics.

In this chapter, we are going to cover the following main topics:

- The need for NAT

- NAT policies, rules, and processing orders

- Automatic NAT use cases and behaviors

- When NAT is not enough

- NAT logging configuration and interpretation

The need for NAT

NAT is most commonly used to circumvent the problems caused by the shortage of public IP addresses. Due to the exhaustion of public IPv4 addresses, private networks predominately use IPv4 addresses defined in **RFC 1918** (**RFC**, or **Request for Comment**, being the form of publication for the development and adoption of standards of communication for the internet). These are comprised of the following ranges:

RFC 1918 name	IP address range	CIDR block and subnet mask	Classful description
24-bit block	`10.0.0.0 - 10.255.255.255`	`10.0.0.0/8` `255.0.0.0`	1 x Class A Network
20-bit block	`172.16.0.0 - 172.31.255.255`	`172.16.0.0/12` `255.240.0.0`	16 x Class B Networks
16-bit block	`192.168.0.0 - 192.168.255.255`	`192.168.0.0/16` `255.255.0.0`	256 x Class C Networks

Table 10.1 – The RFC 1918 IPv4 address ranges and networks

These addresses are reserved for use in private environments that are isolated from one another. Traffic to external services hosted on the internet from these ranges (and the networks they are comprised of) should either be translated into public IP addresses or be discarded by the upstream routers.

Conversely, services hosted in private networks usually reside on hosts with private IP addresses. To make these services accessible from the outside, they are statically translated into specific public IP addresses. Then, these public IP addresses are used in publicly accessible DNS records to direct inbound traffic to the networks that the external interface(s) of your gateways or clusters are connected to. NAT is then used to replace the public IP address with the private IP address assigned to a host (or a load balancer).

Another use case for NAT is the merging of private networks. During mergers and acquisitions, a situation might arise where multiple environments are comprised of either identical or overlapping networks. To merge such infrastructures, these conflicts must be resolved. Either some of those networks must be renumbered (which is a complex and intrusive process with downtime), or NAT can be used between them.

Yet another case that is similar to the previous one is VPN connectivity between organizations with overlapping encryption domains (as encryption domains are often comprised of the networks from RFC 1918). If such situations are encountered, peers can supply each other with non-conflicting addresses or ranges that can be used to translate the sources.

An additional upside to the use of NAT is obfuscation of the private hosts and networks behind the addresses they translate to, allowing for egress traffic to reach its destinations but preventing connectivity in the opposite direction.

With the NAT use cases covered, in the next section, let's take a look at NAT policies, rules, and processing orders.

NAT policies, rules, and processing orders

With each new policy package, a NAT policy is automatically created. When we click on it in the navigation view on the left-hand side, we can see that it is comprised of the following column headers:

- **No.**
- **Name**
- **Original Source**
- **Original Destination**
- **Original Services**
- **Translated Source**
- **Translated Destination**
- **Translated Services**
- **Install On**
- **Comments**

Right-clicking anywhere in the headers area will manifest a column selector menu where we can check an empty checkbox for **Hits** to enable counters for each rule:

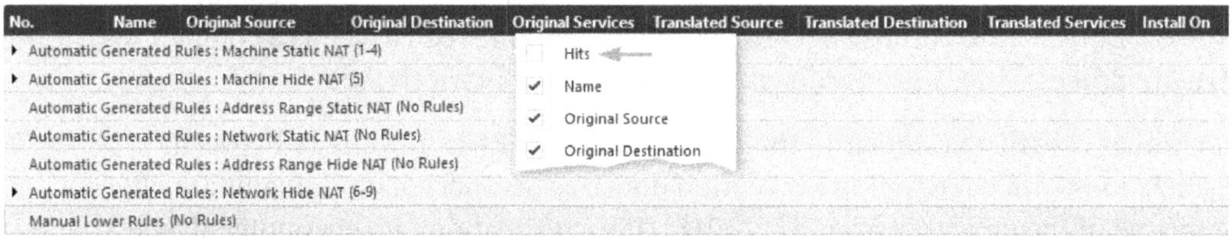

Figure 10.1 – The NAT policy structure

A policy is prepopulated with several section titles prefixed with **Automatic Generated Rules:**, and followed by either **(No Rules)** or the rule numbers within the section.

These are listed as follows:

- **Machine Static NAT**
- **Machine Hide NAT**
- **Address Range Static NAT**
- **Network Static NAT**
- **Address Range Hide NAT**
- **Network Hide NAT**
- **Manual Lower Rules**

Right-clicking on either the top or the bottom one of these section titles will reveal an option to create new rules either above or below the automatically generated rules.

NAT rules are processed top-down, and up to two rules can be applied to traffic; for example, one rule can cover the source and another rule can cover the destination address translation.

Manual rules that are created above the automatic NAT rules should be used for either exceptions from the address translations or for specific address and service translations.

Manual rules located below the automatic rules should be used for cases not covered by the automatically generated rules.

NAT is direction-dependent. Traffic traversing our gateways or clusters is modified either before it is routed or after, to achieve the desired behavior. To visualize the process, let's look at the following diagram. We can see that when packets are moving from left to right [1], **Destination NAT** [2] is applied prior to **Gaia IP Routing** [3] and **Source NAT** [4] is applied after the routing:

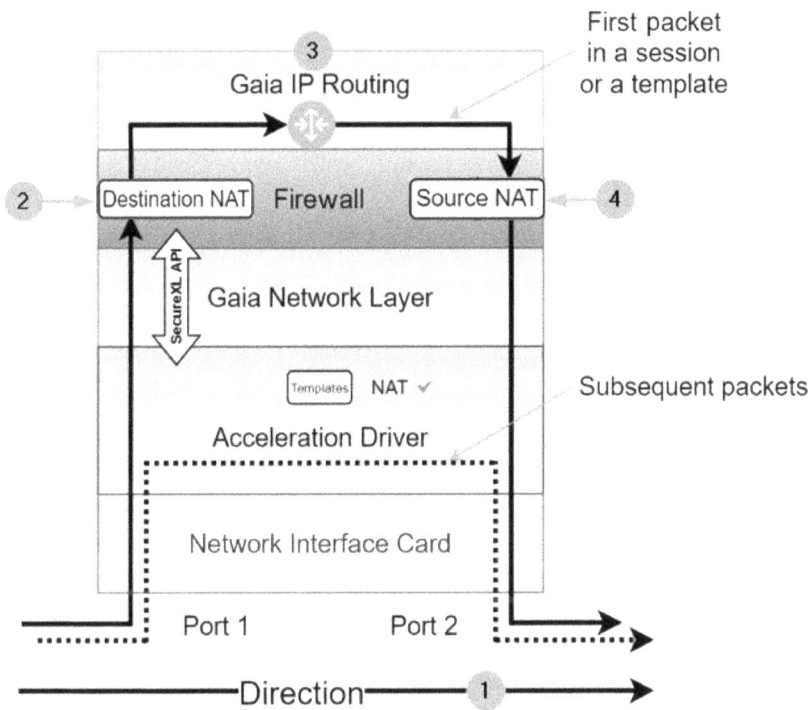

Figure 10.2 – The source and destination NAT options relative to routing

Alternatively, to see where NAT is taking place relative to the access control policy rules, VPNs, and Gaia IP routing, take a look at the following diagram:

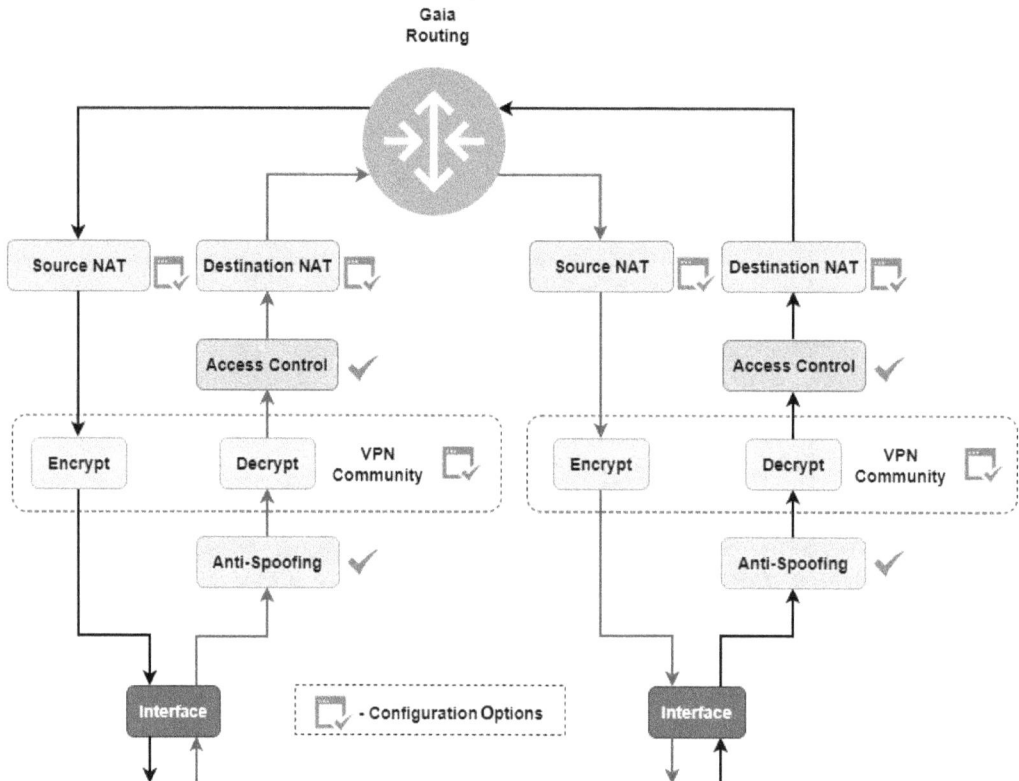

Figure 10.3 – The source and destination NAT options relative to other modules

Note that not all the actions depicted in the preceding diagram are necessarily taking place. Only selected options in a specific direction are enforced. Packets must be allowed by **Anti-Spoofing** and the rule in the **Access Control** policy; hence, they are shown as checked.

The simplest, if somewhat limited in its capabilities, way to implement NAT is to rely on an automatic NAT rule provisioning mechanism.

Automatic NAT

Automatic NAT is configured in the properties of objects. Each time NAT is defined in the properties of a gateway/cluster, management server, host, network, or range, several rules for each object are automatically created in the corresponding **Automatic Generated Rules** sections.

We should bear in mind that since we are working in a single security domain with the common objects database, once NAT has been configured for an object, the corresponding automatic rules will be identical in all the policy packages we might create.

Automatic static NAT

Static NAT refers to the consistent translation of hosts' original IP addresses into specific addresses. Let's take a look at a few example implementations next.

One-to-one

Automatic static NAT is defined in a host object and performs one-to-one NAT, translating the actual IP address of the object into the address we would like it to be accessible by from the networks behind different interface(s) of the firewall.

One common scenario is an assignment to host object [1] with **NAT** properties [2] configured to **Add automatic address translation rules** [3], the **Translation method** option configured as **Static** [4], and the **Translate to IP address** value defined as one of the public IPs from the network on the external interface of the gateway or a cluster [5]. For static NAT, we should choose the target in the **Install on gateway** field [6]:

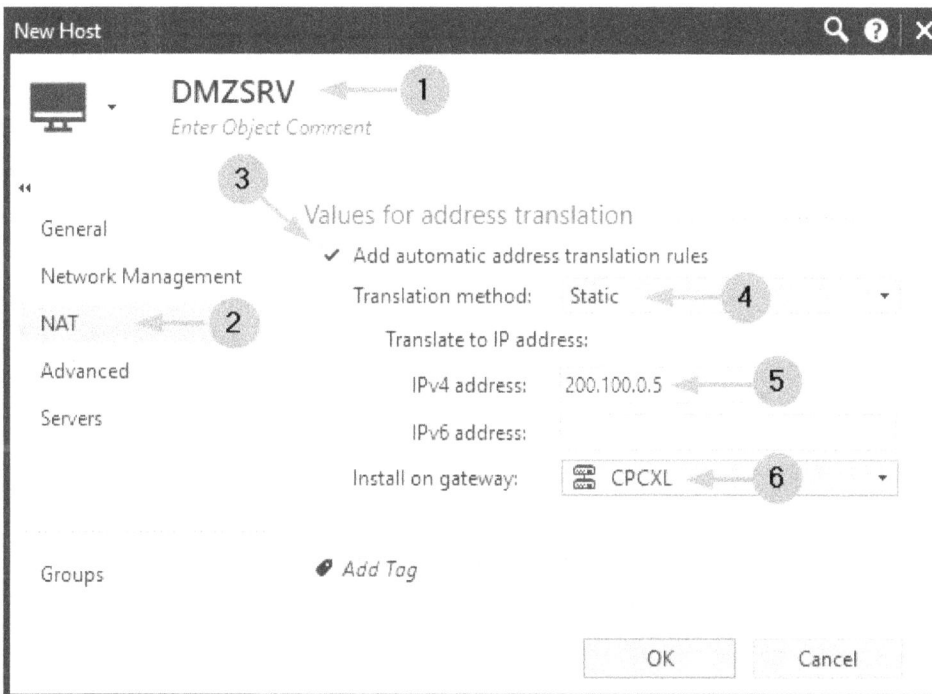

Figure 10.4 – Automatic static NAT for the host object

> **Note**
> Configure the **DMZSRV** object in our lab, as shown in the preceding screenshot.

For the management server(s) to manage remote gateways and receive logs from those devices, we must assign it a public IP address, too.

In the properties of the management server [1], go to the **NAT** section of the navigation tree [2]. With **Add Automatic Address Translation rules** selected [3], choose **Static** [4], and in **Translate to IP Address | IPv4 Address**, type in the public IP address indicated in our lab topology diagram (200.100.0.10) [5]. In **Install on Gateway**, click on the ellipsis (…) and select the **CPCXL** cluster object [6]. Check the box for **Apply for Security Gateway control connections** [7] to ensure that we'll be presenting this public IP to the remotely managed gateways:

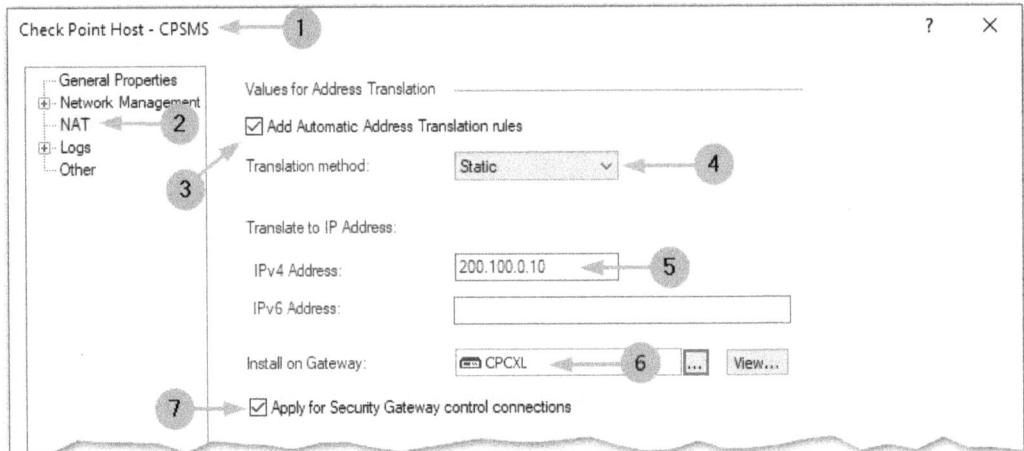

Figure 10.5 – Automatic static NAT for the management server

> **Note**
> Configure the **CPSMS** object in our lab, as shown in the preceding screenshot.

Many-to-many

There are situations where a one-to-one NAT should be performed for either the range of addresses or an entire network to preserve the individual addressability of each NATed host.

If we have to translate an entire network, follow these steps. In the properties of the **Network** object, select the **Static** NAT and specify the corresponding address of the desired NAT network:

1. To achieve this, open the **Network** object [1], and in its **NAT** properties, check the **Add automatic address translation rules** box [2]. Then, choose **Static** as the **Translation method** option [3], and specify the network for correlated many-to-many translation in the **IPv4 address** field [4]. In **Install on gateway**, choose the desired gateway or a cluster [5]:

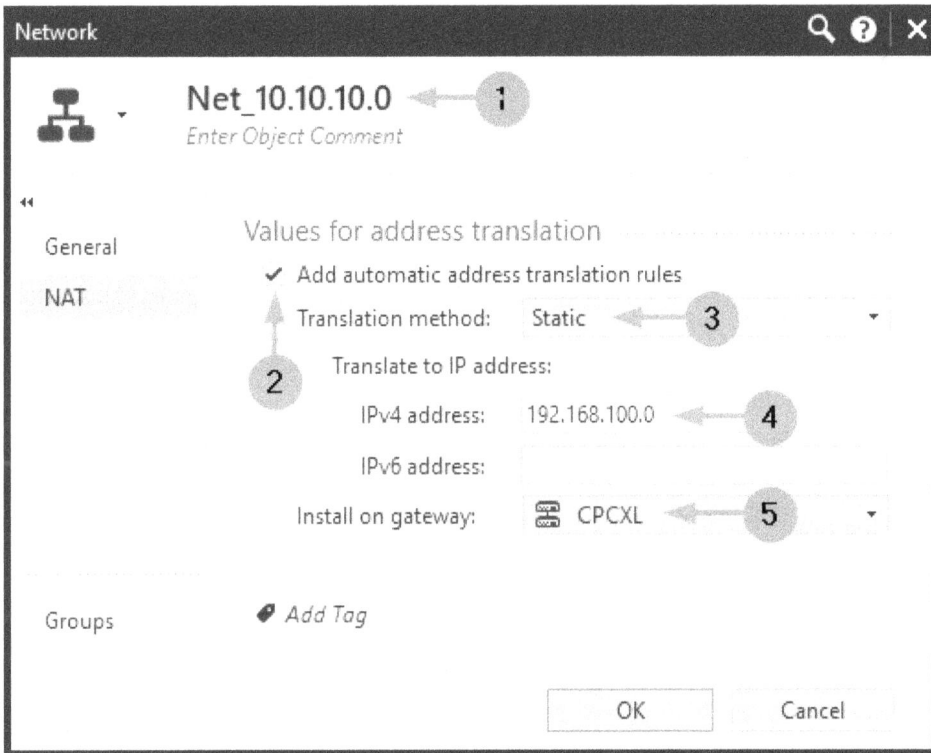

Figure 10.6 – Automatic static NAT for a many-to-many scenario

This will automatically create the corresponding rules and allow automatic static translation of each source IP address to the aligned IP in the target (**Translate to IP address**) network, that is, `10.10.10.1–192.168.100.1` to `10.10.10.254–192.168.100.254`.

> **Note**
> Do *not* configure this in the lab.

If this configuration is already present and active, the resultant traffic logs [1] will indicate the preservation of the last octet [2] when translated [3], along with the current number of the automatically generated NAT rule [4]:

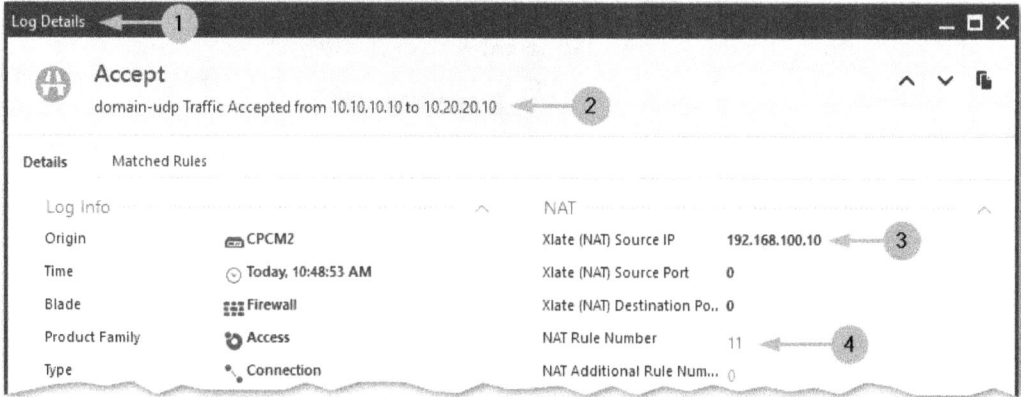

Figure 10.7 – The automatic static NAT log for a many-to-many scenario

2. If a many-to-many static NAT is performed on a range of IP addresses [1], set the **NAT** properties [2] to automatic [3], choose **Static** [4] as the translation method, type in the IPv4 address from which the one-to-one correlation will start [5], and select the installation target [6]:

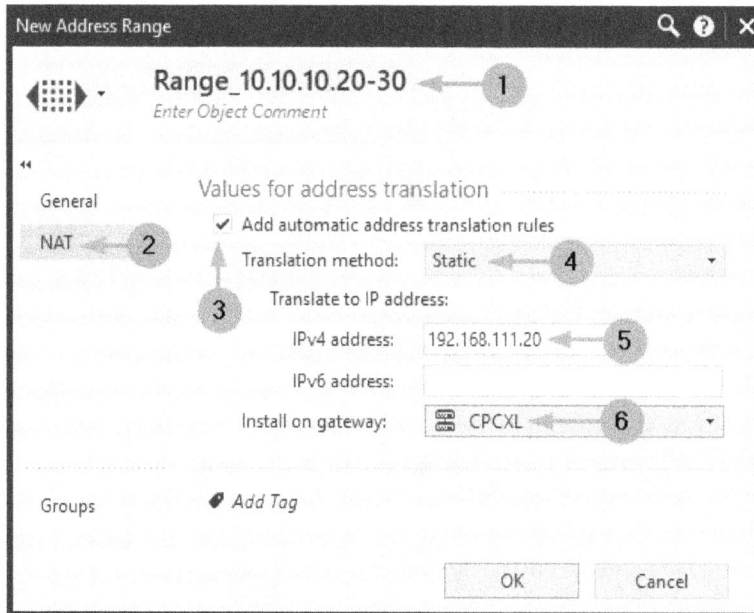

Figure 10.8 – Automatic static NAT for a range of IP addresses

In *Figure 10.8*, the `10.10.10.20` IP address will be translated into `192.168.111.20` and so on.

Note

Do *not* configure this in the lab.

Single host hiding behind a unique IP address

Let's configure a host object [1] called **SmartConsole_VM** [2] with **NAT** properties [3] for automatic NAT [4]. Select the translation method as **Hide** [5], choose **Hide behind IP address** [6], specify the public IP address (**200.100.0.20**) [7], and select our cluster as the installation target [8]:

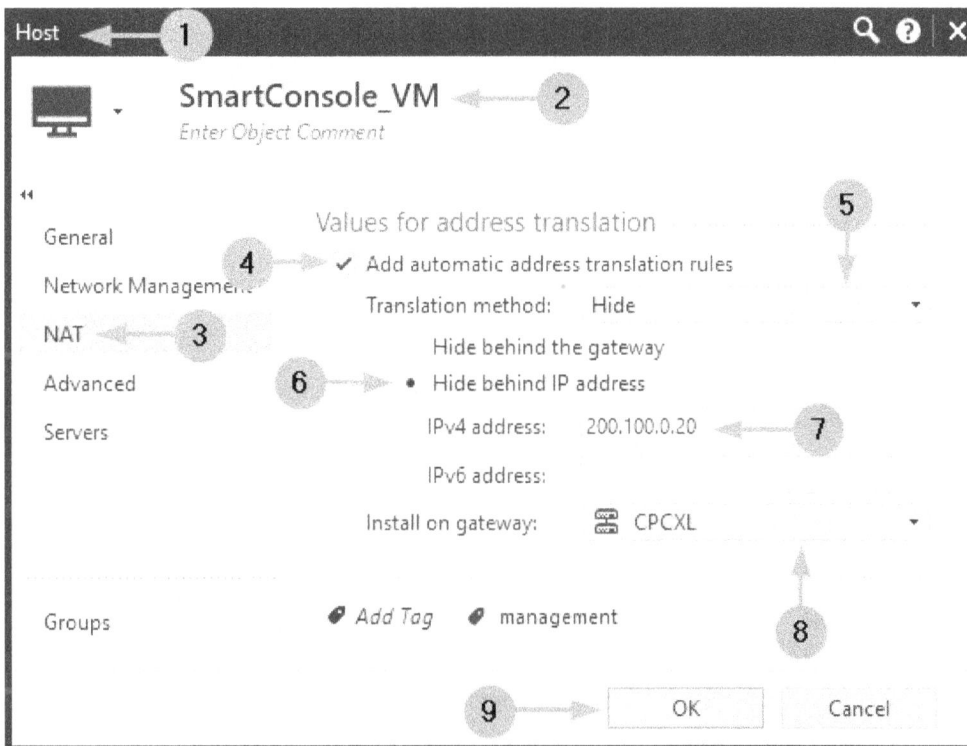

Figure 10.9 – Static hide behind IP address NAT configuration

> **Note**
> Configure the **SmartConsole_VM** object's NAT in our lab, as shown in the preceding screenshot, for consistent egress traffic origin.

Regardless of the destination, whether external (the internet), internal (the networks behind other interfaces), or DMZ, our source IP will always be translated to that which is specified in the **Hide behind IP address** properties.

This might not always be desirable behavior. We'll address how to limit hide NAT to only the external connections shortly.

Automatic dynamic NAT

The most frequently encountered type of address translation that can be found in every home router's out-of-the-box configuration is when all IPs behind the router's internal interfaces are translated into a single public IP address assigned by the ISP.

When we are talking about dynamic NAT, we are indeed touching on the subject of PAT, so the subject of NAT pools should now be mentioned.

Source port pools

Check Point uses the ports **Pool** feature to multiplex the use of identical source ports if any of these parameters in address translation are different: **IP protocol**, **Hide source IP**, or **Destination IP**. To learn more about this feature, see *sk156852 "NAT Hide failure - there are currently no available ports for hide operation. Please refer to sk156852." error in SmartConsole / SmartLog / SmartView Tracker*. In total, Check Point uses approximately 50,000 "high" ports (those between 10,000 and 60,000).

In gateways with less than eight CPU cores or six **CoreXL FW instances** (we learned about **CoreXL SND** and **CoreXL FW instances** in the *Packet flows and acceleration* section of *Chapter 8, Introduction to Policies, Layers, and Rules*), the entire range of ports used for NAT is statically divided between these instances. If we are to execute `cpview` and navigate to **Advanced | NAT | Pool-IPv4** [1], then in the **High port** section [2], we can see the top two used pools [3] and their capacities [4] being close to 50K divided by the number of CoreXL FW instances (our four-core gateway has one **CoreXL SND** instance and three **CoreXL FW** instances):

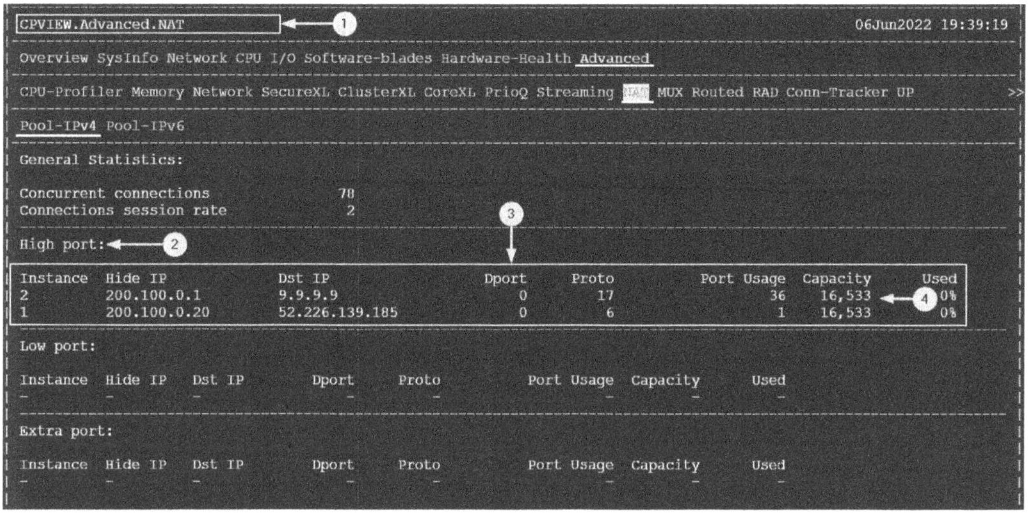

```
┌─────────────────────────────────────────────────────────────────────────────────┐
│ │CPVIEW.Advanced.NAT│  ◄────── ①                                06Jun2022 19:39:19 │
│├─────────────────────────────────────────────────────────────────────────────────│
││ Overview SysInfo Network CPU I/O Software-blades Hardware-Health Advanced         │
│├─────────────────────────────────────────────────────────────────────────────────│
││ CPU-Profiler Memory Network SecureXL ClusterXL CoreXL PrioQ Streaming NAT MUX Routed RAD Conn-Tracker UP  >>│
│├─────────────────────────────────────────────────────────────────────────────────│
││ Pool-IPv4 Pool-IPv6                                                                │
│├─────────────────────────────────────────────────────────────────────────────────│
││ General Statistics:                                                                │
││                                                                                    │
││ Concurrent connections          78                                                 │
││ Connections session rate         2                      ③                          │
│├─────────────────────────────────────────────────────────────────────────────────│
││ High port:  ◄────── ②                                                              │
│├─────────────────────────────────────────────────────────────────────────────────│
││ Instance  Hide IP        Dst IP           Dport   Proto       Port Usage  Capacity     Used │
││ 2         200.100.0.1    9.9.9.9              0      17              36    16,533 ◄──④ 0% │
││ 1         200.100.0.20   52.226.139.185       0       6               1    16,533       0% │
│├─────────────────────────────────────────────────────────────────────────────────│
││ Low port:                                                                          │
││                                                                                    │
││ Instance  Hide IP  Dst IP     Dport    Proto      Port Usage  Capacity    Used     │
││ -         -        -          -        -          -           -           -        │
│├─────────────────────────────────────────────────────────────────────────────────│
││ Extra port:                                                                        │
││                                                                                    │
││ Instance  Hide IP  Dst IP     Dport    Proto      Port Usage  Capacity    Used     │
││ -         -        -          -        -          -           -           -        │
└─────────────────────────────────────────────────────────────────────────────────┘
```

Figure 10.10 – The NAT pool with GNAT disabled (unit with four CPU cores)

Larger gateways with more than eight CPU cores (or more than six **CoreXL FW** instances) are automatically configured to use the **Global NAT** (**GNAT**) feature, which utilizes the global NAT ports allocation table.

For some reason, gateways with less than eight CPU cores have that feature disabled by default. I'm not sure what the reason for that is, since it is explicitly stated in *sk164155 Dynamic Balancing for CoreXL* that on models with fewer than eight cores, you must enable a GNAT port allocation feature.

To do that, let's check whether **GNAT** is already enabled by running `fw ctl get int fwx_gnat_enabled` in expert mode. If it is not running (`fwx_gnat_enabled = 0`), enable **GNAT** by appending the `fwx_gnat_enabled=2` line to `fwkern.conf` located in `$FWDIR/boot/modules/` and rebooting the gateway.

Once rebooted, the size of the pool for each source/destination pair is close to 50,000. If we are to execute `cpview` and navigate to **Advanced** | **NAT** | **Pool-IPv4** [1], in the **High port** section [2], we can see the top two used pools [3] and their capacities [4] being close to 50K:

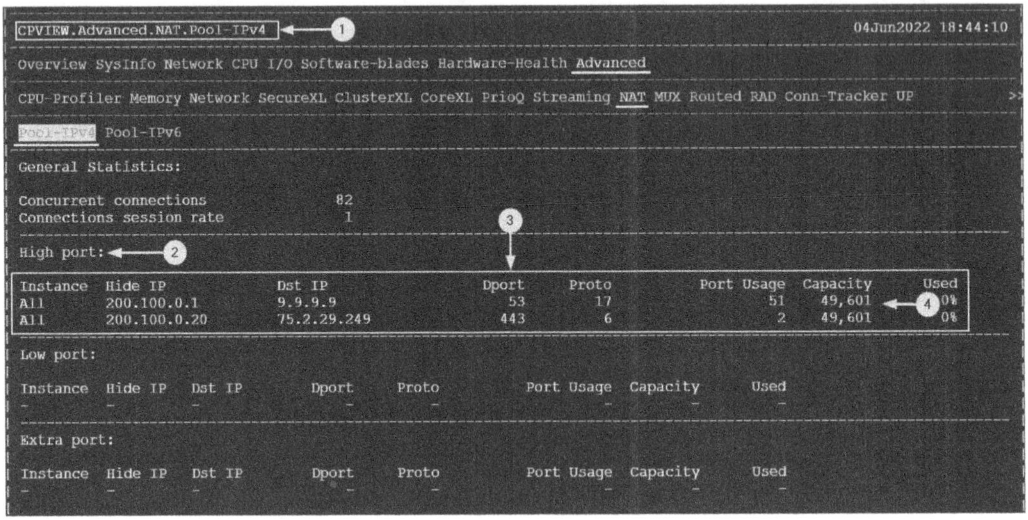

```
CPVIEW.Advanced.NAT.Pool-IPv4  ◄────── 1                                04Jun2022 18:44:10

 Overview SysInfo Network CPU I/O Software-blades Hardware-Health Advanced

 CPU-Profiler Memory Network SecureXL ClusterXL CoreXL PrioQ Streaming NAT MUX Routed RAD Conn-Tracker UP    >>

 Pool-IPv4  Pool-IPv6

 General Statistics:

 Concurrent connections          82
 Connections session rate         1
                                              3
 High port: ◄────── 2

 Instance  Hide IP        Dst IP           Dport    Proto        Port Usage  Capacity       Used
 All       200.100.0.1    9.9.9.9            53       17                 51   49,601  ◄── 4  0%
 All       200.100.0.20   75.2.29.249       443        6                  2   49,601         0%

 Low port:

 Instance  Hide IP  Dst IP     Dport    Proto        Port Usage  Capacity     Used
 -         -        -          -        -            -           -            -

 Extra port:

 Instance  Hide IP  Dst IP     Dport    Proto        Port Usage  Capacity     Used
 -         -        -          -        -            -           -            -
```

Figure 10.11 – Port pools with GNAT enabled

Now that we've seen how the ports are allocated for dynamic NAT, let's look at the number of ways the automatic dynamic NAT can be configured.

Automatic many-to-one (hide behind NAT)

Hide behind, also referred to as many-to-one NAT. In Check Point, we can use three ways of implementing it:

- Hide internal networks *behind the gateway's external IP* in the gateway/cluster properties.

- Hide internal networks, ranges, or hosts *behind the gateway's IP* in the objects' properties.

- Hide internal networks, ranges, or hosts *behind the specific IP* in the objects' properties.

Hiding internal networks behind the gateway's external IP

In most rudimentary environments, private IP addresses located behind internal (or DMZ) interfaces of the firewalls are hidden behind public IP addresses assigned to the external interface of a single gateway or as a virtual IP address of a cluster.

While this is the easiest option to configure, it should be used judiciously. I would recommend it for the smallest environments, such as retail or banking branches with no more than 50 IPs behind the gateway/cluster.

This option is disabled by default. It can be enabled in **Gateway Cluster Properties**, in the **NAT** section, by checking **Hide internal networks behind the Gateway's external IP**:

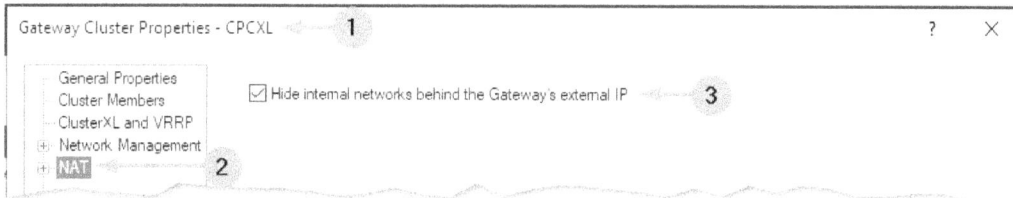

Figure 10.12 – Hide internal networks behind the Gateway's external IP

> **Note**
>
> Do *not* configure this in the lab.

In this case, our internet-bound traffic is NATed, while the traffic between the internal networks and hosts is not. The log details for the outbound connections [1] would indicate that our source IP address is translated into that of our cluster [2] and that the **NAT Rule Number** field is **0** [3]. This rule is invisible in the NAT policy and is superseded by any other more specific rule:

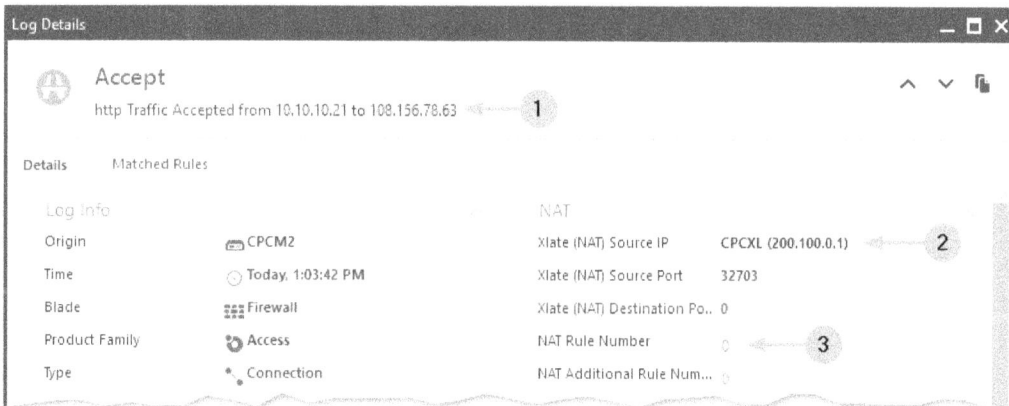

Figure 10.13 – Hiding internal networks behind the gateway's external IP log

Hiding internal networks, ranges, or hosts behind the gateway's IP

This option gives us slightly more flexibility, as we might choose specific networks, ranges, and hosts that require access to the internet and, thus, should be subjected to NAT.

As shown in the following screenshot, in the Net_10.20.20.0 network object [1], select **NAT** [2]. Then, choose automatic address translation [3], select **Hide** for **Translation method** [4], choose **Hide behind the gateway** [5], and select our cluster as the installation target [6]:

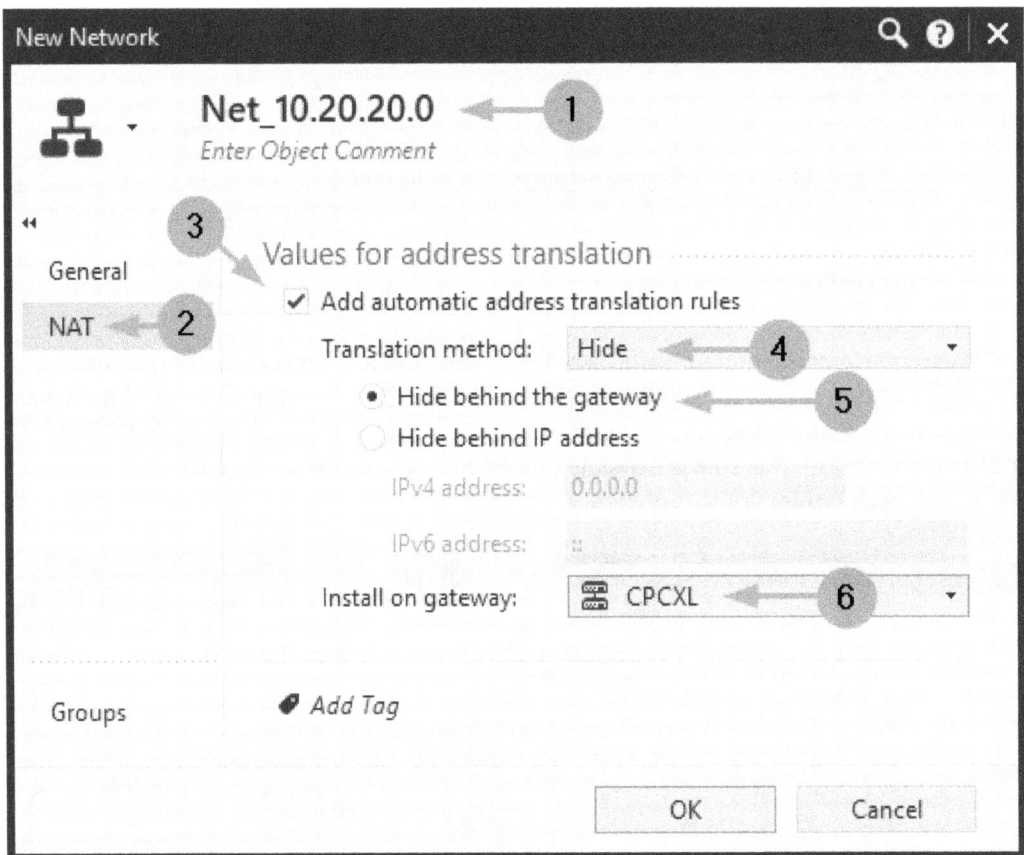

Figure 10.14 – Dynamic NAT for a network (or a range) object; hide behind the gateway

> **Note**
> Configure the **Net_10.20.20.0** and **Net_10.10.10.0** objects in our lab, as shown in the preceding screenshot.

> **Note**
>
> If we are using the **Hide behind the gateway** NAT option, **Install on gateway** might remain at the * **All** default setting. This could be relevant if you are using internal routing for egress traffic redundancy.

In this case, traffic originating from this object will always be translated into the IP address of the gateway/cluster interface closest to the destination. As shown in the following screenshot, when we are accessing resources on a different internal network [1], our source is translated into the virtual IP of the cluster in the destination's range [2]. When we are reaching out to the internet [3], our source IP is translated to the external IP of the cluster [4]. In both cases, the same **NAT Rule Number, 9,** is in effect [5]:

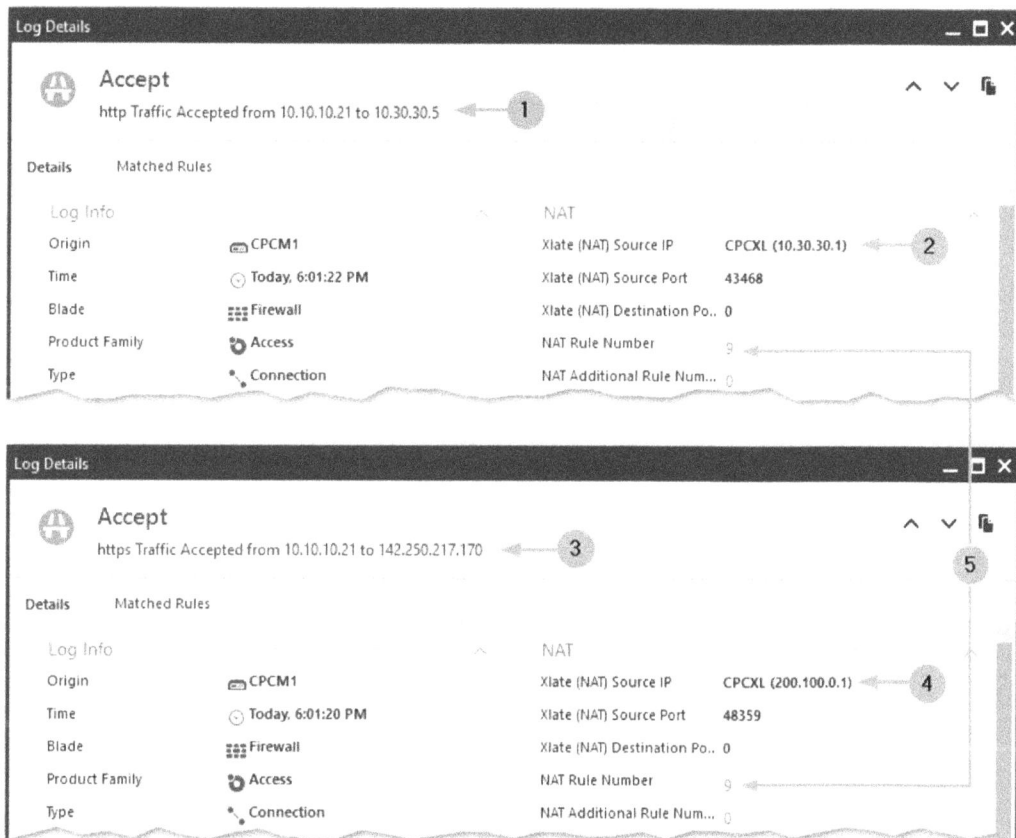

Figure 10.15 – Hiding behind the gateway in the objects' properties logs

Hiding internal networks or ranges behind specific IP addresses

This is similar to the **Hide behind the gateway** option but allows for added flexibility. There is a good reason not to hide your internal networks behind your gateway's or cluster's public IPs. Consider the possibility that either one or several of your hosts have been compromised and, despite your best efforts, generate malicious traffic. In this case, the public IP address associated with it will likely be blocked upstream. If that IP is the same as the external interface of your gateway, its functionality might be negatively affected.

Another compelling reason for this is possible port exhaustion, which happens if the capacity of the **Source Port Pool** instance has been exceeded. If you have a sufficiently large public IP range or a connected network on your external interfaces and if your internal networks are relatively small (like Class C), you can configure each of the internal networks [1] for automatic NAT [2] hiding them [3] behind individual public IP addresses [4] on a specific gateway or a cluster [5], as shown in the following screenshot:

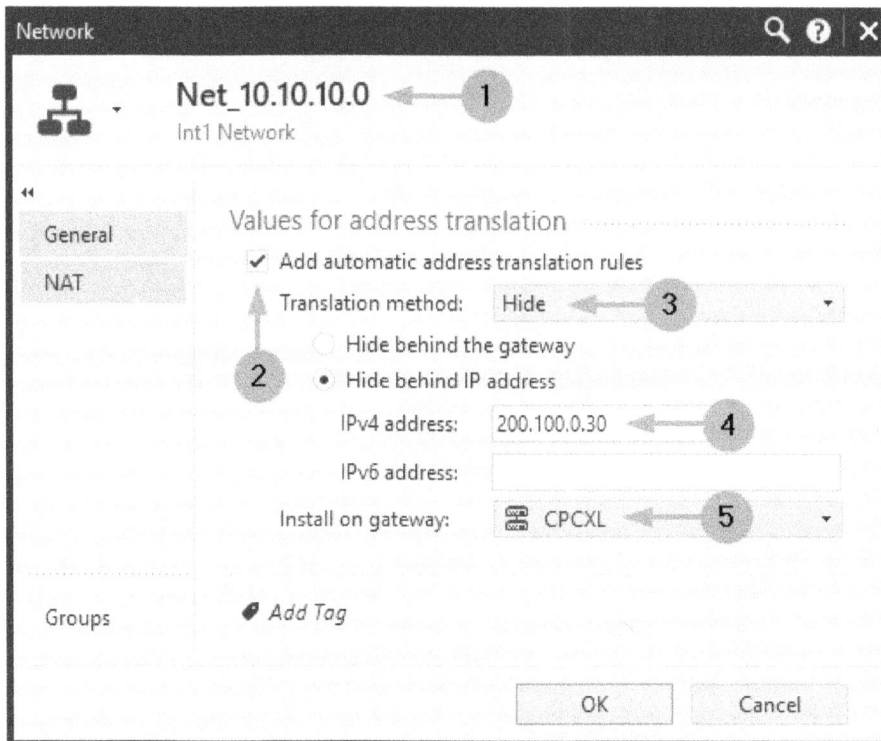

Figure 10.16 – Hiding internal networks or ranges behind specific IP addresses

> **Note**
> Do *not* configure this in the lab.

This will increase the capacity for the dynamically allocated hide NAT ports.

The same approach could be used for the IP ranges' NAT properties.

Preventing unnecessary NAT

While it makes sense to translate addresses for outbound or inbound access to and from the internet, in most cases, traffic inside the environment protected by the gateway or a cluster should remain untranslated.

To prevent unnecessary NAT between the internal and DMZ hosts, we should create a **Network Group** object [1], name it `Int_Nets` [2], and add all our internal and DMZ networks to it [3]. Save the object and, in our NAT policy, create a rule at the top [4] called `No NAT`. Add the newly created **Int_Nets** group to both the **Original Source** and **Original Destination** fields [5]:

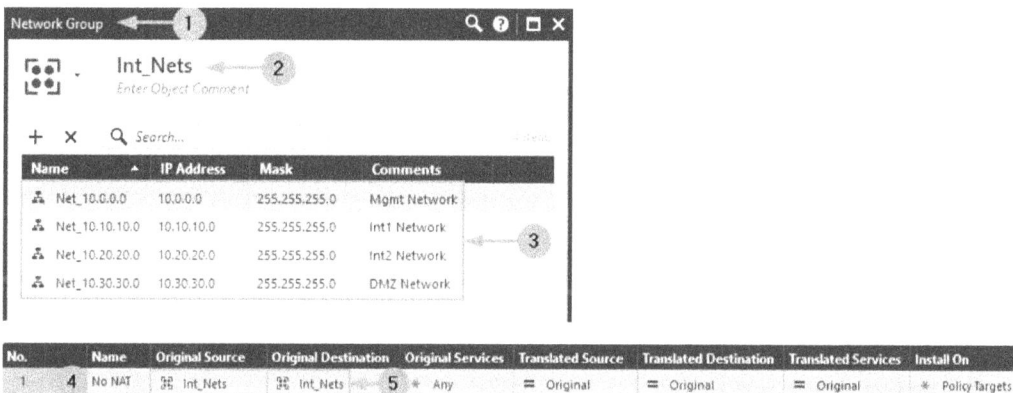

Figure 10.17 – Preventing NAT inside our environment

> **Note**
> Create the **Int_Nets** group, and create the **No NAT** rule in the lab.

This will restrict NATed traffic to the external interface of the gateway using the virtual IP address of the external cluster interfaces, or the specific IP addresses indicated in the **Hide behind IP address** fields.

Now that we have seen the possibilities afforded by automatic NAT defined in the properties of the objects, let's take a look at how more complicated tasks involving NAT and PAT could be accomplished.

When NAT is not enough

While the majority of cases in simple and small environments can be handled by automatic NAT defined in objects' properties, with increasing scale and complexity, we will have to resort to manual NAT rules.

> **Note**
> The information in this section should be used for reference only since it requires the presence of an access control policy, which we have not yet created.

Let's look at a few more complex examples of manual NAT implementation that, in addition to the creation of NAT rules, might require a few Gaia configuration changes.

Many-to-less

If your internal networks are larger than Class C, using the **Hide behind IP address** option described in the previous section might not be feasible. You can still do that by representing your large network as several ranges, each with its own public IP in the NAT properties, but this is a rather tedious process.

An alternative way to handle this issue with greater flexibility is with the following steps:

1. Create a new manual rule at the bottom of the NAT policy under the **Manual Lower Rules** section title.

2. Create a **Network Group** object for your source networks requiring outbound access to the internet (for example, we can use our **Int_Nets** group) and place it in the **Original Source** field of the new manual NAT rule. Create a new **Address Range** object [1], naming it appropriately [2], and define the scope of unused IP addresses from your public range or a network [3]:

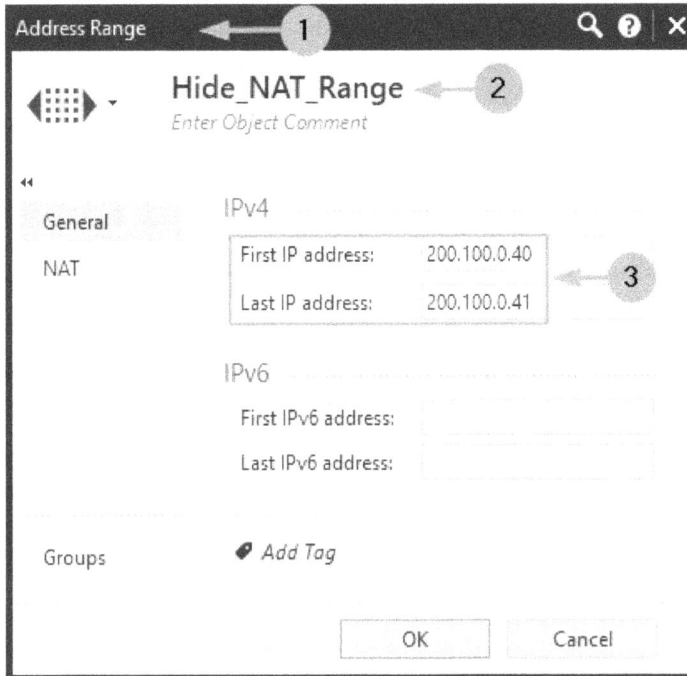

Figure 10.18 – Creating an IP range to use as a hide NAT translated source

3. Place this object in the **Translated Source** field of our rule. You should now see this object with a small red **S** letter, indicating that the preset NAT mode is static [1]. Right-click on it, click on **NAT Method** [2], and click on **Hide** [3]:

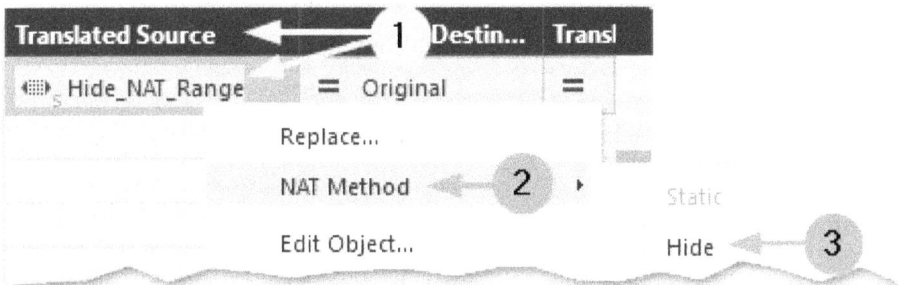

Figure 10.19 – NAT Method change from Static to Hide

4. Once done, the red **S** character is changed to **H**, and the rule should look like this:

Name	Original Source	Original Destination	Original Services	Translated Source	Translated Destination	Translated Services
Hide NAT Scope	Int_Nets	* Any	* Any	Hide_NAT_Range	= Original	= Original

Figure 10.20 – Setting manual NAT rule to Hide behind address range

Now things are about to get a little bit complicated... For each automatically created NAT rule with either **Static** or **Hide** behind the IP address options, the background process creates a **proxy ARP** record. **A proxy ARP** record is an association of the translated IP address with the MAC address of the gateway's (or cluster members') interface, and the actual IP address of that interface. We can look those up for the objects we have configured by executing the `fw ctl arp -n` command in the **Expert** mode of our gateway (in our case, each cluster member). I am using another command, `clish -c "show cluster state" | grep local`, for the compact illustration of active and standby cluster member states.

```
[Expert@CPCM1:0]# clish -c "show cluster state" | grep local
1 (local)   192.168.255.1    100%                  ACTIVE           CPCM1
[Expert@CPCM1:0]# fw ctl arp -n
 (200.100.0.20) at 08-00-27-c8-7b-1d interface 200.100.0.2
 (200.100.0.10) at 08-00-27-c8-7b-1d interface 200.100.0.2
 (200.100.0.5) at 08-00-27-c8-7b-1d interface 200.100.0.2
[Expert@CPCM1:0]#
```

```
[Expert@CPCM2:0]# clish -c "show cluster state" | grep local
2 (local)   192.168.255.2    0%                    STANDBY          CPCM2
[Expert@CPCM2:0]# fw ctl arp -n
 (200.100.0.10) at 08-00-27-94-58-27 interface 200.100.0.3
 (200.100.0.20) at 08-00-27-94-58-27 interface 200.100.0.3
 (200.100.0.5) at 08-00-27-94-58-27 interface 200.100.0.3
[Expert@CPCM2:0]#
```

Figure 10.21 – Automatic proxy ARP records

The output is comprised of the NAT IP address, the MAC address of the interface it is associated with, and the IP address of that interface configured on the gateway (or a cluster member).

For our rule to go into effect, we have to create individual **proxy ARP** records for each IP address in the Hide_NAT_Range object.

To accomplish this, we must perform the following steps:

1. Make a change in the **Global Properties** dashboard's [1] NAT section [2] by checking the **Merge manual proxy ARP configuration** box [3] and clicking on **OK** [4]:

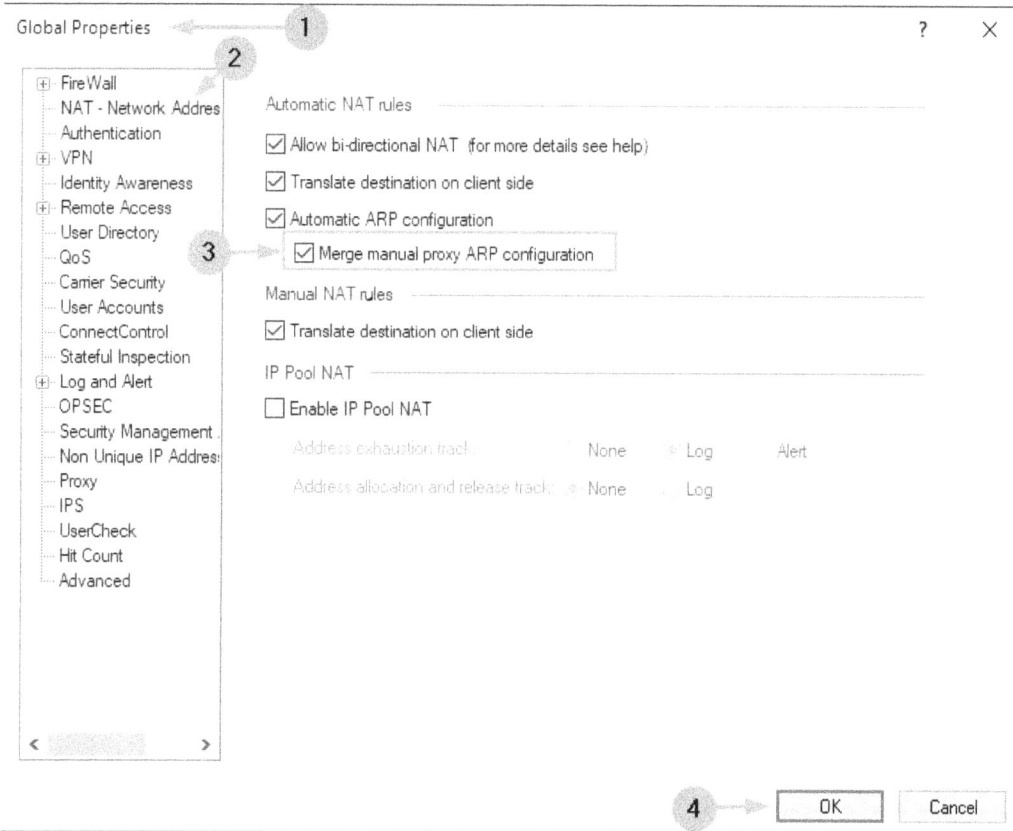

Figure 10.22 – The Global Properties Merge manual proxy ARP configuration option

2. Publish the changes and install the policy.

3. Log into each cluster member via SSH, and in CLISH, enter the following commands:

    - For CPCM1, enter the following:

      ```
      add arp proxy ipv4-address 200.100.0.40 interface eth4
      real-ipv4-address 200.100.0.2
      add arp proxy ipv4-address 200.100.0.41 interface eth4
      real-ipv4-address 200.100.0.2
      save config
      ```

- For CPCM2, enter the following:

```
add arp proxy ipv4-address 200.100.0.40 interface eth4
real-ipv4-address 200.100.0.3
add arp proxy ipv4-address 200.100.0.41 interface eth4
real-ipv4-address 200.100.0.3
save config
```

4. Install the policy again.

Now, we can verify that new proxy ARP entries are in effect:

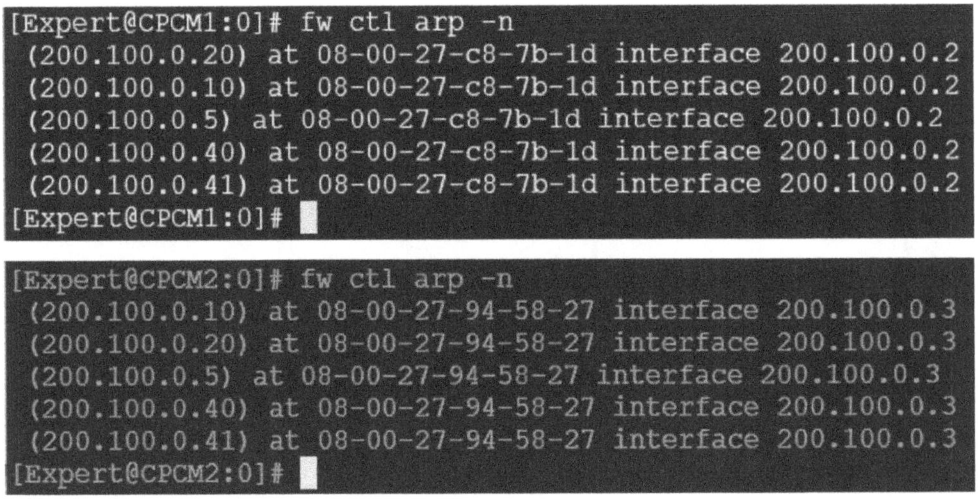

```
[Expert@CPCM1:0]# fw ctl arp -n
(200.100.0.20) at 08-00-27-c8-7b-1d interface 200.100.0.2
(200.100.0.10) at 08-00-27-c8-7b-1d interface 200.100.0.2
(200.100.0.5) at 08-00-27-c8-7b-1d interface 200.100.0.2
(200.100.0.40) at 08-00-27-c8-7b-1d interface 200.100.0.2
(200.100.0.41) at 08-00-27-c8-7b-1d interface 200.100.0.2
[Expert@CPCM1:0]#
```

```
[Expert@CPCM2:0]# fw ctl arp -n
(200.100.0.10) at 08-00-27-94-58-27 interface 200.100.0.3
(200.100.0.20) at 08-00-27-94-58-27 interface 200.100.0.3
(200.100.0.5) at 08-00-27-94-58-27 interface 200.100.0.3
(200.100.0.40) at 08-00-27-94-58-27 interface 200.100.0.3
(200.100.0.41) at 08-00-27-94-58-27 interface 200.100.0.3
[Expert@CPCM2:0]#
```

Figure 10.23 – The proxy ARP entries with IPs from the Hide_NAT_Range object

If we remove the **Hide behind the gateway** settings from our two internal network objects and install the policy again, we'll be using IPs from the `Hide_NAT_Range` object.

Manual static NAT

If (in addition to the IP translation) we must rely on port translation, automatic NAT will be of no help, as the rules created by defining NAT in the properties of the objects are not editable. In this case, we must create an additional object with the public (or translated) IP address and create associated proxy ARP records for it on relevant interfaces*.

> *
>
> Proxy ARP entries are only necessary if we are translating to IP addresses in
> connected networks. If your public IP range does not reside on a connected
> network of your external interfaces (for instance, when the ISP provides you
> with an additional small range used between your gateways and their own
> equipment), proxy ARP entries are not required.

For example, for a `WebSrvPrivate` web server object with an IP of `10.30.30.100`
listening on port `5080`, we should create a `WebSrvPublic` object with an IP address of
`200.100.0.100` and one proxy ARP record added on each cluster member.

On CPCM1, use the following:

```
add arp proxy ipv4-address 200.100.0.100 interface eth4 real-
ipv4-address 200.100.0.2
save config
```

On CPCM2, use the following:

```
add arp proxy ipv4-address 200.100.0.100 interface eth4 real-
ipv4-address 200.100.0.3
save config
```

Then, the pair of NAT rules for outbound and inbound traffic must be created:

Name	Original Source	Original Destination	Original Services	Translated Source	Translated Destination	Translated Services
WebSrv_Out	WebSrvPrivate	* Any	* Any	WebSrvPublic	= Original	= Original
WebSrv_In	* Any	WebSrvPublic	http	= Original	WebSrvPrivate	http_5080

Figure 10.24 – The manual static NAT rules for service translation

Additionally, the accompanying access control rule allowing the **http** traffic from
ExternalZone to **WebSrvPublic** needs to be created:

Name	Source	Destination	Services & Applications	Action	Track
Outside to WebSrv Accept	ExternalZone	WebSrvPublic	http	Accept	Log

Figure 10.25 – The access control policy rule for the manually NATed server

With these steps completed, we should install the policy.

The server is now accessible [1], and traffic from the **External** zone [2] on the HTTP service [3] addressed to the public IP of the server [4] is being translated to the correct destination [5] and port 5080 [6]:

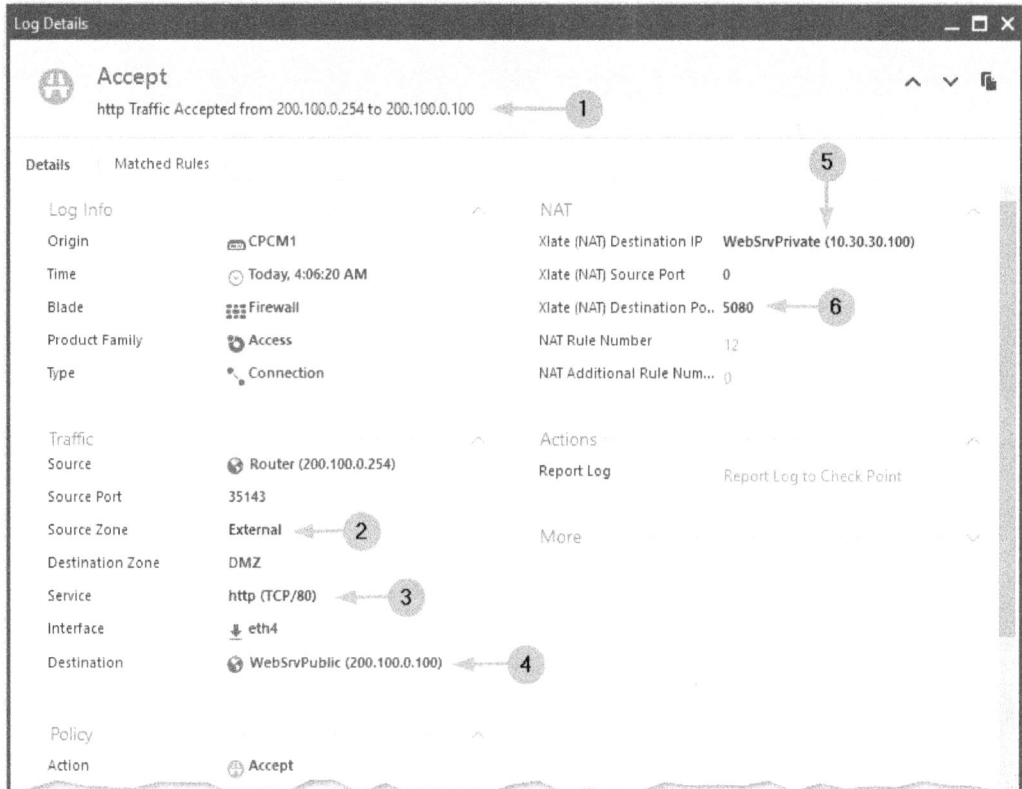

Figure 10.26 – The access control rule log for manual static NAT with service translation

An additional edge case is applicable in very small environments where you have a small gateway with a single public IP address. In this situation, you can create an object with the same public IP address as the one assigned to the external interface of your gateway, use public DNS to redirect traffic to the high port, and use manual static NAT to translate it to the service running on the desired port on internal or DMZ networks.

In this case, no proxy ARP configuration is required, as the IP address being used is already associated with the interface.

NAT pools

Occasionally, you might have to deal with situations where some of your peer organizations require you to source your connections from specific IPs or networks. To accomplish this, you can either create an actual network to accommodate the peer or you can use NAT pools.

Individual NAT pools are created in Gaia WebUI **Advanced Routing | NAT Pools**.

The **Add NAT Pool** [1] window contains the following fields: a **Destination** field [2] and a **Subnet mask** field [3]. We can use networks or individual hosts (with the /32 mask):

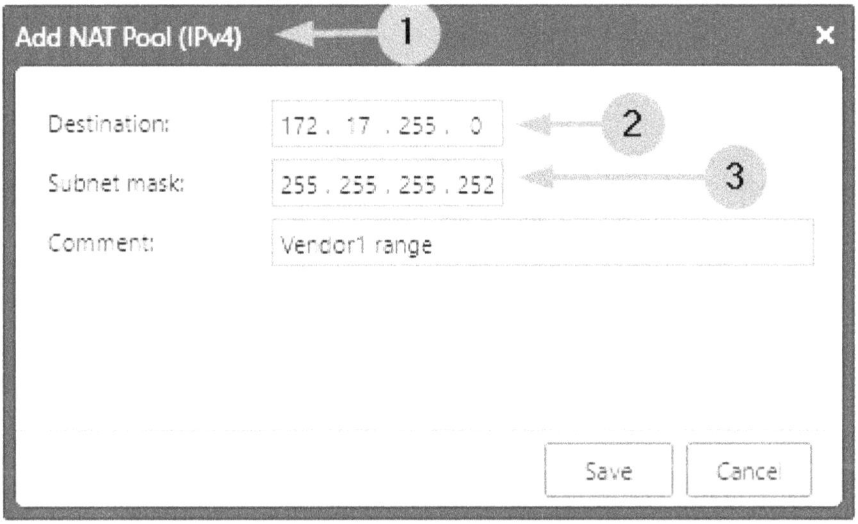

Figure 10.27 – NAT pools

In a cluster, identical pools must be defined for all members. These pools can then be advertised as regular networks via routing protocols to the upstream router or routers might be preconfigured with static routes for specific NAT pools using our firewalls or clusters' actual IP as the gateway.

We can now use addresses from **NAT Pool** in the **Hide behind IP address** fields for our hosts or networks initiating a connection to that peer.

Bells and whistles

So far, all the cases we've looked at are relatively simple but necessary to understand. To get a better idea about Check Point's NAT policy capabilities, let's look at the NAT policy in the demo mode:

No.	Name	Original Source	Original Destination	Original Services	Translated Source	Translated Destination	Translated Services
1	Dynamic Object - No NAT - DYN_ADMINS to DYN_ADMINS	DNS Server	DYN_ADMINS	* Any	= Original	= Original	= Original
2	Access Role - No NAT - Sales to Sales	Sales	Sales	* Any	= Original	= Original	= Original
3	Network - No NAT - HR LAN to HR Server	HR LAN	HR Server	* Any	= Original	= Original	= Original
Automatic Generated Rules : Machine Static NAT (No Rules)							
Automatic Generated Rules : Machine Hide NAT (No Rules)							
Automatic Generated Rules : Address Range Static NAT (No Rules)							
Automatic Generated Rules : Network Static NAT (No Rules)							
Automatic Generated Rules : Address Range Hide NAT (No Rules)							
▶ Automatic Generated Rules : Network Hide NAT (4-5)							
▼ Manual Lower Rules (6-14)							
6	Network - HR LAN to ANY	HR LAN	* Any	* Any	Corporate-GW	= Original	= Original
7	Security Zone - Exchange to External	Exchange	ExternalZone	* Any	EXCHANGE_EXT	= Original	= Original
8	Domain Object - Guests to Corp	Wireless Guests Network	.corp.com	* Any	Guests Corp Srv	= Original	= Original
9	Dynamic Object - DYN_ADMINS to ANY	DYN_ADMINS	* Any	* Any	Corporate-GW	= Original	= Original
10	Translated Dynamic Object - SRV_EXT to SRV_INT	* Any	SRV_EXT	http	= Original	SRV_INT	= Original
11	Security Zone - Internal to DMZ	InternalZone	DMZZone	* Any	Remote-4-gw	= Original	= Original
12	Updatable Object - TDF Lab to France	TDF Lab Range	France	* Any	Remote-2-gw	= Original	= Original
13	Updatable Object - DSW Lab to United States	DSW Lab Range	United States	* Any	Remote-3-gw	= Original	= Original
14	Updatable Object - Internal Lab to Skype	Internal Lab Net	Skype for Business Services	* Any	Corporate-GW	= Original	= Original

Figure 10.28 – The NAT policy in demo mode

As we can see, in addition to the examples of hosts, networks, and ranges that we had to work with, dynamic objects [1], access roles [2], FQDN domains [3], zones [4], and updatable objects [5] can be used in manual rules, providing great flexibility in traffic handling. Services can be conditionally redirected [6] and, of course, PAT can be used with all of these by populating the **Original Services** and **Translated Services** fields in the rules.

While we have discussed logging in general and have mentioned NAT logging in passing, let's spend a minute on NAT-specific log entries in the event log cards.

NAT logging

NAT is logged in rules where **Log Generation | Per Connection** are enabled. As per the recommendations made in *Chapter 8, Introduction to Policies, Layers, and Rules*, this should be the default setting for the firewall-only rules.

The NAT portion of the log card is really easy to understand, but I have seen repeated questions in forums regarding these two fields: **NAT Rule Number** and **NAT Additional Rule Number**.

To illustrate, let's take a look at the following log card:

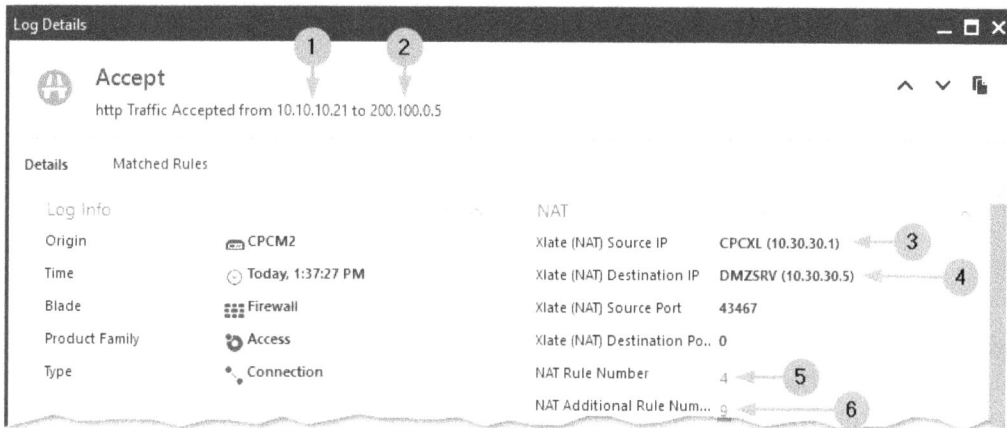

Figure 10.29 – The NAT Rule Number and NAT Additional Rule Number log fields

In the preceding screenshot, we can see the following:

- Traffic is accepted from host `10.10.10.21` [1] to host `200.100.0.5` [2].

- The translated source IP is `10.30.30.1` [3], which belongs to one of the virtual IPs of our cluster, and the translated destination is `10.30.30.5` [4].

- We can see that **NAT Rule Number** is **4** [5] and **NAT Additional Rule Number** is **9** [6].

This log describes the scenario where the `DMZSRV` object is being accessed from inside by its public IP from the network configured to **Hide behind the gateway**. NAT is performed on both; the destination [1], source [2], and corresponding rule numbers are shown here:

No.	Name	Original Source	Original Destination	Original Services	Translated Source	Translated Destination	Translated Services	Install On
▼ Automatic Generated Rules : Machine Static NAT (1-4)								
1	Automatic Rule: CPSMS	CPSMS	✳ Any	✳ Any	CPSMS (Valid Address)	= Original	= Original	CPCXL
2	Automatic Rule: CPSMS	✳ Any	CPSMS (Valid Add	✳ Any	= Original	CPSMS (Valid Address)	= Original	CPCXL
3	Automatic Rule: DMZSRV	DMZSRV	✳ Any	✳ Any	DMZSRV (Valid Address)	= Original	= Original	CPCXL
4 (1)	Automatic Rule: DMZSRV	✳ Any	DMZSRV (Valid Ac	✳ Any	= Original	DMZSRV (Valid Address)	= Original	CPCXL
▶ Automatic Generated Rules : Machine Hide NAT (5)								
Automatic Generated Rules : Address Range Static NAT (No Rules)								
Automatic Generated Rules : Network Static NAT (No Rules)								
Automatic Generated Rules : Address Range Hide NAT (No Rules)								
▼ Automatic Generated Rules : Network Hide NAT (6-9)								
6	Automatic Rule: CP_default_Offi ce_Mode_addre sses_pool	CP_default_Offi...	CP_default_Office	✳ Any	= Original	= Original	= Original	✳ All
7	Automatic Rule: CP_default_Offi ce_Mode_addre sses_pool	CP_default_Offi...	✳ Any	✳ Any	CP_default_Office_Mode_addre	= Original	= Original	✳ All
8	Automatic Rule: Net_10.10.10.0	Net_10.10.10.0	Net_10.10.10.0	✳ Any	= Original	= Original	= Original	✳ All
9 (2)	Automatic Rule: Net_10.10.10.0	Net_10.10.10.0	✳ Any	✳ Any	Net_10.10.10.0 (Hiding Address)	= Original	= Original	✳ All
Manual Lower Rules (No Rules)								

Figure 10.30 – The NAT Rule Number and NAT Additional Rule Number rules

If only one translation is taking place, we will see either the **Xlate (NAT) Source IP** field or the **Xlate (NAT) Destination IP** field. The other one will be suppressed. In such a case, only **Nat Rule Number** will have the rule number value and **NAT Additional Rule Number** will have **0** in it.

Summary

In this chapter, we learned how to work with NAT policies, rules, and methods. We covered the automatic static and dynamic NAT capabilities of different objects and saw how they could be applied and logged. We learned about cases requiring manual NAT rules alongside additional Gaia configurations that might be necessary to accommodate more complex NAT scenarios in our data centers. As we went through this chapter, we configured NAT properties for the objects in our lab.

In the next chapter, using objects created in this and the previous chapters, we'll be creating and working with our first access control policy.

Part 3: Introduction to Practical Administration for Achieving Common Objectives

In this portion of the book, you will use the knowledge from previous chapters to create a functioning access control policy, gradually increasing its granularity and complexity. You will learn how to implement site-to-site and remote access VPNs, work with high-availability clusters, and learn about the methodology and tools for performing basic troubleshooting. You will also learn about Check Point licensing.

The following chapters will be covered in this section:

- *Chapter 11, Building Your First Policy*
- *Chapter 12, Configuring Site-to-Site and Remote Access VPNs*
- *Chapter 13, Introduction to Logging and SmartEvent*
- *Chapter 14, Working with ClusterXL High Availability*
- *Chapter 15, Performing Basic Troubleshooting*
- *Appendix, Licensing*

11

Building Your First Policy

We have learned a lot over the course of the previous chapters. Now, it's time to put that knowledge to the test and construct our first fully functional access control policy.

We'll be following the process of defining a policy structure, trying to account for the most common scenarios likely to be encountered in any infrastructure. Once that is done, we'll proceed with the creation of the rules and, when necessary, additional objects. As we progress, we'll be periodically testing the behavior of our policy to ascertain that it is performing as intended.

Additionally, we will gradually expand its capabilities with HTTPS Inspection, application control, URL filtering, and Identity Awareness to experience and gain an understanding of most aspects of policy building.

In this chapter, we are going to cover the following main topics:

- Defining the access control policy structure
- Creating rules for the firewall/networking layer
- Creating the APCL/URLF layer and rules
- Using access roles for identity-specific access control

Defining the access control policy structure

From *Chapter 8, Introduction to Policies, Layers, and Rules*, you might recall the requirement to have three rules present in the policy to ensure that you can access gateways via SSH and WebUI, drop any unsanctioned connections to the gateways as early as possible, and drop/log all other unsanctioned traffic.

With this in mind, let's create section titles and include these rules from the start. For now, ignore the rule numbers depicted in the **No.** column of the following screenshot. These will eventually align when the rest of your rules have been created:

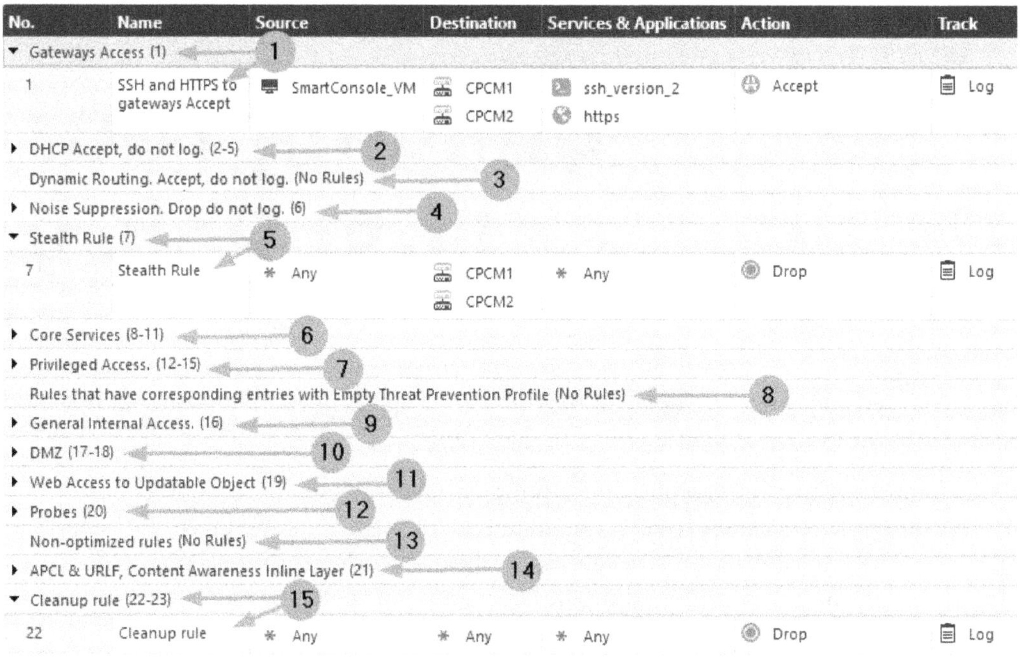

No.	Name	Source	Destination	Services & Applications	Action	Track
▼ Gateways Access (1)						
1	SSH and HTTPS to gateways Accept	SmartConsole_VM	CPCM1 CPCM2	ssh_version_2 https	Accept	Log
▶ DHCP Accept, do not log. (2-5)						
Dynamic Routing. Accept, do not log. (No Rules)						
▶ Noise Suppression. Drop do not log. (6)						
▼ Stealth Rule (7)						
7	Stealth Rule	Any	CPCM1 CPCM2	Any	Drop	Log
▶ Core Services (8-11)						
▶ Privileged Access. (12-15)						
Rules that have corresponding entries with Empty Threat Prevention Profile (No Rules)						
▶ General Internal Access. (16)						
▶ DMZ (17-18)						
▶ Web Access to Updatable Object (19)						
▶ Probes (20)						
Non-optimized rules (No Rules)						
▶ APCL & URLF, Content Awareness Inline Layer (21)						
▼ Cleanup rule (22-23)						
22	Cleanup rule	Any	Any	Any	Drop	Log

Figure 11.1 – Defining a policy structure using section titles and mandatory rules

While naming the section titles, you might optionally include action and tracking notes in some of them, as shown in the preceding screenshot.

With our policy structure defined, let's go from the top-down, learning the purpose of each section and, when necessary, creating rules in them. We will be omitting the three rules already created.

Creating rules for the firewall/networking layer

Since a **Gateway Access** rule is already present, the next rules are listed as follows:

- **DHCP**
- **Dynamic Routing**
- **Noise Suppression**

These three are interconnected in that all of them are related to traffic addressed to the gateways. Because of this, these three sections are located higher up in the policy than the **Stealth Rule** section.

The **Noise Suppression** section and rules are there to drop and avoid logging useless traffic, specifically, the broadcast traffic from connected networks and the general broadcast. However, DHCP relies on it, so the relevant section must be positioned higher. Additionally, we are supposed to define objects representing the broadcast addresses for connected networks and general broadcasts.

Defining hosts for broadcast addresses

To handle broadcast traffic, we have to create a number of dummy host objects with broadcast addresses. In SmartConsole, launch the **COMMAND LINE** window from the bottom-left corner of the UI and paste into it the following block of code (using the *Paste* icon):

```
add host name BCast_10.0.0.255 ip-address 10.0.0.255
add host name BCast_10.10.10.255 ip-address 10.10.10.255
add host name BCast_10.20.20.255 ip-address 10.20.20.255
add host name BCast_10.30.30.255 ip-address 10.30.30.255
add host name BCast_255.255.255.255 ip-address 255.255.255.255
```

The **BCast** hosts have now appeared in our **Objects | Hosts** list. You can close the **COMMAND LINE** window.

Creating rules for DHCP traffic

Now we can use these BCast objects in our DHCP and **Noise Suppression** rules. For this purpose, let's consider that the **Internal1** network might use DHCP services running on the **Internal2** host, **ADDCDNS**. In this case, the rules required to accommodate DHCP functionality will look like this:

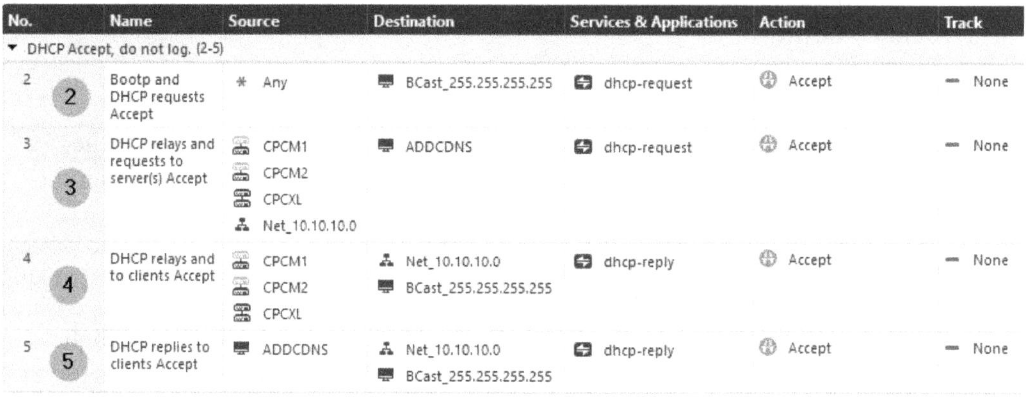

No.		Name	Source	Destination	Services & Applications	Action	Track
▼ DHCP Accept, do not log. (2-5)							
2	2	Bootp and DHCP requests Accept	✻ Any	🖥 BCast_255.255.255.255	🔁 dhcp-request	⊕ Accept	— None
3	3	DHCP relays and requests to server(s) Accept	CPCM1 CPCM2 CPCXL Net_10.10.10.0	🖥 ADDCDNS	🔁 dhcp-request	⊕ Accept	— None
4	4	DHCP relays and to clients Accept	CPCM1 CPCM2 CPCXL	Net_10.10.10.0 🖥 BCast_255.255.255.255	🔁 dhcp-reply	⊕ Accept	— None
5	5	DHCP replies to clients Accept	🖥 ADDCDNS	Net_10.10.10.0 🖥 BCast_255.255.255.255	🔁 dhcp-reply	⊕ Accept	— None

Figure 11.2 – The DHCP rules requiring directional services

In general, the four rules addressing the DHCP needs are as follows:

No.	Source	Destination	Services and Applications	Action	Track
1	Any	BCast_255.255.255.255	dhcp-request	Accept	None
2	Cluster members Cluster object Client network(s)	DHCP server(s)	dhcp-request	Accept	None
3	Cluster members Cluster object	Client network(s) BCast_255.255.255.255	dhcp-reply	Accept	None
4	DHCP server(s)	Client network(s) BCast_255.255.255.255	dhcp-reply	Accept	None

Table 11.1 – General DHCP rule requirements

Besides policy rules for DHCP traffic, additional OS-level configuration is required.

Gaia configuration for BOOTP/DHCP Relay

In addition to the rules within the policy, we should also define the Gaia parameters for handling DHCP traffic across interfaces. This is similar to Cisco's DHCP Relay Agent and associated IP-helper-address configuration. Note that the `eth1` interface is where our client network is connected, and `10.20.20.10` is the IP address of our **ADDCDNS** server.

DHCP relays can be configured in Gaia WebUI, **Advanced Routing | DHCP Relay | Add**:

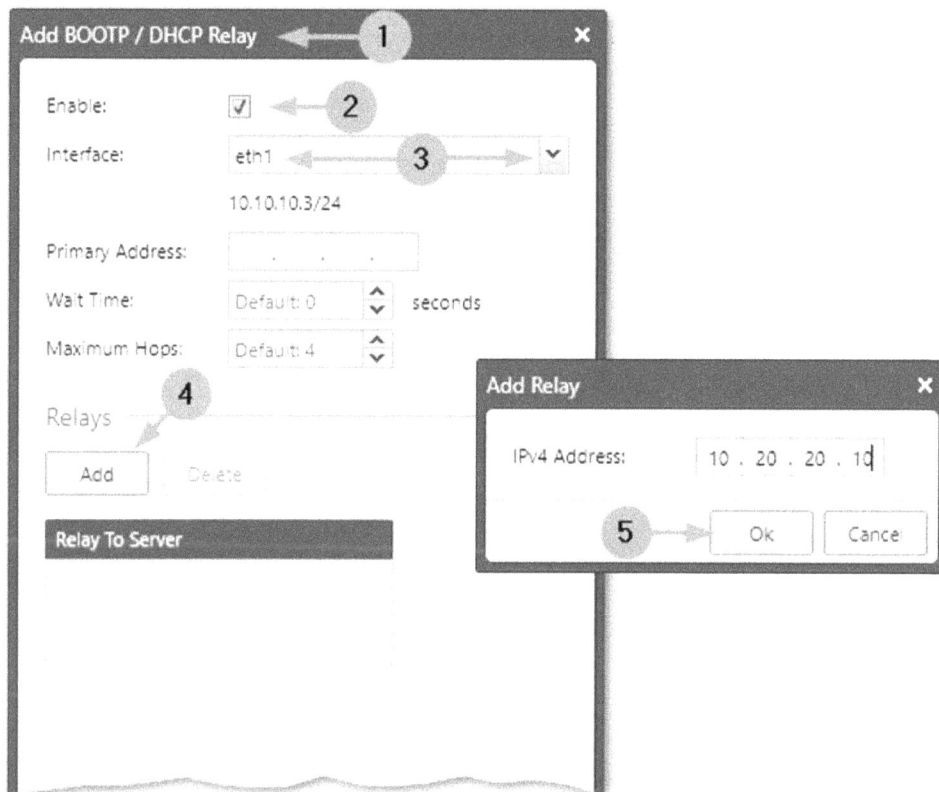

Figure 11.3 – Configuring BOOTP/DHCP Relay in Gaia WebUI

Click on **Ok** and then **Save** (this is hidden in the **Add BOOTP /DHCP Relay** window)

Alternatively, you can configure them via CLISH, using identical commands on both cluster members:

```
set bootp interface eth1 relay-to 10.20.20.10 on
save config
```

Configuring DHCP on Windows Server

To test DHCP's functionality, add the DHCP Server role to **ADDCDNS**, create a **Scope** with the 10.10.10.0/24 network, and define an **Address Pool** that could be used for a dynamic assignment, such as 10.10.10.51–10.10.10.60:

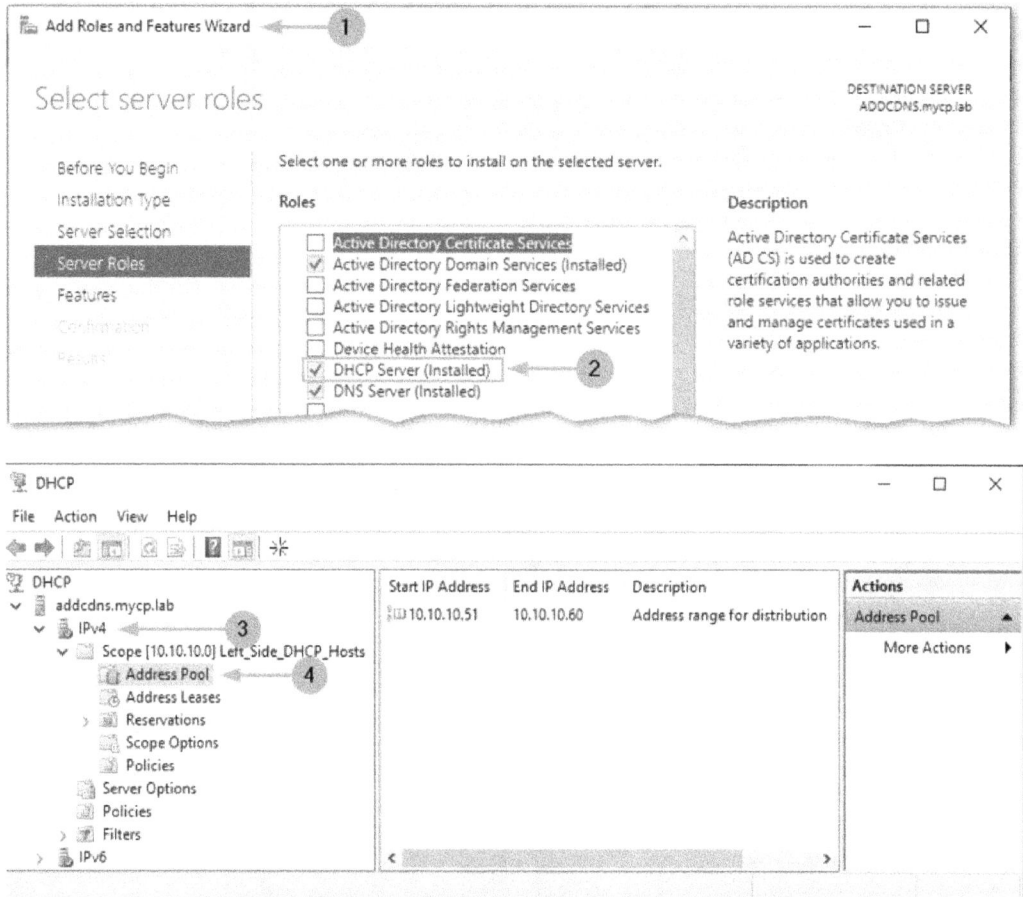

Figure 11.4 – Enabling the DHCP server and defining the scope and address pool (optional figure)

Note that while it might not make sense to log DHCP traffic on a regular basis, you can turn it on during implementation and change its tracking to **None** afterward.

Configuring rules for noise suppression

With DHCP taken care of, we can now create a rule to drop and avoid logging otherwise meaningless broadcasts using the objects created earlier in this chapter:

No.	Name	Source	Destination	Services & Applications	Action	Track
▼ Noise Suppression. Drop do not log. (6)						
6	Broadcasts Drop. Do not log.	✳ Any	🖥 BCast_10.0.0.255	✳ Any	◉ Drop	━ None
			🖥 BCast_10.10.10.255			
			🖥 BCast_10.20.20.255			
			🖥 BCast_10.30.30.255			
			🖥 BCast_255.255.255.255			

Figure 11.5 – Broadcast noise suppression rule

Stealth Rule and its section title are already in place, so we can move on to the core services.

Configuring rules for core services

These are the rules that, in addition to the ones we have already created, ensure the basic functionality of your environment. Where the rest of the rules are (or should be) governed by your company's written security policy, these rules should reflect the prerequisite services and communication channels. At a minimum, these will address the domain name services and the network time services traffic, but each environment is different, and you might find other dependencies that should be addressed in your **Core Services** section. You can call this the *rest of the core services* section, as your infrastructure could very well depend on DHCP and dynamic routing for basic functionality.

Let's take a look at the rules required in our lab:

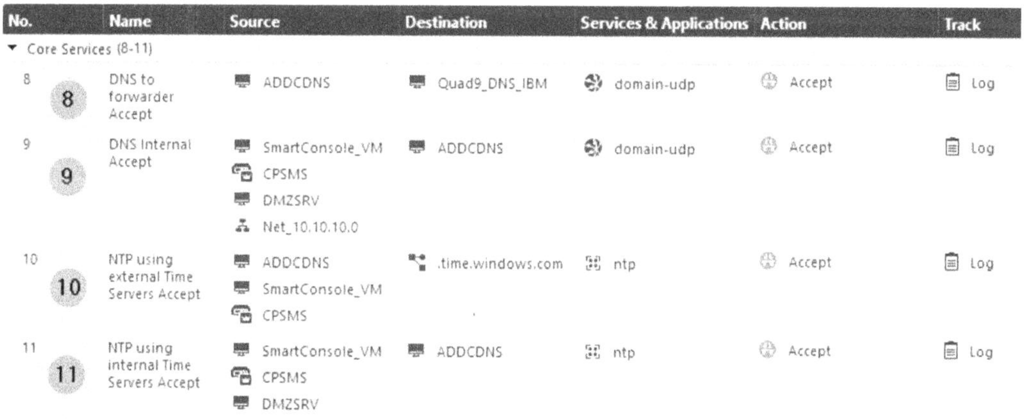

No.	Name	Source	Destination	Services & Applications	Action	Track
▼ Core Services (8-11)						
8	DNS to forwarder Accept	🖥 ADDCDNS	🖥 Quad9_DNS_IBM	🌐 domain-udp	⏱ Accept	📄 Log
9	DNS Internal Accept	🖥 SmartConsole_VM	🖥 ADDCDNS	🌐 domain-udp	⏱ Accept	📄 Log
		🖥 CPSMS				
		🖥 DMZSRV				
		⛢ Net_10.10.10.0				
10	NTP using external Time Servers Accept	🖥 ADDCDNS	📡 .time.windows.com	🕒 ntp	⏱ Accept	📄 Log
		🖥 SmartConsole_VM				
		🖥 CPSMS				
11	NTP using internal Time Servers Accept	🖥 SmartConsole_VM	🖥 ADDCDNS	🕒 ntp	⏱ Accept	📄 Log
		🖥 CPSMS				
		🖥 DMZSRV				

Figure 11.6 – The Core Services rules for the lab

When creating these rules, it helps to walk through the logic of requirements; for instance, we need DNS to resolve anything, so let's configure the rules for that first. Which of our servers are allowed to communicate with which specific public DNS servers? Do we have an internal time server? If the answer is *yes*, which hosts should be able to use it? Let's define the missing objects if any are present, and use them in rules to be as specific as possible.

I'd like to use the public DNS service provided by IBM, due to its security filtering and privacy policy. So, let's create a `Quad9_DNS_IBM` host using the command line:

```
add host name Quad9_DNS_IBM ip-address 9.9.9.9
```

Windows servers are preconfigured to sync the time to `time.windows.com` servers, so it makes sense to allow our hosts that might rely on external time sources to reach it. Let's create the FQDN domain object, `.time.windows.com`, using the command line:

```
add dns-domain name ".time.windows.com"  is-sub-domain false
```

In addition to the rules allowing hosts to sync time with external time servers, let's create a rule for our infrastructure's components to use the internal NTP server(s), too.

The rest of the objects depicted in *Figure 11.6* are already available, so we can assemble our rules.

If you have to use DNS zone replication, add the rules for the primary and secondary DNS servers and `domain-tcp` services.

Configuring rules for privileged access

It is a good idea to define rules for privileged access of static assets, such as administrative workstations or a dedicated **privileged access management** (**PAM**) solution for the crown jewels of your infrastructure. To illustrate them, let's create rules for the SSH, RDP, and LDAP access to some of the lab components:

No.		Name	Source	Destination	Services & Applications	Action	Track
▼ Privileged Access. (12-15)							
12	12	SSH acccess to Router Accept	SmartConsole_VM	Router	ssh_version_2	Accept	Log
13	13	Admins RDP to all Accept.	SmartConsole_VM	DMZZone InternalZone	Remote_Desktop_...	Accept	Log
14	14	Admins RDP to all Accept.	*Negated* SmartConsole_VM	DMZZone InternalZone	Remote_Desktop_...	Drop	Log Alert
15	15	LDAP Access	SmartConsole_VM CPSMS	ADDCDNS	ldap ldap-ssl	Accept	Log

Figure 11.7 – Privileged access rules for the lab

Note the use of the negated source cell in rule **14** with the corresponding **Alert** in its **Action** cell. This is a simple way to ensure that at least straightforward attempts at unsanctioned access are investigated immediately if you configure appropriate alert notifications.

Rules that have corresponding entries with an empty threat prevention profile

We do not have any of these in our policy, but it is nice to have a placeholder for the rules that will be used for latency-sensitive traffic. Those are the rules that we discussed in *Chapter 8*, *Introduction to Policies, Layers, and Rules*, which are exempt from content inspection using an empty threat prevention profile assigned to the same sources, destinations, and services.

Configuring internal access rules

This is a placeholder rule for internal communication. It is suitable for new implementations when dataflows and services that are in use are not yet documented. Having it in place will allow you to log all the traffic between the internal segments of your infrastructure, document the observed behavior, and work with IT and business units to craft the rules that are aligned with your written security policy.

At the very least, you'll have the ability to enforce threat prevention on all internal communication while working on the refinement of the policy:

No.	Name	Source	Destination	Services & Applications	Action	Track
▼ General Internal Access. (16)						
16	Internal communication Accept (used until specific rules are created)	InternalZone	InternalZone	＊ Any	⊕ Accept	🖹 Log

Figure 11.8 – Internal communication rule(s)

Recall that we have assigned **InternalZone** to two clustered interfaces: eth1 and eth2. The LeftHost VM, representing a typical network client, is behind eth1, while **ADDCDNS**, which hosts all the core services and represents one of the server-side networks, is behind eth2. Also, recall that even though these interfaces belong to the same zone, communication between the networks associated with the same zone(s) is not permitted unless specific rule(s) are created.

Configuring DMZ access rules

Let's configure our DMZSRV host as the web server to make the experience more believable. Log in to DMZSRV, start PowerShell, and paste into it the following string. Then, press *Enter*:

```
Install-WindowsFeature -name Web-Server -IncludeManagementTools
```

Typically, the installation process completes in under a minute, and you have an **Internet Information Server** (**IIS**) running and serving the default page over HTTP.

Now, let's get back to our `SmartConsole_VM`, and within the SmartConsole application, create the following rules:

No.	Name	Source	Destination	Services & Applications	Action	Track
▼ DMZ (17-18)						
17	Internal2 network to DMZ Accept	Net_10.10.10.0	DMZSRV	http https	Accept	Log
18	Outside to WebSrv Accept	ExternalZone	DMZSRV	http https	Accept	Log

Figure 11.9 – DMZ access rules

Note that our **DMZSRV** web server cannot serve pages via HTTPS, as we have not made any provisions for it yet. However, it makes sense to include HTTPS in the rules that would eventually rely on it.

Configuring rules for access to updatable objects

Where it is necessary to allow connection to dynamic cloud-based resources that might be comprised of multiple domains and IP addresses or ranges and that are trusted, let's configure a rule for access to the updatable object. The reason we are separating these rules from the remainder that will belong in application control and URL filtering is that we are explicitly trying to avoid unnecessary content inspection.

Remember that, to include an updatable object, we must hover over the cell, click on the + sign that appears in it, click on **Import** and select **Updatable Objects**. In the **Updatable Objects** browser, check the **Check Point Services** box and click on **OK** to include it in the **Destination** cell:

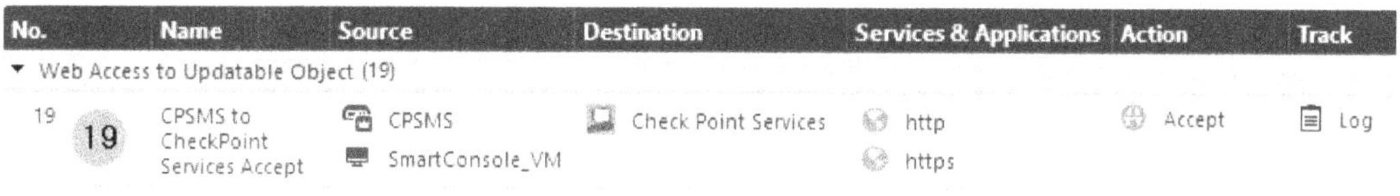

No.	Name	Source	Destination	Services & Applications	Action	Track
▼ Web Access to Updatable Object (19)						
19 19	CPSMS to CheckPoint Services Accept	CPSMS SmartConsole_VM	Check Point Services	http https	Accept	Log

Figure 11.10 – Rules for access to updatable objects

Configuring rules for probes

Some of our applications, appliances, or services rely on the ability to determine the availability of resources outside of our environment. A case in point is Microsoft Windows' internet connectivity test. Unless it succeeds, users are presented with a warning in the system tray claiming that the internet is not reachable.

In production environments, resources hosted in a public cloud are frequently polled to either trigger notifications or to take automated scripted actions if these are not available.

Let's define Microsoft's internet connectivity test targets and create a rule allowing probes to succeed. To do that, we must create two FQDN domain objects by performing a copy and paste of these commands into the **COMMAND LINE** window of SmartConsole:

```
add dns-domain name ".www.msftncsi.com"  is-sub-domain false
add dns-domain name ".www.msftconnecttest.com"  is-sub-domain false
```

We can use them in the **Destination** field of the rule:

No.	Name	Source	Destination	Services & Applications	Action	Track
▼ Probes (20)						
20 20	MS Internet connectivity probe Accept	Int_Nets	.www.msftncsi.com .www.msftconnecttest.com	http	Accept	None

Figure 11.11 – Rules for service probes

Non-optimized rules

These are the rules that we have referred to as non-optimized in *Chapter 8, Introduction to Policies, Layers, and Rules*, as described in the *sk32578*. Services defined in these rules either cannot be accelerated themselves or prevent the **Accept** template creation for all subsequent rules (a subject we have discussed in the *Best practices for access control rules* section of *Chapter 8, Introduction to Policies, Layers, and Rules*). It might actually make sense to experiment with these rules by placing them right above the **Cleanup** rule that is located at the very bottom of the policy if you are trying to squeeze maximum performance from your appliances.

This pretty much covers the structure of our lab's firewall/network access control policy. Next, let's continue our work by adding an inline APCL/URLF and content awareness layer.

Creating the APCL/URLF layer and rules

If we were to create a rule containing internal networks as sources, **ExternalZone** as a destination, and **Accept** as an action, we would be free to roam the internet from any internal host subjected to NAT.

Since it would be an exceptionally bad idea without the appropriate access controls, let's change the action to **Inline Layer | New Layer**. In **Layer Editor** [1], name it APCL_URLF_ Layer [2], and check the **Applications & URL Filtering** box [3] in its properties. Also, check the box for **Multiple policies and rules can use this layer** [4]:

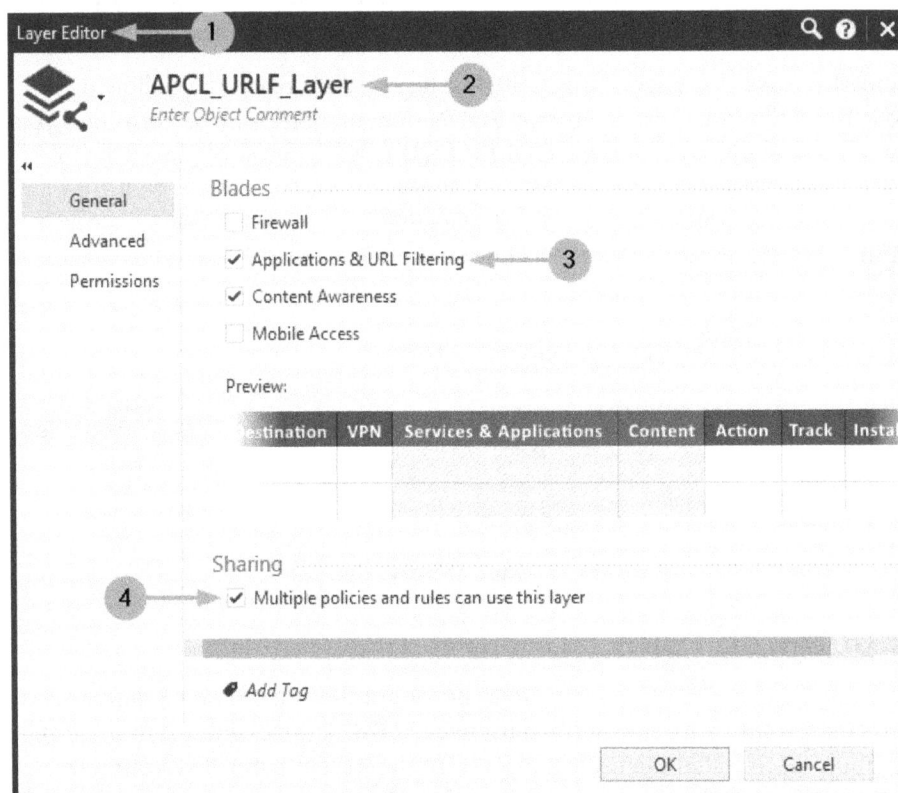

Figure 11.12 – Creating APCL_URLF_Layer for the lab

Choose the **Advanced** [1] section on the left-hand side, and change the **Implicit Cleanup Action** option to **Drop** [2]. Check the box for **Detect users located behind http proxy with X-Forward-For** [3], and click on **OK** [4]:

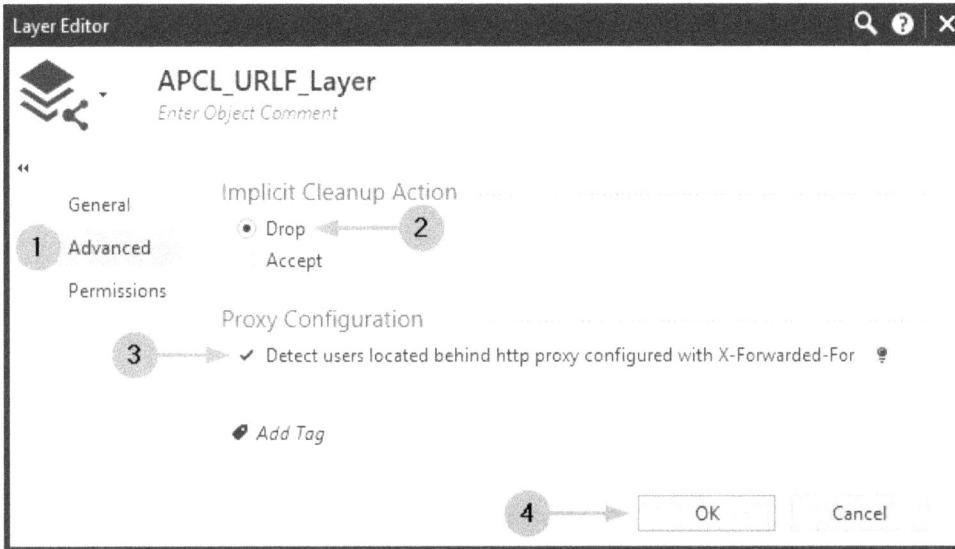

Figure 11.13 – The advanced properties of APCL_URLF_Layer

Now, let's populate this newly created layer according to the following screenshot:

Figure 11.14 – The APCL_URLF_Layer rules

Now, we must remember, that in addition to enabling the **Applications & URL Filtering** box in the layer's properties, we must perform a few additional steps before we start using these rules. Namely, we should do the following:

1. Enable APCL/URLF in the properties of the cluster.
2. Create an outbound CA certificate for HTTPS inspection, and enable HTTPS Inspection in the properties of the cluster.
3. Configure the HTTPS Inspection policy.
4. Distribute and install the outbound CA and ICA certificates to the client machines.
5. Change the website categorization to **Hold** mode.

So, let's get on with these tasks.

Enabling APCL/URLF in the properties of the cluster

To enable the features/blades in the properties of the cluster, double-click on the **CPCXL** cluster object [1] anywhere it can be found. Then, check the **Application Control** and **URL Filtering** boxes [2]:

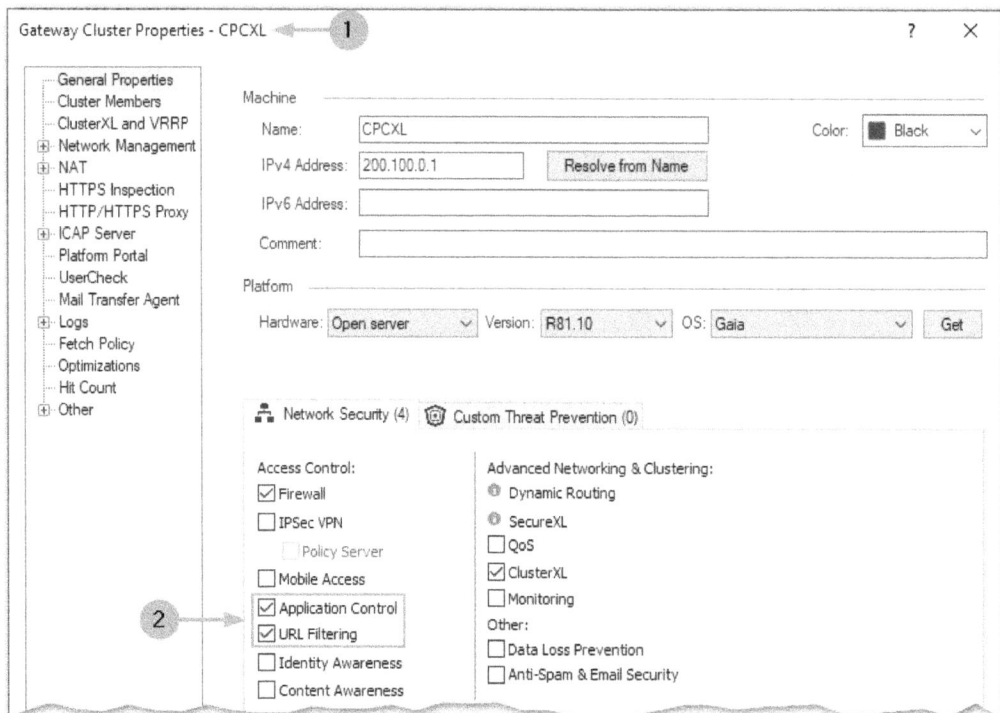

Figure 11.15 – Enabling Application Control and URL Filtering in the cluster's properties

Creating an outbound CA certificate for HTTPS inspection and enabling HTTPS Inspection in the properties of the cluster

Our gateway's ability to recognize applications and the content they are exchanging reliably depends on HTTPS inspection. This, in turn, requires that our gateways or clusters act as trusted **Man-In-The-Middle** (**MITM**) devices. For this to work, an outbound CA certificate must be created by the first gateway or cluster in our environment. If your environment consists of multiple gateways or clusters administered by the same management server, the same certificate will be used by the rest:

1. Continuing work on the properties of our cluster [1], click on **HTTPS Inspection** [2]. In **Step 1**, click on **Create** [3]:

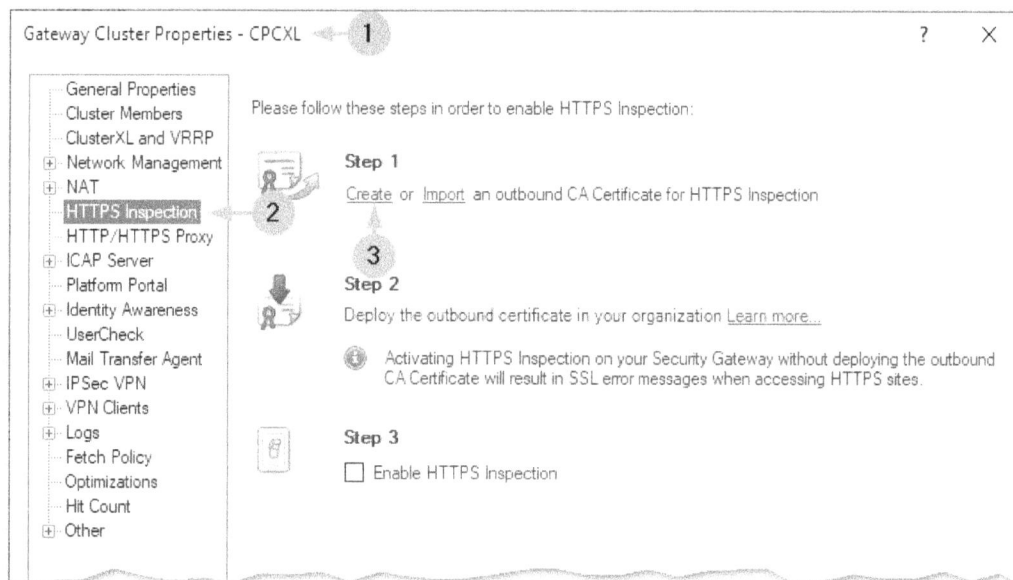

Figure 11.16 – Initiating the creation of an outbound CA certificate

2. Now, you will be prompted with the **Create** window [1]. In **Issued by (DN)**, type in our domain name [2]. You can use the same password that we adopted elsewhere in the lab, CPL@b8110 [3]. Click on **OK** [4] to complete the process:

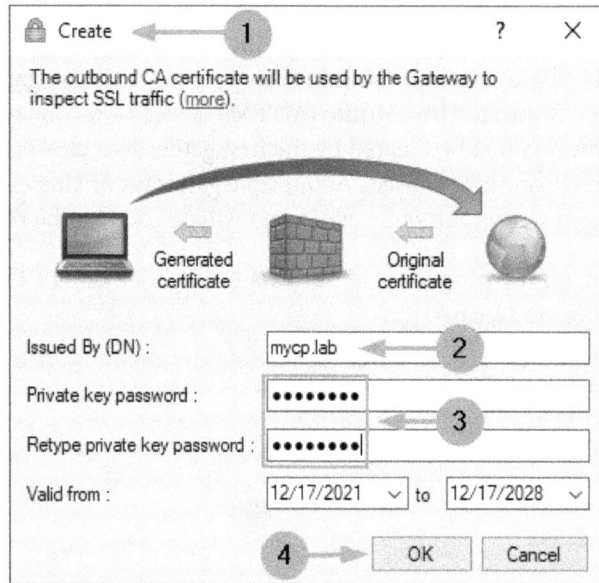

Figure 11.17 – Creating an outbound CA certificate

3. You can now view the created certificate [1] and learn about the distribution of this certificate in the production infrastructure [2]. Now, we must export this certificate for distribution to the machines in our environment by clicking on **Export certificate** [3]:

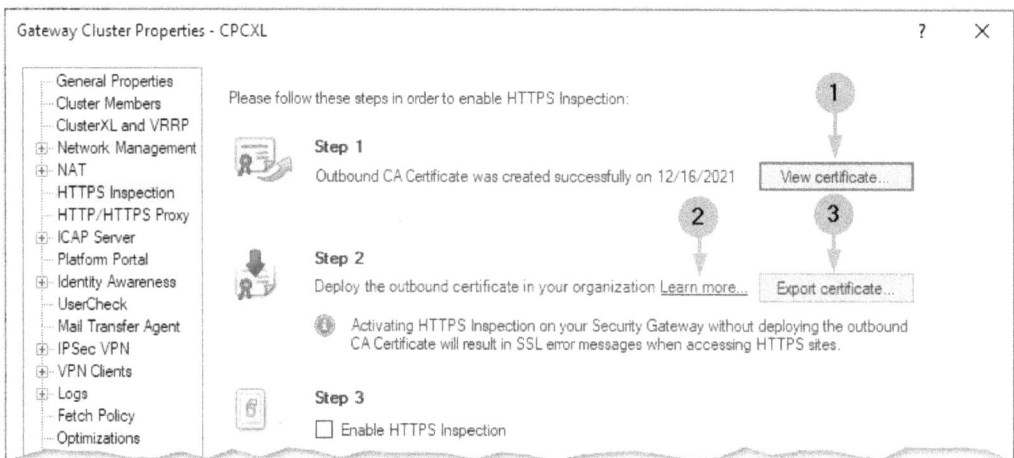

Figure 11.18 – Exporting the outbound CA certificate in SmartConsole

4. In the **Save Certificate** window [1], browse to LabShare (mapped drive F:) [2]. Then, in **File name**, enter a recognizable name, such as outbound.mycp.lab [3], leave the default **Save as type** value of **Certificate file (*.cer)** [4] unchanged, and click on **Save** [5]:

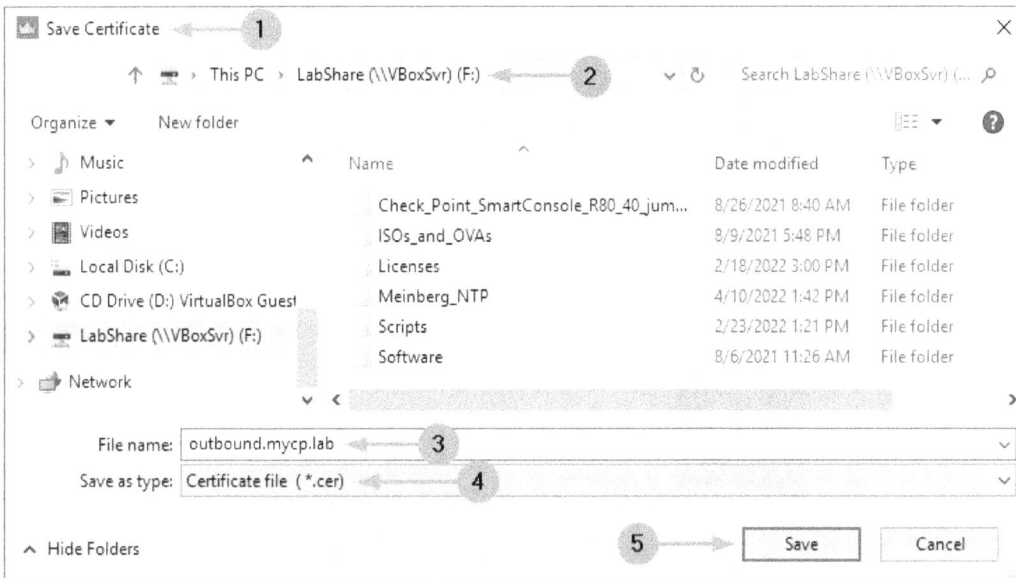

Figure 11.19 – Saving the exported outbound CA certificate for the distribution

5. Let's check the **Enable HTTPS Inspection** box [1] and click on **OK** at the bottom of the **Gateway Cluster Properties** window (not pictured):

Figure 11.20 – Enabling HTTPS Inspection in the gateway cluster properties

Configuring the HTTPS Inspection policy

As soon as HTTPS Inspection has been enabled, we should define at least a rudimentary HTTPS Inspection policy to prevent the violation of privacy regulations. In the **SECURITY POLICIES** view [1], under the **HTTPS Inspection** section, click on **Policy** [2]:

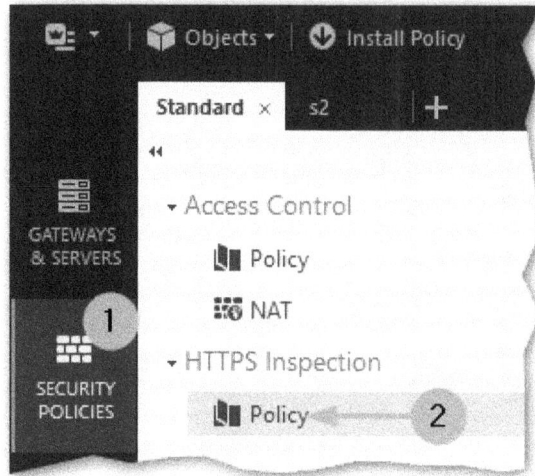

Figure 11.21 – Opening the HTTPS Inspection policy for editing

When we enable HTTPS Inspection, the **Predefined Rule** [1] is created automatically. Let's add the rule [2] for the **Financial Services** and **Health** categories [4] as well as an additional rule [3] with the **Internet** destination replaced with **Updatable Object HTTPS services – recommended bypass** [5]. Remember that to add that object to the destination, you'll have to hover in the cell, click on +, then click on **Import | Updatable Objects**, expand the **HTTPS services – bypass** section, and select the recommended bypass group. Take a moment to examine the content of this group. Now, change the **Action** fields for our two new rules to **Bypass** [6]:

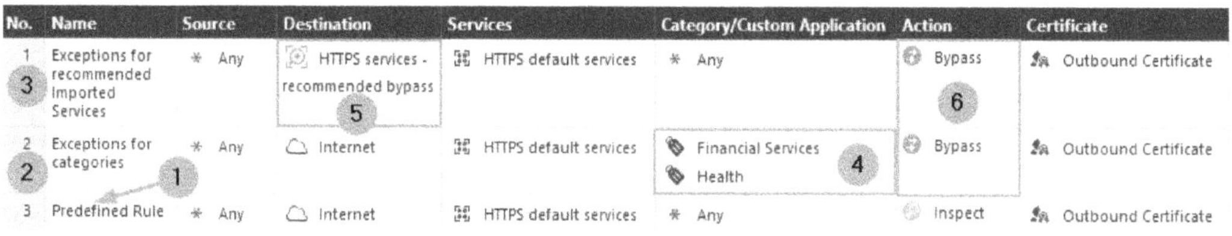

No.	Name	Source	Destination	Services	Category/Custom Application	Action	Certificate
1	Exceptions for recommended Imported Services	* Any	HTTPS services - recommended bypass	HTTPS default services	* Any	Bypass	Outbound Certificate
2	Exceptions for categories	* Any	Internet	HTTPS default services	Financial Services / Health	Bypass	Outbound Certificate
3	Predefined Rule	* Any	Internet	HTTPS default services	* Any	Inspect	Outbound Certificate

Figure 11.22 – Configuring the HTTPS Inspection policy

Now is a good time to publish the pending changes and install the policy before moving on to the next section.

Distributing and installing the outbound CA and ICA certificates to the client machines

The Check Point outbound CA, which was generated earlier in this chapter, can be distributed to your domain member machines using **Group Policy Objects** (**GPOs**). For installation on non-domain member machines and appliances, use a manual or scripted installation.

For our lab, we can simply install the certificates manually on machines that we are experimenting with. Since we'll need more than this single certificate installed on our internal clients, let's install this one just on SmartConsole_VM for now.

Navigate to LabShare and filter it out using cer to see the certificate we saved earlier:

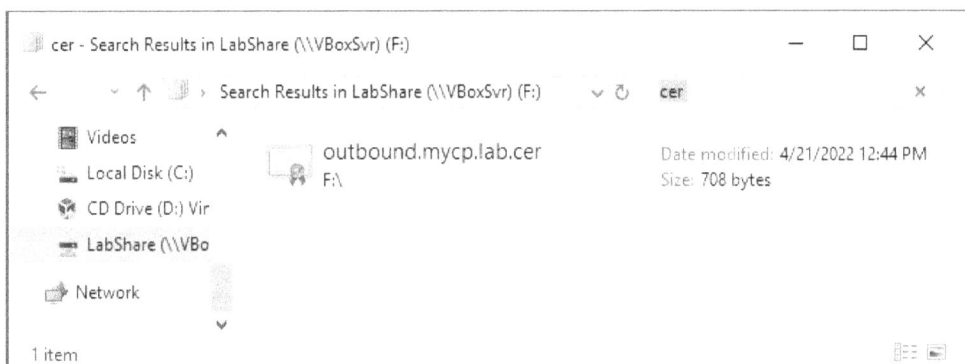

Figure 11.23 – The outbound certificate installation on the client VMs, part one

Double-click on the outbound.mycp.lab.cer file. In the **Open File – Security Warning** window [1], click on **Open** [2]:

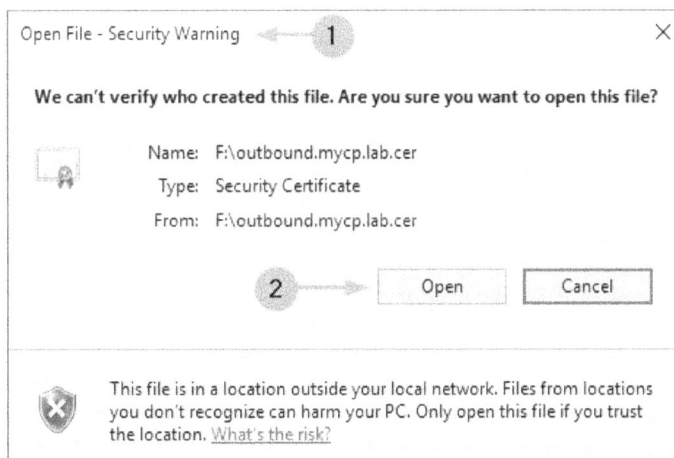

Figure 11.24 – The outbound certificate installation on the client VMs, part two

In the **Certificate** window [1], read the note about the current state of trust in this certificate and the necessary steps to change it [2]. Verify the issuance data [3], and click on **Install Certificate** [4]:

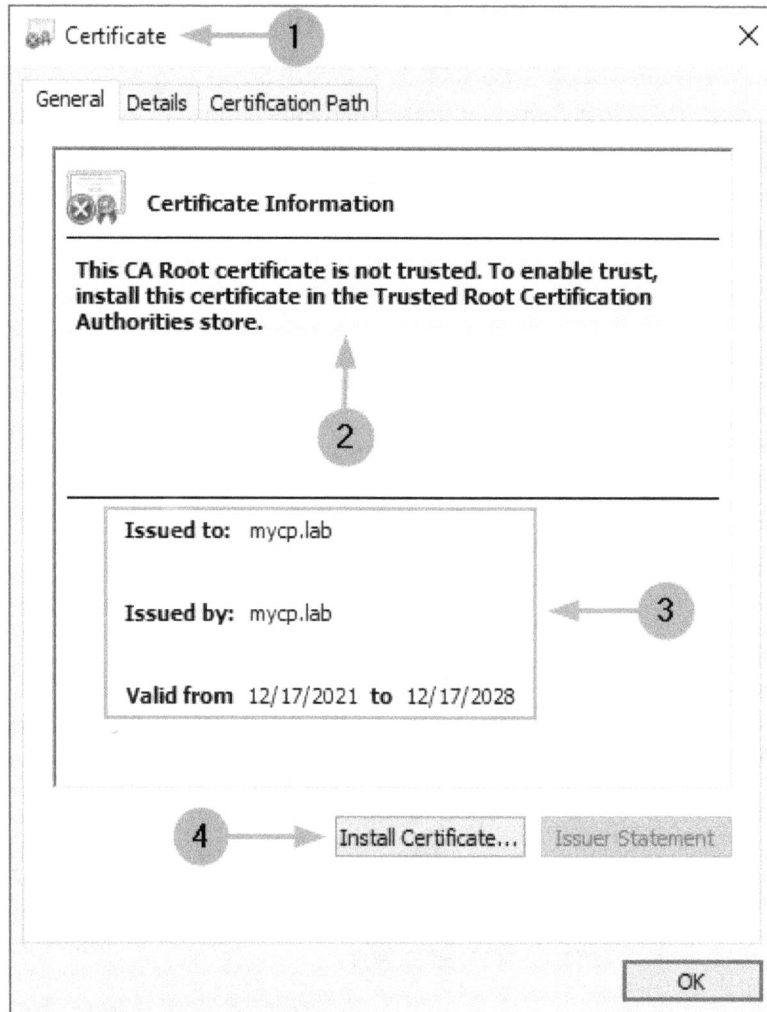

Figure 11.25 – The outbound certificate installation on the client VMs, part three

In the **Certificate Import Wizard** window [1], change the store location to **Local Machine** [2], and click on **Next** [3]:

Figure 11.26 – The outbound certificate installation on the client VMs, part four

In the next window of **Certificate Import Wizard**, change the selection to **Place all certificates in the following store** [1] and click **Browse** [2]. In the **Select certificate Store** window, select **Trusted Root Certification Authorities** [3] and click on **OK** [4]. Then, click on **Next** [5]:

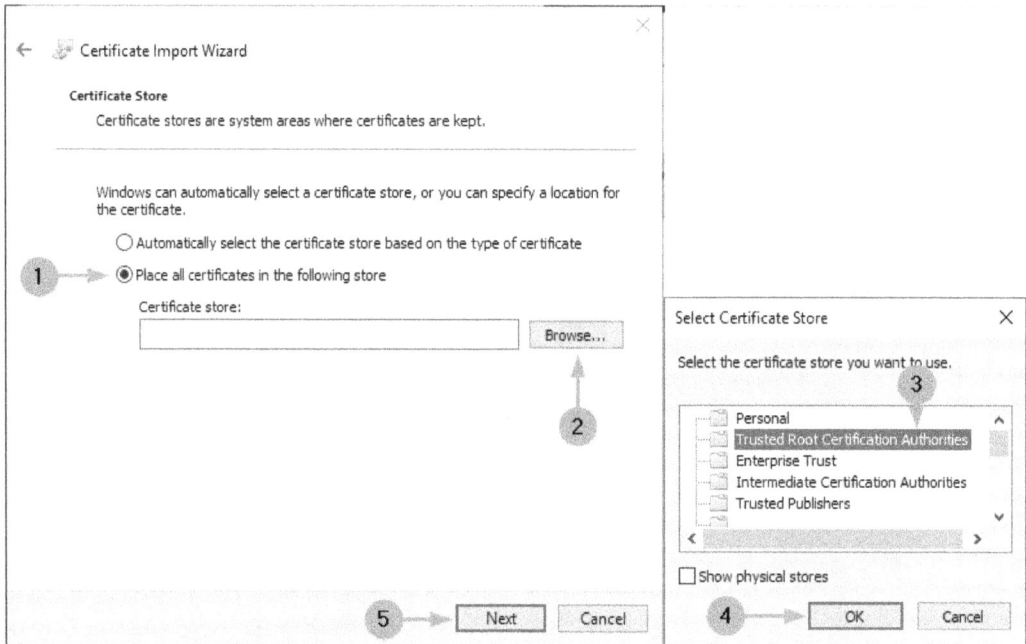

Figure 11.27 – The outbound certificate installation on the client VMs; part five

In the next window, click on **Finish**. Then, when presented with the **Certificate Import was successful** message, click on **OK**. Click on **OK** again in the certificate window to dismiss it. Close the **Windows Explorer** window.

Once done, open Google Chrome on `SmartConsole_VM` and launch a new Incognito mode session to avoid caching. In the browser's address bar, type in `https://www.google.com` and press *Enter*. You should see the familiar search portal.

Click on the lock icon on the left-hand side of the `google.com` address [1], click on the expansion arrow next to the **Connection is secure** status [2], and then click on the **Certificate is valid** option [3]:

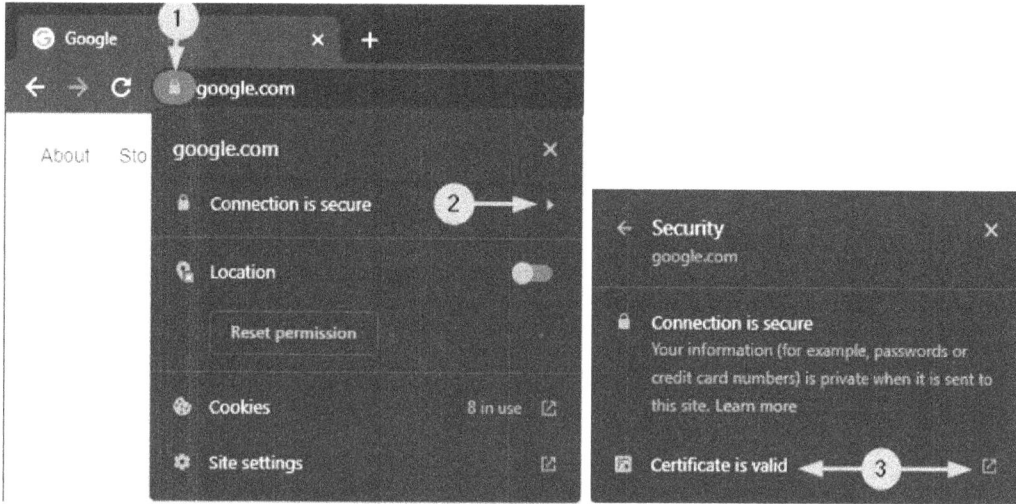

Figure 11.28 – Validation of the outbound CA certificate

We can examine the certificate properties to see that it was issued to www.google.com by our mycp.lab environment [2]. Click on **OK** [3] to dismiss the window:

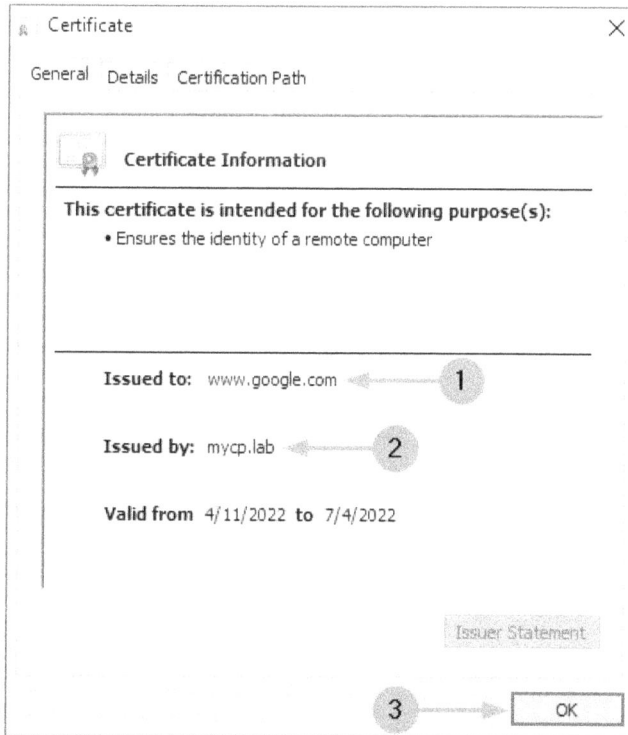

Figure 11.29 – The outbound CA certificate in action

In the same browser window, open a new tab, type in www.facebook.com, and press *Enter*. If you are prompted with a **Your connection is not private** warning, ignore it, and click **Advanced**. Then, click on **Proceed to 200.100.0.1 (unsafe)**. This is the IP address of our cluster. You will now be greeted with a **Page Blocked** UserCheck message:

Access to facebook.com is blocked according to the organization security policy.

And in our browser's address bar, we can see the **Not Secure** warning with the HTTPS strike-through. Click on the **Not Secure** message in the address bar [1], and then click on the **Certificate is not valid** status [2]. In the **Certificate** properties, we can see that it could not be verified by a trusted certification authority. That's because our own Check Point ICA certificate has not yet been added to the **Trusted Root Certification Authorities** list:

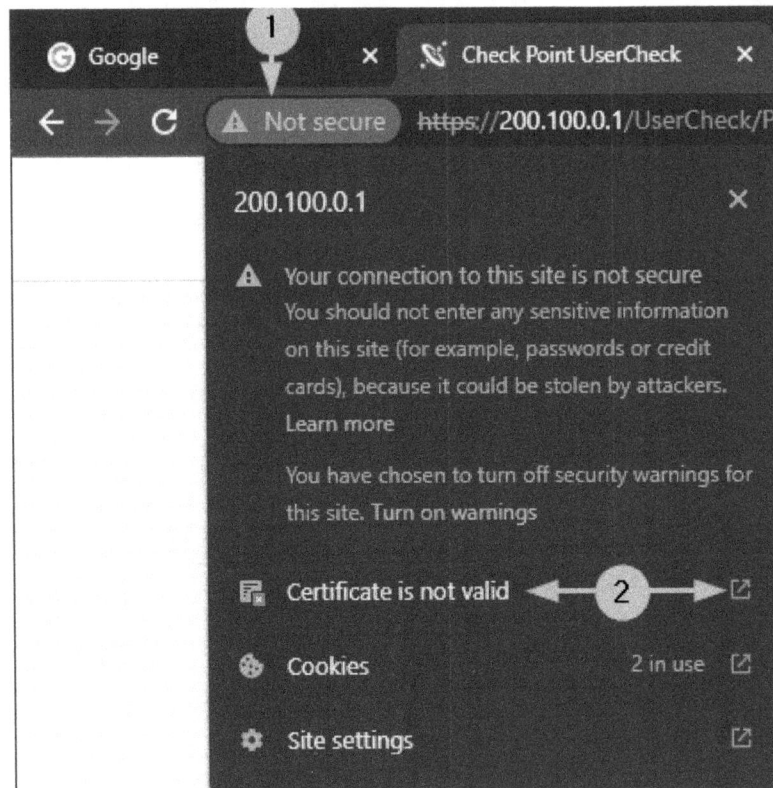

Figure 11.30 – Using the security warning from UserCheck to extract the ICA certificate

Click on the **Not Secure** warning [1] (not shown), click on the **Certification Path** tab [2], and select the **CPSMS.mycp.lab.7kr2b2** certificate from the top of the certification path [3]. Then, click on **View Certificate** [4]. In a new CPSMS certificate window, click on the **Details** tab [5]. Click on the **Copy to file** button [6]:

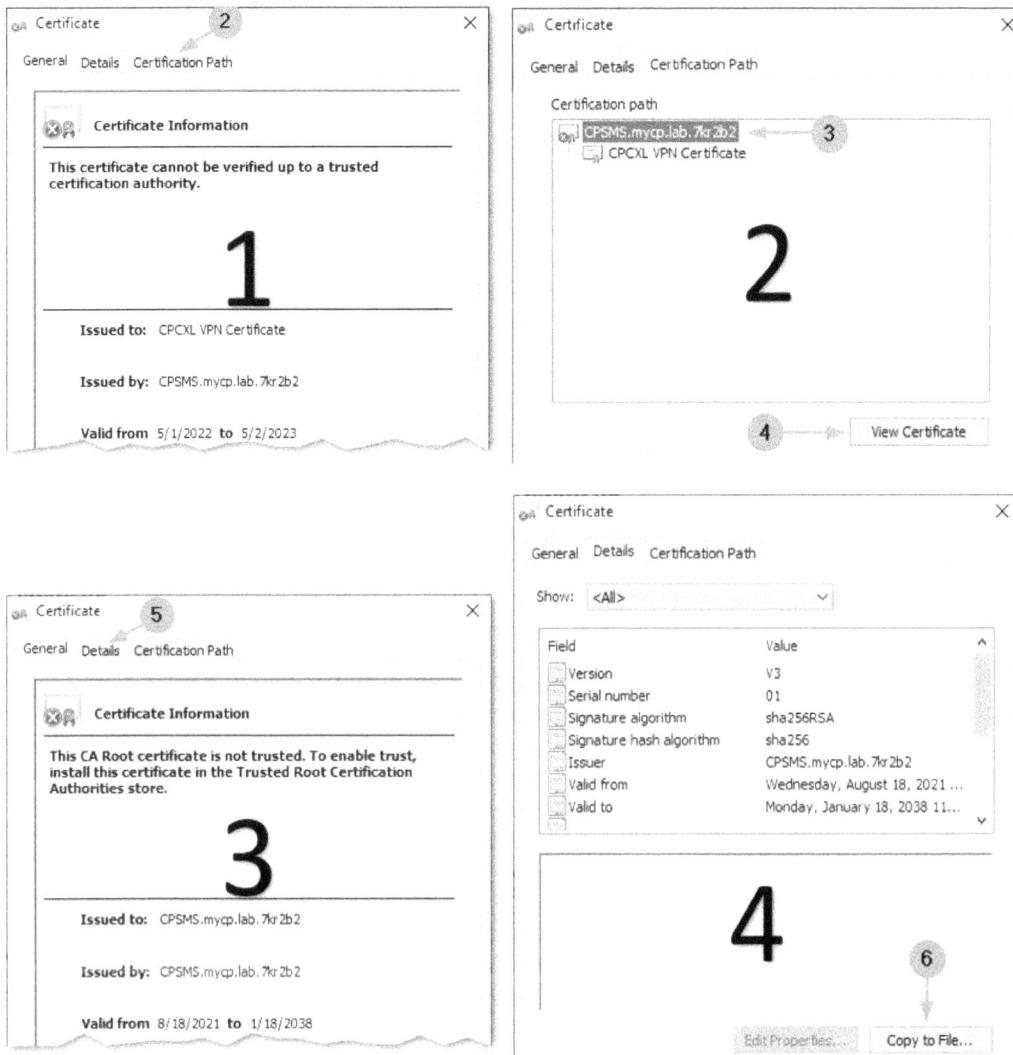

Figure 11.31 – Preparing the management server's certificate for distribution

In **Certificate Export Wizard**, click on **Next | Next | Browse**, and navigate to **LabShare**. In **File name**, type in `cpsms.mycp.lab`, leave the **Save as type** default setting of **DER Encoded Binary X.509 (*.cer)**, and click on **Save**. Click on **Next** and then click on **Finish**. Click on **OK** three times to dismiss all the certificate windows.

Now, open LabShare and repeat the installation of this certificate for **Local Machine** in the **Trusted Root Certification Authorities** location, which is the same as the one we have used for the outbound CA we imported earlier.

In our Incognito browser tab, navigate to another social networking site (for instance, Instagram), and you should be greeted with UserCheck's **Page Blocked** notification but no more security warnings. To verify that our HTTPS bypass rules are working, visit one of the banking sites, such as www.citi.com or www.jpmorgan.com, and check the certificates' properties for them. You will see that they present their own certificates and not the mycp.lab certificate used for the inspected traffic.

Repeat the installation of both certificates on the rest of the Windows VMs in the lab.

Changing the website categorization to Hold mode

By default, URL categorization occurs in the background. What this means is that any first attempts to connect to a previously unknown URL will succeed, until Check Point ThreatCloud decides it is used by bad actors; only then will it be blocked. This approach is only acceptable if you are exercising a solid defense-in-depth strategy with other means of preventing the compromise.

If Check Point gateways are your primary (or the most advanced) line of defense, we should change the website categorization to hold (requests are blocked until the categorization is complete). To do that, in SmartConsole, click on **MANAGE & SETTINGS** [1], and click on **Blades** [2]. Then, under the **Application Control & URL Filtering** section, click on **Advanced Settings** [3]. In the **Application Control & URL Filtering Settings** window [4], click on **Check Point online web services** [5] and select the **Hold** option [6]. Click on **OK** [7]:

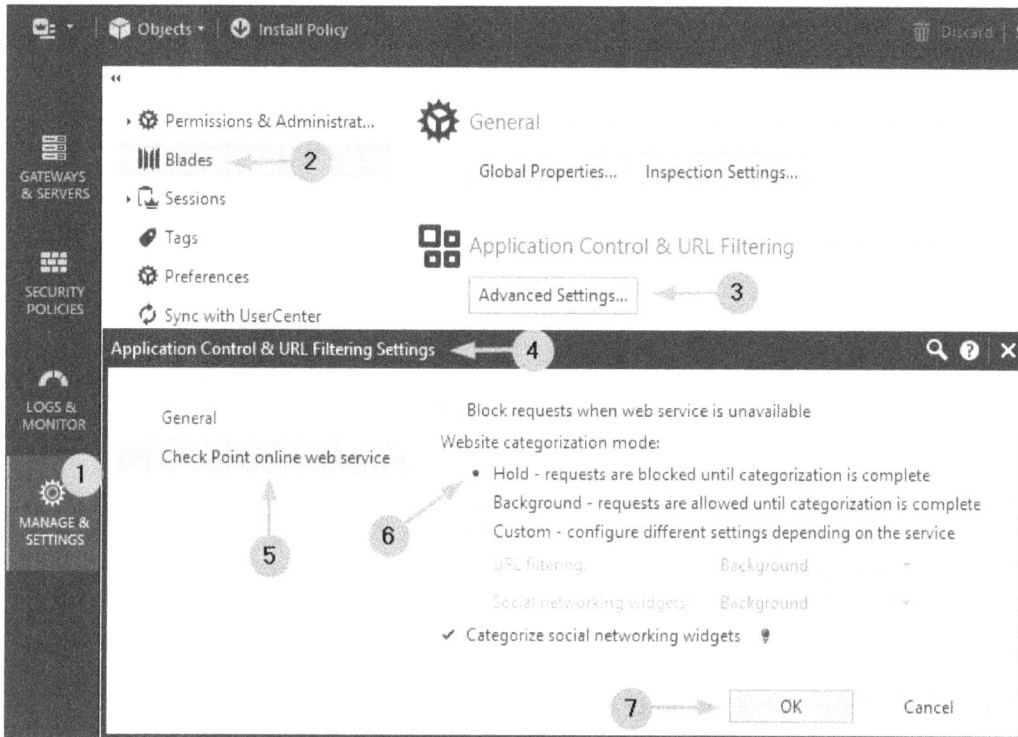

Figure 11.32 – Changing the website categorization mode to Hold

Once the changes have been published and the policy is installed, we are in business. So far, the only caveat is that we are dealing with statically defined internal assets. This is great for the core components side of things, but not so much for the users. We should have the ability to dynamically adapt access control based on user identity. Read on to see how this could be accomplished.

Using Identity Awareness and access roles

Prior to 2021, the easiest way to demonstrate Identity Awareness with an external user directory was to use the **AD Query** method. Unfortunately, due to the discovery of Microsoft's vulnerability *CVE-2021-26414*, Check Point had to deprecate this feature. It is still available to accommodate existing customers that are slow to migrate to a better solution, but being bad practice, I'd like to avoid it. Now, the appropriate way to implement Identity Awareness is to deploy **Identity Collector**. This is a more serious undertaking that we do not have space for and is a slightly more advanced subject. If you'd like to learn more about it, see *sk108235 Identity Collector – Technical Overview*. The other way for me to demonstrate it here is to use **Browser-Based Authentication**.

> **Note**
>
> To learn more about the reasons for AD Query deprecation and Microsoft's deadline for the retirement of WMI features, read the *sk176148, Check Point response to CVE-2021-26414 - "Windows DCOM Server Security Feature Bypass"*

Preparing Active Directory for integration with Identity Awareness

Let's create a simple domain user account that Check Point will be using for authentication attempts alongside a couple of test accounts and security groups. Log in to **ADDCDDNS** and open **PowerShell**. Copy and paste the following code block into **PowerShell**:

```
New-ADUser -Name "cpauth" -UserPrincipalName "cpauth@mycp.lab"
-AccountPassword(Read-Host -AsSecureString "Input Password")
-Enabled $true

New-ADUser -Name "itadmin" -UserPrincipalName "itadmin@
mycp.lab" -AccountPassword(Read-Host -AsSecureString "Input
Password") -Enabled $true

New-ADUser -Name "hruser" -UserPrincipalName "hruser@mycp.lab"
-AccountPassword(Read-Host -AsSecureString "Input Password")
-Enabled $true

New-ADGroup -Name "IT_Admins" -SamAccountName IT_Admins
-GroupCategory Security -GroupScope Global -DisplayName "IT_
Admins" -Path "CN=Users,DC=mycp,DC=lab"

New-ADGroup -Name "HR" -SamAccountName HR -GroupCategory
Security -GroupScope Global -DisplayName "HR" -Path
"CN=Users,DC=mycp,DC=lab"

Add-ADGroupMember -Identity IT_Admins -Members itadmin

add-adgroupMember -Identity HR -Members hruser
```

You are prompted for passwords three times. Each time, type in `CPL@b8110` and press *Enter* to continue. Press *Enter* at the end to ensure that the last line has been executed.

Enabling Identity Awareness and browser-based authentication

Return to `SmartConsole_VM`. In **Gateway Cluster Properties** [1], check **Identity Awareness** [2]:

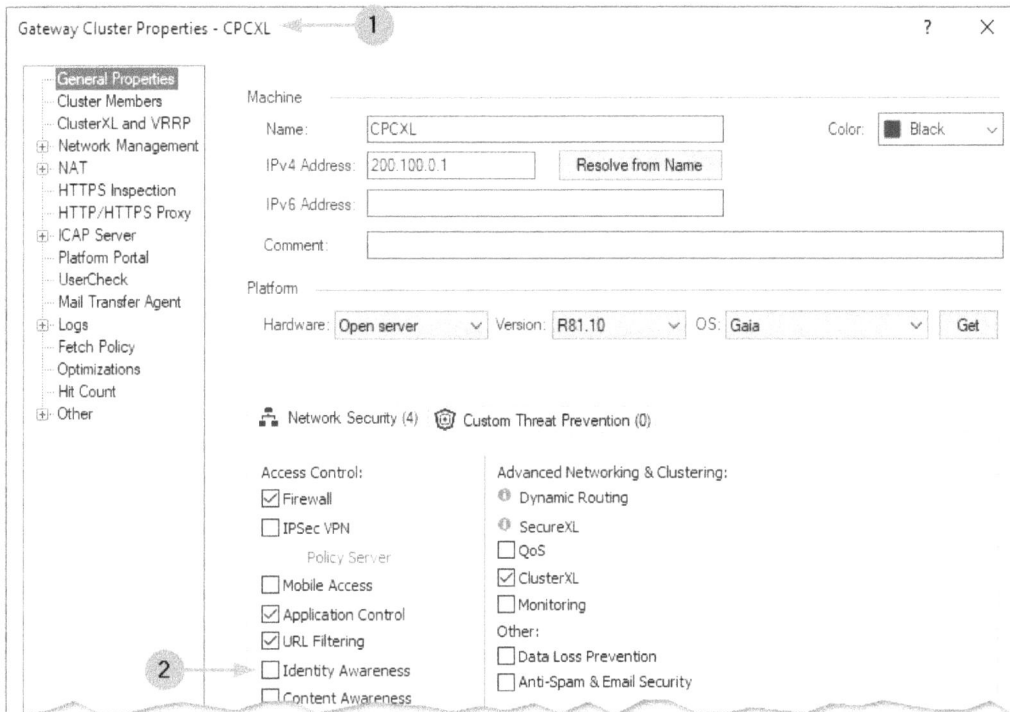

Figure 11.33 – Enabling Identity Awareness

In **Methods for Acquiring Identity**, uncheck the **AD Query** box [1] and check the **Browser-Based Authentication** box [2]. Click on **Next**:

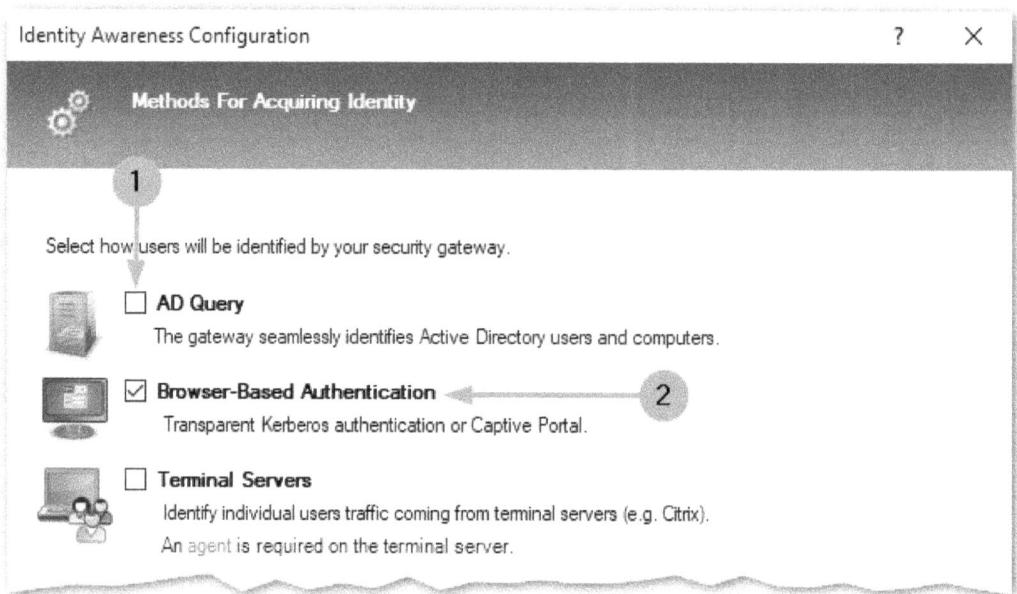

Figure 11.34 – Choosing Browser-Based Authentication

In the **Select an Active Directory** section of the next window, choose **Create new Active Directory** [1]. Then, fill out the **Domain Name** (mycp.lab), **Username** (cpauth), **Password** (CPL@b8110), and **Domain Controller** (ADDCDNS.mycp.lab) fields [2]. Click on **Connect** [3]:

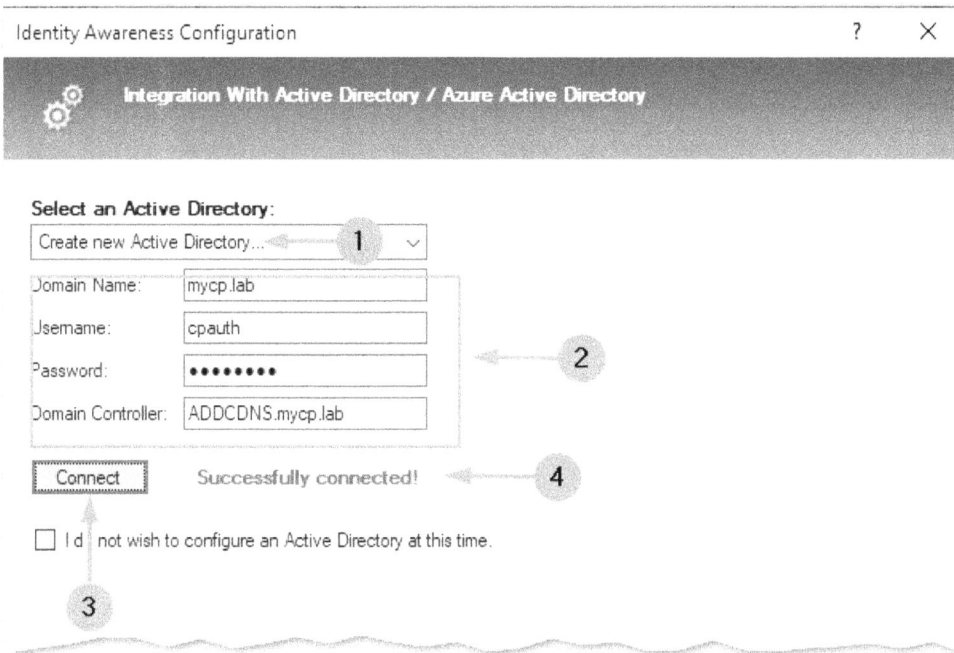

Figure 11.35 – Connecting Identity Awareness to Active Directory

Read through the information in the **Browser-Based Authentication Settings** window and copy the **Main URL** address (`https://200.100.0.1/connect`):

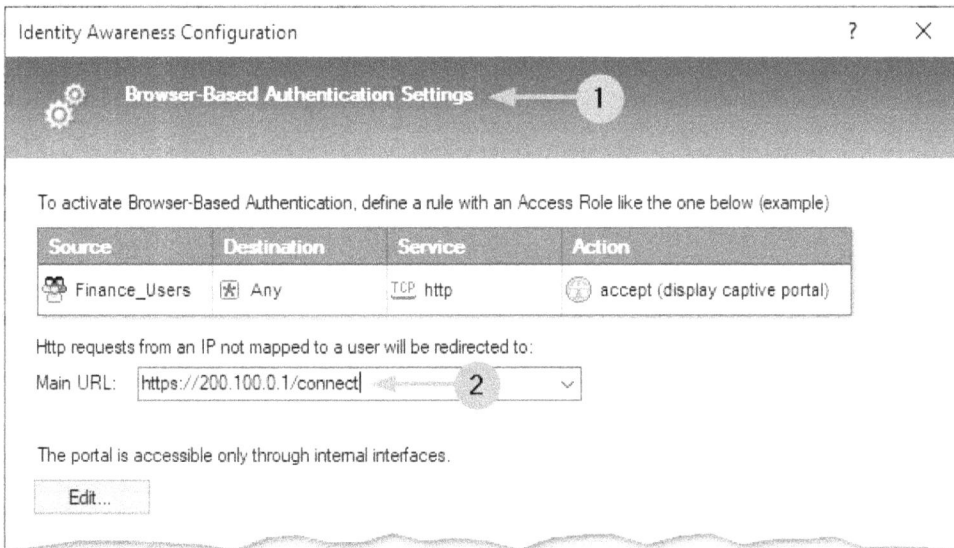

Figure 11.36 – Completing the Identity Awareness configuration

Click on **Next** and, when presented with the **Identity Awareness is Now Active!** window, click on **Finish**.

In **Gateway Cluster Properties**, click on the **Identity Awareness** navigation tree option [1]. You will see the **Browser-Based Authentication** box checked. Click on the **Settings** button next to it [2]. In the new **Portal Settings** window [3], click on **Edit** under **Access Settings** [4]. In the **Portal Access Settings** sub-window, note two buttons, **Aliases** [5] and **Import**, under the **Certificate** section [6].

I am drawing your attention to these two options [5] and [6] to point out that if you would like to use **Browser-Based Authentication** for the guests' networks, where you cannot enforce self-signed ICA certificate installation, you have an option to use a certificate issued by a publicly trusted certificate authority.

This is relevant to the **Unregistered guest login** option [7] in the **Portal Settings | User Access** section. Check the **Log out users when they close the portal browser** box [8]:

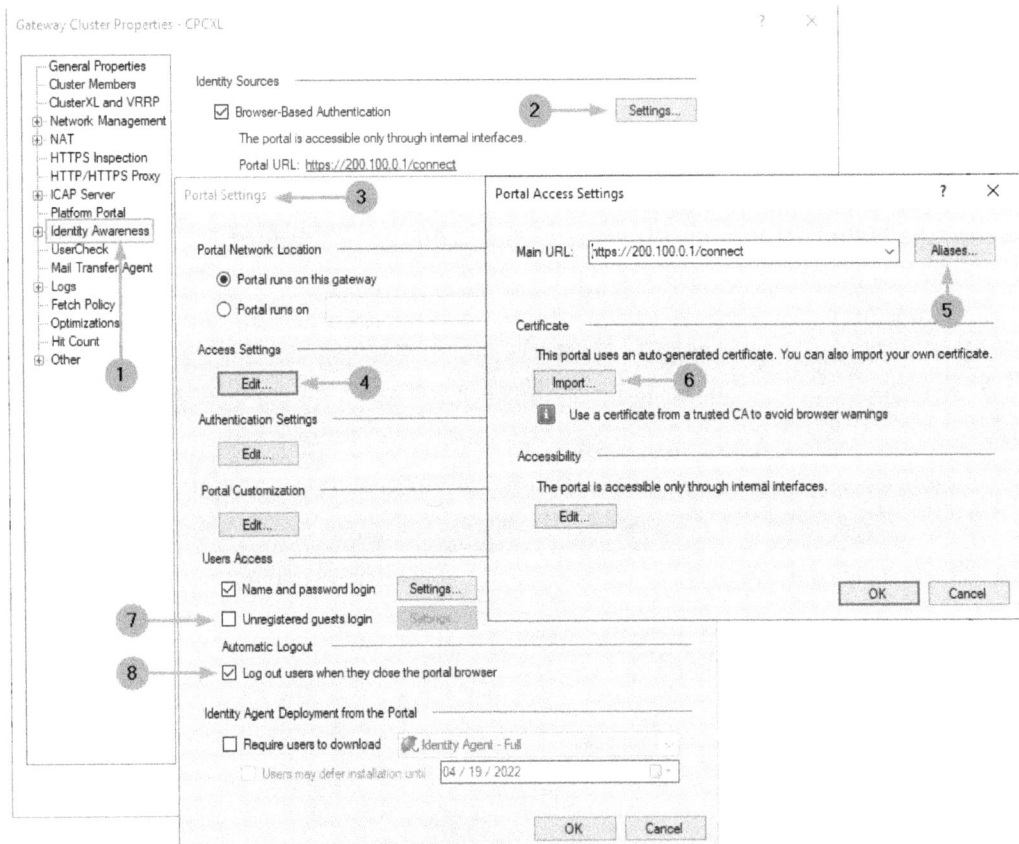

Figure 11.37 – Configuring and exploring the Browser-Based Authentication portal settings

Click on **OK** three times to exit **Gateway Cluster Properties**.

Now, let's paste the **Main URL** address for **Browser-Based Authentication** that we copied during the Identity Awareness configuration into the Chrome browser's address bar on SmartConsole_VM and on LeftHost https://200.100.0.1/connect, press **Enter**, and you will be presented with the **Check Point Network Access Login** screen. Add it to the bookmarks section of the browser on each VM.

Creating and using access roles

Let's return to SmartConsole and our policy. Hover over the **Source** field in rule **21.6**. Click on the + sign, click on the new icon, and then click on **Access Role**:

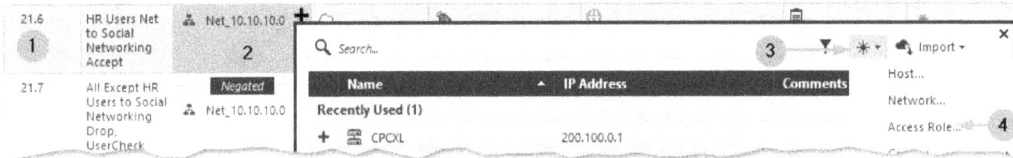

Figure 11.38 – Creating an Access Role object

Type HR into the **Enter Object Name** field. Leave the **Networks** setting as **Any Network**. Click on **Users**:

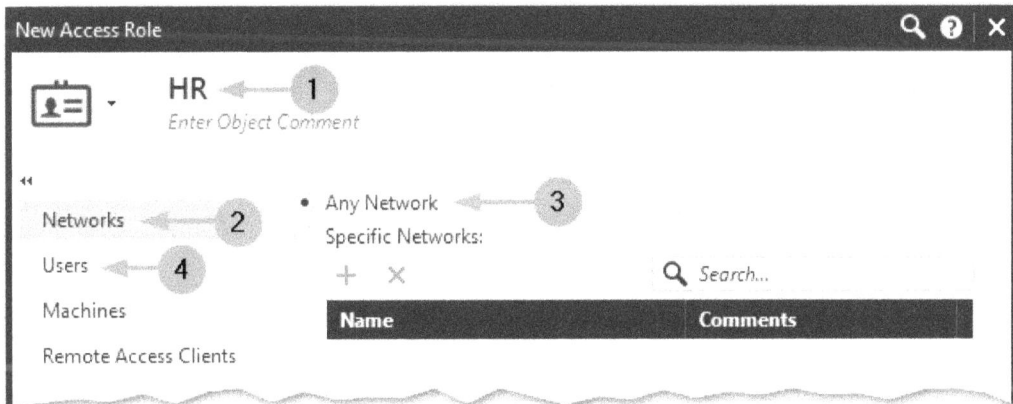

Figure 11.39 – Configuring the access role networks

While in the **Users** section [1], select the **Specific users/groups** option [2]. Click on the large + sign under it [3]. Note that the mycp.lab object [4] is shown in the upper-left corner of the child window. This is an LDAP_Account_Unit object that is automatically created when we have set up Identity Awareness. After a few seconds, the content of your Active Directory | **Users** container is shown in the window. Type HR into the search bar [5] to filter it out. Hover your cursor over HR in the remaining options, and click on the small + sign on the left-hand side of it [6] to turn it into a checkmark. Exit the sub-window:

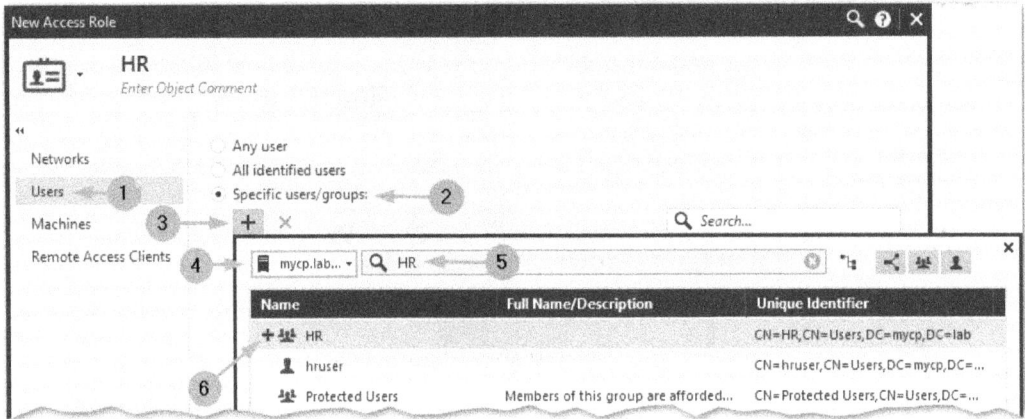

Figure 11.40 – Configuring the access role users

Click on **OK**.

In the **Source** field of the **21.6** rule, right-click on Net_10.10.10.0 and click on **Remove**.

In the **Source** field of rule **21.7**, remove the Net_10.10.10.0 address and the cell negation option. Replace them with **Any**.

Your resultant rules should look like this:

Figure 11.41 – Changing APCL_URLF_Layer's rules for access roles

Publish the changes, and install the policy.

Testing access role-based rules

Open the Chrome browser on `SmartConsole_VM`, and click on the **Network Login** bookmark:

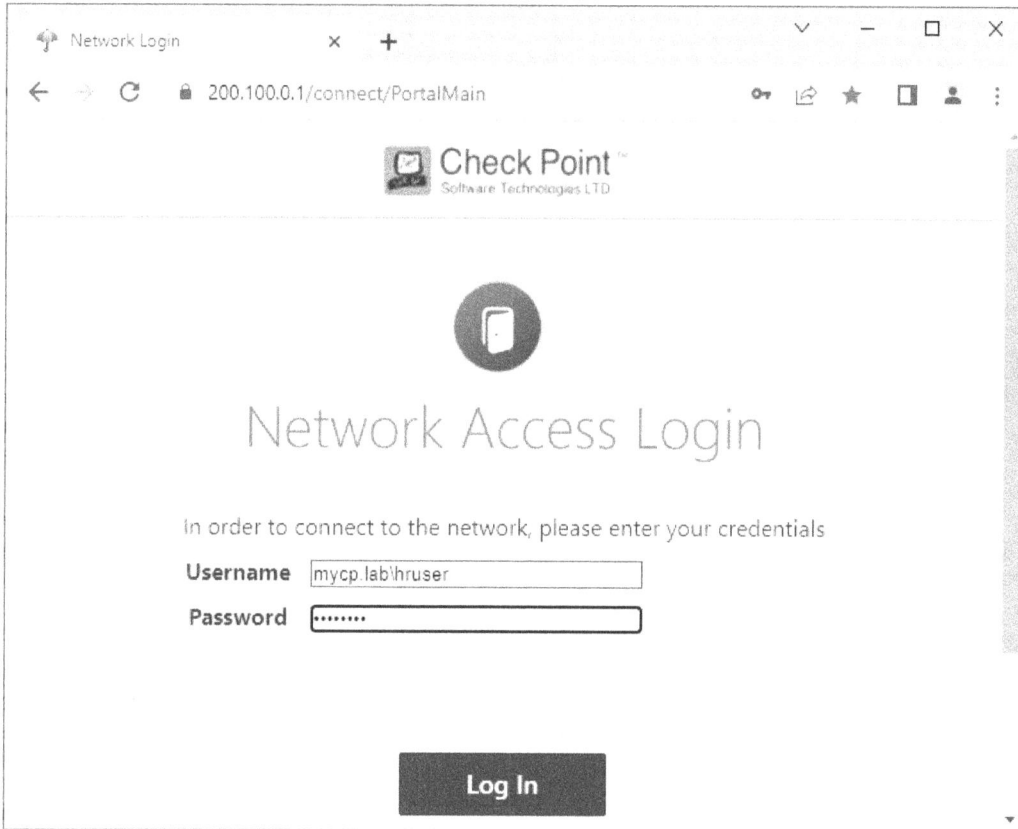

Figure 11.42 – The browser-based authentication Network Access Login portal

Enter **Username** as `mycp.lab\itadmin` and **Password** as `CPL@b8110`. Click on **Log In**. You will see the **Network Access Granted** screen with the **Log Out** button. Leave it as it is, and open another tab in the browser. Try accessing `facebook.com` – you will be greeted with a **Page Blocked UserCheck** message. Now try accessing `reuters.com` – you should be able to get to the site. Close this tab and, in the remaining **Network Logon** tab, click on **Log Out**.

Now, enter **Username** as `mycp.lab\hruser` and **Password** as `CPL@b8110`. Click on **Log In**. After seeing the **Network Access Granted** screen, open another tab in the browser. Attempt to access `facebook.com`, and you should see the Facebook Log In prompt. Close this tab, and click on **Log Out** on the remaining tab.

Repeat the exercise on the LeftHost VM, using the `mycp.lab\itadmin` account to reach the following accounts:

```
facebook.com      Outcome: Page Block
google.com        Outcome: Page Block
cnn.com           Outcome: success
```

Close the three open tabs and click on **Log Out**.

Click on **Regain access to the network**, and log in using the `mycp.lab\hruser` credentials.

Now try accessing the following accounts:

```
facebook.com      Outcome: success
google.com        Outcome: Page Block
apnews.com        Outcome: success
```

We can now see that the access is being granted based on the AD group membership, rather than the IP address.

Browser-based authentication for Identity Awareness is more suitable for guest or contractor access control. In the rules of our access control policies, for any **Action** field, we can right-click on **Accept**, click **More,** and check **Enable Identity Captive Portal** to automatically redirect users to it.

If you are using **Enable Identity Captive Portal**, in the access role of that rule, choose specific internal networks. Additionally, in the preceding rule, using the same access role with the same networks, either accept some innocuous application or drop one that should not be allowed anyway. This last step is not strictly necessary if **X-Forwarded-For (XFF)** is enabled in the layer's advanced properties, which we did earlier in this chapter (*Figure 11.13*). This is done to avoid the situation described in *sk162856, HTTPS traffic can not be redirected to Captive Portal if matched on first packet.*

Internal User Groups (as created in **New | More | User/Identity | User** and **User Group**) can be used in access roles instead of, or in addition to, those from the external directories (not in the same rules though).

With HTTPS Inspection and Identity Awareness enabled, our logs are now enriched with information relevant to both:

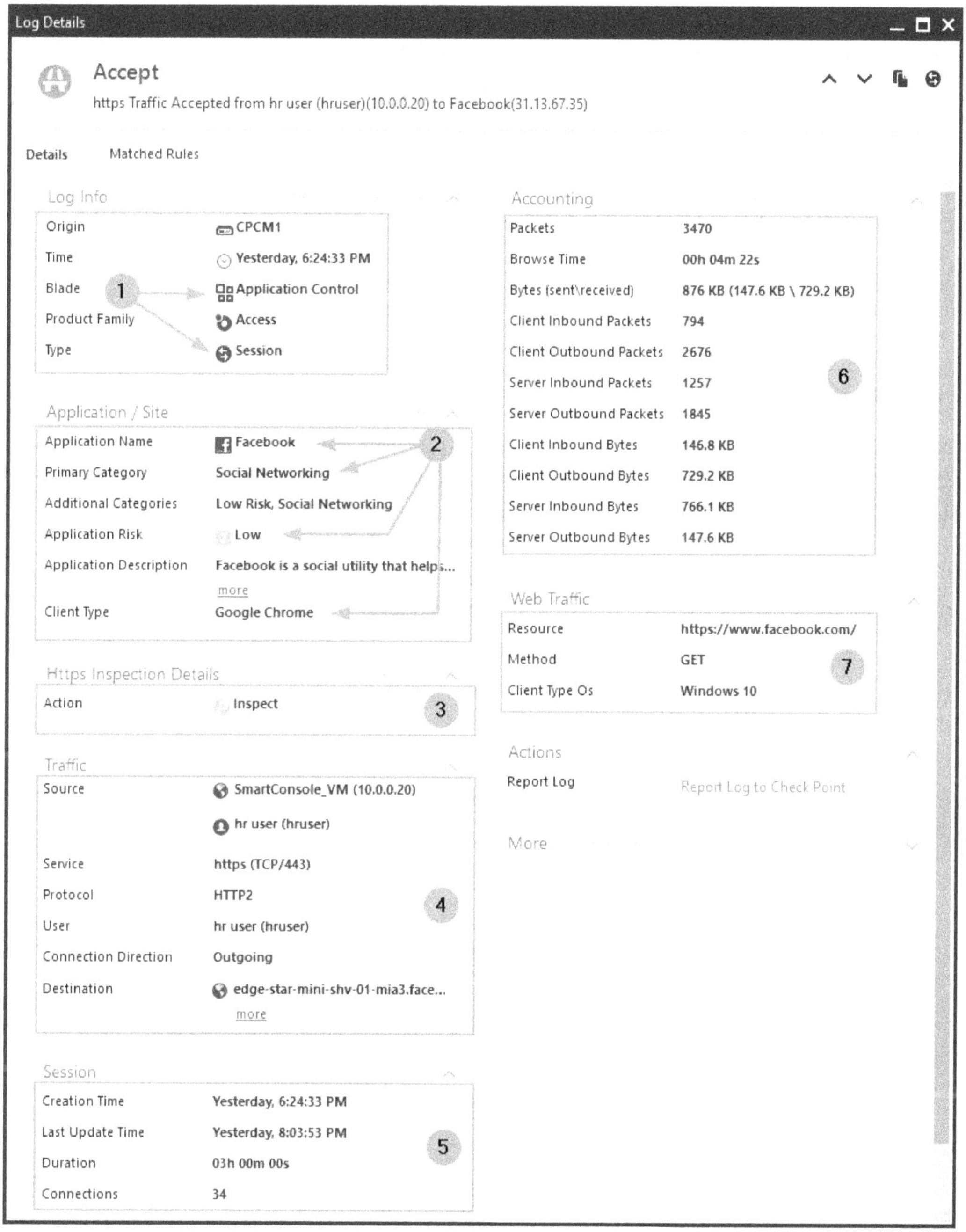

Figure 11.43 – The log for the event with APCL/URLF HTTPS Inspection and Identity Awareness

Looking at the preceding example, we can see that the **Application Control** setting for this **Session** [1] is working, that the application we were accessing was **Facebook**, categorized as **Social Networking** with **Low Risk**, and that the client type was **Chrome** [2]. We can see that **HTTPS Inspection** has taken place [3] and that the traffic **Source** is identified by the hostname, the IP, and the user [4]. We can see the **Session Duration** and the number of **Connections** within the session [5], and gain access to extensive **Accounting** information [6], along with **Resource, Method,** and **Client type Os** [7].

I highly recommend investing the time to implement Identity Collector(s) instead of, or in addition to browser-based authentication for a seamless Identity Awareness experience.

Summary

It looks like we accomplished quite a bit in this chapter. We defined the structure of our access control policy and created rules for the firewall/networking layer. We created the APCL/URLF inline layer and rules for categories and applications using different actions. To properly process the application-specific rules, we implemented an HTTPS Inspection policy and performed a number of operations with certificates. To round up this chapter, we implemented Identity Awareness using an external directory service and were able to use access roles to replace static objects in our rules.

In the next chapter, we'll cover the basics of VPNs on the Check Point platform and build both site-to-site and remote access VPN communities.

12
Configuring Site-to-Site and Remote Access VPNs

One of the most common tasks that firewall and gateway administrators must perform is configuring **Virtual Private Networks** (**VPNs**) for communicating with peers, data, or service providers, as well as implementing remote access solutions.

It is time for us to cover the introduction to these concepts and their capabilities in Check Point environments, and to implement rudimentary site-to-site and remote access VPNs in our lab.

As we go through the lab exercises, we'll cover some additional subjects (such as local users, templates, and groups) and their use in the access roles, as well as the changes in gateway certificates necessary to accommodate UserCheck and browser-based authentication for remote users.

In this chapter, we are going to cover the following main topics:

- An introduction to site-to-site VPN capabilities
- Configuring a remote gateway and creating its policy

- Building site-to-site VPNs using gateways managed by the same management server
- An introduction to Check Point remote access VPN solutions
- Configuring a remote access IPSec VPN

An introduction to site-to-site VPN capabilities

Check Point site-to-site VPN topologies are defined in **VPN Communities** objects and can be a mesh, a star (hub and spoke), or a combination of the two.

Meshed VPN communities are used to interconnect equally important locations. In a mesh, all members have tunnels established between each pair of gateways.

Star, or hub-and-spoke, communities are used for branches or satellite offices connected to central locations. Star VPN communities can be used in one of three modes:

- Split tunnel, with local internet egress and a VPN connection to the networks behind the hub
- The same, but with access to other spokes' networks
- All traffic routed through the hub

Multiple Entry Point (**MEP**) configuration options are available to connect spokes to one of several hub gateways using a variety of preference methods, based on the hubs' availability. A hub's availability is continually assessed using a proprietary probing protocol. MEP is considered a high-availability option but without state synchronization, meaning that all sessions must be re-established anew once MEP failover has taken place. **Route Injection Mechanism** (**RIM**) can be integrated with MEP to populate the routing tables on the hub gateways that the spokes are connected to with their encryption domain networks, if the **Permanent Tunnels** option is enabled for the community. In this case, keepalive traffic is used to maintain a tunnel state.

Site-to-site VPNs can be either domain-based (encrypted traffic routing defined on participating gateways) or route-based (encrypted traffic routing defined by **Virtual Tunnel Interfaces** (**VTIs**)).

Domain-based VPNs are mostly used between sites that do not require dynamic routing and are prevalent in corporate VPN-based intranets and the majority of VPNs with peers.

Route-based VPNs are more suitable when encrypted traffic relies on dynamic routing, where gateways are one of the routing nodes. These are prevalent in VPNs with cloud-hosted environments.

A **wire mode VPN** is also available, in which interfaces are declared as **trusted** in gateways participating in one or more communities. This mode allows a VPN between ports on different gateways to act as a virtual Ethernet cable and is used for transmission of uninspected traffic. The firewall functionality for traffic traversing these interfaces is completely bypassed. Wire mode can be used for dynamic routing through a VPN where gateways are used as transient nodes.

To illustrate a site-to-site VPN in our environment, we need a second gateway. Since we pre-provisioned one earlier, let's finish its configuration in the next section.

Configuring a remote gateway and creating its policy

When we were creating our lab, the gateway object for the right side of our environment (the CPGW) was just cloned, but we never completed its configuration. It is time to do so now. Since we've already used the Gaia First Time Configuration Wizard a number of times, no screenshots will be provided for FTW, but step-by-step instructions for it are provided:

1. From VirtualBox, start CPGW [1]. In its console, note that the hostname [2] is that of the linked clone. We are also seeing reminders that the FTW was not yet completed:

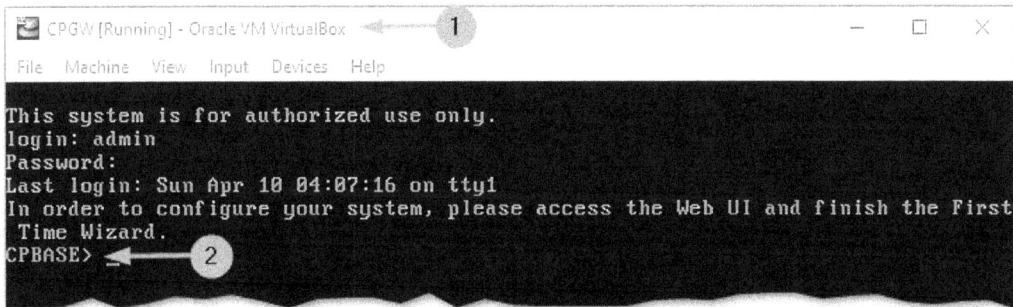

Figure 12.1 – The unconfigured CPGW VM

Before we run the FTW, let's change its hostname and the IP addresses of its interfaces, lock down the Gaia access to only that of our `SmartConsole_VM`, and set its expert password.

2. Type in the following commands, pressing *Enter* after each line:

    ```
    set hostname CPGW
    set interface eth0 ipv4-address 200.200.0.1 subnet-mask
    255.255.255.0
    ```

```
set interface eth0 state on
set interface eth1 ipv4-address 172.16.16.1 subnet-mask
255.255.255.0
set interface eth1 state on
set static-route default nexthop gateway address
200.200.0.254 on
add allowed-client host ipv4-address 10.0.0.20
add allowed-client host ipv4-address 200.100.0.20
delete allowed-client host any-host
set expert-password
```

Type CPL@b8110 and press *Enter*.

Type CPL@b8110 again and press *Enter*.

Type save config and press *Enter*.

3. Execute ping 200.200.0.1 to confirm that you can reach the router
   VM's interface.

Now, we have to allow our SmartConsole VM to access this new remote gateway
on HTTPS:

1. In SmartConsole, in the **SECURITY POLICY | Access Control | Policy Rule
   1 Destination** field, create and add a new temporary host, CPGW_tmp, with a
   200.200.0.1 IP address. Publish the changes and install the policy. This is done
   to give us access to WebUI and to permit communication between the management
   server and this new gateway. Otherwise, it will be blocked by the APCL/URLF
   Uncategorized Drop/Block Message rule.

2. Open WebUI from SmartConsole VM's browser at https://200.200.0.1,
   click on **Advanced** when presented with the warning, and click **Proceed to
   200.200.0.1 (unsafe)**.

3. Log in using the CPL@b8110 admin credentials.

4. At **Welcome to the Check Point FTCW**, click **Next**.

5. At **Deployment Options**, click **Next**.

6. At **Management Connection**, click **Next**.

7. At **Internet Connection**, click **Next**.

8. At **Device Information**, enter 9.9.9.9 as the primary DNS server and click **Next**.

9. At **Date and Time Settings**, check **Use Network Time Protocol (NTP)**, select a time
   zone identical to the one used in CPCM1, CPCM2, and CPSMS, and click **Next**.

10. At **Installation Type**, click **Next**.

11. At **Products**, uncheck **Security Management** and click **Next**.

12. At **Dynamically Assigned IP**, click **Next**.

13. At **Secure Communication to Management Server**, enter CPL@b8110 twice and click **Next**.

14. At **First Time Configuration Summary**, click **Finish** and then **Yes** to confirm.

15. When prompted to restart the system, click **OK**.

16. Wait until the **LOGIN** prompt reappears. If it does not, hit **Reload this page** in your browser. When you see it, return to SmartConsole.

Let's continue our work in SmartConsole, following these steps:

1. Remove the CPGW_tmp object from Rule 1.

2. Go to **GATEWAYS & SERVERS**.

3. Click **New | Gateway | Classic Mode**.

4. Enter **Name** as CPGW [1], **IPv4 Address** as 200.200.0.1 [2], check **IPSec VPN** [3], and click **Communication…** [4]:

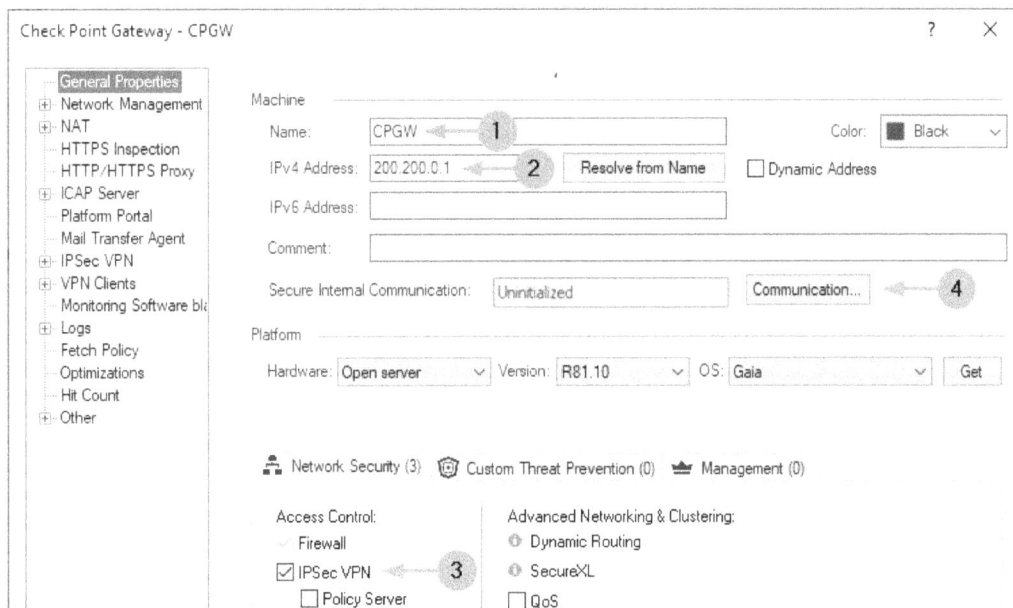

Figure 12.2 – Configuring a new gateway object

5. Enter CPL@b8110 twice and click **Initialize**. If you see a warning about an object with a duplicate IP, ignore it.

6. After seeing **Trust established** [1], click **OK** [2]:

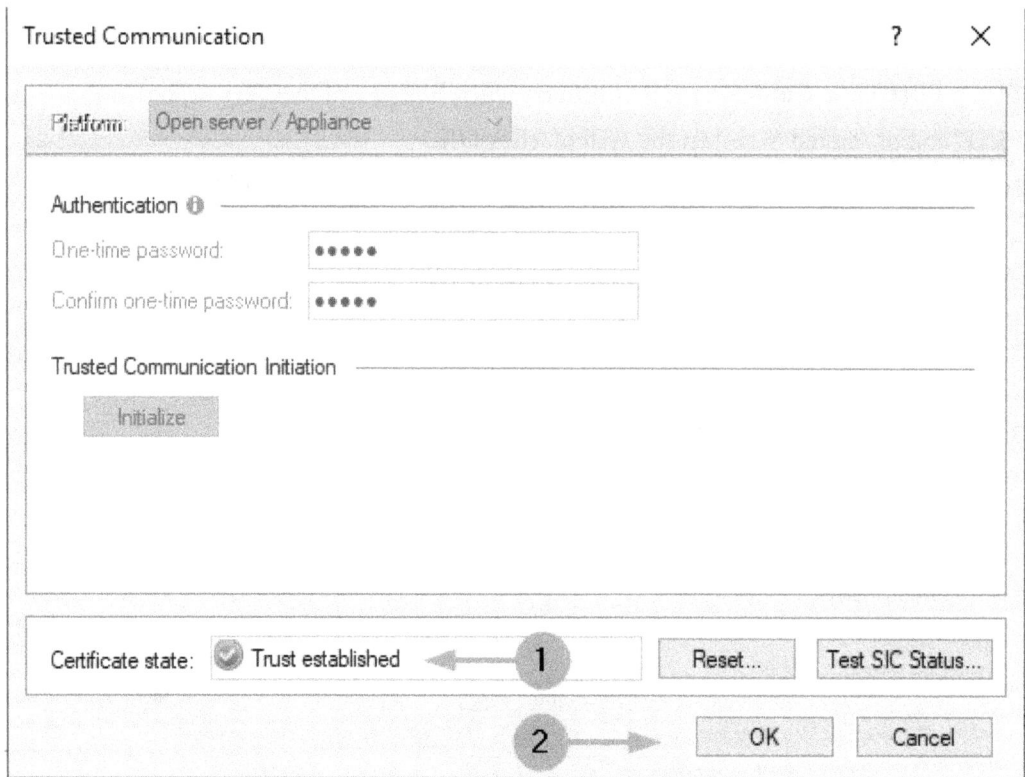

Figure 12.3 – Trust established with CPGW

7. You will be presented with the **Get Topology Results** window. Verify that both interfaces have the correct IPs [1] and click **Accept** [2]:

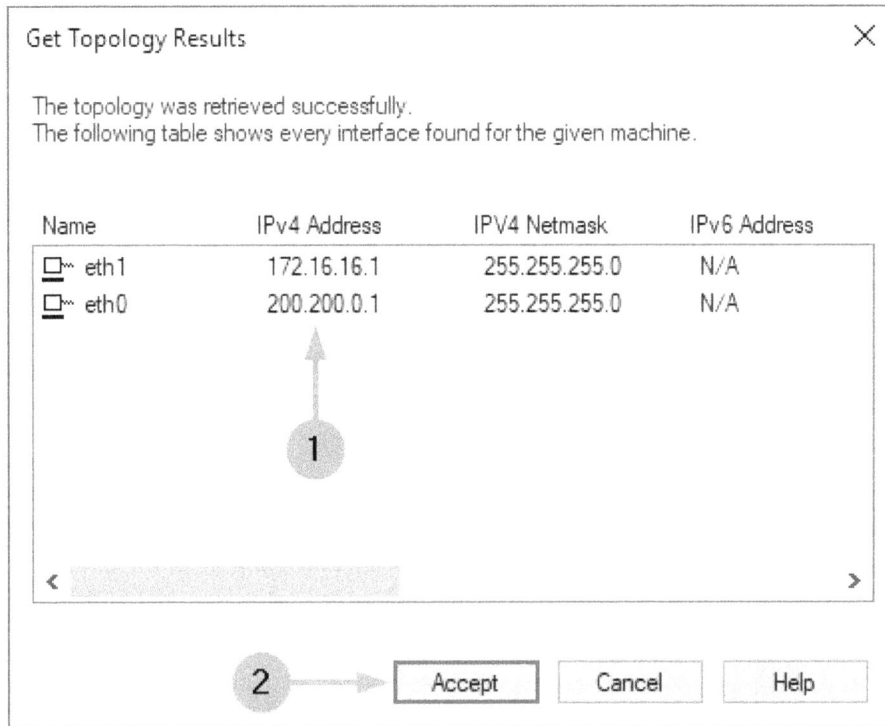

Figure 12.4 – The automatic topology retrieval process

8. Now, click on **Network Management** [1]. You will see the **eth0** interface, identified as **External** [2] in the **Topology** field, but the **eth1** interface is listed as **Undefined** [3]:

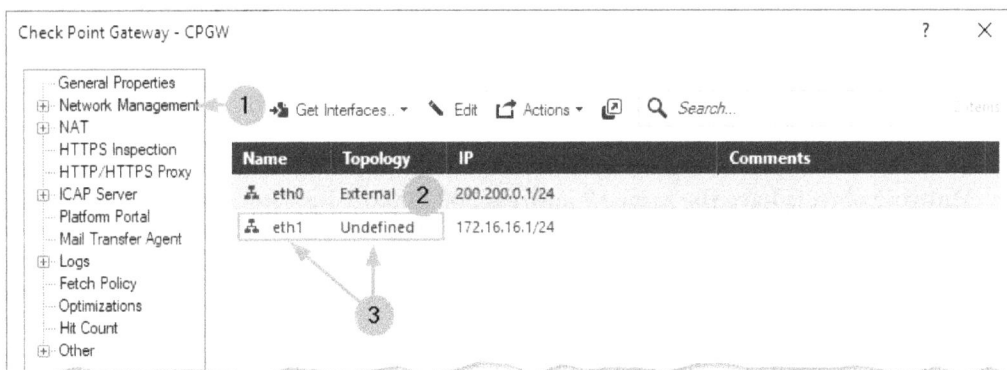

Figure 12.5 – Defining the topology of the remote gateway

> **Important Note**
>
> If you see interface names as anything other than **eth0** and **eth1**, click on **Get Interfaces | Get Interfaces without topology** to re-enumerate them.

9. Double-click the **eth0** interface and click on the **Modify** button in the **Topology** section. In the **Security Zone** section, select **According to topology: ExternalZone**. Click **OK** twice to return to the **Network Management** view.

10. Double-click the **eth1** interface and click on the **Modify** button in the **Topology** section. In the **Topology Settings** window's **Leads To** section, select **Override** and then **Network Defined by the interface IP and Net Mask**. In the **Anti-Spoofing** section, check **Perform Anti-Spoofing based on interface topology**.

11. Click **OK** twice to commit the topology configuration.

12. Now, let's define a specific VPN domain for this gateway. Expand the **Network Management** item in the navigation tree [1] and click **VPN Domain** [2]. In the **VPN Domain** view on the right, select **User defined** [3], and using the browse function (**...**), select an internal network on the right side of our lab – **Net_172.16.16.0** [4]:

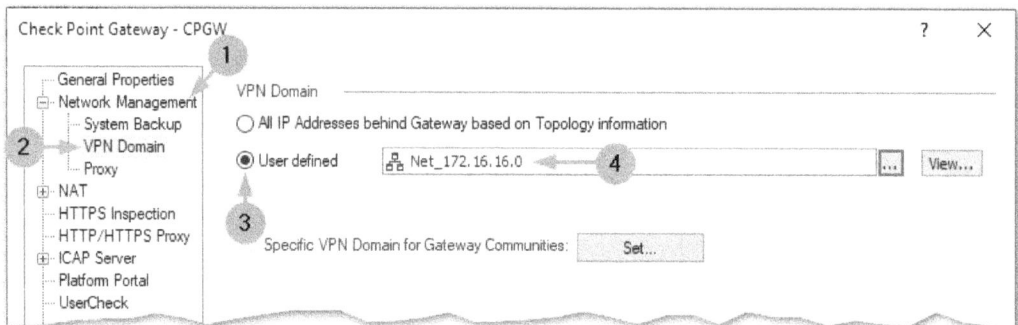

Figure 12.6 – Configuring the user-defined VPN domain

13. Click **OK** to complete the gateway creation process. If presented with another **Multiple objects have the same IP address 200.200.0.1** warning, choose to keep the changes by clicking **Yes**.

   The **CPGW** gateway is now added to your inventory.

14. Now, in the **Standard** policy, add the **CPGW** object to the **Destination** field of `rule 1` and delete the temporary object, `CPGW_Temp`, that we created earlier, from **Object Explorer**.

   Since we only have a single policy package called **Standard**, now is a good time to change that. Right-click on **Access Control | Policy** and click **Edit Policy**.

15. In the **Policy** properties [1], rename the policy package from **Standard** to LeftSide_S2S [2]. We should also restrict it to the **CPCXL** cluster – click on **Installation Targets** [3], select **Specific gateways** [4], click on the + sign [5], and click on **CPCXL**. Your **Policy** window should now look like this:

Figure 12.7 – Changing the Standard policy and specifying its installation target

16. Click **OK** (not shown), publish changes, and install the policy. Now, if you look in **GATEWAYS & SERVERS**, you should see the **CPGW** status and CPU utilization.

Let's create a second policy package for our new gateway:

1. Click the + sign to the right of our **LeftSide-S2S** policy.

2. In the **Manage Policies** tab, click the **New** icon. Name the policy package RightSide.

3. While in the **General** section of the navigation tree of that policy, uncheck the **Threat Prevention** checkbox.

4. Click on **Installation Targets**, select **Specific gateways**, click the + sign, and click on **CPGW**. Click **OK**.

The tab name has changed from **Managed Policies** to **RightSide**, and a new policy with a single **Cleanup** rule is shown.

Let's create some objects and throw together a rudimentary policy for the right side of our lab environment:

1. Create a host object, BCast_172.16.16.255, with an IP address of 172.16.16.255.

2. Create a host object, `SmartConsole_VM_NAT`, with an IP address of `200.100.0.20`.

3. In a network, `Net_172.16.16.0`, set its NAT translation method to **Hide behind the gateway**, and set **Install on gateway** to *__*All__*.

The reason we are creating a **SmartConsole_VM_NAT** object is that the original **SmartConsole_VM** is NATed to the address that belongs to the external network of the **CPCXL** cluster. Our new gateway, **CPGW**, should be able to accept inbound management traffic from that public IP if the VPN is not operational.

With all the new objects in place, let's create a rudimentary access control policy. It will look like this:

No.	Name	Source	Destination	VPN	Services & Applications	Action	Track
▼ Gateways Access (1)							
1	SSH and HTTPS to gateways Accept	SmartConsole_VM SmartConsole_VM_NAT	CPGW	* Any	ssh_version_2 https	Accept	Log
▼ Noise suppression (2)							
2	Broadcasts Drop Do not log	* Any	BCast_172.16.16.255 BCast_255.255.255.255	* Any	* Any	Drop	— None
▼ Stealth Rule (3)							
3	Stealth Rule	* Any	CPGW	* Any	* Any	Drop	Log
▼ Internet Access (4)							
4	Internet Access Accept	Net_172.16.16.0	* Any	* Any	* Any	Accept	Log
▼ Cleanup Rule (5-6)							
5	Cleanup rule	* Any	* Any	* Any	* Any	Drop	Log

Figure 12.8 – The RightSide access control policy

4. Publish your changes and install the policy.

5. Test the internet connectivity from **RightHost** to confirm its functionality.

At this point, we are ready to configure the VPN.

Building a site-to-site VPN using gateways managed by the same management server

In this section, we will examine two popular choices for star VPN topologies:

- **To center only**

- **To center or through the center to other satellites, to Internet and other VPN targets**

Star community – To center only

This VPN topology (selected by default) is used where independent access control is preferred at each satellite location. When implementing it in production, you need to do the following:

1. Expand the basic access control policy we created earlier to contain specific rules.

2. Enable HTTPS inspection on CPGW and configure the local HTTPS inspection policy.

3. Either reuse the existing shared APCL_URLF layer or create a new one for this policy.

For our purposes in the lab, the basic policy, already in place, will suffice.

Regardless of which policy you are in, follow these steps:

1. Click on **VPN Communities** [1] under **Access Tools**:

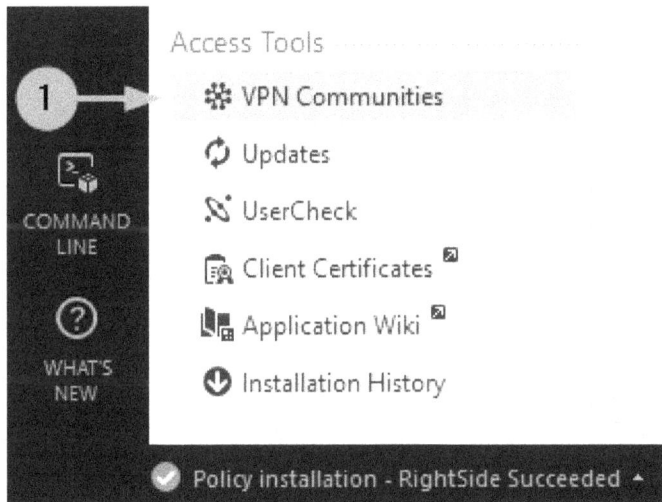

Figure 12.9 – The VPN Communities editor

2. In the action menu of **VPN Communities**, click the **New** icon and then **Star Community**.

3. In **New Star Community** [1], enter the community's name, `Branches` [2], and while in the **Gateways** section of the navigation tree [3], add **CPCXL** to **Center Gateways** [4] and **CPGW** to **Satellite Gateways** [5]:

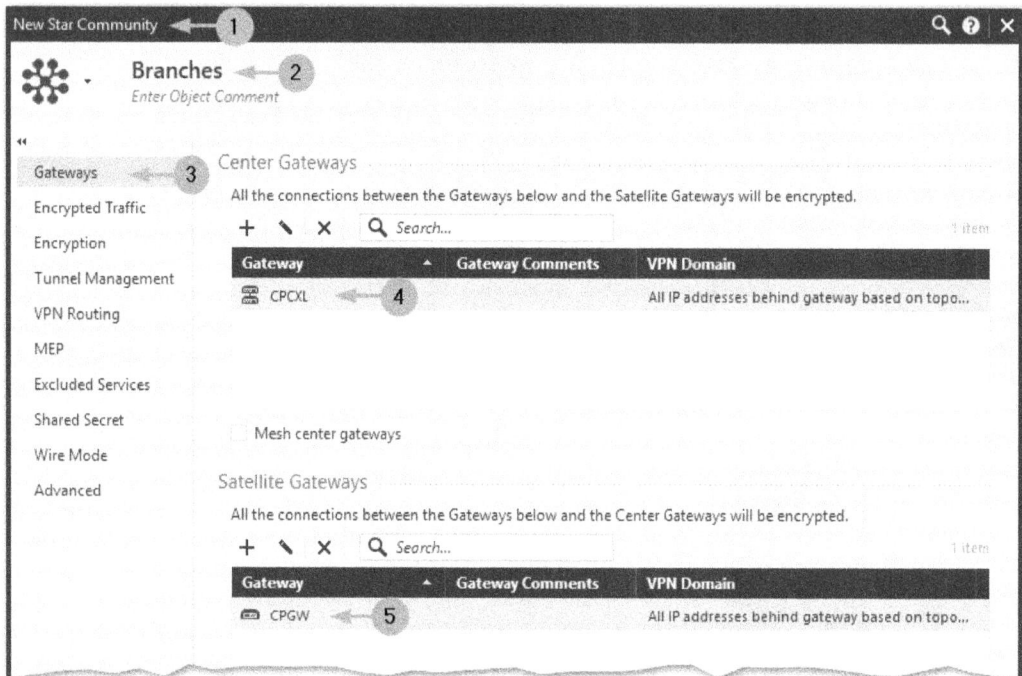

Figure 12.10 – The New Star Community center and satellite members

4. In the **Advanced** section of the navigation tree [1], check **Disable NAT inside the VPN community** [2] and click **OK**.

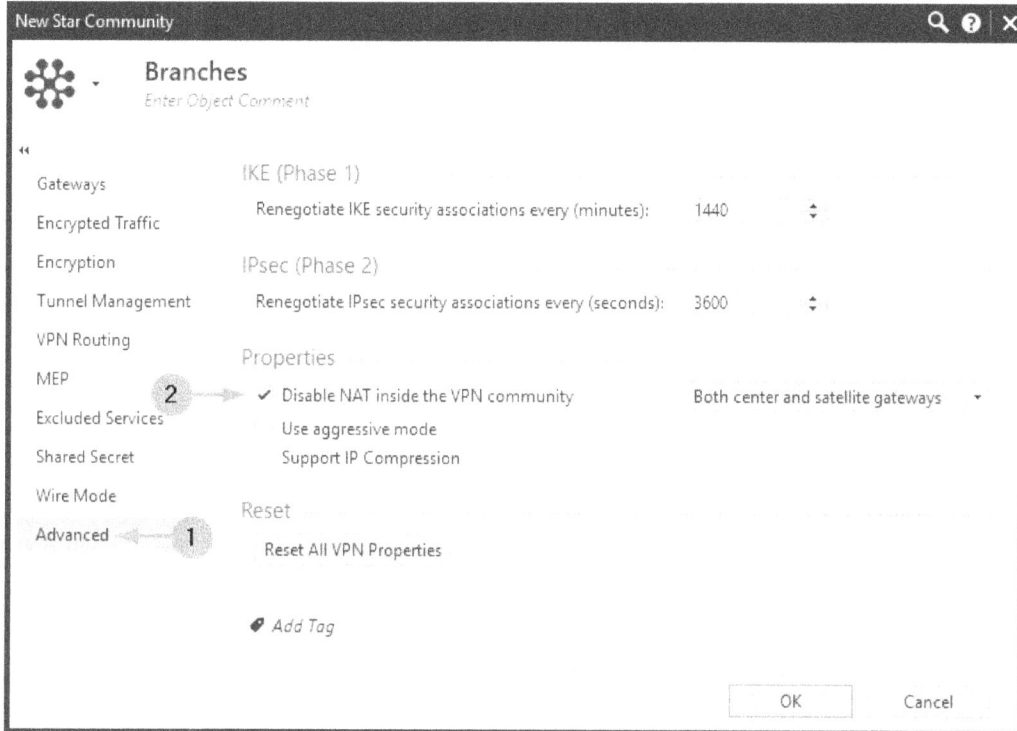

Figure 12.11 – Disable NAT inside the VPN community

If we check the **VPN Routing** section, we'll see that it is set as **To center only**.

Select the **LeftSide_S2S** policy. Let's make some changes to a few rules to allow our VPN traffic to access some of the destinations:

1. Add the **Net_172.16.16.0** object to the **Source** fields of these rules:

    - In rule no. 9 (**DNS Internal Accept**) [1]

    - In rule no. 13 (**Admins RDP to all Accept.**) [2]

    - In rule no. 17 (**Internal2 network to DMZ Accept**) [3]

    - In rule no. 20 (**MS Internet connectivity probe Accept**) [4]

The results should look like this:

9	1	DNS Internal Accept	SmartConsole_VM CPSMS DMZSRV Net_10.10.10.0 Net_172.16.16.0	ADDCDNS	* Any	dns	Accept
13	2	Admins RDP to all Accept.	SmartConsole_VM Net_172.16.16.0	InternalZone DMZZone	* Any	Remote_Desktop_Pr...	Accept
17	3	Internal2 network to DMZ Accept	Net_10.10.10.0 Net_172.16.16.0	DMZSRV	* Any	http https	Accept
20	4	MS Internet connectivity probe Accept	Int_Nets Net_172.16.16.0	.www.msftncsi.com .www.msftconnecttest.com	* Any	* Any	Accept

Figure 12.12 – The modified rules for site-to-site VPN access from the remote side

2. Publish the changes and install both policies.

3. On **RightHost**, change the DNS settings by adding `10.20.20.10` (**ADDCDNS**) as a secondary DNS server.

4. Initiate traffic from **RightHost** to any host on the left side (e.g., RDP or HTTP to `DMZSRV.mycp.lab`).

5. Open **LOGS & MONITOR | New Tab | Logs**. In the search bar, type `blade:VPN` and press *Enter*. Click on the refresh icon, and the logs will indicate that traffic is encrypted (*closed lock icon*) by **Origin CPGW** and decrypted by **Origin CPCM1** (*open lock icon*). Open one of the log cards for decrypted traffic, where you will find that, despite us selecting **No NAT inside VPN Community**, traffic with a `172.16.16.10` source is translated to the **CPXL** virtual cluster IP addresses closest to the destination you were trying to reach (i.e., `10.30.30.1` when accessing **DMZSRV** or **10.20.20.1** when accessing **ADDCDNS**).

While you may be puzzled by this behavior, it is correct, as this translation is happening after packets are decrypted (outside of the VPN). We'll address how to avoid this situation in the next subsection. Remember to use `<hostname>.mycp.lab` if using hostnames for access, or use their IP addresses for the tests.

6. Open another **Logs** tab and enter the `src:172.16.16.10` query. We are running a split-tunnel configuration, so you will see internet-bound unencrypted traffic exiting CPGW.

Star community – To center or through the center to other satellites, to Internet and other VPN targets

This configuration is more suitable when you need to route all traffic from a satellite site to, or through, a central hub. In this case, our rudimentary **RightSide** policy is sufficient for production (except when multiple networks are present in the satellite site, and these require specific rules for interaction between the hosts that belong to them). There is no need for local HTTPS inspection or application control, as all the relevant resources are located at or accessed through the hub.

Let's follow these steps to set it up:

1. Double-click the **Branches** community and select **VPN Routing** [1] in the navigation tree. Select the third option, **To center or through the center to other satellites, to Internet and other VPN targets** [2]:

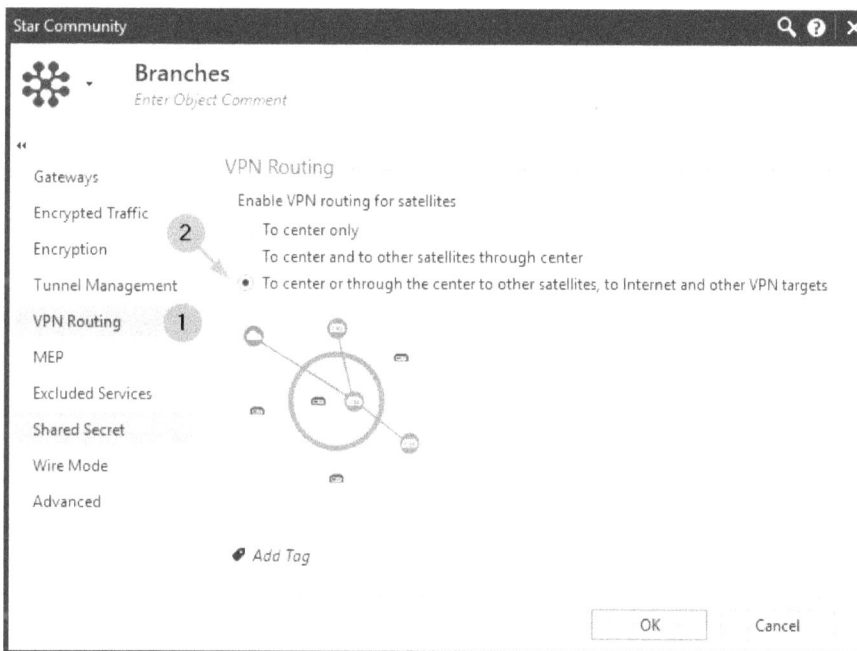

Figure 12.13 – Selecting the To center or through the center to other satellites, to Internet and other VPN targets option

2. Publish the changes and install both policies, since the **Branches** VPN community is used by both the **CPCXL** cluster and the **CPGW** gateway.

3. On **RightHost**, change the DNS server settings, leaving a single DNS server, 10.20.20.10, to use its internal resolver.

4. Initiate traffic from **RightHost** to the internet – we are running "through the center" configuration, so the internet traffic is now subjected to the `APCL_URLF` layer of **LeftSide-S2S**.

5. Observe logs with **Source** as `172.16.16.10` [1]. You will see that traffic is encrypted on **CPGW**, decrypted on the **CPCXL** cluster member, **CPCM1**, and forwarded to the destinations [2]. That said, even with the **Disable NAT inside the VPN community** setting selected, we are still seeing traffic addressed to internal hosts on the left side, such as DNS or HTTP to DMZ, being hidden behind the **CPCXL** cluster's vIPs [3]:

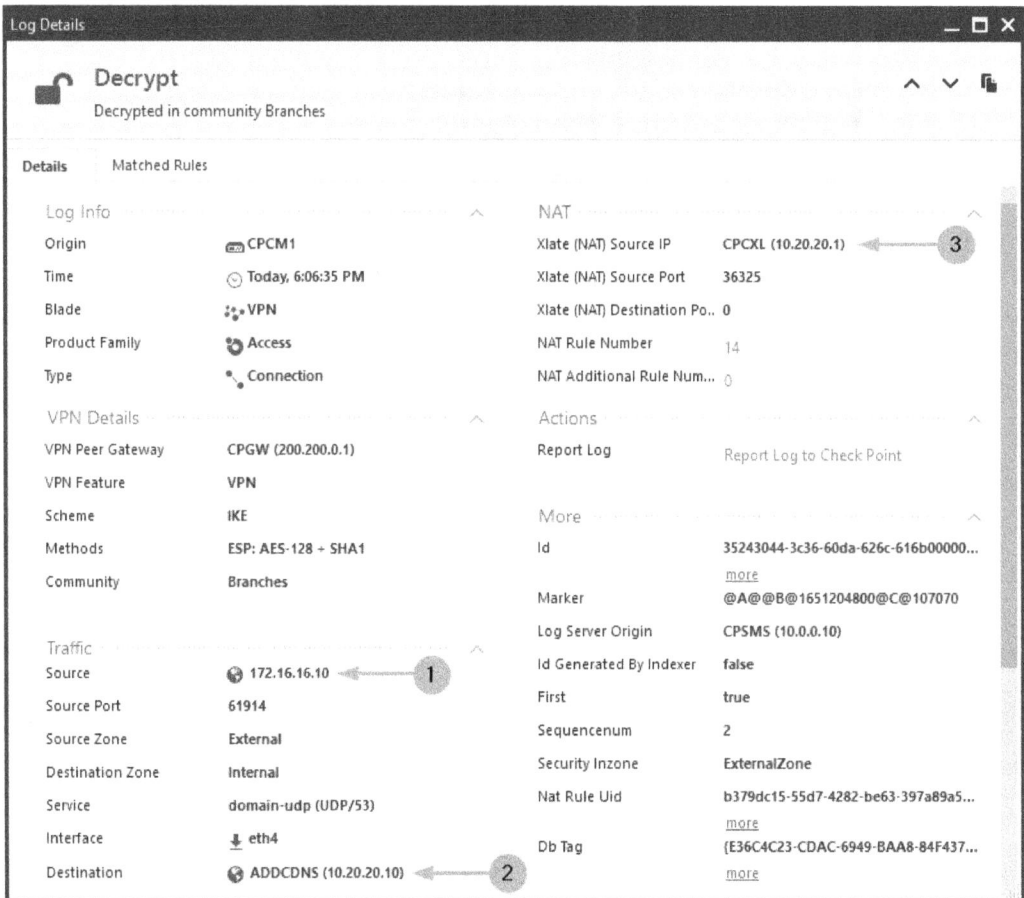

Figure 12.14 – Hide behind the gateway in action with Disable NAT inside the VPN community

6. To avoid NAT being enforced on our VPN traffic, add two rules to the top of the **LeftSide-S2S** NAT policy under the previously created rule, **No NAT1**, preventing automatic NAT between local networks:

No.	Name	Original Source	Original Destination	Original Services	Translated Source	Translated Destin...	Translated Services
1	No NAT1	Int_Nets	Int_Nets	* Any	= Original	= Original	= Original
2	No NAT2	Net_172.16.16.0	Int_Nets	* Any	= Original	= Original	= Original
3	No NAT3	Int_Nets	Net_172.16.16.0	* Any	= Original	= Original	= Original

Figure 12.15 – No NAT between the VPN domains with the No NAT2 and No NAT3 rules

7. Initiate traffic from **LeftSide-S2S** to **RightHost**.

8. Observe **Logs** with the `blade:VPN  dst:172.16.16.10` filter. Since we do not have any services listening on **RightHost**, we should not be able to connect to it, but we should still see the encrypted and decrypted packets in the logs.

Excluding a satellite gateway's external IP from the VPN

Now, while the VPN seems to be working fine, there is one more item left to complete that, in my opinion, is exceptionally annoying. It is specific to **star communities** configured as **To center or through the center to other satellites, to Internet and other VPN targets**.

By default, Check Point includes the public IP of gateways, clusters, and cluster members in its VPN domain. What that means for us is that traffic generated by CPGW itself suddenly fails. Namely, DNS, NTP, and ICMP from CPGW to public IPs are encrypted and sent over the VPN. As there are no rules in the access control and NAT policies on the left side to account for it, this traffic is either dropped by the rules or forwarded, if allowed, but the replies come back to the origin IP of `200.200.0.1` outside the VPN and are discarded because they are not part of a known session.

Additionally, in situations where your satellite gateway has only a single public IP available, you may not be able to generate any internet-bound traffic locally by using hide behind IP address. This is a problem, since you may need to allow local guest users on a different network hiding behind the same external IP address to access the internet.

Try the following exercise: SSH into CPGW and ping our router's IP, `200.200.0.254`. In the SmartConsole logs, use the `src:200.200.0.1 AND service:icmp-proto` filter, and you will see the encrypted [1] and decrypted [2] traffic from CPGW traversing CPCXL member CPCM1:

★ Queries	<	>	C	C_A	Q	⌚ Last 24 Hours ▾	src:200.200.0.1 AND service:icmp-proto

Found 71 results (39 ms)

Time	..	..	..	..	Origin	Source	Destination	Service
Today, 8:52:29 PM ①					CPGW	CPGW (200.200.0.1)	200.200.0.254	echo-request (ICMP)
Today, 8:52:29 PM ②					CPCM1	CPGW (200.200.0.1)	200.200.0.254	echo-request (ICMP)

Figure 12.16 – Unexpected encrypted traffic

To address this issue using recommendations in *sk25675 Customizing VPN Domain to exclude IP Address and allow clear text*, take the following steps:

1. SSH into our management server, **CPSMS**. Enter the Expert mode.

2. In it, create a backup copy of the `crypt.def` file and open it in the editor:

   ```
   cp  $FWDIR/lib/crypt.def  $FWDIR/lib/crypt.def_BKP
   vi  $FWDIR/lib/crypt.def
   ```

3. For non-Linux users or those not used to the vi editor, follow these steps:

   A. Press *Esc*.

   B. Press *Shift + G* to go to the bottom of the file.

   C. Find the `#define NON_VPN_TRAFFIC_RULES 0` line.

   D. Press *I*.

   E. Using the arrow keys, move the cursor to the start of that line.

   F. Press *Enter* to insert a new line.

   > **Important Note**
   > Do not use the numpad of your keyboard; use the number keys of the top row.

   G. Type (or paste) `vpn_exclude={200.200.0.1};` and replace 0 in `#define NON_VPN_TRAFFIC_RULES 0` with `(src in vpn_exclude)`. The block of lines at the end of the file should now look like this:

   ```
   #ifndef NON_VPN_TRAFFIC_RULES
   vpn_exclude={200.200.0.1};
   #define NON_VPN_TRAFFIC_RULES (src in vpn_exclude)
   #endif

   #endif /* __crypt_def__ */
   ```

   Figure 12.17 – Excluding the public IP of the satellite gateway from VPN sources

   H. Press *Esc*.

   I. Type `:wq`.

   J. Press *Enter*.

4. Install both policies.

5. If you find yourself messing up the file in the editor, exit any time without saving changes by doing the following:

    A. Pressing *Esc*

    B. Typing `:q`

    C. Pressing *Enter*

> **Note**
>
> If you are seeing policy installation errors, you have probably made a mistake while editing the `crypt.def` file.
>
> To fall back, use the following:
>
> **`cp $FWDIR/lib/crypt.def_BKP $FWDIR/lib/crypt.def`**
>
> ... and install the policies again.

6. With the file successfully edited and policies installed, SSH to CPGW again, ping the router at `200.200.0.254`, and perform `nslookup` of the known internet domain. These should now work.

The rest of the traffic is still routed to and through the CPCXL hub.

Note that we did not remove `200.200.0.1` from the encryption domain of CPGW but simply restricted traffic sourced from it from being encrypted.

Changing portals' URLs and renewing a gateway cluster certificate

There are two more items that we should take care of at this point – the portals for browser-based authentication and UserCheck, which still contain CPCXL's public IP addresses, `https://200.100.0.1/connect` and `https://200.100.0.1/UserCheck` respectively.

These should now be changed to the **virtual IP (vIP)** of one of the internal cluster interfaces, since we will be accessing them from the VPN:

1. Open the **CPCXL** cluster properties [1] and click on **Identity Awareness** [2]. Click on the **Settings** button next to **Browser-Based authentication** (not shown). In **Portal Settings** [3], under **Access Settings**, click **Edit...** [4]. In **Portal Access Settings** [5], change **Main URL** to `https://10.20.20.1/connect` [6]. Under

the **Accessibility** section of the same view, click **Edit…** [7]. In the **Accessibility** window, check the box for **Including VPN encrypted interfaces** [8]. Click **OK** three times to close these windows:

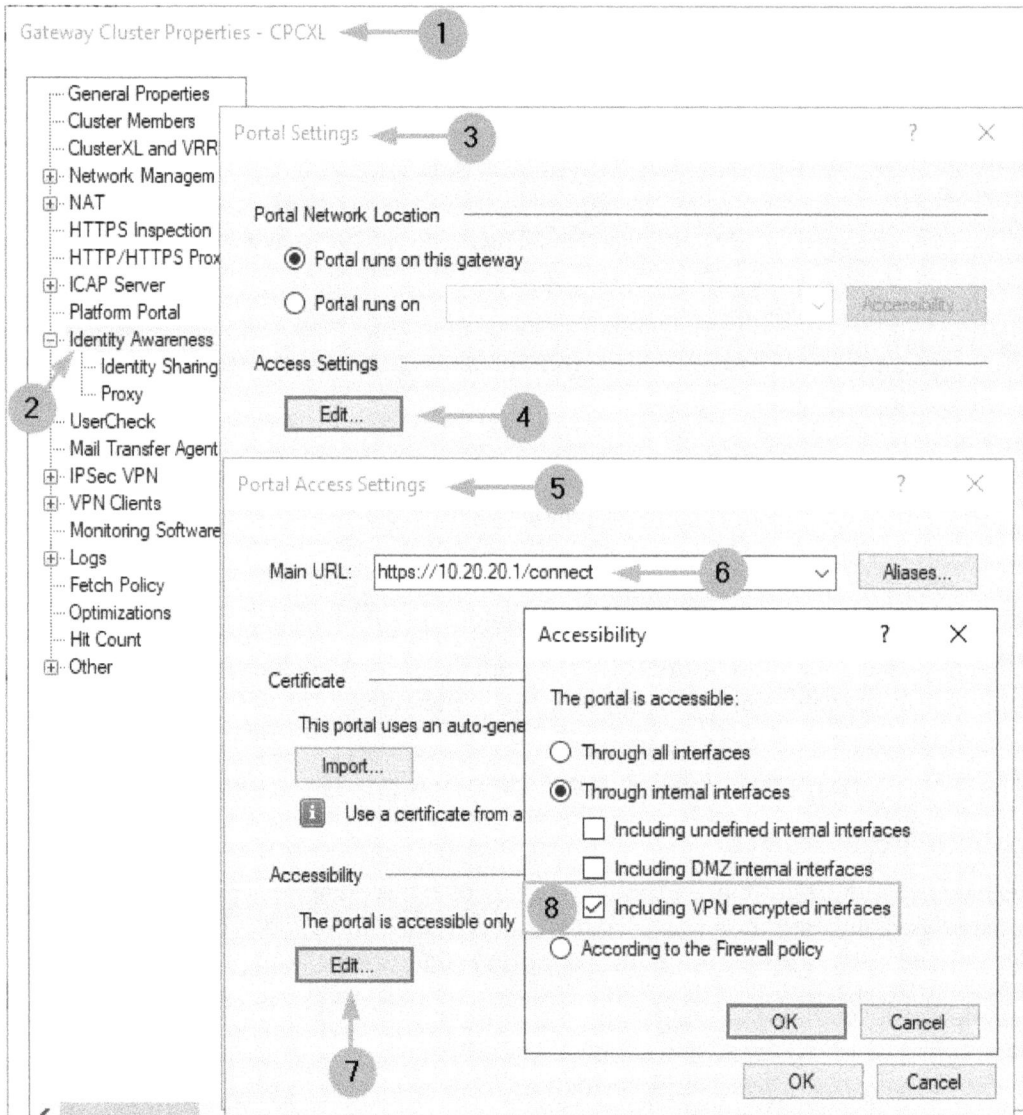

Figure 12.18 – Changing the Identity Awareness browser-based authentication portal URL

2. Now, click on the **UserCheck** item in the cluster properties navigation tree
    [1]. Change **Main URL** to `https://10.20.20.1/UserCheck` [2]. Under
    the **Accessibility** section, click **Edit…** [3]. In the **Accessibility** window, check
    Including VPN encrypted interfaces. Click **OK**:

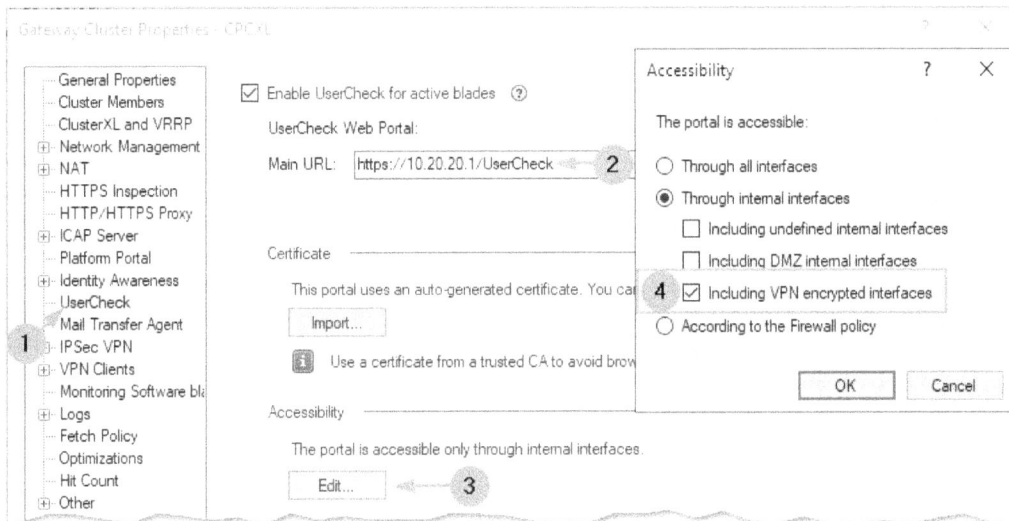

Figure 12.19 – Changing the UserCheck web portal URL

3. While still in the **CPCXL** properties [1], click **IPSec VPN** [2]. On the right side
    under **Repository of Certificates Available to the Gateway**, click **Renew…** [3]. In
    the **Generate Keys and Get Internal CA Certificate** window, click **Add…** [4]. In
    the **Add Subject Alternate Name** window, change **Type** [5] to **IP Address**, and in
    the **Alternate Name** field, type `10.20.20.1` [6]. Click **OK** [7] to see it added to
    the list of the alternate names [8] and click **OK** twice [9 and 10] to commit.

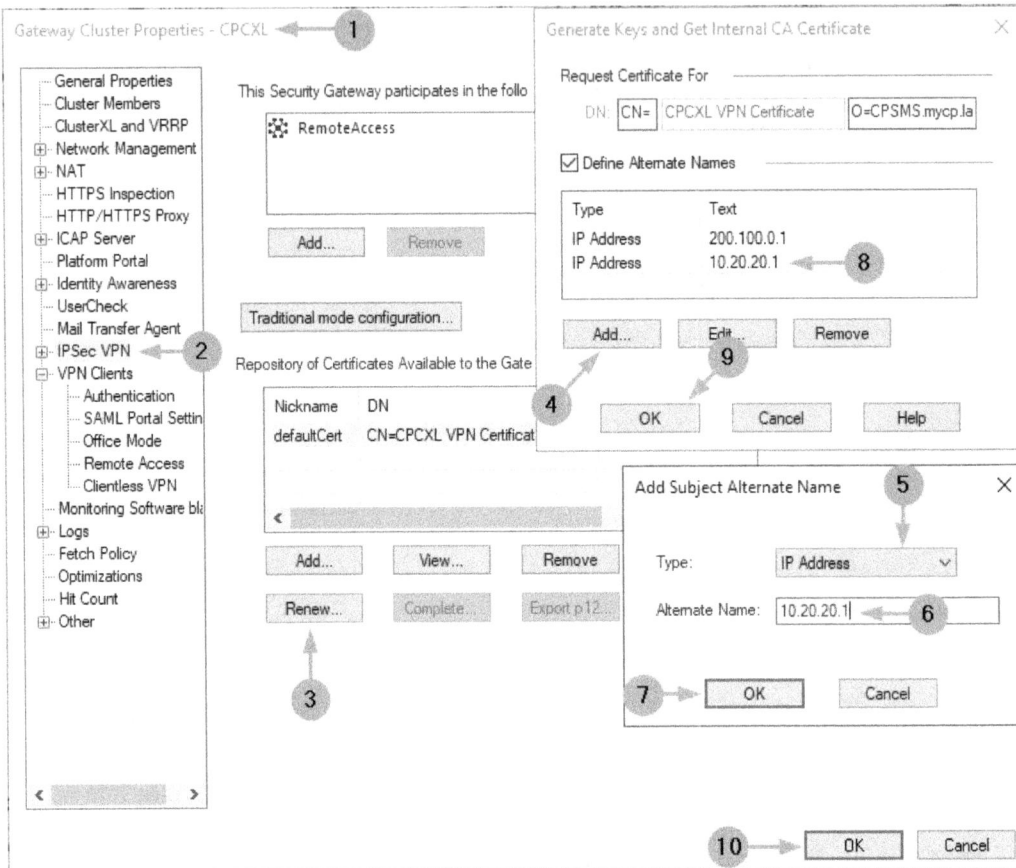

Figure 12.20 – Changing the gateway certificate to include subject alternate names

4.   Publish the changes and install the policy.

Now, should you encounter **UserCheck** when accessing a site subject to this action from a remote VPN location, you will see it presented in your browser. There is no need to redistribute certificates for internal users at both locations, as your ICA certificate, as well as the HTTPS outbound certificate, remained the same. The validity of new gateways' certificates will be confirmed by ICA. If you open the certificate properties of the **UserCheck** page on **RightHost**, you will see that it is issued to the **CPCXL** VPN certificate and issued by **CPSMS.mycp.lab**, and in **Details | Subject Alternate Names**, both IP addresses are now shown.

This will also aid us in ensuring **Portal** accessibility for a remote access VPN using IPSec clients, a subject we will examine next.

An introduction to Check Point remote access VPN solutions

Check Point has a rather large number of options for remote access. Some of them focus strictly on connectivity; others include further options for client security and compliance. You can read about most of the available options here: `https://sc1.checkpoint.com/documents/R81/WebAdminGuides/EN/CP_R81_RemoteAccessVPN_AdminGuide/Topics-VPNRG/Remote-Access-Solutions.htm`.

In addition to those listed in the **Remote Access VPN R81 Administration Guide**, there are a few more options: **Harmony Endpoint**, a complete EDR with an IPSec VPN (read more about it here: `https://www.checkpoint.com/harmony/advanced-endpoint-protection/`), as well as **Harmony Connect**, a browser-based, clientless, zero-trust application and network-secure remote access – a VPN as a service that you can host in your own infrastructure (read more about it here: `https://www.checkpoint.com/harmony/connect-sase/`).

Since our goal, for now, is simply to implement remote access in our lab, we'll be using a product called Endpoint Security VPN. Let's move on to the next section to see how it is done.

Configuring a remote access IPSec VPN

Before we begin working on the remote access IPSec VPN, let's clear out the **Branches VPN community**: open the **VPN community** called **Branches**, and remove CPCXL from **Center Gateways** and CPGW from **Satellite Gateways**. Click **OK** to close this VPN community. Publish the changes and install both policies, **LeftSide_S2S** and **RightSide**.

Cloning a policy

Let's create a clone of our **LeftSide_S2S** policy, in case you want to experiment with it later:

1. Click on the main menu (in the top-left corner of SmartConsole) | **Manage policies and layers**. Click on the **LeftSide_S2S** policy, **Actions**, and then **Clone**.

2. In the **Clone Object** dialog, type in `LeftSide_RA` (for remote access) and click **OK**.

3. A new policy tab called **LeftSide_RA** is now opened.

4. Close the **LeftSide_S2S** policy tab.

Creating local user templates, groups, users, and access roles

Let's configure local user templates, groups, and users. When local user groups are used in firewall/networking rules' sources, they are referred to as **Legacy User Access**. The same groups can be added to the access roles to work in layers with application control enabled (or unified policies). In a smaller environment, or when explicitly trying to decouple some of the remote access VPNs from dependency on external directory services, these are still a good choice:

1. Click on **Objects** (at the top left of SmartConsole) | **More Object Types** | **User/Identity** | **New User Template**.

2. In **New User Template** | **General**, name the object RA_Users_Template.

3. Click on the **Groups** option on the left. Under **Add user to groups**, click +. In the pop-up window, click the **New** icon, and then **User Groups**.

4. Name the new user group RA_Users_Group and click **OK**.

5. Click on the **Authentication** option. Under the **Authentication** section on the right, use the drop-down menu to select **Check Point Password**. Ignore the **Password is not defined** warning and click **OK**.

6. Click on **Objects** (at the top left of SmartConsole) | **More Object Types** | **User/Identity** | **New User**.

7. In the **New User** window | the **Choose template** field, use the drop-down menu to select **RA_Users_Template** and click **OK**.

8. Name the new user Ra_user1 and click the **Authentication** option on the left.

9. Click **Set New Password**. When prompted, enter and confirm our lab password, CPL@b8110, and click **OK** twice to complete the user creation process.

10. Click on **Objects** (at the top left of SmartConsole) | **More Object Types** | **User/Identity** | **New Access Role**.

11. Name the new access role [1] RA_Role [2] and click on **Users** [3]. Select **Specific users/groups** [4] and click the + sign [5]. In the drop-down menu, select **Internal User Groups** [6]. Click another + sign to the left of **RA_Users_Group**, which will change into a checkmark [7]. Close the group selection window and click **OK** (not shown):

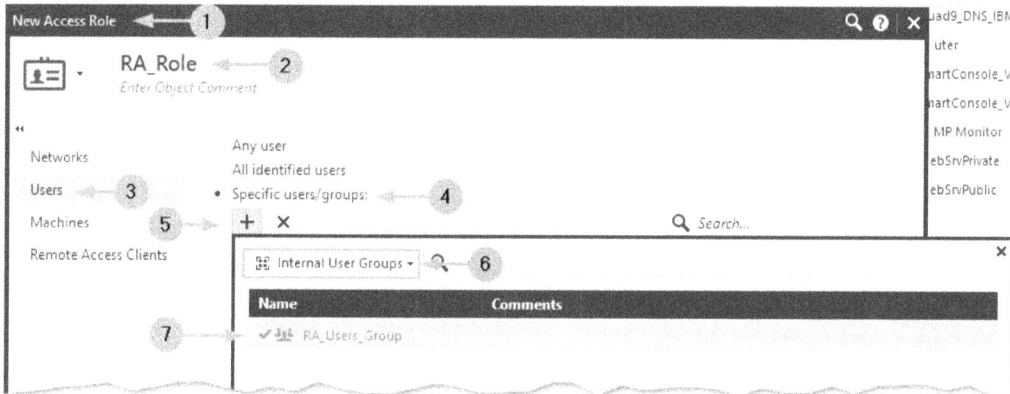

Figure 12.21 – Creating an access role for the remote access VPN

We have the option to assign IP addresses to remote clients once they are connected. To do that, we must define the network that will serve as a basis of the pool of addresses available to us and specify how these are leased to the clients.

Let's create a new network object, Net_192.168.254.0, with a 192.168.254.0 address and a 255.255.255.0 mask.

Configuring a gateway or cluster for remote access

Now, let's configure our cluster object for remote access:

1. Double-click the **CPCXL** object.

2. Select **Identity Awareness** in the navigation tree [1].

3.   Check the **Remote Access** checkbox [2]:

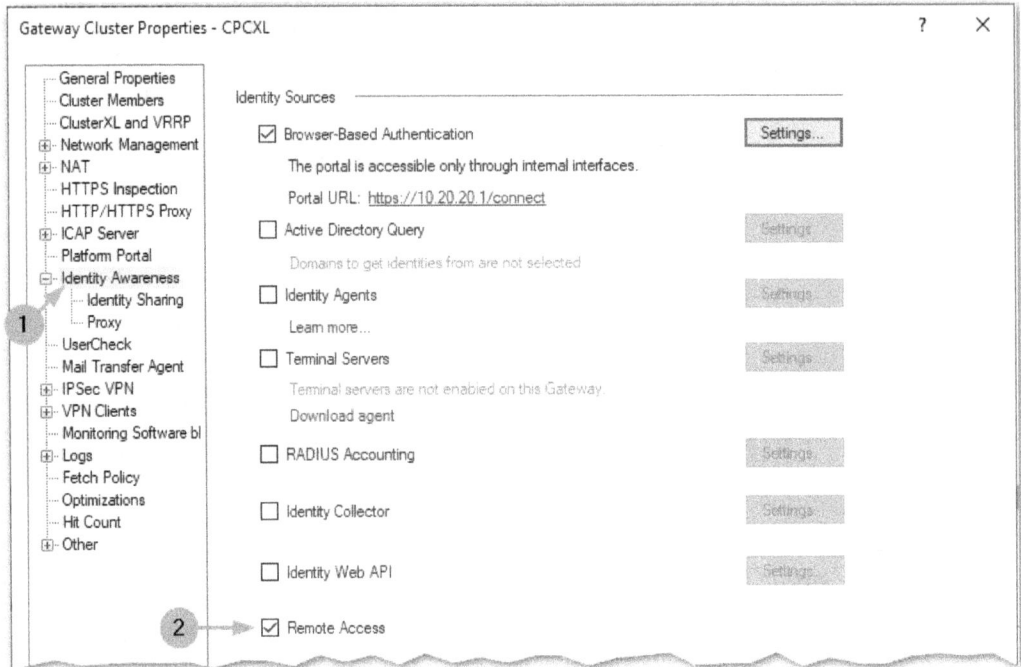

Figure 12.22 – Remote access as an identity source

4.   Select **VPN Clients** in the navigation tree [1].

5.   Uncheck everything except **Endpoint Security VPN** [2]:

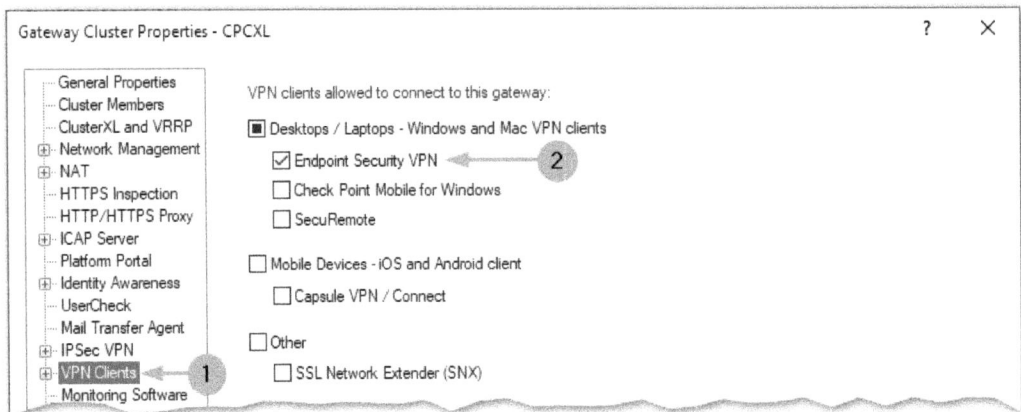

Figure 12.23 – The VPN client types in the gateway's properties

6. Expand **VPN Clients** by clicking on the + icon and then **Office Mode**.

7. In **Office Mode** [1], select **Allow Office Mode to all users** [2], check **Using one of the following methods** [3], and select **Automatic (using DHCP)** [4]. In the **Use specific DHCP server** drop-down menu, select **ADDCDNS** [5]. In the **Virtual IP address for DHCP server replies** field, we can choose any IP from the network we have dedicated to it. Let's set it to `192.168.254.1` [6]. Check **Perform Anti-Spoofing on Office Mode addresses** [7], and in its drop-down menu, select our network, **Net_192.168.254.0** [8].

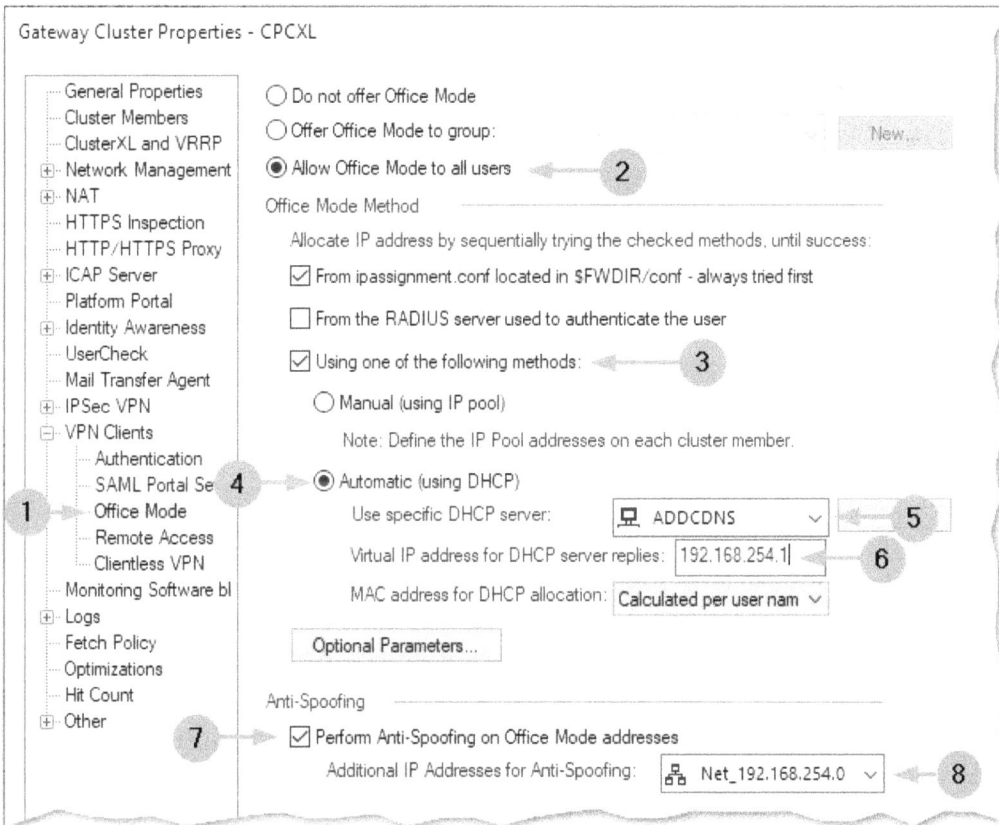

Figure 12.24 – The Office Mode section of VPN Clients in the gateway's properties

8. Now, click on the **Remote Access** item in the **VPN Clients** section of the navigation tree [1] and verify that **Allow VPN clients to route traffic through this gateway** [2] and **Support Visitor Mode** are checked [3]. Leave all other settings in this window as default and click **OK** to complete gateway cluster configuration:

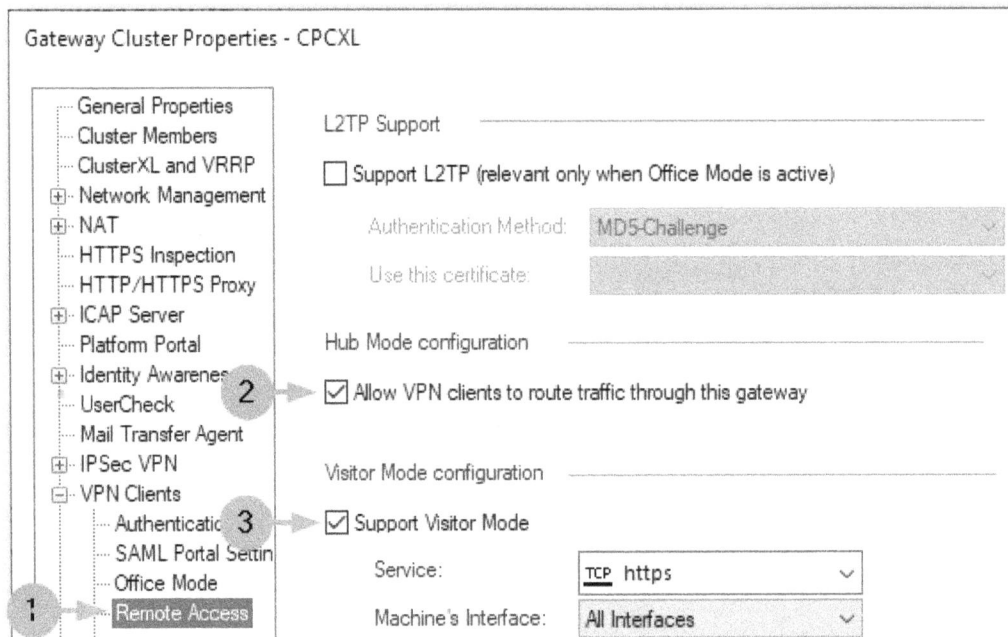

Figure 12.25 – The Remote Access section of VPN Clients in the gateway's properties

The **Allow VPN clients to route traffic through this gateway** option is necessary if we intend to deny the use of a split tunnel on clients' PCs. Additional changes are required in **Global Properties** to enforce these settings. We will cover those shortly.

The purpose of **Visitor Mode** is to allow connectivity from clients in restrictive environments. If you are using public or private Wi-Fi that restricts you to HTTP and HTTPS only and does not allow **Internet Key Exchange (IKE)** or IPSec, TCP tunneling is used to encapsulate all connecting traffic on port 443.

Configuring global properties for remote access

Now, let's click on the main menu and click on **Global Properties** [1] Then take the following steps:

1. In the **FireWall** view [2], we are presented with configurable implied rules parameters. Check the **Accept ICMP requests** box and verify that the requests are set to **Before Last** [3]. This will be useful for a variety of tests we'll be running once remote access is configured. Let's also check the **Log Implied Rules** box [4] located at the bottom of the screen under the **Track** section:

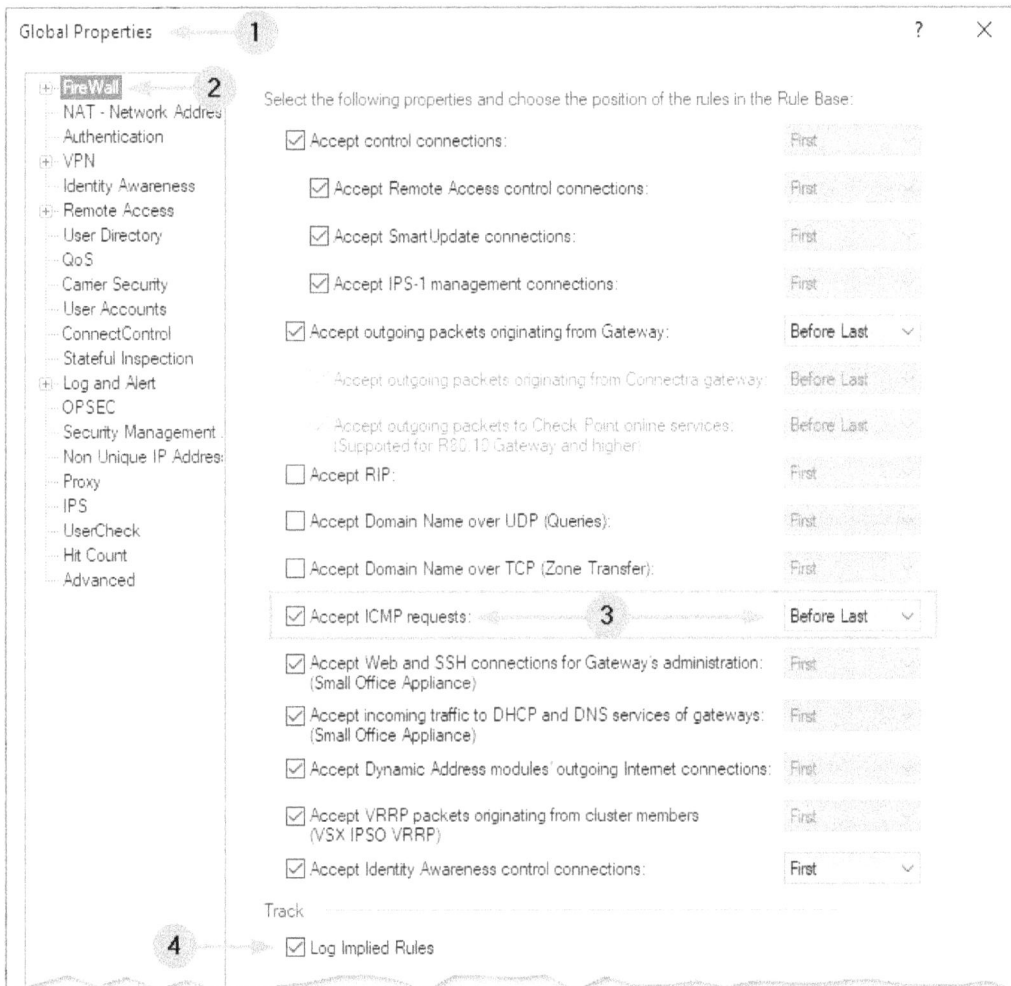

Figure 12.26 – Global Properties, FireWall for ICMP, and implied rules tracking

> **Important Note**
>
> In production environments, it is not a good idea to enable ICMP globally. Instead, rules for specific network monitoring solutions and, perhaps, certain administrative workstations should be created.

Remember that **Log Implied Rules** is a global parameter. If left enabled, later installation of all policies will enforce that on all target gateways. It is useful if you are not seeing anticipated traffic in your logs and must determine and document the root cause, using them instead of relying on low-level CLI troubleshooting techniques. Set a reminder to disable **Log Implied Rules** and install the policy on affected gateways to discontinue.

2. Now, let's select **Remote Access** [1] and check (or verify) that the **Enable Back Connections (from gateway to client)** [2] and **Encrypt DNS traffic** [3] boxes are checked. The back connections will allow for connections to the VPN client that originated from behind the gateway/cluster. Not encrypting DNS traffic will force you to do one of the following: refer to corporate resources by IP addresses, have local hosts' files on remote clients, or serve private IPs from your public DNS servers.

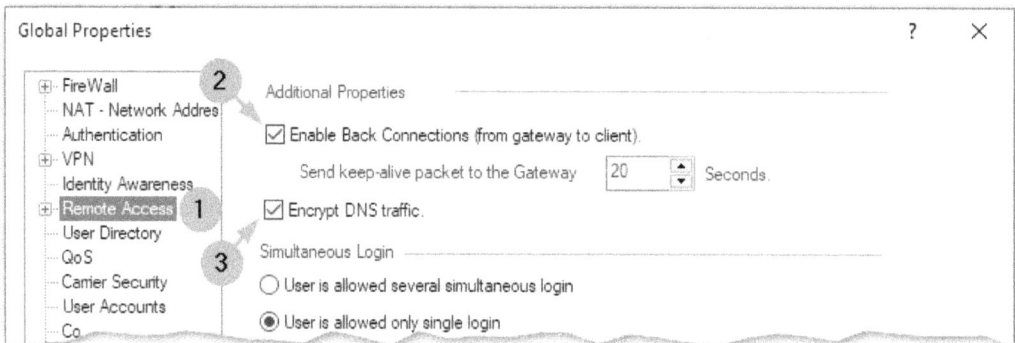

Figure 12.27 – Global Properties, Remote Access back connections, and DNS encryption

3. Let's expand the **Remote Access** navigation tree branch and move on to **Endpoint Connect** [1]. In the **Security Settings** section, set **Route all traffic to gateway** to **Yes** [2]. In the **Connectivity Settings** section, set **Network Location Awareness** to **Yes** [3], and examine the options that become available when the **Configure...** button to the right of this field becomes active [4].

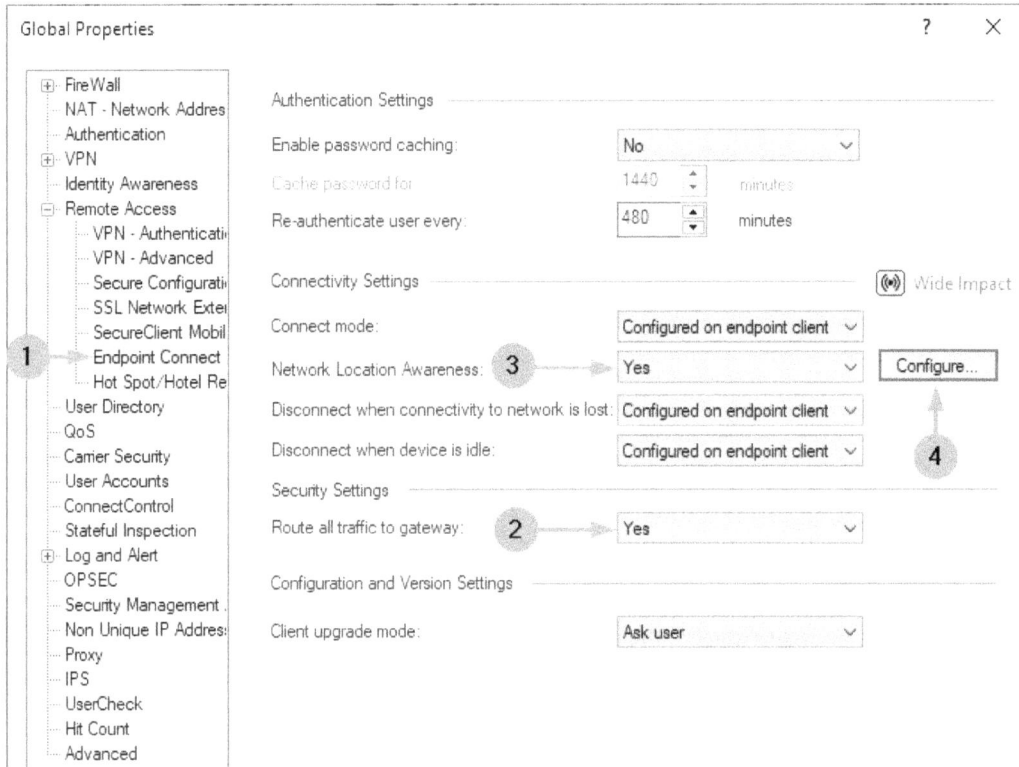

Figure 12.28 – Global Properties, Remote Access, and Endpoint Connect

4. Click it to see that it is used to terminate the VPN when the same client, such as a laptop, connects to corporate resources from the inside of the corporate LAN:

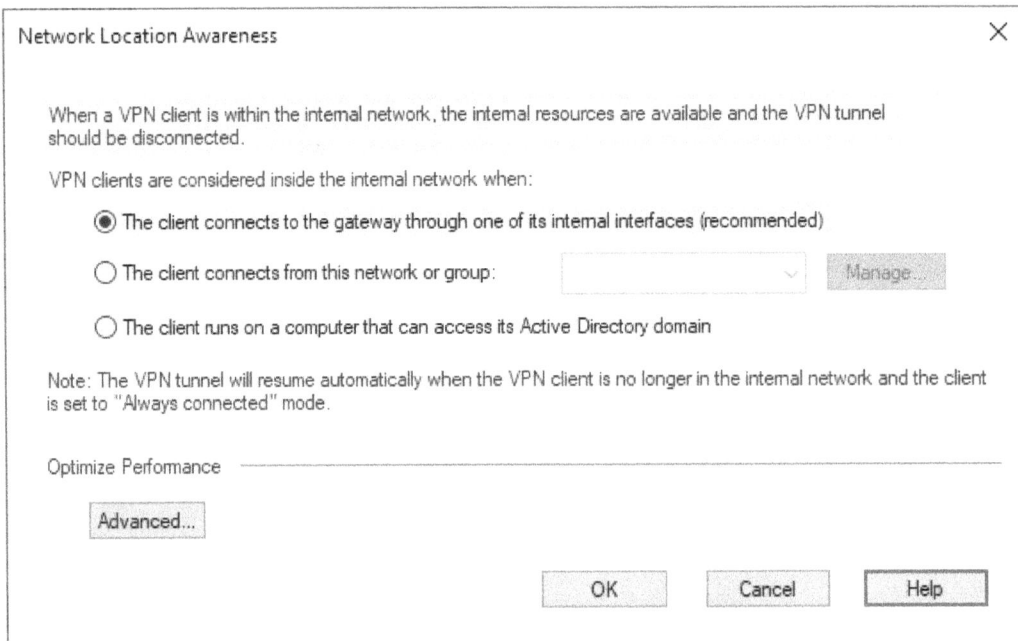

Network Location Awareness   ✕

When a VPN client is within the internal network, the internal resources are available and the VPN tunnel should be disconnected.

VPN clients are considered inside the internal network when:

⦿ The client connects to the gateway through one of its internal interfaces (recommended)

◯ The client connects from this network or group: [▽] [Manage...]

◯ The client runs on a computer that can access its Active Directory domain

Note: The VPN tunnel will resume automatically when the VPN client is no longer in the internal network and the client is set to "Always connected" mode.

Optimize Performance ───

[Advanced...]

                                          [OK   ]   [Cancel]   [Help   ]

Figure 12.29 – Network Location Awareness for Endpoint Connect clients

The **Advanced...** configuration options under the **Optimize Performance** section allow you to quickly make a decision about a client by looking at the Wi-Fi **Service Set Identifiers** (**SSIDs**), organization DNS suffixes, and client-side caching of detected external networks.

Now is not a bad time to examine all other available options under **Remote Access** in **Global Properties** (as long as you are not changing anything else). Click **OK** to complete the process and return to our LeftSide_RA security policy.

Configuring a VPN community for remote access

Under **Access Tools** in SmartConsole, click on **VPN Communities** and open a pre-defined VPN community called **Remote Access**. In the **Participating Gateways** section, click the + sign, select **CPCXL**, and exit the selector window. In the participating user groups, add **RA_User_Group** and click **OK**.

Now, we must decide what our remote access clients will actually be able to access by adding or modifying rules in our policy.

> **Important Note**
>
> In the firewall/networking layer, we can use both the user groups and the IP addresses (or a network object) that we decided to allocate to the remote access users in **Office Mode** as sources (e.g., addresses from `Net_192.168.254.0/24`).
>
> Also, note that we must use the access role containing that user group in our application control and URL filtering rules.

Configuring access control policy rules for remote access

With this in mind, in the **Core Services** section of our policy, let's do the following:

1. Add `Net_192.168.254.0` to rule 9 (**DNS Internal Accept**).

2. In the **Privileged Access** section of the policy, let's add a new rule below rule 13 (**Admins RDP to all Accept**).

3. Name it `RA_Users RDP to InternalZone Accept`.

4. In a source field of that rule, right-click and select **Add Legacy User Access**.

5. In **New Legacy User At Location** [1] | **User Group**, select **RA_Users_Group** [2]. Leave **Location** at **Any** and click **OK**:

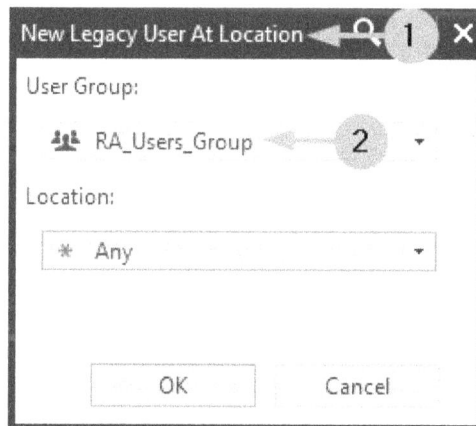

Figure 12.30 – Adding a legacy user access source to a rule

6. Add **InternalZone** to the **Destination** field.

7. In the **VPN** field, right-click and click on **Specific VPN Communities**. Select **Remote Access** and exit the selector dialog.

8. In **Services & Applications**, add **Remote_Desktop_Protocol**.

9. In **Actions**, select **Accept**.

10. In **Track**, select **Log** plus **Accounting**:

No.	Name	Source	Destination	VPN	Services & Applications	Action	Track
14	RA_Users RDP to InternalZone Accept	RA_Users_Group@Any	InternalZone	RemoteAccess	Remote_Desktop_...	Accept	Log Accounting

Figure 12.31 – A remote access rule with legacy user groups and a VPN community

Add another rule at the top of the **DMZ** section of the policy, allowing the same user group access to **DMZZone** via the VPN using HTTP and HTTPS services:

No.	Name	Source	Destination	VPN	Services & Applications	Action	Track
▼ DMZ (18-20)							
18	RA_Users to DMZ Accept	RA_Users_Group@Any	DMZZone	RemoteAccess	http https	Accept	Log Accounting

Figure 12.32 – A remote access rule with legacy user groups and a VPN community

Now, let's amend our APCL and URLF layer to account for remote access.

There are a few things to remember for rules in layers with blades other than **Firewall** enabled:

- We cannot combine access roles and network objects in the same cell. So, when it is necessary to grant or deny the same access based on identities, we must use duplicate rules specifying access roles in sources. Rules with access roles take priority if contending with rules where sources contain simple network objects or **Any.**

- We cannot use **Legacy Users** or **Legacy User Groups** in layers with **Application Control and URL Filtering**. This is why we have created the access role containing the legacy users group.

- When using local legacy users and groups, we only have group-level granularity in rules and access roles – that is, if you must grant an individual locally defined user access to resources, you'll have to create a separate local group and place that user into it. For layers with **Application Control and URL Filtering,** use either that group for **Firewall** rules or a separate access role containing that group.

With the preceding points in mind, let's do the following:

1. Right-click the rule (`All Users to News Accept`) in its **Rule No.** field and click **Copy**. Right-click it again in the same field and click **Paste | Above**.

    Rename the newly pasted rule and replace the `Int_Nets` object in the **Source** field with the `RA_Role` access role (both the original and new rules are shown) [1].

2. Right-click the rule (`All except IT Admins to Search Drop`) in its **Rule No.** field and click **Copy**. Right-click it again in the same field and click **Paste | Above**.

    Rename the newly pasted rule, deselect the **Negate** option for the **Source**, and replace `Net_10.0.0.0` with `RA_Role` (both the original and new rules are shown) [2].

3. Add `RA_Role` in the **Source** field of `HR and Remote Users to Social Networking Accept` [3].

    The following labeled rules represent the new or changed rules:

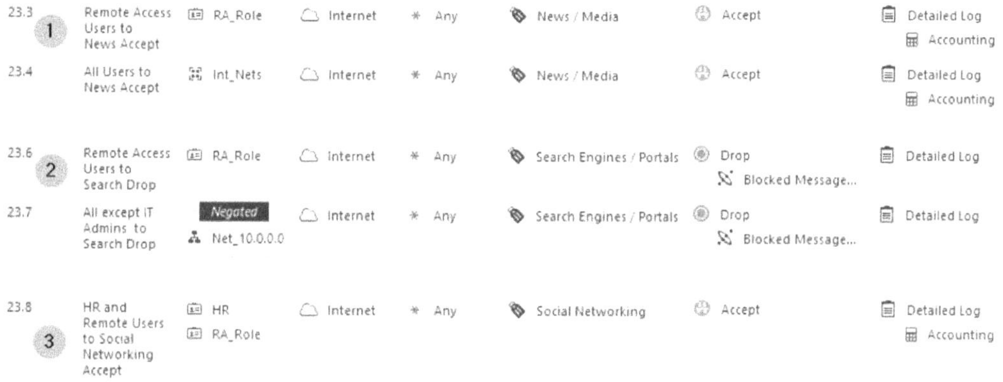

Figure 12.33 – The application control layer rules created or modified for the lab

4. Publish the changes and install the policy.

Configuring a DHCP server for a remote access Office Mode IP range

We must provision the `192.168.254.0/24` network, its associated DHCP range, and its parameters in the ADDCDNS server; otherwise, we cannot serve the DHCP leases for remote access clients' Office Mode.

Let's log in to **ADDCDNS**, start PowerShell, and paste the following code block into it. Press *Enter* to execute the last line:

```
New-ADReplicationSubnet -Name "192.168.254.0/24"

Add-DnsServerPrimaryZone -NetworkID "192.168.254.0/24"
-ReplicationScope "Forest"

Add-DhcpServerv4Scope -Name 'CP_RA_Hosts' -StartRange
192.168.254.100 -EndRange 192.168.254.150 -SubnetMask
255.255.255.0

Set-DhcpServerv4OptionValue -ScopeId 192.168.254.0 -DnsServer
10.20.20.10 -DnsDomain "mycp.lab"
```

With the DHCP server ready to supply dynamic addresses for this range, let's move on to remote client configuration.

Preparing remote client

Now, let's prepare **RightHost** to become a remote access client.

From our previous lab, where we were working on a site-to-site VPN, our DNS server was set to the `10.20.20.10` IP address; change it to `9.9.9.9`. Try accessing any safe websites and look at their certificates to confirm that we are connecting directly.

Also, examine the logs of the **Internet Access Accept** rule in the **RightSide** policy using the `service:https` filter query. You should see the unencrypted traffic to public IPs.

From the **RightHost** browser, go to *Endpoint Security Homepage sk117536* located at `https://supportcenter.checkpoint.com/supportcenter/portal?eventSubmit_doGoviewsolutiondetails=&solutionid=sk117536`.

Scroll down and download the latest release of **Remote Access clients for Windows**. At the time of writing, it is `E86.20`. This is a universal installer, allowing you to choose the type of client at runtime. Double-click it to start the installation process.

Let's run through the installation process, showing only a selection of important screenshots:

1. At **Welcome to Check Point VPN Installation Wizard**, click **Next**.

2. At **Client Products**, leave the default selection, **Endpoint Security VPN**, in place and read the description of the other two options. For our purposes, **Check Point Mobile** is also an acceptable choice that requires a corresponding checkbox to be selected in the gateways' VPN clients' properties. **SecuRemote** is a no-cost option for remote access, but it comes with limitations that make it ill-suited for enterprises (it does not support Office Mode).

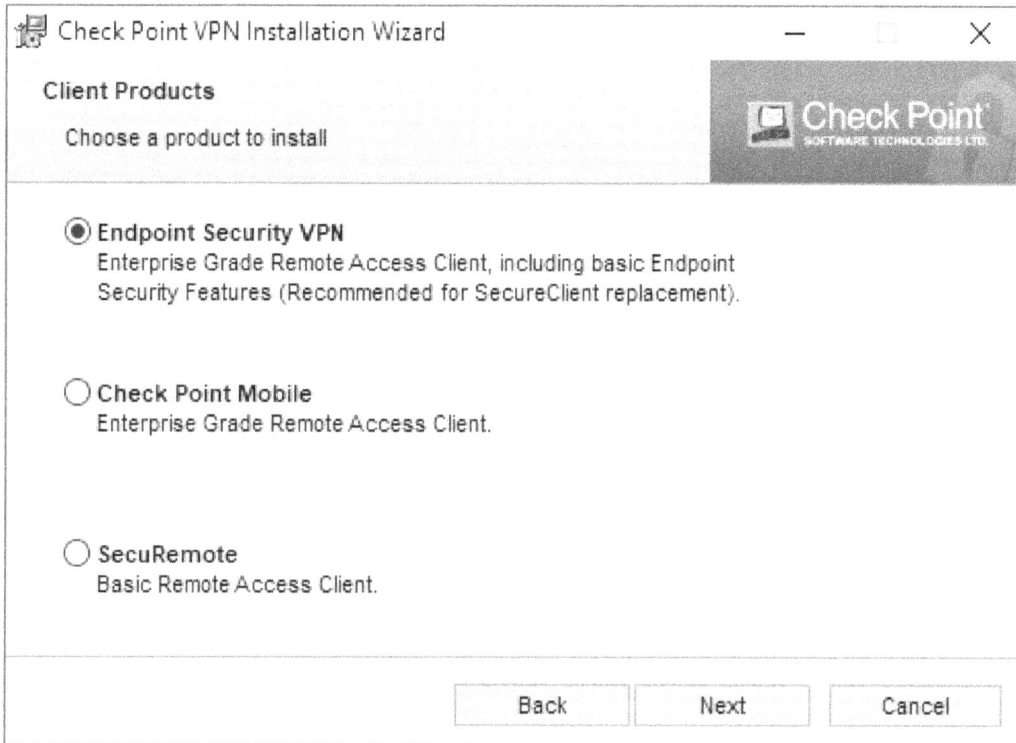

Figure 12.34 – The VPN Installation Wizard Client Products selector

3. At **License Agreement**, read the agreement, accept the terms, and click **Next**.

4. At **Destination Folder**, note that you can change it, but let's not. Click **Install**.

5. At **Installation Wizard Completed**, click **Finish**.

6. In the **RightHost** system tray, you can see the icon of the running **Check Point Endpoint Security** client. Double-click it:

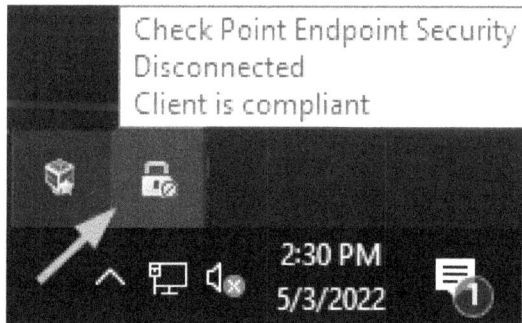

Figure 12.35 – The disconnected VPN client icon in the system tray

7. At **No site is configured. Would you like to configure a new site?**, click **Yes**.

8. At **Welcome to the Site Wizard**, click **Next**.

9. In the **Server address or Name** field, enter the public IP of our CPCXL cluster [1], check the **Display name** box [2], and enter a name associated with the site. Let's call it LeftSide [3]. Click **Next**:

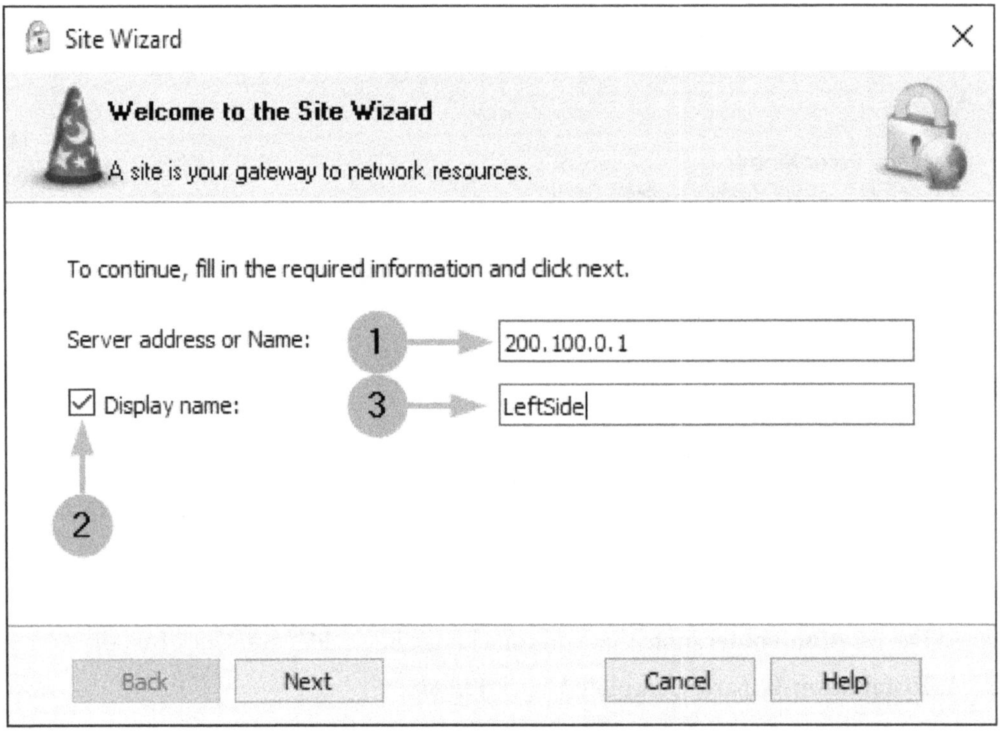

Figure 12.36 – Defining the VPN server and site

10. At **Login Option Selection**, leave it as **Standard (Default)** and click **Next**:

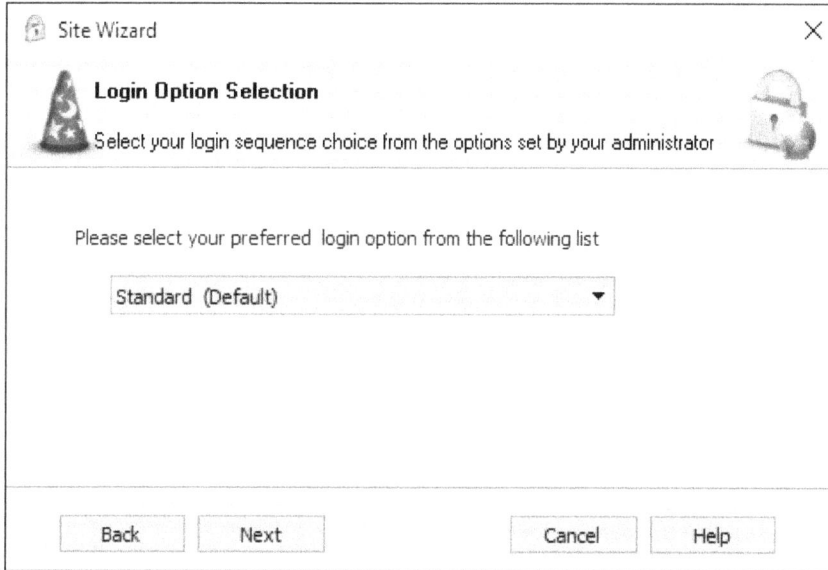

Figure 12.37 – Remote access VPN login option selection

11. Note that the login options available here are limited to those specified in the **User Template | Authentication** option and are defined in the **Gateway** [1] | **VPN Clients | Authentication** [2] view. If **Allow older clients to connect to this gateway** [3] is checked and **Settings…** [4] is configured with **Allow newer clients that support Multiple Login Options to use this authentication methods** checked [5], **Display Name** is then set as **Standard** [6] and **Authentication method** is set as **Defined On User Record (Legacy)** [7]:

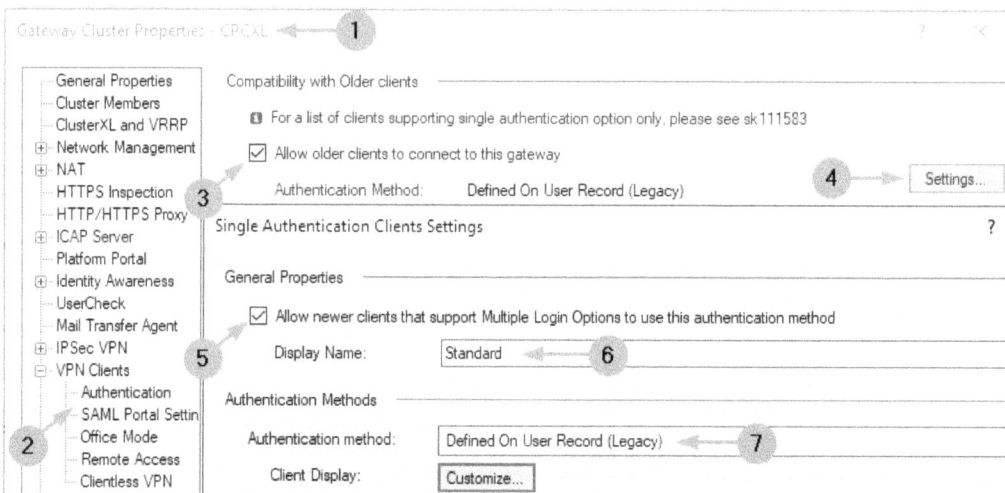

Figure 12.38 – Where the Standard authentication method is defined

12. When presented with the **Authentication Method** selector, read through the available options, leave the option as **Username and Password**, and click **Next**:

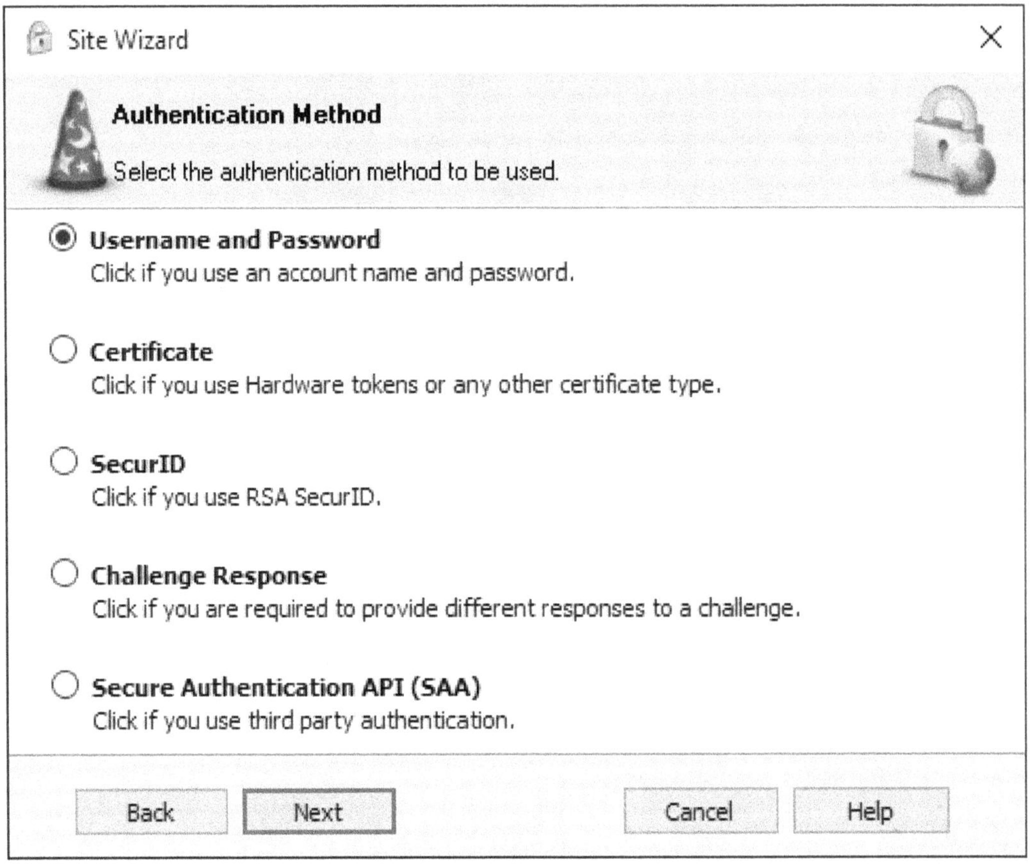

Figure 12.39 – The Endpoint Security Authentication Method selector

13. You will now be presented with a **Site created successfully** message. Click **Finish**.

14. When presented with **Would you like to connect?**, click **Yes**.

15. On the **Endpoint Security** screen, enter the RA_user1 username and the CPL@b8110 password that we used when creating this user in SmartConsole, and click **Connect**:

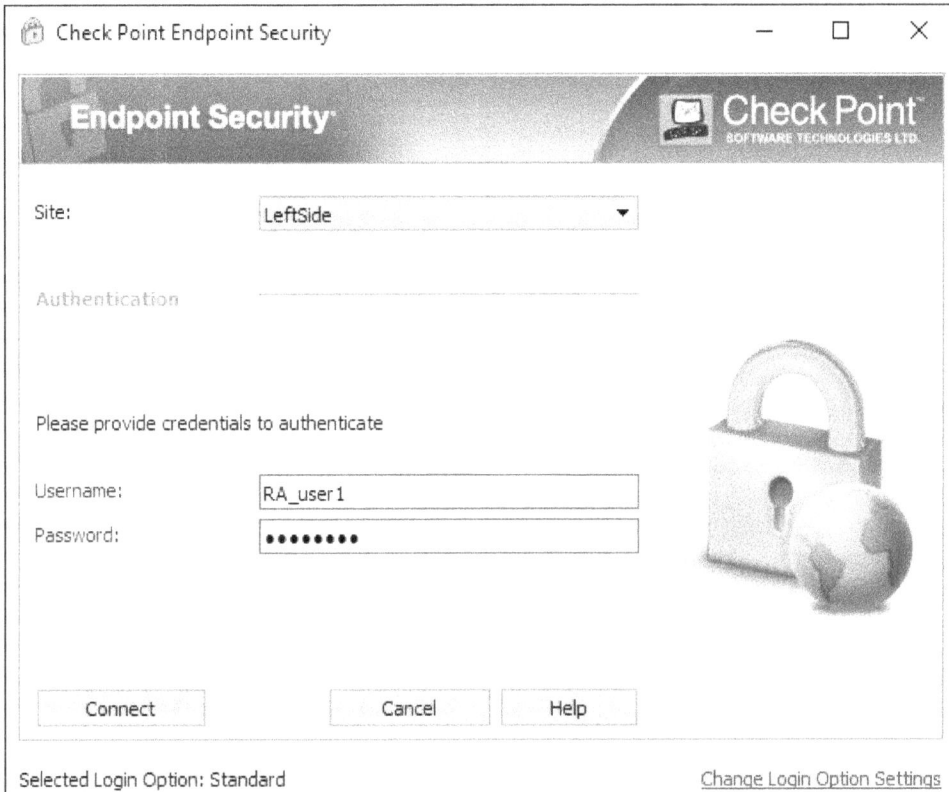

Figure 12.40 – The Endpoint Security authentication and connection prompt

16. You will be briefly presented with the **Connecting to <Site Name>** screen and the progress bar [1]. If you want to see the actual progress steps, click on **Details >>** [2]:

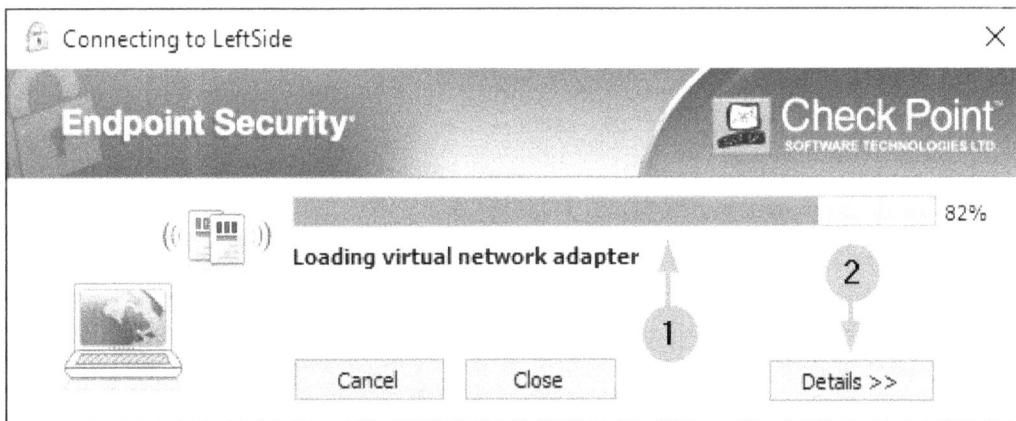

Figure 12.41 – The remote access VPN connection progress view

If you missed an opportunity to do that while the connection was being established, this dialog box will disappear. You will then see the system tray icon with a green light. When you right-click it, the following action options are present: **Disconnect**, **VPN Options**, **Help**, **Show Client**, and **Shutdown Client**:

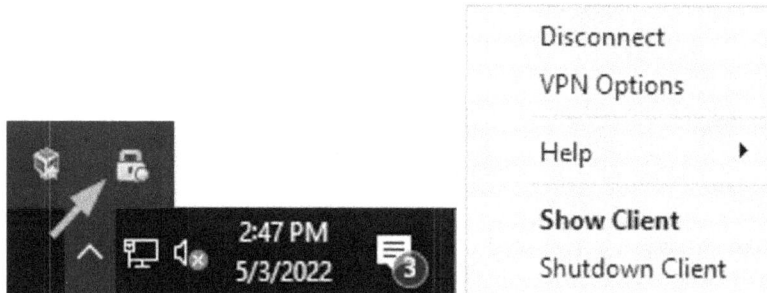

Figure 12.42 – The connected VPN system tray status icon and action options

Testing a remote access VPN

Now, with the endpoint VPN established, let's perform some tests to verify that it is working as expected:

1. Using Chrome, access these hosts and compare your outcomes and site certificates:

Destination	Outcome	Certificate	Issued by
dmzsrv.mycp.lab	Success	Not secure	NA
google.com	Page block	CPCXL VPN certificate	CPSMS.mycp.lab
bbc.com	Success	www.bbc.com	mycp.lab

Table 12.1 – The endpoint security VPN-connected state tests and outcomes

2. Check the logs for each rule in the **LeftSide_RA** policy that uses either **RA_Users_Group@Any**, **Net_192.168.254.0**, or **RA_Role** in its sources to observe their behavior.

3. Initiate the RDP session from **RightHost** to 10.20.20.10. Check the logs.

Let's take a look at the log card for decrypted connection, where we can see the **Decrypt** action in a specific community [1] performed by the **VPN** blade [2], **VPN Details** [3], the **Traffic** source (which is identified as the IP assigned to the client by DHCP) [4], **Policy**, including **Access Rule Number** [5], as well as the applied **NAT Rule Number** [6]:

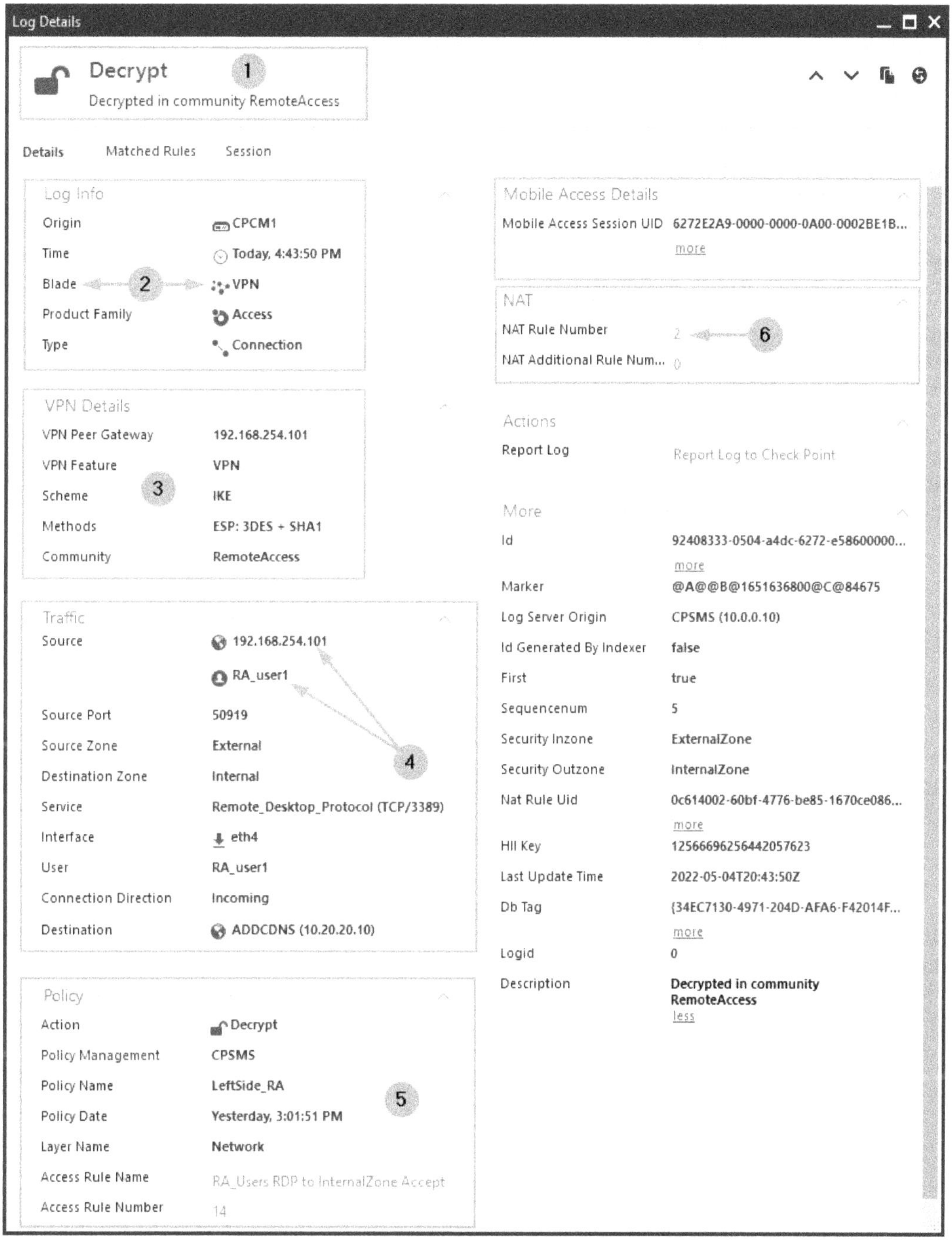

Figure 12.43 – The remote access VPN decrypt log card

Note that connections to the internet-based applications are treated as if they originated locally – there is no encryption data in these logs; they are simply identified as originating from the 192.168.254.0/24 network by a particular user.

There are indeed corresponding decryption logs, but they do not carry application information:

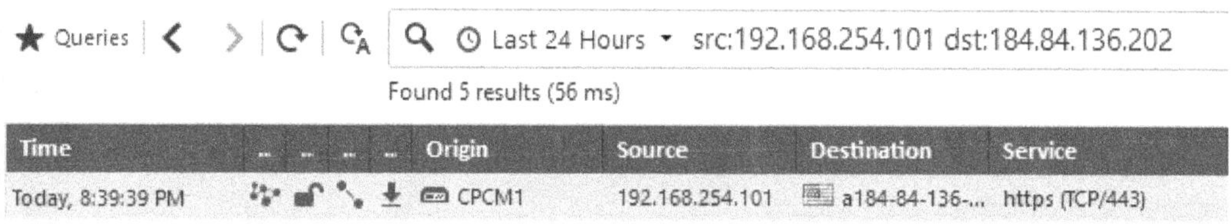

Figure 12.44 – A decrypted connection associated with a local egress traffic log for an application

This is pretty much it for the introduction to remote access VPNs. You can obviously see that it is a large subject, and depending on your infrastructure requirements, you may end up using multiple solutions to address different security and connectivity scenarios.

Summary

In this chapter, we learned about Check Point VPN capabilities. We configured and tested site-to-site VPNs for split tunnel and "to and throughout the center" gateway topologies. Additionally, we covered remote access IPSec VPNs using one of Check Point's endpoint clients. We were also able to incorporate the creation and use of locally defined users, groups, and access roles based on those in our policies. Additionally, we addressed changes to some of the portals and associated changes to gateway cluster certificates, allowing interaction from either users in remote sites or remote access clients.

In the next chapter, we'll address logging into a single security domain and go further into SmartEvent's views, reporting, and policy capabilities.

13
Introduction to Logging and SmartEvent

As we've been working with logs for a while, now is a good time to learn a bit more about how logging works in Check Point, along with how to use different configuration options to better address your infrastructure requirements.

In addition, we will be introduced to SmartEvent, a feature that simplifies the work of Check Point administrators by providing enhanced views, reporting capabilities, and automated responses to events.

In this chapter, we are going to cover the following main topics:

- Logging into a single security domain
- Introduction to SmartEvent

Logging into a single security domain

Logging is available on all Check Point components – that is, gateways, management servers, dedicated log servers, and SmartEvent servers.

Security logs created by gateways are sent to either management servers (if they are acting as log servers), dedicated log servers, or both if so configured. Additionally, logs can be stored locally and forwarded to the management/log servers on schedule.

Audit logs are created by management servers and are stored locally. They can be forwarded to the designated log servers on schedule, too.

Logs are indexed by log servers and are accessible via SmartConsole, SmartView (browser-based access), an API, or in a raw form via the CLI using either the `fw log` command or the `CPLogFilePrint` command in Expert mode. The `CPLogFilePrint` command, although unwieldy, returns more information. It is officially unsupported beyond logging troubleshooting but might come in handy. See the *sk153972 CPLogFilePrint utility* section for references. Use this command without any parameters to see the extended list of available options.

SmartEvent correlation units perform automated analysis of the logs residing on log servers and forward events to the SmartEvent server. The events are then accessible in either actionable views or reports in the SmartConsole or SmartView web application. Events can also be used as triggers for SmartEvent Automated Reactions defined in its policy. SmartEvent stores events in the Events database.

Configuring logging on gateways or clusters

Typically, we would be sending logs to either a management server(s) or a dedicated log server(s). That being said, gateways and clusters can and do log locally under the following circumstances:

- They are configured to do that (for clusters, this option is shown when only a single management server is available).
- When connectivity with the designated management server(s) or log server(s) has been lost.
- When the designated target management server(s) or log server(s) is under a heavy load.

Let's take a look at the various logging configurations, storage, and other options in the following subsections.

Logs configuration options for gateways with a single management server

As defined in the properties of our cluster's [1] main **Logs** configuration section [2], our management server is specified and selected by default [3]. The **Save logs locally, on this machine (CPCXL)** option is available [4]:

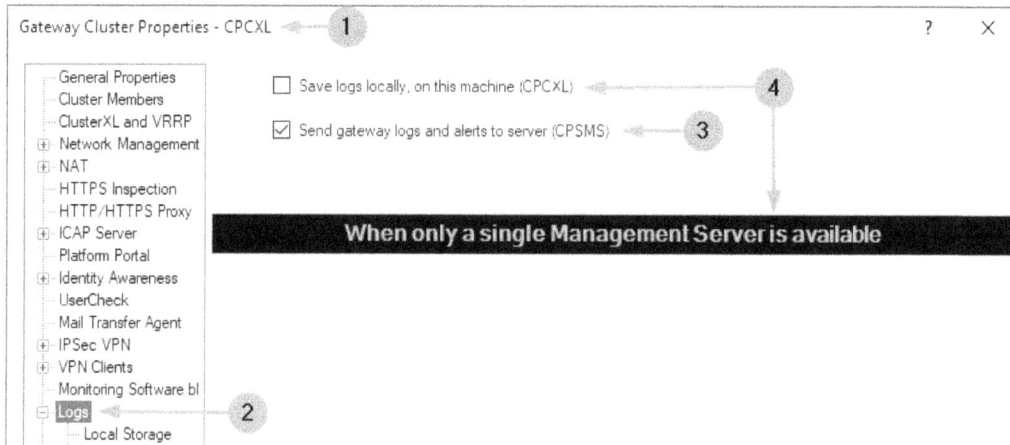

Figure 13.1 – Gateway or cluster logging with a single management server

The **Save logs locally, on this machine** option will *rotate* the log files at midnight by closing the currently active instance of a `fw.log` log file and naming it using the `YYYY-MM-DD_HHMMSS.log` format. The **Additional Logging** options, which we will discuss later in this chapter, might result in multiple files created daily if preset file size limits are met. The **Local Storage** configuration section is where the options for disk utilization by logs are defined.

Local Storage configuration options

Local Storage for gateways or clusters is where the **Disk Space Management** options are defined, such as alerts when running below specified values, and the watermarks for deletion of the older local log files. Additionally, we can also specify a script to execute on the gateway before deleting old files.

Note that, in clusters, local traffic logs are created only on the active cluster members. If you experience a massive outage that might have resulted in a cluster failover and are looking for local logs on the gateways, check all of the cluster members.

Additional Logging configuration options

If we expand the **Logs** option in the navigation tree of the gateway properties, additional options become visible: **Local Storage** and **Additional Logging**.

Additional Logging [1] presents more interesting options for configuration. In **Log Forwarding Settings** [A], we can specify a designated log server to send locally accumulated log files to [2] and **Log forwarding schedule** [3]. There are a number of predefined schedules available, but we might also use the **Manage... | New | Scheduled Event** [4] sequence to create additional schedules and then choose them from the drop-down list. Logs forwarded to a management server or a log server are prepended with the hostname to the file like this: CPCM1__2022-05-14_000000.log.

The **Log Files** section [B] allows us to limit the size of log files [5] and automatically rotate the log files at defined intervals [6]. If you have log forwarding configured, I suggest not doing this [6] on gateways, as it will cause the creation and subsequent forwarding of empty log files if there are no daily or multi-day outages. Additionally, even when you define a periodic local log creation schedule, logs are automatically rotated at midnight.

The **Advanced Settings** [C] section contains circuit-breakers to either stop logging [7] and/or reject all connections [8] if the disk space dips below a certain level. I suspect the **Update Account Log every** frequency has been misnamed and is supposed to refer to **Accounting Log** [9]. Options [10] and [11] are available when corresponding functions are enabled and configured. The **Include TCP state information** option [12] is interesting, but unless you have an explicit need (troubleshooting), leave it at **Never** or your logging volume will go through the roof:

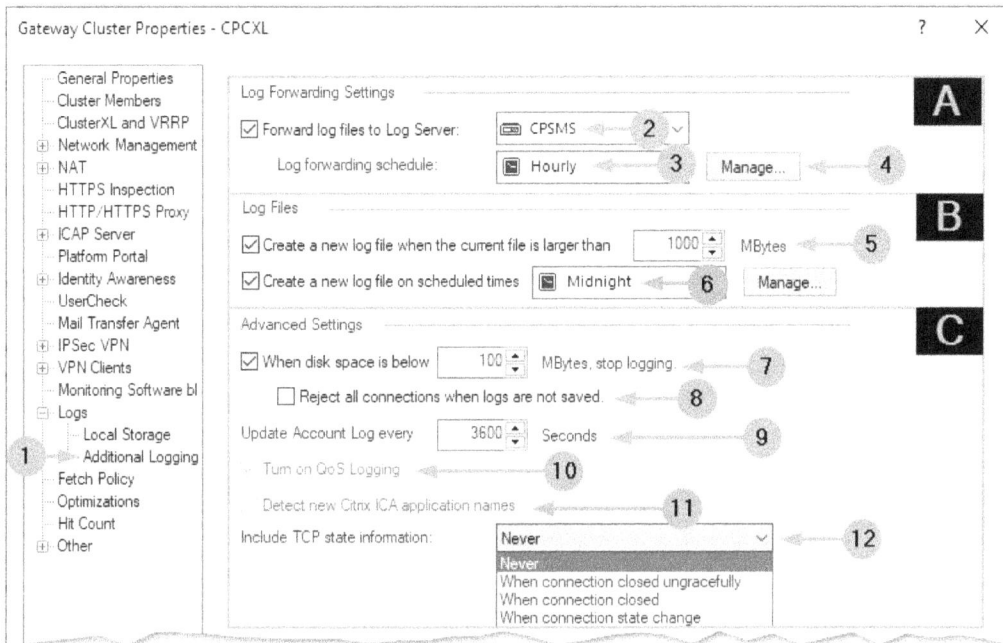

Figure 13.2 – Additional logging configuration options for the gateways

Note that if you have configured the gateways or clusters for local logging, when a log forwarding event is triggered, by default, all the local logs are shipped to the target server and are deleted from the gateways. If you are using **Log locally** as the only means of log redundancy, use the directions found in *sk106039, How to forward log file from Security Management Server to a Log Server without deleting the local log file.*

Your **Log forwarding schedule** section should reflect the utilization of your network and appliances and take the size of the logs under consideration.

If the log server configured in the properties of the gateways becomes unavailable, either due to the loss of connectivity, a heavy load, or scheduled or unexpected downtime, local logging kicks in. Once that server becomes available, you can either wait for **Log forwarding schedule** to deliver the logs to the server or use the `fw fetchlogs <IP of the Gateway>` command executed on that management server or a log server to get the logs immediately.

Once logging resumes, we will see the following relevant **System Monitor** logs:

1. This indicates that logging has resumed, along with the number of logs stuck on the gateway. Note that it claims the reason is a high load, even if it was a loss of connectivity, so it cannot be used as a reliable cause indicator.

2. This indicates that the gateway started logging (I am only showing a **Sys Message** clip from that log in the orange frame):

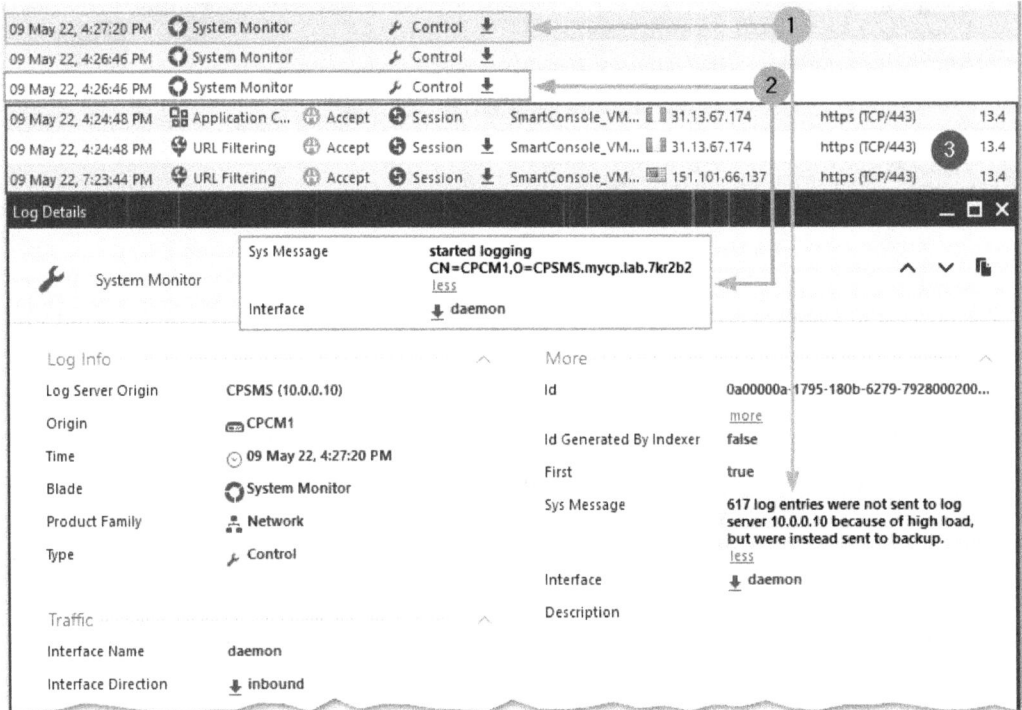

Figure 13.3 – Logs indicating logging resumption and the number of logs on the gateways

In the following screenshot, you can see the process of getting the logs from the gateway using `fw fetchlog`. The top section [1] shows the gateway/cluster member containing a number of local logs. We can see an abbreviated log list using the `fw lslogs` command. The lower section [2] shows the execution and output of the `fw fetchlog` command. [3] shows that the local logs, except current active `fw.log` file, are no longer present on the gateway after `fetchlog` was completed:

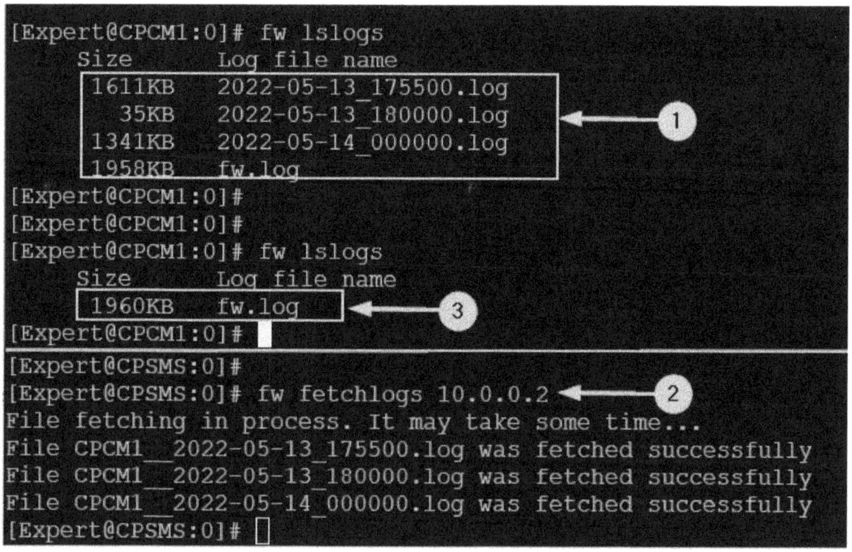

```
[Expert@CPCM1:0]# fw lslogs
     Size         Log file name
     1611KB       2022-05-13_175500.log
       35KB       2022-05-13_180000.log
     1341KB       2022-05-14_000000.log
     1958KB       fw.log
[Expert@CPCM1:0]#
[Expert@CPCM1:0]#
[Expert@CPCM1:0]# fw lslogs
     Size         Log file name
     1960KB       fw.log
[Expert@CPCM1:0]#
[Expert@CPSMS:0]#
[Expert@CPSMS:0]# fw fetchlogs 10.0.0.2
File fetching in process. It may take some time...
File CPCM1__2022-05-13_175500.log was fetched successfully
File CPCM1__2022-05-13_180000.log was fetched successfully
File CPCM1__2022-05-14_000000.log was fetched successfully
[Expert@CPSMS:0]#
```

Figure 13.4 – Fetching local logs manually

Note the size of the current log file, `fw.log`, before and after `fw fetchlogs` is executed. If you see it growing, this is an indication that until it is rotated, you might still be missing logs on your management server for the time interval between the last log's rotation and the management server coming back online. To get the logs from an active log file, use the `fw logswitch` command on the gateways to close and rename the current active log file and to create a new active log file before running `fw fetchlogs`.

Security management servers or log servers

Logs and nested configuration options for all management and log servers are identical and are covered next.

Logs configuration options

In the **CPSMS** [1] | **Logs** configuration menu [2], we have the option of disabling log indexing [3] (it is enabled by default for all appliances running the **Logging & Status** blade with more than four CPU cores). Additionally, we see all the gateways and clusters configured to send logs to it [4]:

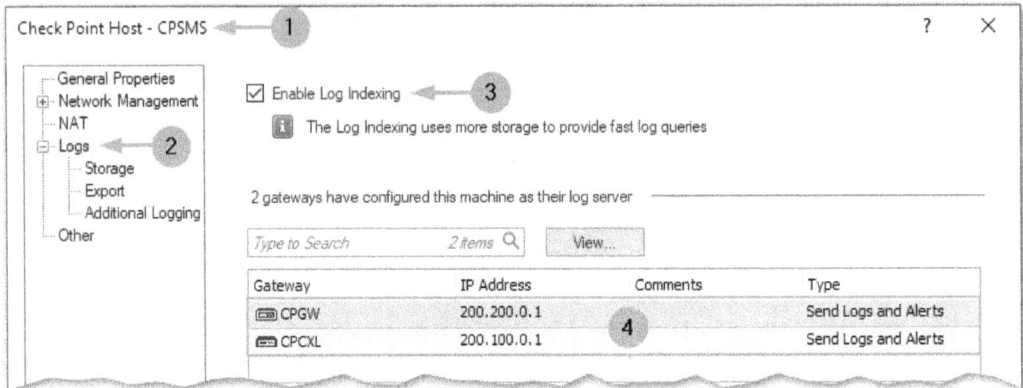

Figure 13.5 – The management server logs configuration option

The **Storage** configuration [1] contains two sections: **Disk Space Management** [2], which is identical to that of the gateways, and **Daily Logs Retention Configuration** [3]. The last one will be present on all log servers and is the way we control the life cycles of the logs and log indexes:

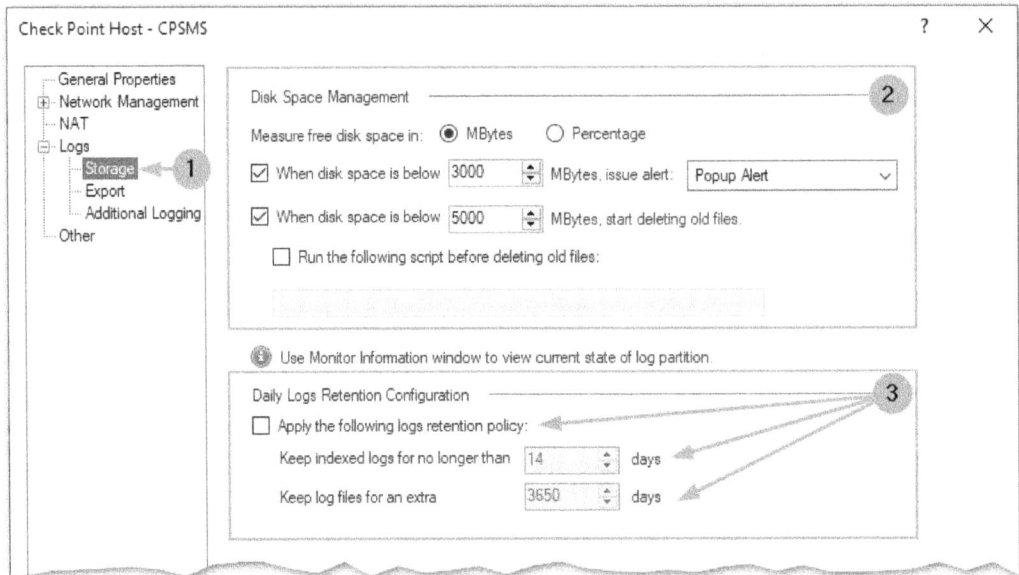

Figure 13.6 – The management or log servers' storage configuration options

By default, indexed log files are only retained for 14 days, even if the **Apply the following logs retention policy** checkbox is unchecked. Do not forget to set the retention of indexed logs and the log files to the values that suit your company's requirements while keeping available space in mind.

> **Note**
> Increasing this value will not re-index logs older than the previous limit retroactively. The change will be applied to the newer logs moving forward.

Unindexed logs are viewable and searchable by a limited number of fields. They can be manually re-indexed using instructions provided in *sk164553 How to re-index a log file.*

There is a difference in search results for indexed and unindexed log files. Let's look at the indexed log search results [1] and compare them to the same search in the unindexed log file [2] and the actual content of the file filtered by `blade` [3]:

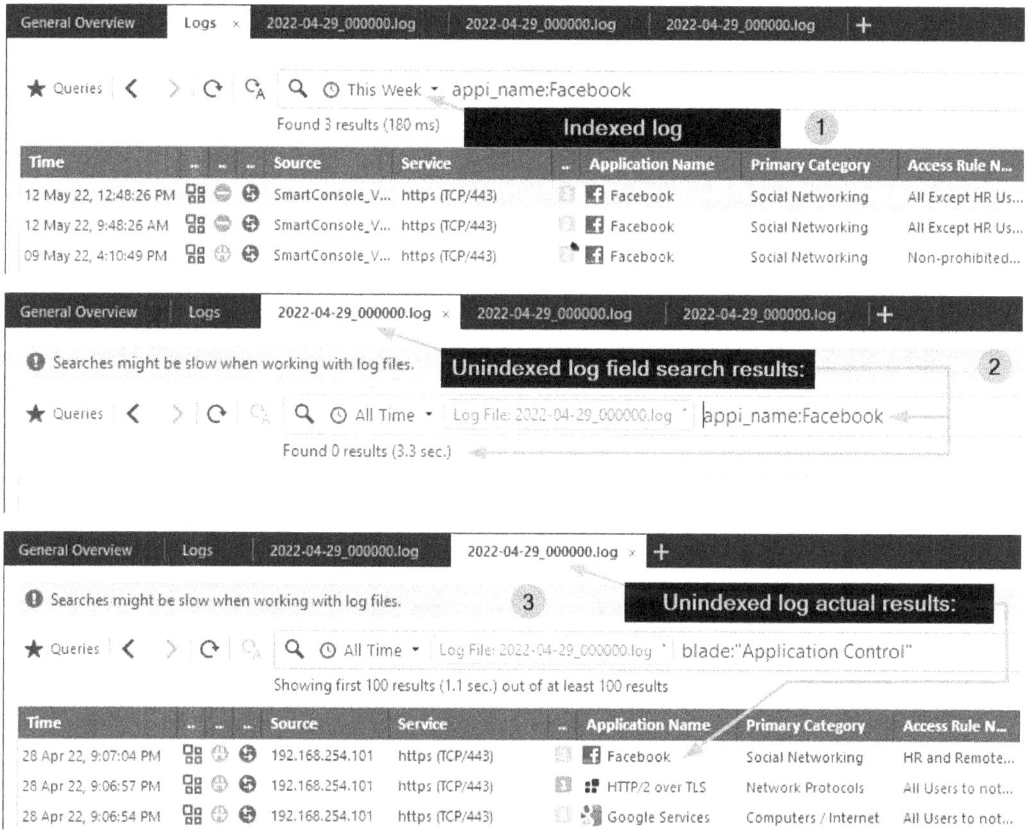

Figure 13.7 – The indexed and unindexed log searches

I am uncertain whether this is a bug or whether we are limited to the searches of unindexed logs for specific fields. This issue has been reported to Check Point.

Export configuration option

The **Export** option allows us to define an external SIEM or a Syslog server(s) to propagate our logs to.

When selected [1], we can click on + [2], the new icon [3], and **Log Exporter/SIEM...** [4] to configure the target:

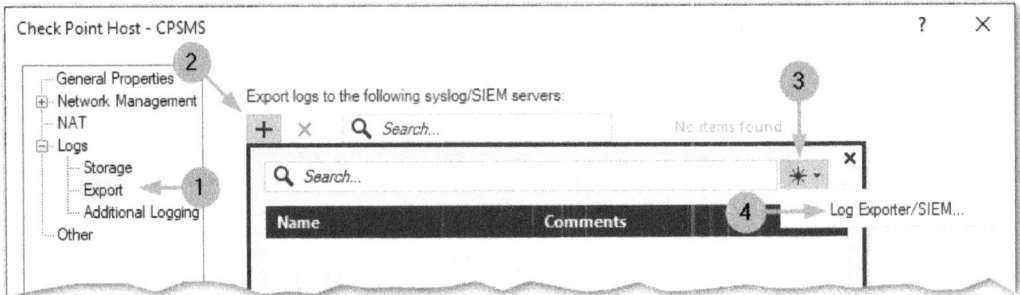

Figure 13.8 – Adding the log exporter/SIEM

Name the new **Log Exporter/SIEM** instance [1], select **General** [2], click on **Enabled** [3], and then specify its IP address [4], port, and protocol [5]:

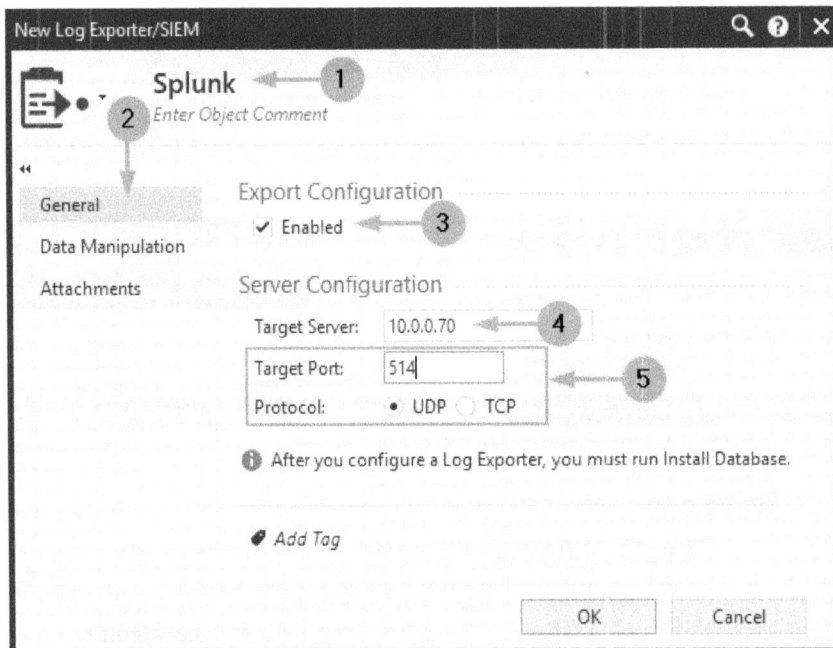

Figure 13.9 – The new Log Exporter/SIEM's General properties

Click on **Data Manipulation** [1] and, in the **Format Configuration** section, using the drop-down menu [2], choose one of the available **Format** options [3]:

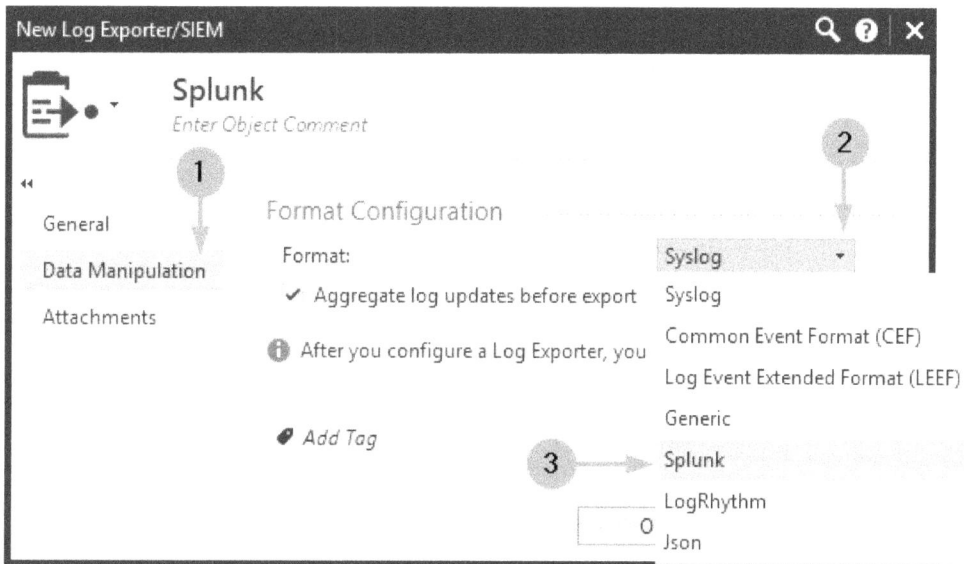

Figure 13.10 – The new Log Exporter/SIEM Data Manipulation format

Select **Attachments** [1] and choose the desired **Attachments Configuration** setting [2]:

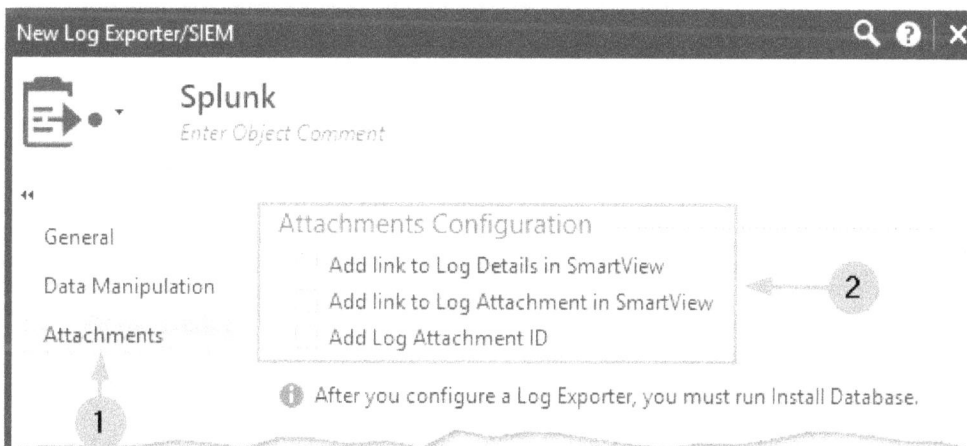

Figure 13.11 – New Log Exporter/SIEM attachments configuration

Click on **OK** (not shown) to complete the process.

If you need more information on **Log Exporter**, see *sk122323 Log Exporter - Check Point Log Export*.

Additional Logging configuration option on management servers

Additional Logging on management or log servers [1] is similar to the same options we saw on the gateways earlier, but with the addition of **Accept Syslog messages** [2] and **SmartEvent Intro Correlation Unit** [4]:

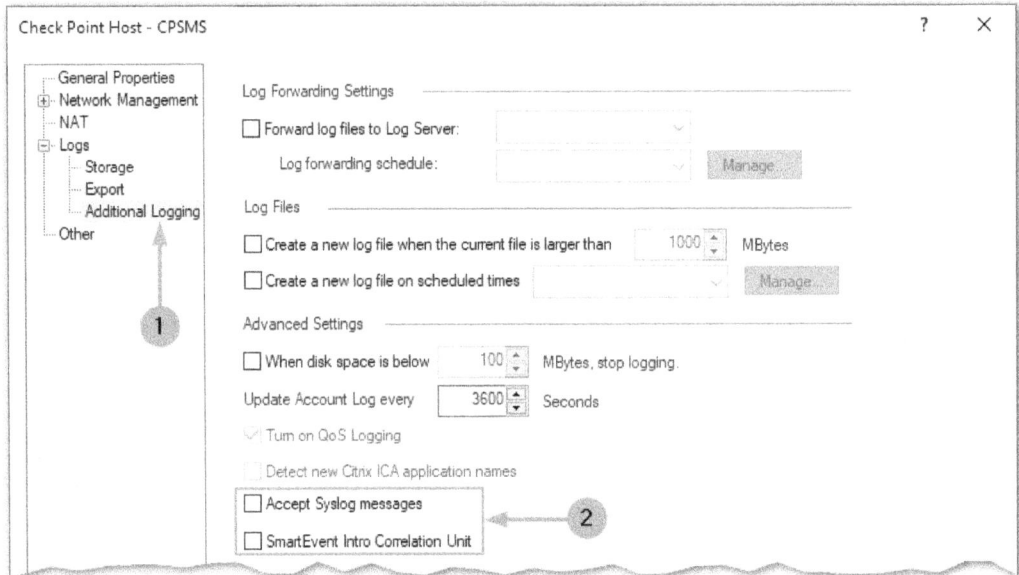

Figure 13.12 – The management or log servers' Additional Logging configuration options

Note
For some reason, the **SmartEvent Intro Correlation Unit** box is checked by default in every secondary management or log server object you create. I suggest unchecking it as soon as you have named the object and specified its IP address.

While it is possible to use Check Point log servers as Syslog servers (there is even a **Log Parsing Editor** option for creating custom parsers; please refer to *sk55020 How to generate a log parser for a third-party Syslog server*), in my opinion, it is more trouble than it's worth. There are better dedicated Syslog servers with parser libraries that could be used for this purpose.

The **SmartEvent Intro Correlation Unit** checkbox is likely an artifact that should be removed by developers, as it was discontinued in version R80, and I do not see any references to it in later versions.

Logging with management high availability or log servers

If you are expanding your management infrastructure by either adding a secondary management server and/or additional dedicated log servers, the **Logs** options in the properties of your gateways and clusters will change.

The **Logs** configuration section [1] will look quite different. We will now see a **Log Distribution** section [2], and the **Log Servers** section contains **Primary log servers** [3] and **Backup log servers** [4]. Our original management server (**CPSMS**) will be listed as the only primary log server by default [5]. Additional log servers (if available and configured) can be added to each section. Once the primary and backup log servers are defined, we can choose the **Log Distribution mechanism** option for either availability [6] or performance [7]:

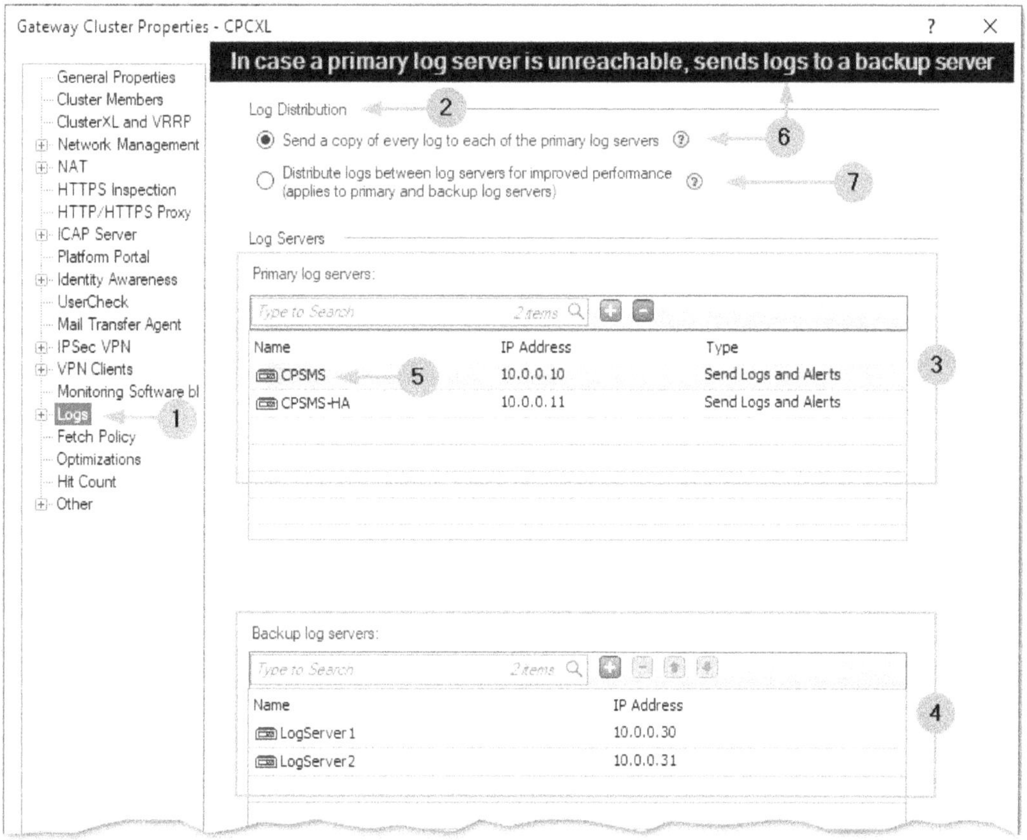

Figure 13.13 – Gateway or cluster logging with multiple log servers

> **Note**
>
> At the time of writing, there is a bug that shows a **Log locally** checkbox for the **Logs** properties of single gateways, but not clusters when more than a single management/log server is available.

If we are choosing the availability option by specifying multiple primary servers, we will see duplicate log entries in the **Logs** view in SmartConsole unless we uncheck one of the redundant log servers in the **Logs** view:

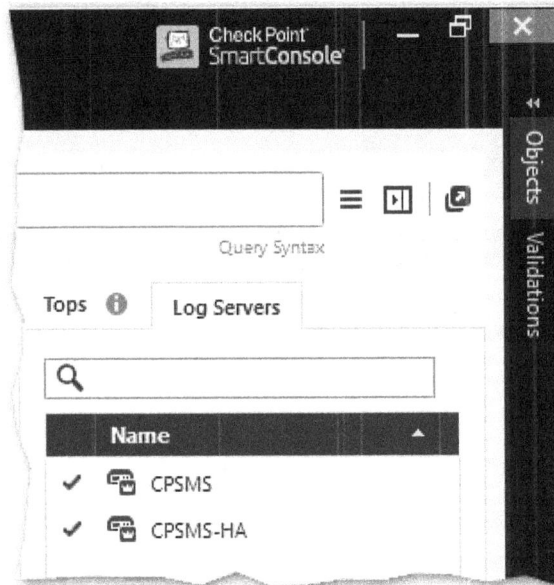

Figure 13.14 – Selecting the Log Servers sources

If you have log forwarding configured on your gateways, only one of the primary log servers will receive locally accumulated logs in the case of the loss of connectivity or scheduled maintenance. To ensure that both primary servers have identical logs, use the procedure described in *sk106039 How to forward log file from Security Management Server to a Log Server without deleting the local log file* on your gateways. Then, after connectivity has been restored, the current active log is rotated, and the logs are forwarded, perform `fw fetchlogs` from the second primary log server. This will sync the logs on both servers and remove the local logs from the gateways.

Strategies for the effective use of management high availability and log servers

Consider how to best utilize the available management and log servers. For instance, if we have two management servers, primary and secondary, we could elect to log to both simultaneously if latency and bandwidth are acceptable, or we could opt to log from geo-collocated gateways to the nearest server.

If we are logging into both management servers, we can use a secondary, standby management server as a log exporter, relieving the load on the primary.

If our management servers are physical Check Point appliances and we are seeing spikes in utilization that impact our management activities, additional dedicated log servers might be needed. In this case, configure the gateways to send logs to log the new dedicated log servers, specified as primary, and use old management servers for administration and as secondary log servers.

Check Point log servers do not have a built-in log archival mechanism capable of offloading older logs to external storage, but there are plenty of scripts written by community members that could be adopted for this purpose.

Additional considerations should be given to the options of using hardware or virtual appliances. If you are mandated to keep Check Point separate from the rest of the infrastructure, then you are limited to the hardware appliances and should endeavor to size those appropriately.

If you do not have these restrictions, you might be better off using open server licenses and having either or both management and log servers virtualized.

In this case, you can utilize hypervisor-specific backups and replication solutions for redundancy and have more flexibility in allocating additional storage, memory, and compute resources to your servers.

Smart-1 Cloud

Another viable solution for high-availability logging and management is the use of the Smart-1 Cloud product. If your company has embraced the cloud-first or hybrid cloud architecture, has less than 100 gateways, and is comfortable with an SLA of 99.9%, run a total cost of ownership analysis. If the numbers are in its favor, you could save a lot of time and effort by using cloud-based management. It is exceptionally easy to implement and is using the same SmartConsole delivered as a web application. A short introductory video, datasheet, and solution brief are available at `https://www.checkpoint.com/quantum/unified-cyber-security-platform/smart-1-cloud/`.

In a nutshell, you do not have to worry about maintaining the underlying management infrastructure beyond licensing for the necessary capacity and features.

This pretty much covers the logging in a single security domain subject. As you might have noticed, there is a multitude of ways to implement and scale logging. When designing your own implementation architecture, it is a good idea to model it in the lab to work out the flows, retention, and accessibility of the logs and to document the maintenance and troubleshooting procedures. We have seen that Check Point can produce copious amounts of detailed logs. Now, let's take a look at another tool, designed to generate custom views to alert us of actual security events, define automated actions for them, and produce highly customizable reports.

Introduction to SmartEvent

We were able to see how extensive Check Point logging capabilities are and how we can filter the logs to focus on traffic flows, connections, and actions. That said, there are a lot of logs being generated even by modest infrastructures. While a periodic manual log review is still a good practice for abnormal traffic detection, it is not a realistic approach for real-time reactions to security events by administrators.

If your company has implemented a SIEM and SOAR and have those configured to consume Check Point logs, generate actionable intelligence, and trigger automated responses, that is great. However, the likelihood of that happening for the majority of implementations is not very high.

In this case, the solution you are looking for is Check Point's own SmartEvent server. Essentially, it is a native SIEM and SOAR product that is primarily focused on working with Check Point gateways and endpoint clients that is also capable of integrating with a range of additional security solutions.

Even if we limit its application for integration with Check Point gateways and log servers, the benefits are tremendous.

SmartEvent provides real-time, centralized correlation of log data into events, assigning them different priorities as well as performing statistical correlation. Events are then presented to administrators in a number of different, easily customizable views that allow for rapid drill-downs into specific underlying logs. Capable of processing millions of logs per day, it is aggregating, normalizing, and correlating those to a manageable number of items of interest. Additionally, it is a reporting tool that can generate them on-demand and on schedule.

SmartEvent could coexist with a single management and log server or scale out to include dedicated event correlation units, with each connected to a number of log servers in a very large environment along with a dedicated SmartEvent server for aggregation, presentation, reporting, and automated actions based on combined events.

Initial configuration

Let's go over the installation and initial configuration of SmartEvent in our environment:

1. Open the properties of the **CPSMS** server, and click on the **SmartEvent Server** option [1]. You will notice that the **SmartEvent Correlation Unit** box has been automatically selected. If it had been a separate **SmartEvent Server** option, the **Logging & Status** checkbox would be checked for us, too. Click on **OK** (not shown):

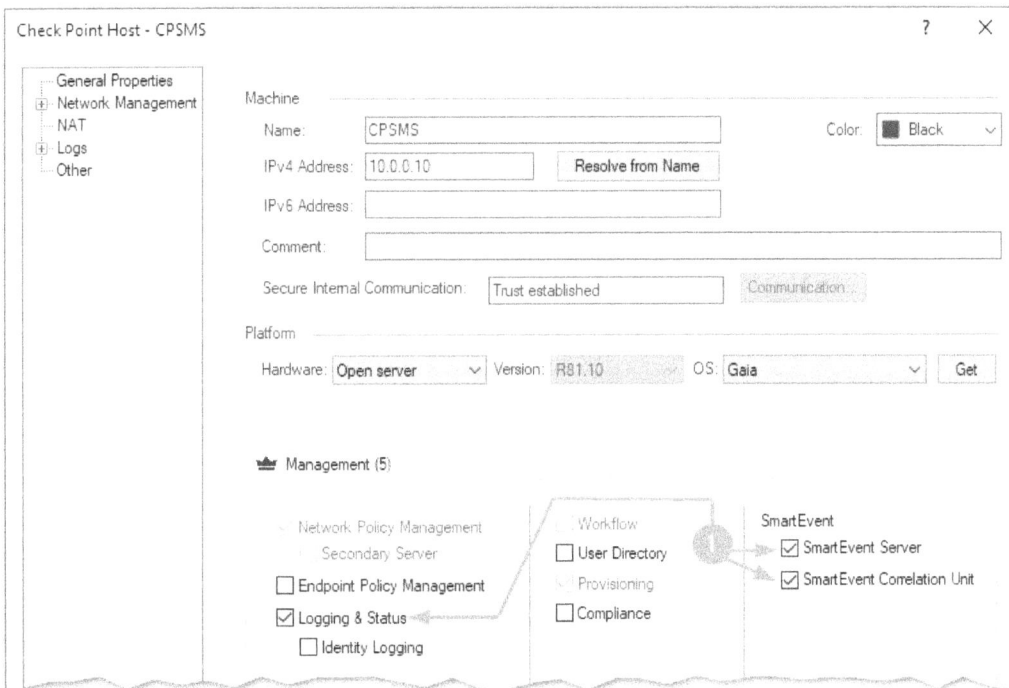

Figure 13.15 – Enabling SmartEvent

2. In the **Main Menu** window, click on **Install Database**, and in the **Install Database** confirmation window, click on **Install**.

3. Wait to see the installation status: **Install Database on CPSMS Succeeded**.

4. In **SmartConsole**, click on **LOGS & MONITOR** [1]. Once you are in it, open a **New Tab** window by clicking on the + sign on the right-hand side of the last open tab [2]. Under the **External Apps** section, click on **SmartEvent Settings & Policy** [3]:

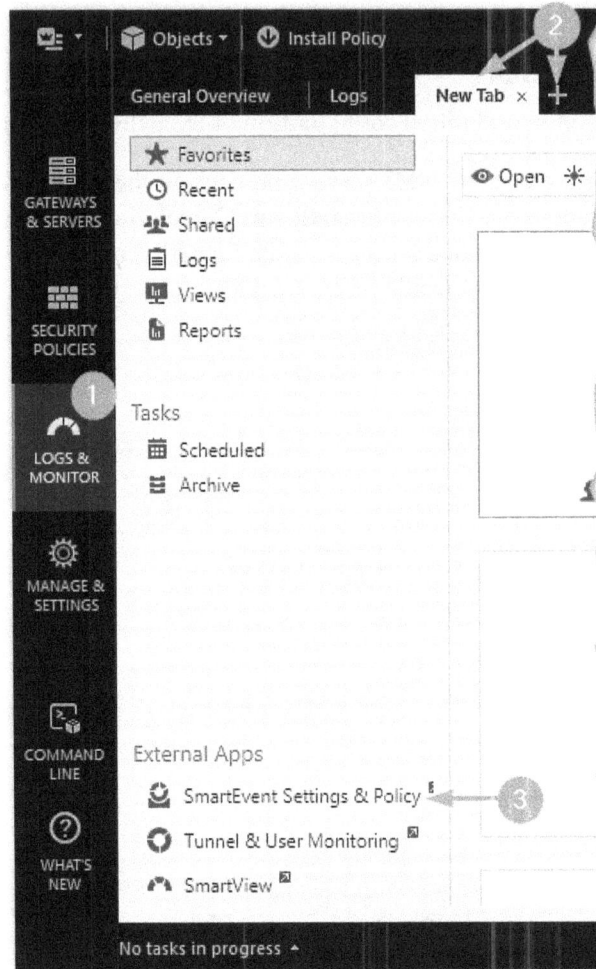

Figure 13.16 – Accessing the SmartEvent application

5. This will open a dedicated legacy application for the **SmartEvent** configuration [1]. Under **General Settings** [2], click on **Correlation Units** [3]. Unless **CPSMS** is already listed in **Correlation Units**, click on **Add...** [4], select **CPSMS** (*the only available option in the lab at this point*), and click on **OK** to see it in the **Correlation Unit** column [5]. Items from [6] to [9] are there to illustrate how additional log servers, if present, could be added to the correlation unit. When done, click on the install SmartEvent policy icon [10]:

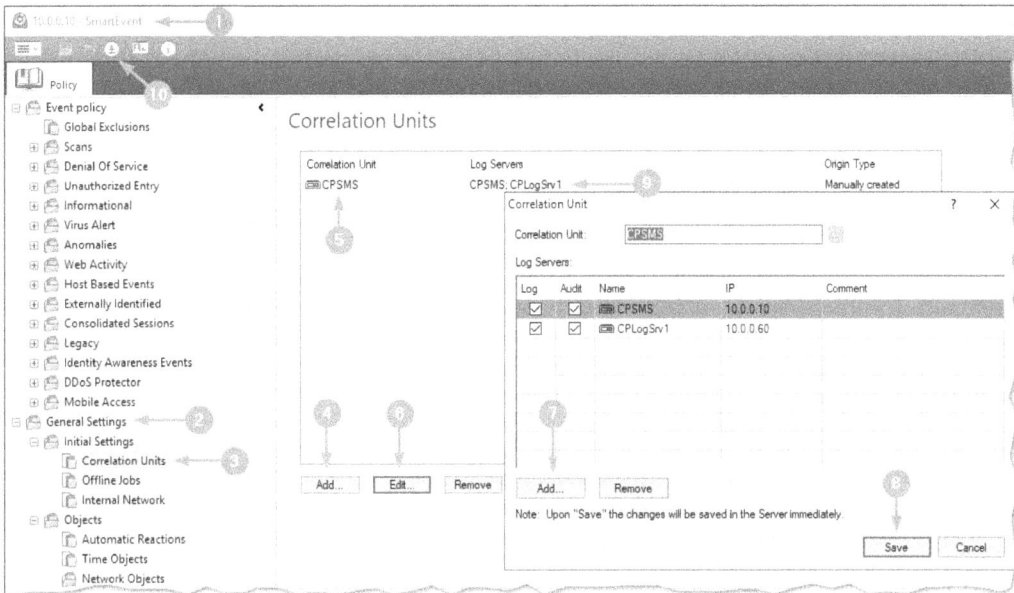

Figure 13.17 – Adding correlation units and log servers

6. While still in **Initial Settings** [1], click on **Internal Network** [2]. In the **Not in Internal Network** list [3], select all the networks in your environment and click on **Add** [4] to move those to the **In Internal Network** list [5]. Repeat the same process with the gateways, management servers, and hosts that have static NAT configured to move those to In Internal **Network**. Once done, click on the install Event Policy icon [6]:

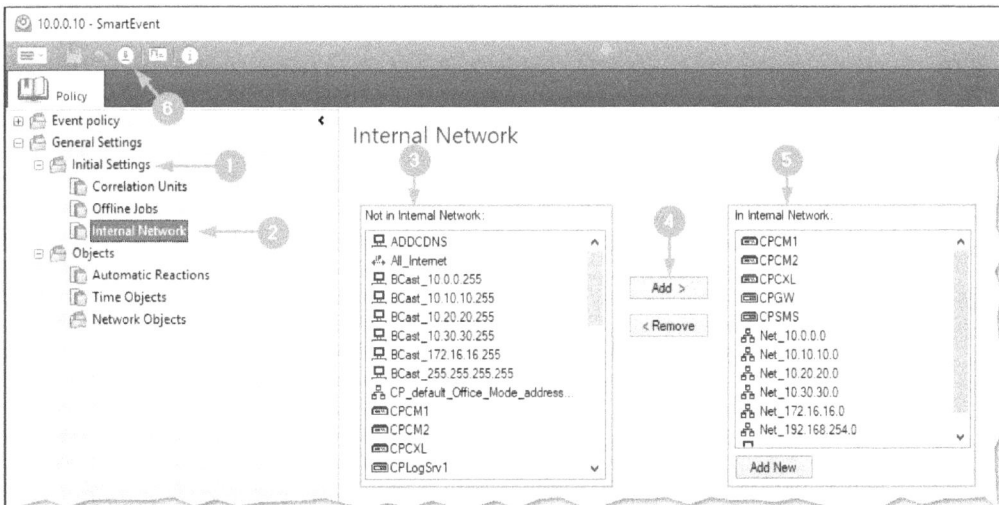

Figure 13.18 – Defining internal networks and installing an event policy

7. When prompted, confirm that you want to proceed with installation, and observe the **Install policy report** pop-up window to see the progress and confirmation of successful installation. Click on **Close** when completed.

8. Close and re-open SmartConsole, and log in again.

The three pillars of SmartEvent are policies (comprised of events and reactions), views, and reports. There is a predefined standard policy that we installed in the previous section, and we will return to it later to use a few of the event options to demonstrate its capabilities. For the remaining options, we are using a SmartConsole or SmartView web application.

Views

Let's head to **LOGS & MONITOR** [1], **New Tab** [2], where additional items, such as **Shared**, **Views**, **Reports**, **Scheduled**, and **Archive** [3], have become accessible. When we click on **Views** [4], we see a long list of the predefined views (or dashboards) that either focus on a specific functionality (such as access control or applications and sites) or broader subjects, such as **General Overview** or **MITRE ATT&CK** for threat prevention:

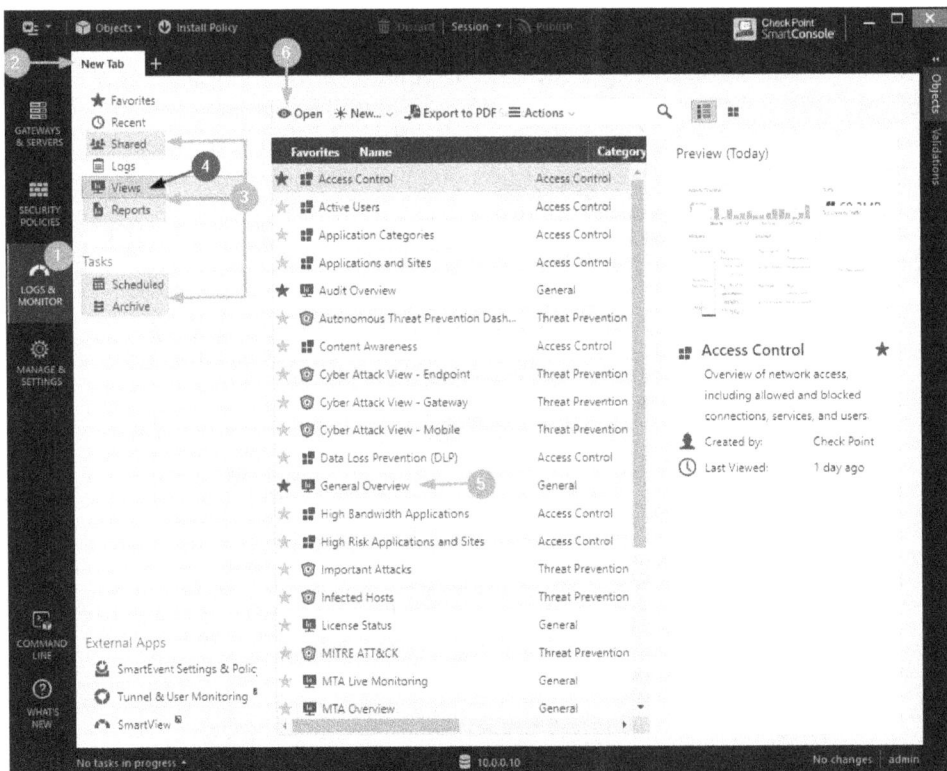

Figure 13.19 – Accessing the expanded capabilities in LOGS & MONITOR

Select **General Overview** [5], and click on **Open** [6].

You are presented with the dashboard containing multiple sections referred to as widgets. In the bottom-left corner, there is one called **Timeline**.

Now, on the same SmartConsole VM, open Chrome and access the following:

- `https://10.20.20.1/connect/PortalMain` using `mycp.lab\hruser` and `CPL@b8110`
- `https://edition.cnn.com/`
- `https://www.facebook.com/`
- `https://www.google.com/`

Change the VM to **RightHost**, start your VPN client, and connect to the **LeftSide** site using the `RA_user1` and `CPL@b8110` credentials. Once connected, open your browser, and try accessing the following:

- `https://google.com`
- `https://pinterest.com`
- `https://3proxy.com`
- `https://ninjaproxy.com`
- `https://mirc.com`

Return to the **SmartConsole** application and **General Overview** [1]. Change the time range in the search query to **Last Hour** [2] and press *Enter*. Your **General Overview** dashboard will look similar to the following screenshot. Let's go over the remaining numbered labels next.

Statistics [3] gives us a high-level overview of the number of **Critical Attack Types** and **Infected Hosts (With bots)** [4].

> **Note**
> I would suggest changing this name to `Affected Hosts`, as it'll list all the hosts that have attempted to communicate with the bot, even if the attempt was unsuccessful and is not caused by persistent compromise.

Software Blades [5] will indicate the number of logs generated by each feature. For instance, if you did not have any VPN clients connecting, your **Mobile Access** blade will not be shown. **Timelines** [6] contains two horizontal sections, with only the bottom one, **Applications and URL Filtering (by Logs)**, being active [7].

The top section of the **Timelines** blade, along with **Attack Prevention by Policy** and **Critical Attacks Allowed by Policy** [8], is showing **No data found** due to us not running any of the **Threat Prevention** blades. The **Allowed High Risk Applications** widget [9] contains a single instance of **mIRC**:

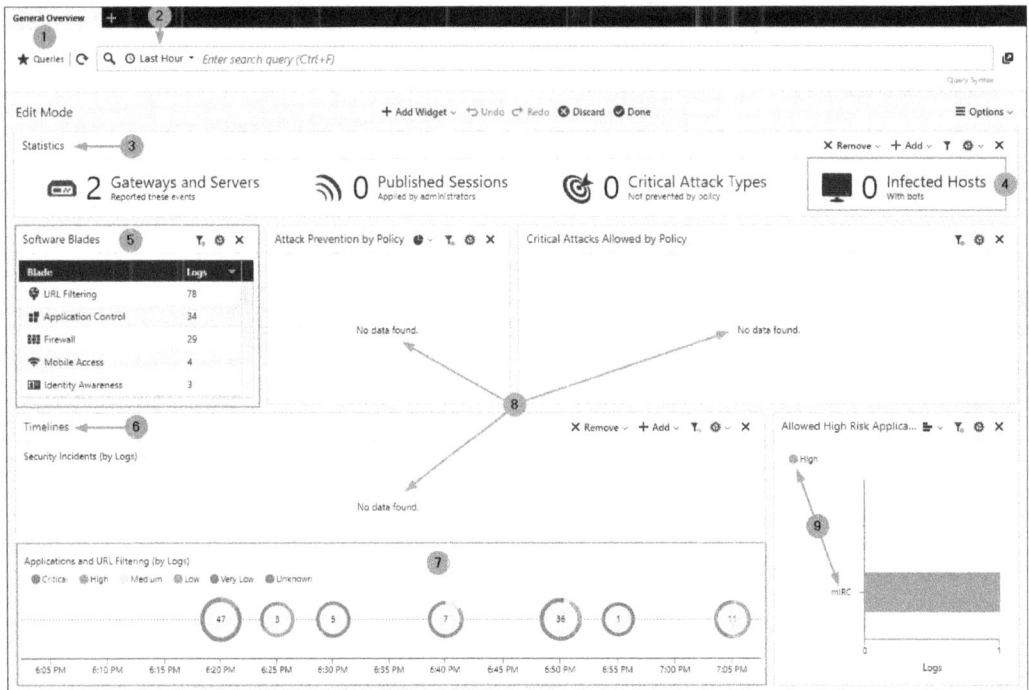

Figure 13.20 – The General Overview dashboard for SmartEvent

If we take a closer look at **Timelines | Applications and URL Filtering (by Logs)**, we can de-select individual categories by clicking on the individual **Risk** categories [1], leaving us with only two items that have a **Risk** rating of **Critical**:

Figure 13.21 – Narrowing down the scope of investigation in Timelines

Double-clicking on each opens the time-corresponding session logs. At the end of the prepopulated query, change the `app_risk:*` setting to the desired risk level, in this case, `risk:Critical`. The resultant filtered logs for `3proxy.com` [1] and `Ninjaproxy.com` [2], both known anonymizers, are shown:

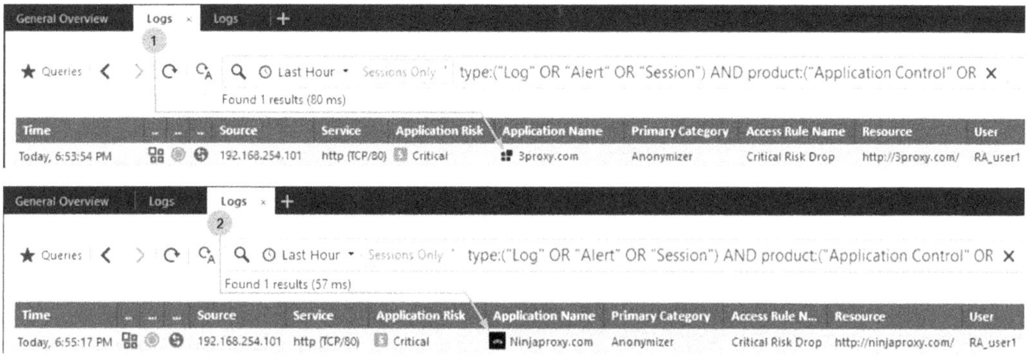

Figure 13.22 – The rapid filtering of logs or events by criticality

You could justly remark that we are just looking at our logs, not events. Well, the interaction principles remain the same, once events are present. With the **Threat Prevention** blades enabled, it will happen automatically for either predefined events or those that we configure manually.

Do not do this in the lab, as we have limited storage space in our virtual appliances; just read on as I illustrate how it is done. To simulate an event, we'll have to create a security omission in our access control policy, so let's add a rule at the bottom of the **Privileged Access** section, allowing **SmartConsole_VM** unrestricted access to **ADDCDNS**:

No.	Name	Source	Destination	VPN	Services & Applications	Action	Track
17	Temp-SmartEvent PortScan test	SmartConsole_VM	ADDCDNS	* Any	* Any	Accept	Log

Figure 13.23 – A faulty rule with wide-open access

Open the **CPCXL** cluster object, and in **General Properties**, select **Custom Threat Prevention** [1]. Under **Threat Prevention**, check **IPS** (which stands for intrusion prevention system) [2]. Click on **OK**:

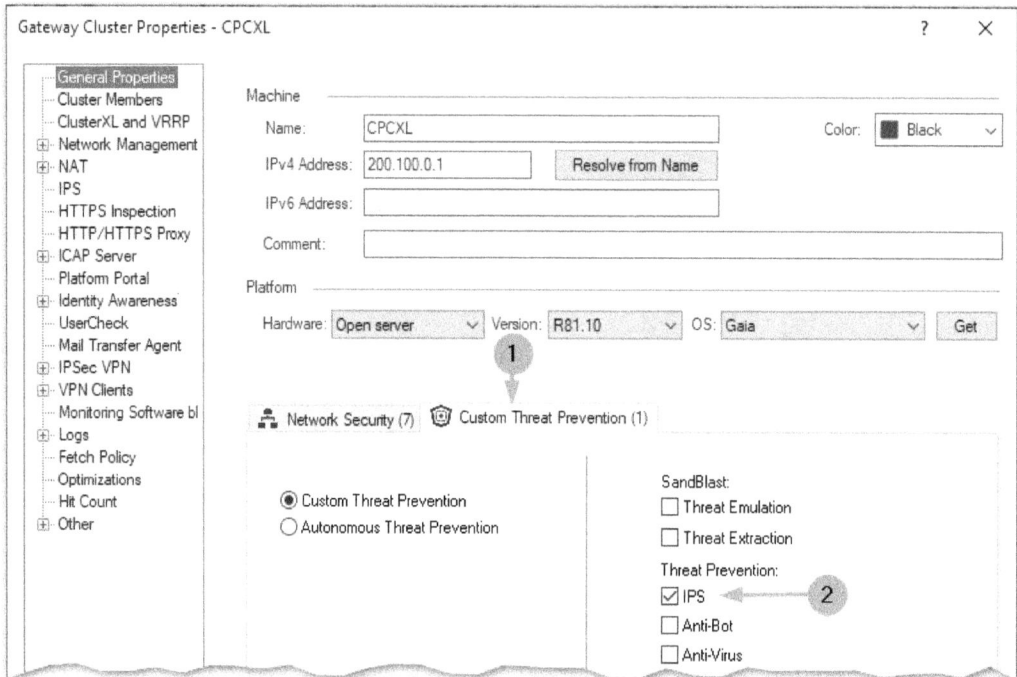

Figure 13.24 – Enabling the Custom Threat Prevention IPS setting

In **SECURITY POLICIES | LeftSide_RA | Access Control**, right-click on **Policy** and click on **Edit Policy**.

In the properties of the **LeftSide_RA** policy [1], check the **Threat Prevention** checkbox [2]:

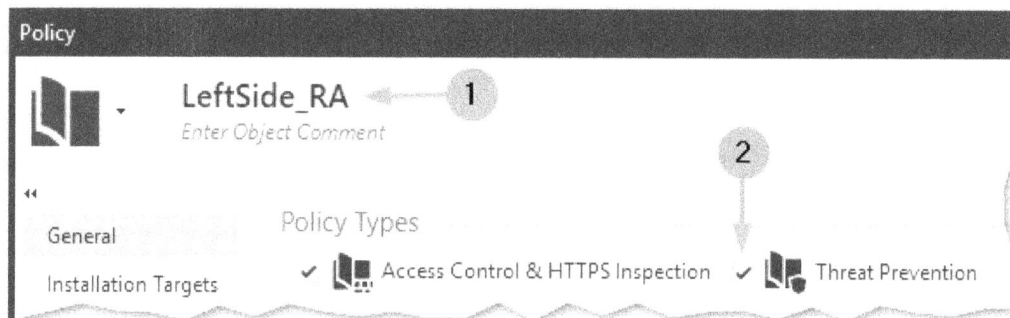

Figure 13.25 – Enabling Threat Prevention in the policy

Click on **OK**. Publish the changes, and install the policy.

Events

Return to the minimized **SmartEvent** application. In **Event policy** [1] | **Scans** [2], check **Port scan from internal network** [3]. In **Severity**, choose **Critical** [4]. Note the other available options [5], and install the policy [6]:

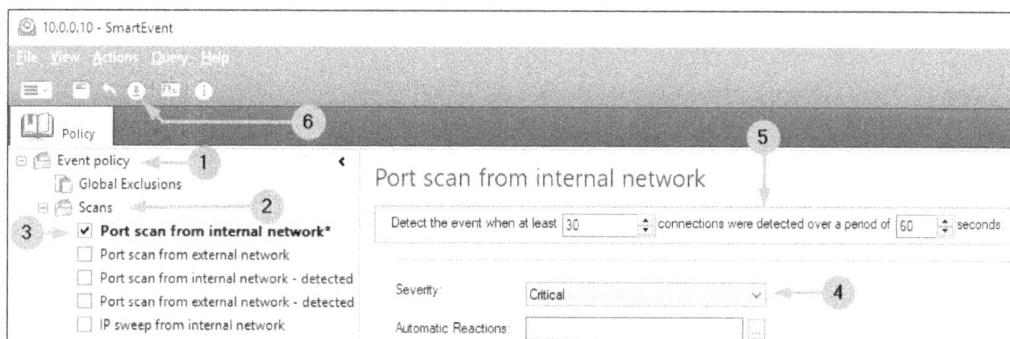

Figure 13.26 – Configuring an event and defining criticality

In **SmartConsole_VM**, install **Nmap** and execute **Quick scan** and the **Intense Scan** of the **ADDCDNS** host, 10.20.20.10, about 10 minutes apart.

Security incidents

Now, when we look at the **General Overview** dashboard, we can see that the **IPS** blade is working [1], **Attack Prevention by Policy** is happening [2], and in the **Timeline** blade, we can see **Security Incidents (by Logs)** numbered **5** for each of the port scans [3]:

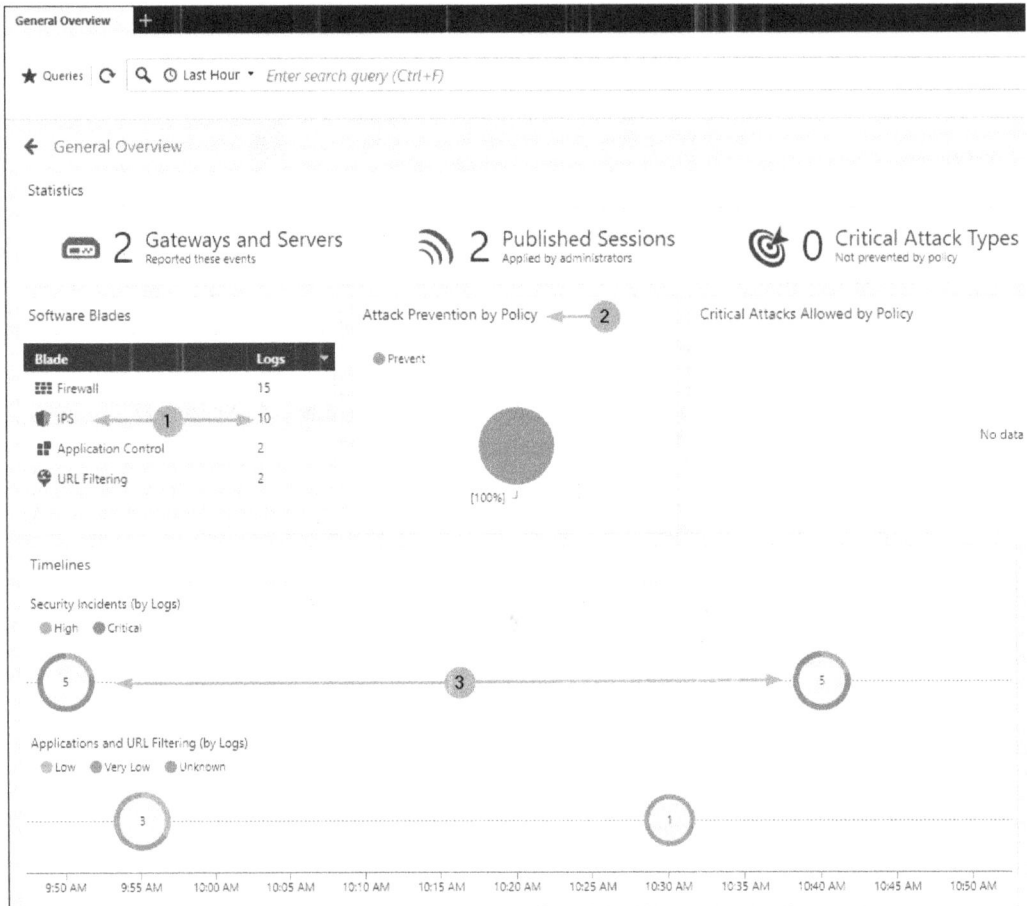

Figure 13.27 – Security incidents reported as events

Now, consider that the actual port scan is comprised of the hundreds or thousands of connections that are logged, but those logs are now distilled into a meaningful number of events.

Furthermore, the relative size of the circles over time can draw your attention to anomalous traffic volumes. If you change the time parameters in the search bar from **Last Hour** to **Last 7 Days**, for instance, you'll be able to spot the deviations from the trends.

> **Note**
> Multiple graph options are available to choose from with widget editing.

Using **Options** [1], the **views** are editable [2], allowing us to select **Hide Identities** (for applicable views). **Views** can be created for individual administrators or shared. The **views** and their current states are exportable [3] and are completely customizable:

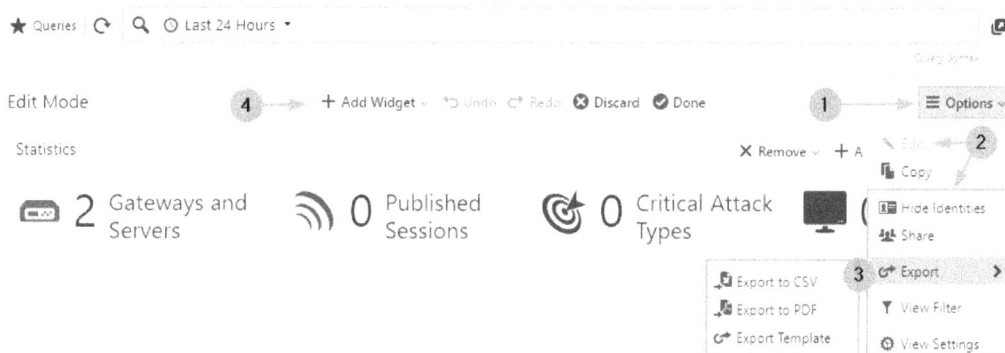

Figure 13.28 – The SmartEvent Views editing options

Depending on the screen space available, you can **Add Widget**(s) [4] that you are interested in, which makes it a great tool for security operations centers. It is especially convenient since you can use a browser-based **SmartView** application, accessible via `https://<IP_of_SmartEvent_Server>/smartview/`, set to auto-refresh, and run in full-screen mode (triggered by the *F11* key in most browsers).

Reports

Reports provide functionality similar to that of **views**, but they allow for automatic generation, archiving, and forwarding. Accessible via **LOGS & MONITOR | Reports**, they are a collection of **pages**, each comprised of **views containing widgets**. You can create custom reports by adding pages and organizing either existing predefined widgets on them or by creating custom widgets. An excellent demonstration of how this is done can be found at `https://sc1.checkpoint.com/documents/R81/ WebAdminGuides/EN/CP_R81_LoggingAndMonitoring_AdminGuide/ Topics-LMG/Making-Custom-Report-Video.htm`.

Once created, reports can be scheduled to run automatically and be delivered via email in either PDF or Excel formats. Once a report has been executed, it is also accessible via the **Archived** option in **LOGS & MONITOR | New Tab**.

An additional benefit of the reports is your ability, as an administrator, to filter them in real time using normal Check Point search queries in SmartConsole or a SmartView. See the following screenshot where a company-wide **Application and URL Filtering** report [1] is filtered by a single user [2] on-demand:

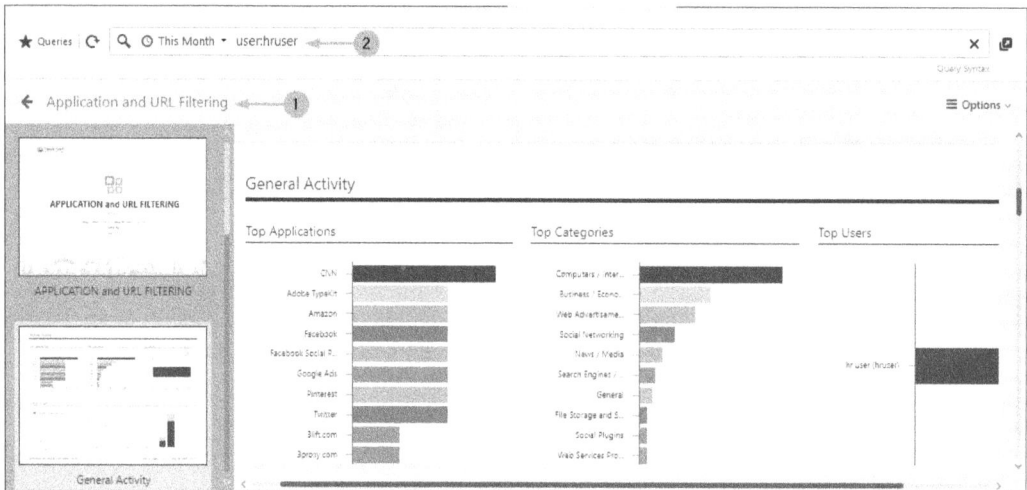

Figure 13.29 – The reports' filtering capabilities

Note that there are no report life cycle management options, and you'll have to either manually delete older reports or script their periodic purging.

Automatic reactions

Last but not least, I'd like to mention that SmartEvent should not only be used as a passive monitoring tool. For example, if we are to open our SmartEvent application and navigate to the same **Port scan from internal network** [1] window we were working with earlier, we could choose **automatic reactions** [2] for this event. The **Add new...** button [3] shows the range of available options, which vary from notifications to blocking the activity or the source or execution of external scripts [4]. Additional flexibility is provided by the options for creating exceptions for specific sources and destinations [5] and/or defining exceptions to the event definition [6]:

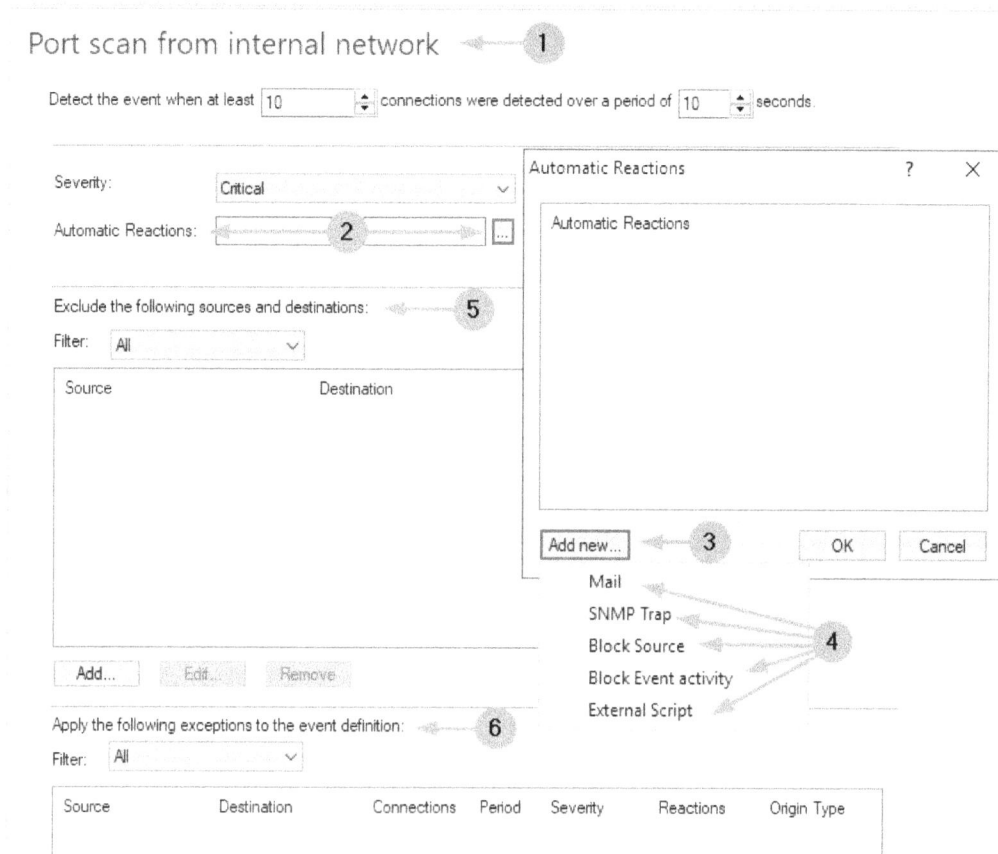

Figure 13.30 – SmartEvent automatic reactions

As you can see, there is a lot that could be accomplished with SmartEvent. I can only introduce you to it and make you aware of the possibilities, as well as vouch for much happier administrators, CISOs, HR, boards, and auditors that have this tool in their environment and have invested time and effort into tuning it.

Summary

In this chapter, we learned about Check Point logging, some of the ways it can be configured, and the different approaches to logging in distributed management architectures. Additionally, we were introduced to SmartEvent, Check Point's **SIEM**, and some of its monitoring and reporting capabilities alongside the possibilities of using automatic reactions.

In the next chapter, we are going to learn about Check Point clustering options and, specifically, focus on operating a ClusterXL High Availability cluster.

14
Working with ClusterXL High Availability

One of the first objects we created in our lab was a cluster consisting of two gateways. Generally, the term **cluster** is applied to a group of computers performing identical tasks. **High-availability** (**HA**) clusters are those that are designed to provide fault tolerance, maintaining uninterrupted functionality in the case of a failure of one or more of their members. Our cluster is running in HA mode, and it is time for us to take a closer look at how it works and at a few tools that we can use for its routine maintenance, diagnostics, and operation.

Considering the importance of firewalls and threat prevention gateways in any environment, often, it is imperative to assure their survivability in case of hardware failure, partial power loss, or connectivity loss. This is where fault tolerance in the form of HA is required.

Check Point offers several ways to achieve HA. After seeing how **ClusterXL High Availability** mode works, we'll briefly discuss these other options.

The ClusterXL functionality and diagnostics commands are available for both CLISH and expert mode, and both have different syntaxes. I'll be using both CLISH and expert mode commands selectively, but references to all of them are available in the *ClusterXL Administration Guide*, which is located at `https://sc1.checkpoint.com/documents/R81.10/WebAdminGuides/EN/CP_R81.10_ClusterXL_AdminGuide/Default.htm`.

In this chapter, we are going to cover the following main topics:

- ClusterXL in HA mode
- ClusterXL HA failover simulations
- Alternative preferred HA options

ClusterXL in HA mode

The prerequisites for an HA cluster are that it should comprise identical hardware (either physical or virtual), which, in its stable production state, runs identical versions of Check Point software. The caveat of a stable production state is that it is possible to use **Multi-Version Cluster** (**MVC**) mode during its transition to R81.10 or higher versions. This is not without certain limitations, but it allows for more gradual upgrades to verify new version stability and provide an easier fallback to earlier versions.

HA mode is defined in the **ClusterXL and VRRP** section of the cluster properties [1]. Under **Select the cluster mode and configuration** [2], we have two options, **ClusterXL** and **VRRP**. **ClusterXL** is selected by default [3] and, unless you have a very specific reason to use **VRRP**, **ClusterXL** should be your choice, as it provides the best failover experience. Under **Tracking**, the default selected option is to **Log** with additional alerting options available in the drop-down menu [4]. In **Advanced Settings**, **Use State Synchronization** with a 3-second delay is preselected [5]. This is the function that is responsible for the seamless failover between cluster members. If session states are not synchronized, in the case of failover, all existing connections are reset. **Use Virtual MAC** [6] is an option that I would recommend enabling in most production environments. I'll add a few words about it in the *Virtual MAC* section. In **Upon cluster member recovery** [7], we have the option to either **Maintain current active Cluster Member** or **Switch to higher priority Cluster Member**. We will discuss this further in the *Cluster member priority* section:

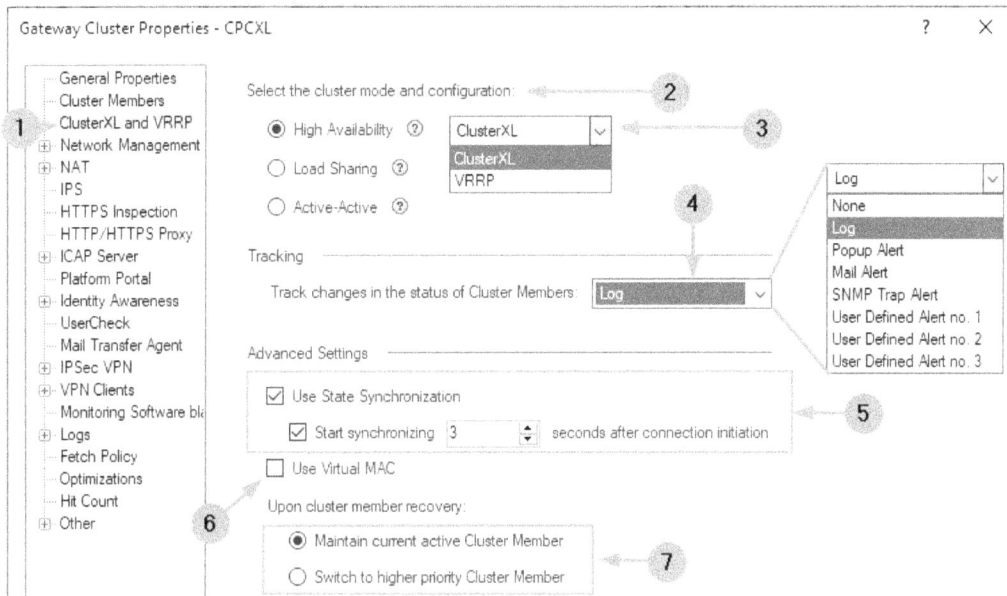

Figure 14.1 – The ClusterXL HA configuration options

Let's go over some of the cluster's configuration options, functionality, control and synchronization protocol, and states.

Virtual MAC

Virtual MAC (**VMAC**) assigns identical MAC addresses to all cluster virtual interfaces.

In Check Point's description of the **VMAC** functionality, it states *"VMAC that is advertised by the cluster members (via Gratuitous ARP Requests) keeps the real MAC address of each member and adds another Virtual MAC address on top of it."*

So, when a failover does occur, the **G-ARP** packet is the only packet sourced from the VMAC address. This alerts the connected switch about the (physical or virtual) port that the new active member is connected to via the **MAC Learning** process, by updating the **Content Addressable Memory** (**CAM**) tables. This is a lot faster and less prone to issues than some L3 switches or routers might experience with traditional ARP table updates, especially if you have a lot of statically NATed hosts.

We cannot use this feature in our lab, as VirtualBox and the majority of other Level-2 hypervisors do not have fully functional virtual switches. The networks they simulate either act as hubs or switches with limited functionality. In physical infrastructure or if your environment resides on a Level-1 (bare metal) hypervisor, such as VMware ESXi or Microsoft Hyper-V, you could safely use VMAC.

Cluster member priority

Cluster member priority is configured in the **Cluster Members** section of **Gateway Cluster Properties** [1], in the ordered list [2] where the priority of individual members can be adjusted using the **Increase Priority** and **Decrease Priority** buttons [3]:

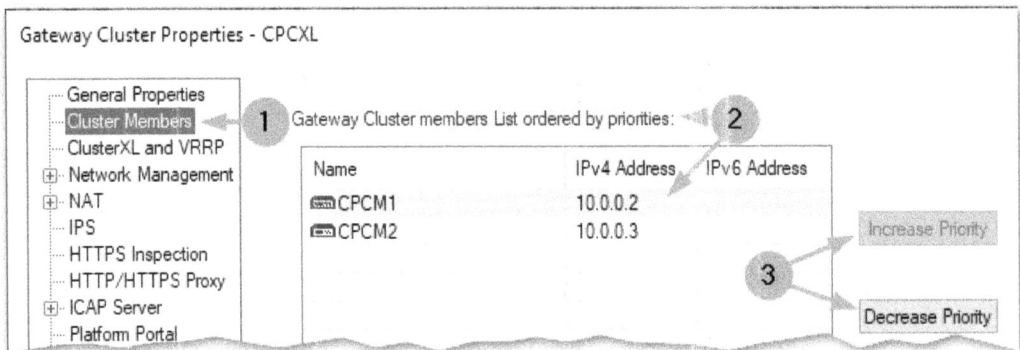

Figure 14.2 – Cluster member priority

The cluster member priority order can be a factor if, for instance, your architecture requires a full rack failover, and you have one of the cluster members in each rack. Additionally, it might be relevant if you are using a stretched VLANs cluster in the primary and disaster recovery sites and do not want to unnecessarily saturate the site-to-site link with traffic once the unit at the primary site has recovered.

Network interfaces

Network Types for interfaces are defined in **Gateway Cluster Properties | Network Management**. They can be designated as **Cluster, Sync, Cluster + Sync**, and **Private**. Let's take a closer look at what these terms are referring to.

Cluster interfaces

Cluster interfaces are defined with virtual IPs and will remain available if at least one of the cluster members is in an active state. Generally, a virtual IP and the corresponding IPs of addresses assigned to the cluster member interface should belong to the same network. However, there is a provision to use a virtual IP from a different network. For instance, when you might only have a single public IP address but are still interested in implementing fault-tolerant gateways. If that is the case within your environment, see *sk32073 Configuring Cluster Addresses on Different Subnets*. A major caveat of this solution is that it is only applicable to either locally managed clusters or remote clusters that are not managed over the internet (for instance, if you have WAN connectivity between the sites). This is because you cannot establish **SIC** with individual cluster members over the internet if they have private IPs on their external interfaces.

In most modern physical implementations, cluster interfaces are implemented as VLANs on bonded interfaces.

> **Important Note**
>
> The parent interface, bonded or standalone, containing VLAN sub-interfaces, should not have an IP address assigned. Also, note that VLAN ID 1 cannot be defined as a sub-interface. These are not ClusterXL-specific requirements but universal.

The cluster interfaces configured on individual NIC ports are all monitored. Of the VLAN cluster interfaces residing on the same parent port or bond, only those with the lowest and highest VLAN IDs are monitored by default. You can change this configuration to monitor all VLANs if necessary.

Private interfaces

Private interfaces are used with asymmetric clusters. Asymmetric clusters are those that, in addition to performing fault tolerance functions for the majority of the traffic, use either some or all cluster members to handle non-fault tolerant connections. Private interfaces created in the **Network Management** section [1]| **Actions** [2] | **New Interface** (not shown)| **General properties** section [3], defined as **Private** [4] and in **Members IPs | Modify** [5] are configured either on one of the cluster members [6] while left undefined on others [7], or when IP addresses that belong to unrelated networks are assigned to each member:

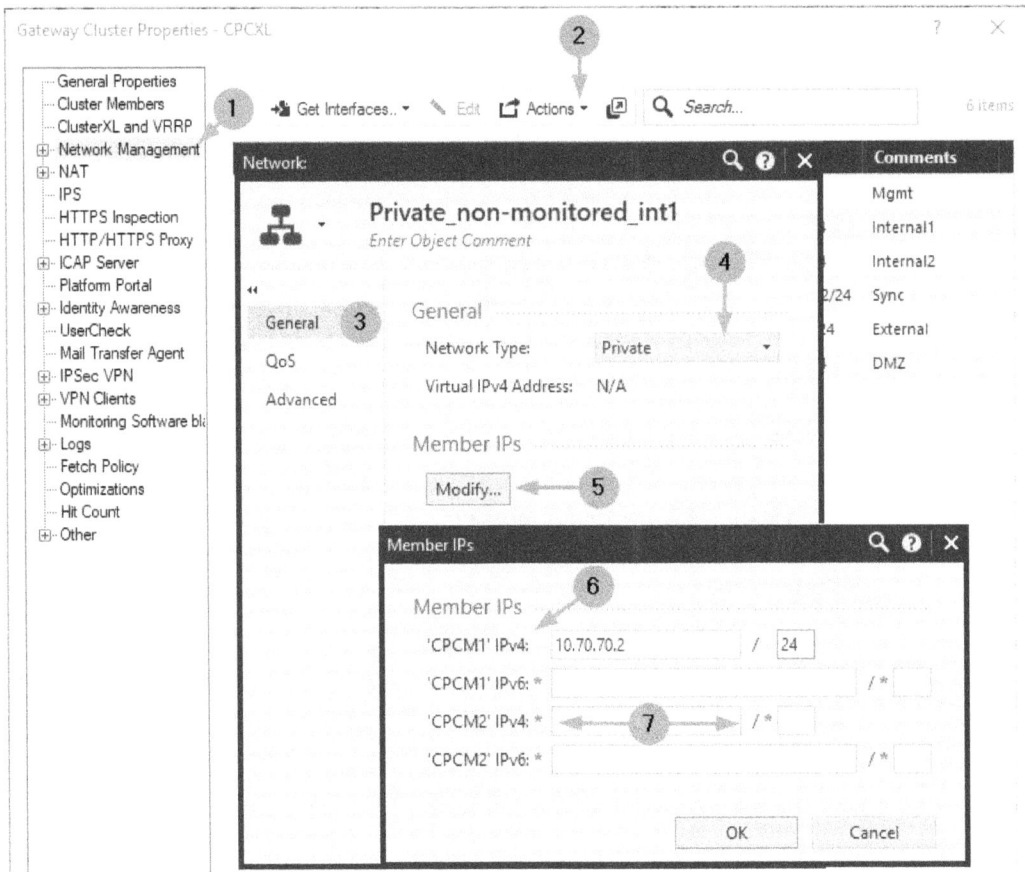

Figure 14.3 – Private interfaces in the gateway clusters

Such interfaces can be configured in the interface's **Advanced** section [1] as **Monitored Interface** [2], with faults triggering a cluster failover, or non-monitored, where faults will be ignored:

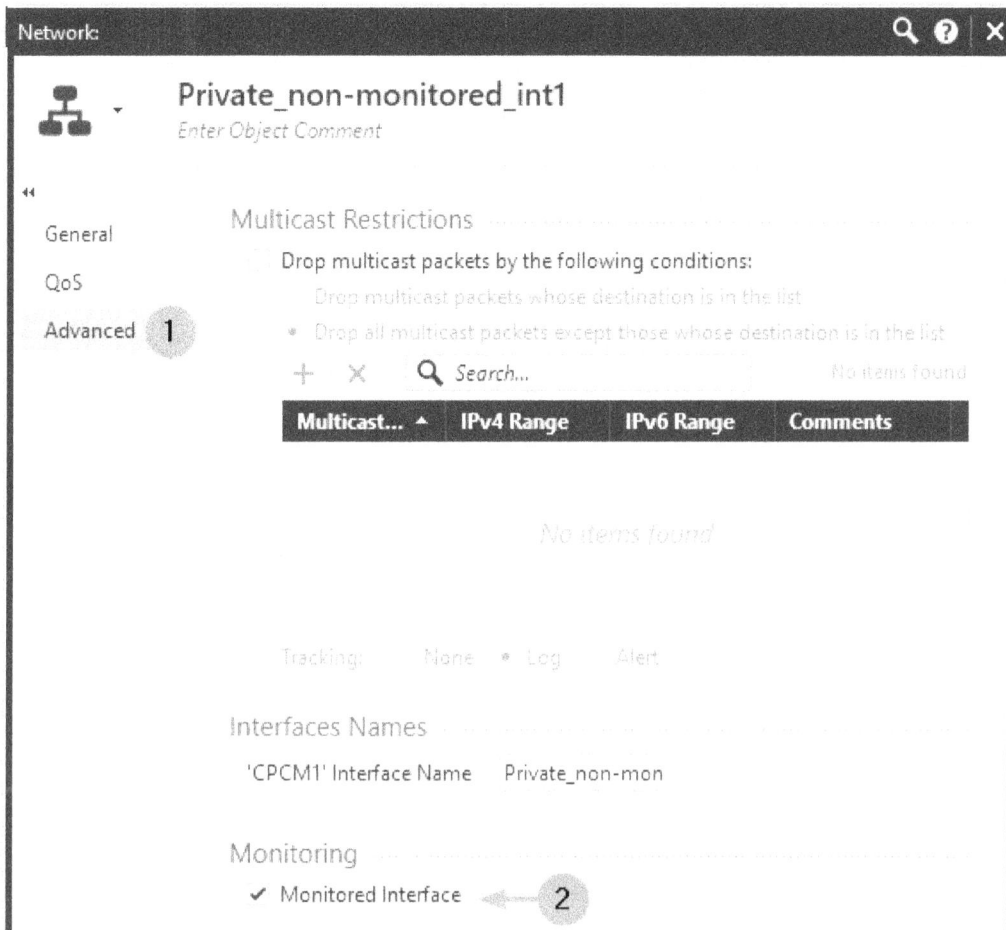

Figure 14.4 – Interface monitoring in the gateway cluster

We could assign different unrelated IP addresses to the same private cluster interfaces, but we cannot define virtual IP addresses for them.

Sync interfaces

Sync interfaces are used for the state synchronization of cluster members. In a two-member cluster situated at the same physical location, they are, typically, interconnected with a direct patch cable. In more complex environments with more than two cluster members, they are terminated on a dedicated VLAN with IP addresses assigned to VLAN interfaces.

While there are no specific requirements for their capacity, Check Point recommends using interfaces that are comparable to the largest one assigned to carry data and, in cases where uptime is the absolute priority, use bonded interfaces with each child interface connected to a separate switch with IP addresses assigned to the VLAN interfaces.

An additional argument in favor of using switches with bonded **Sync** interfaces is that you will not get alerts from the remaining **Active** member in the logs pertaining to the state of the interface.

If your **Sync** interface is defined as one of the multiple VLANs, it must have the lowest VLAN ID number on a parent interface.

Cluster + Sync interfaces

Cluster + Sync interfaces are seldom used in physical implementations but might be encountered in some cloud service deployments.

Critical devices

Certain processes running on a cluster member are predefined as **critical devices**. We could register additional critical devices (processes) manually, if needed, as long as they are configured on all cluster members and are numbered no more than 16. Critical devices are, somewhat counterintuitively, referred to as **Problem Notifications** (or **PNOTEs**). We can list current PNOTEs on a cluster member using `show cluster members pnotes all` to see all the registered critical devices or `show cluster members pnotes problem` to see those that are causing issues. When any of the cluster member's critical devices are in either the **init** or **problem** states, the cluster member changes its state to **DOWN** and triggers a failover.

When any of the **Cluster**, **Sync**, and **Private-Monitored** interfaces fail, either due to the link state or an inability to receive CCP traffic (discussed in the next section), a critical device named **Local Probing** changes its state from **OK** to **problem**.

Cluster Control Protocol, Full Sync, and routing synchronization

Check Point clustering relies on a combination of services to maintain the cluster's functionality. When a new member joins a cluster or an existing member either recovers from a **DOWN** state or is rebooted, the **Cluster Control Protocol** (**CCP**) keepalive traffic on all the cluster member interfaces will report the state of the members to each other. A **Full Sync** takes place and all connections from the active cluster member are replicated. Additionally, if dynamic routing is used by the cluster, routing synchronization using the **Forwarding Information Base** (**FIB**) manager process ensures that the OSPF and BGP routes are propagated to the standby members. To see whether the cluster member has dynamic routing configured and is associated with its interfaces, use the `cphaprob routedifcs` expert mode command. When we run it on one of our cluster members, we see that no such interfaces are configured:

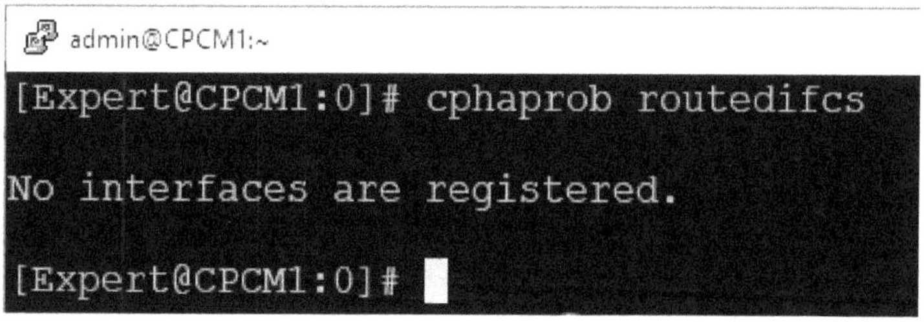

```
admin@CPCM1:~
[Expert@CPCM1:0]# cphaprob routedifcs

No interfaces are registered.

[Expert@CPCM1:0]#
```

Figure 14.5 – Dynamic routing and associated interfaces

From there on, a CCP is responsible for continuous state synchronization (Delta Sync) on the network and the exchange of keepalive packets containing interface status information between cluster members. By default, CCP traffic is encrypted, but we have the option of turning off the encryption (do not do this unless directed by TAC) and seeing under the hood.

For keepalives, the traffic consists of packets with the following OpCodes:

- Normal advertisements from the **ACTIVE** or **STANDBY** cluster members, `FWHA_MY_STATE - Report source machine's state`
- If one of the corresponding interfaces on other cluster members is down, `FWHA_IF_PROBE_REQ - Interface active check request`
- When such an interface comes back online, `FWHA_IF_PROBE_REPLY - Interface active check reply`

The preceding list refers to the numbered lines in the following Wireshark screenshot:

Figure 14.6 – CCP keepalive traffic

For sync interfaces, in addition to all of the preceding information, we will see the packets with OpCode `FWHA_SYNC - New Sync packet`:

Figure 14.7 – CCP Delta Sync traffic

We might see the current state and statistics of the synchronization using the show `cluster statistics sync` command. This presents us with a comprehensive synchronization diagnostics screen. **Sync status** [1] is the most informative section as it is capable of displaying a large number of specific error messages. **Link usage** can help us to spot-check the capacity of the **Sync** interface to determine whether it is sufficient or whether we should migrate it to a different interface [2]. What is also helpful is the presence of **Delta Sync Reset** data, as it shows us the last time the reset was triggered either manually or **triggered by fullsync** [3]:

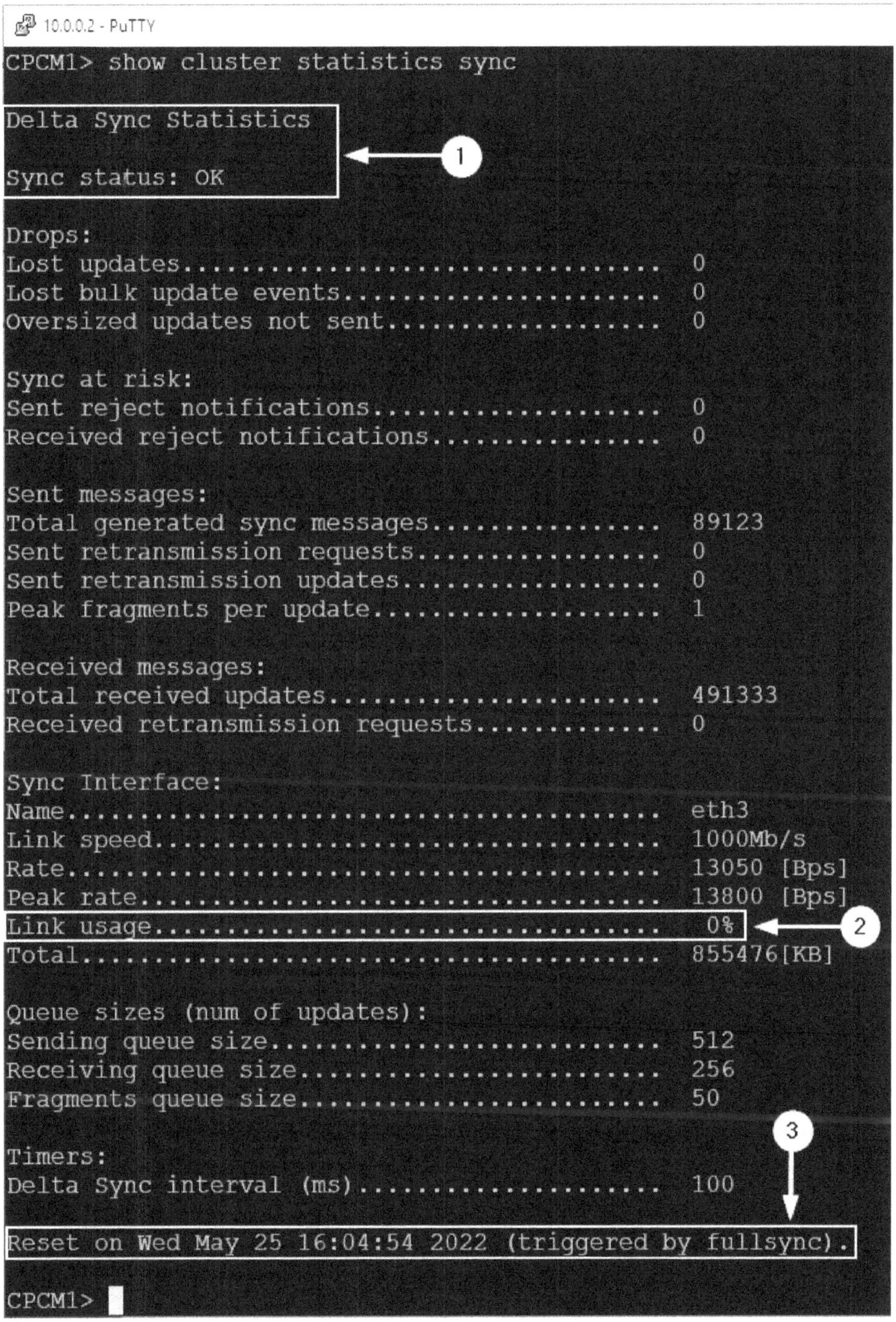

Figure 14.8 – The Delta Sync statistics and status

It's worth noting that state synchronization is optional in cluster HA, although it is enabled by default, and I have never encountered environments where it was disabled. Also, note that state synchronization can be configured for individual services but is enabled for all predefined TCP and UDP services by default. In ClusterXL HA with state synchronization enabled, existing connections for all services (with state synchronization enabled in the service properties) will continue to operate uninterrupted.

Cluster member states

There are five cluster member states:

- **ACTIVE**: This is the normal state of the currently active member in a healthy cluster.

- **ACTIVE(!)**: This indicates a problem with this member, or the entire cluster, while the current member is still in an active state.

- **STANDBY**: This is the normal state of a healthy standby member.

- **DOWN**: This is the state of a standby member with a problem.

- **LOST**: This is the state of an unreachable member (from the point of view of the member unable to communicate with it).

Let's take a look at the HA in action during failover.

Failover

In HA mode, failover is the process of an alternative member becoming active when one of these two scenarios plays out:

- A current active member detects a problem with one of its critical devices and (using CCP) notifies other members of the HA cluster about it, changing its status to **DOWN**. The standby member with the next highest priority becomes **ACTIVE(!)**.

- Standby members stop receiving CCP packets from the active cluster member. Again, the standby member with the highest priority becomes active.

If OSPF or BGP are configured on the cluster members, the *graceful restart* functions are used to re-establish communication between the connected routers and the cluster without triggering an upstream propagation of routing updates.

Manual failover can be performed using the `set cluster member admin down` CLISH command, which triggers a script registering a **Critical Device**, questionably named `admin_down`, and reports its state as a *problem*. The `set cluster member admin up` CLISH command will return that cluster member to a **STANDBY** state if no active PNOTEs are present, or to **ACTIVE** if your preferences are so configured. To survive a reboot and not become a valid **STANDBY** or **ACTIVE** member without your express wishes, use the `clusterXL_admin down -p` expert mode command. Use `clusterXL_admin up -p` to reverse the state.

If you would like to be able to perform a manual failover from SmartConsole, in **GATEWAYS & SERVERS** [1], select the cluster [2], click on **Scripts | Run One Time Script** [3], choose the cluster member [4], enter the desired failover command in the **Script Body** box, and click on **Run** [5]:

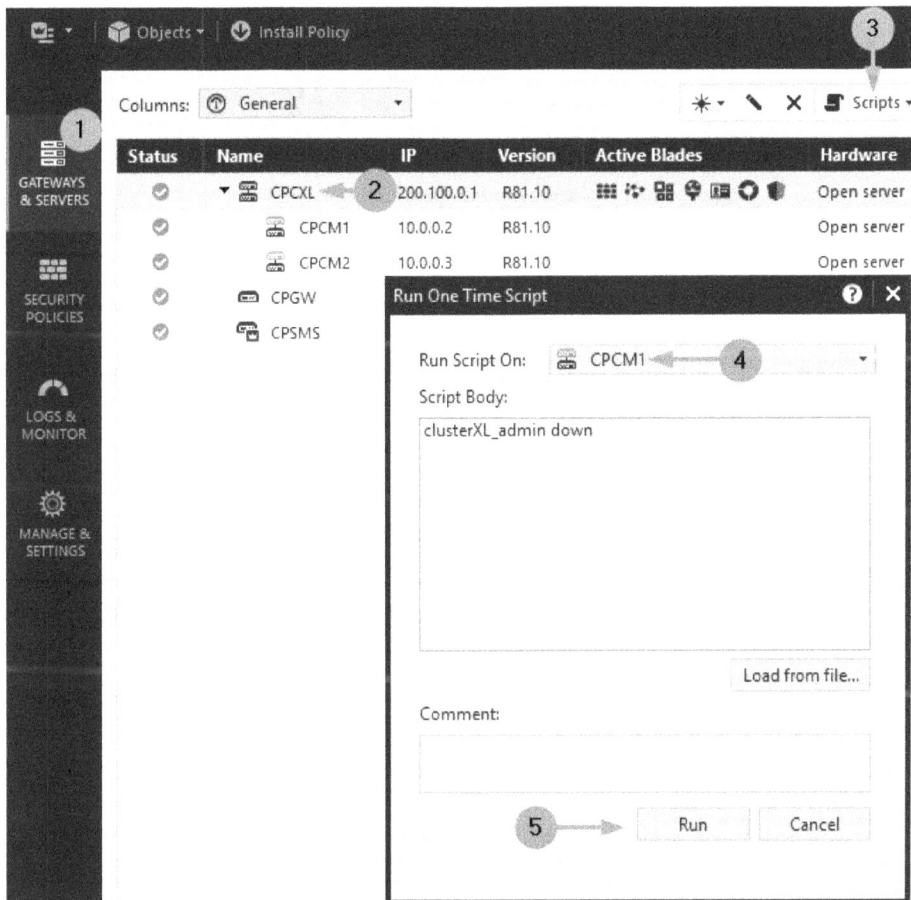

Figure 14.9 – Manual failover in SmartConsole

We can use the `show cluster failover` command on any member to see the details of the most recent failover [1], the overall failover counter since the last reboot and its time [2], and the cluster failover history for the last 20 instances [3]:

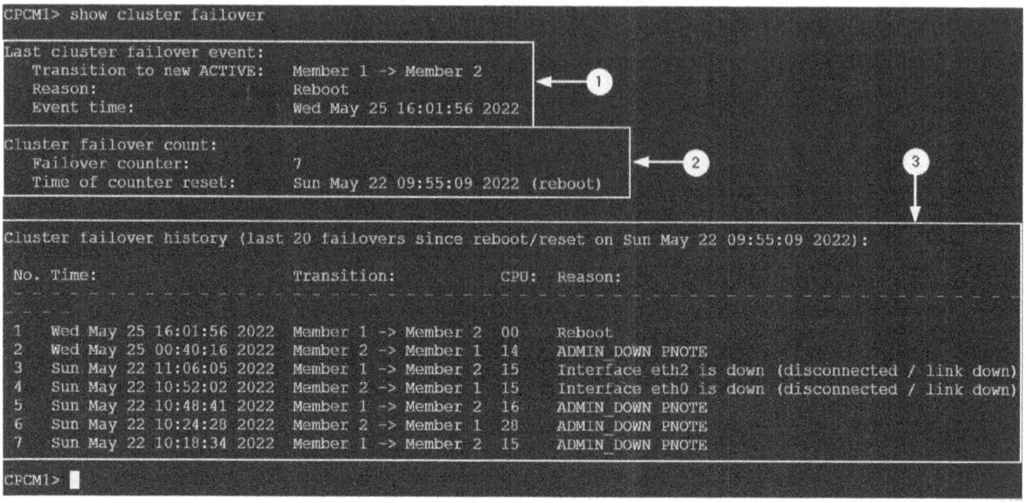

```
CPCM1> show cluster failover

Last cluster failover event:
    Transition to new ACTIVE:    Member 1 -> Member 2          ◄──── 1
    Reason:                      Reboot
    Event time:                  Wed May 25 16:01:56 2022

Cluster failover count:
    Failover counter:            7                             ◄──── 2
    Time of counter reset:       Sun May 22 09:55:09 2022 (reboot)
                                                                              3

Cluster failover history (last 20 failovers since reboot/reset on Sun May 22 09:55:09 2022):

No. Time:                        Transition:           CPU:   Reason:

1    Wed May 25 16:01:56 2022    Member 1 -> Member 2  00     Reboot
2    Wed May 25 00:40:16 2022    Member 2 -> Member 1  14     ADMIN_DOWN PNOTE
3    Sun May 22 11:06:05 2022    Member 1 -> Member 2  15     Interface eth2 is down (disconnected / link down)
4    Sun May 22 10:52:02 2022    Member 2 -> Member 1  15     Interface eth0 is down (disconnected / link down)
5    Sun May 22 10:48:41 2022    Member 1 -> Member 2  16     ADMIN_DOWN PNOTE
6    Sun May 22 10:24:28 2022    Member 2 -> Member 1  28     ADMIN_DOWN PNOTE
7    Sun May 22 10:18:34 2022    Member 1 -> Member 2  15     ADMIN_DOWN PNOTE

CPCM1>
```

Figure 14.10 – The cluster failover details and history

Since not every HA transition is orderly and it could be a result of a much larger infrastructure problem, let's take a look at how our cluster behaves in these instances.

Edge cases

There are a few edge cases that should be considered:

- When all cluster members experience an identical problem but can communicate with each other (for example, a switch used by some of the active and standby cluster members' interfaces is down). In such a case, identical PNOTEs are registered as active on all cluster members. Active members will remain active (with a problem) and will be displayed as **ACTIVE (!)**, but standby members will change their state to **DOWN**.

- When all connections between the HA cluster members are interrupted (for example, a WAN link between sites carrying multiple VLANs that all the gateways' interfaces are a part of). In such a case, former active members will change to **ACTIVE(!)** with a problem, but one of the higher-priority **STANDBY** members that lost connectivity with the rest of the cluster will also become **ACTIVE (!)** and begin to process traffic using its last installed policy. Both members in the **ACTIVE(!)** state will list the unreachable members as **LOST**.

- When a cluster member in a degraded state listed as **DOWN** becomes a sole, partially functioning unit, losing its ability to communicate with remaining cluster members. In this case, it will change its state from **DOWN** to **ACTIVE(!)** and resume executing the currently installed policy on the traffic traversing it.

Now that we have seen what is happening during a failover event, let's take a look at the process of the cluster's recovery to a normal operational state.

Recovery

When ClusterXL members are recovered and/or connectivity between them and the rest of the cluster has been restored, some or all of the following takes place:

1. Full Sync
2. Delta Sync
3. Routing synchronization
4. A comparison of the last installed policy and the fetching of the latest version from other cluster members or, if none are available, the management server
5. Failover to the member with fewer or no active PNOTEs (if warranted)
6. State changes to one of the following:

   A. **DOWN**, if any active PNOTEs are present

   B. **STANDBY**, if no active PNOTEs are present

   C. **ACTIVE**, if so configured in the **Upon cluster member recovery** settings

Unless you have set the option to **Switch to higher priority Cluster Member**, it will remain a standby member until it is forced to become active by another failover event.

So, let's take our HA cluster for a ride to see how it actually performs during different failover scenarios.

ClusterXL HA failover simulations

While knowing the theory is all well and good, there is nothing better for gaining confidence in the reliability of fault-tolerant solutions than practice. Let's conduct a few tests in our lab that will illustrate the functionality of ClusterXL in HA mode using two exercises, one for orderly failover and another one for *real-life* simulation.

Manual failover test

Let's introduce slight modifications to the security policy:

1. In the SmartConsole application, navigate to the **SECURITY POLICIES | LeftSide_ RA** policy and disable the rule named **SSH access to Router Accept**:

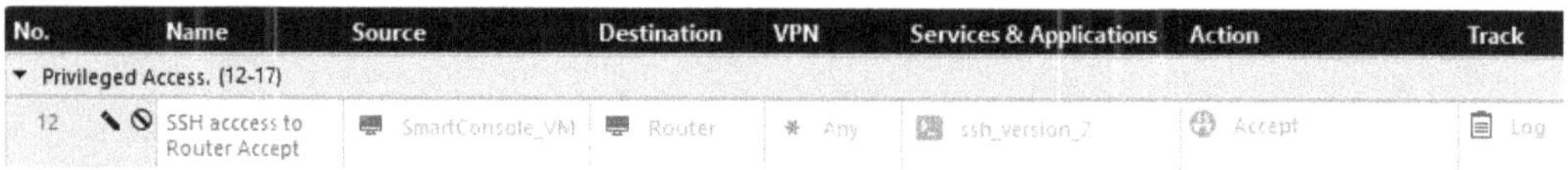

No.	Name	Source	Destination	VPN	Services & Applications	Action	Track
▼ Privileged Access. (12-17)							
12	SSH acccess to Router Accept	SmartConsole_VM	Router	* Any	ssh_version_2	Accept	Log

Figure 14.11 – Disabling the existing firewall layer privileged access rule

2. Scroll down to the **APCL & URLF, Content Awareness Inline Layer** policy section and expand its parent rule.

3. Add two new rules above the existing rule for HTTP and HTTPS [1] (an **APCL/ URLF Layer cleanup** rule [2] is shown for reference):

   A. One for **HR** access role access to the internet over SSH.

   B. Another for **RA_Role** access to the internet over SSH. Disable this rule by right-clicking on its **No.** field and checking the **Disable** checkbox:

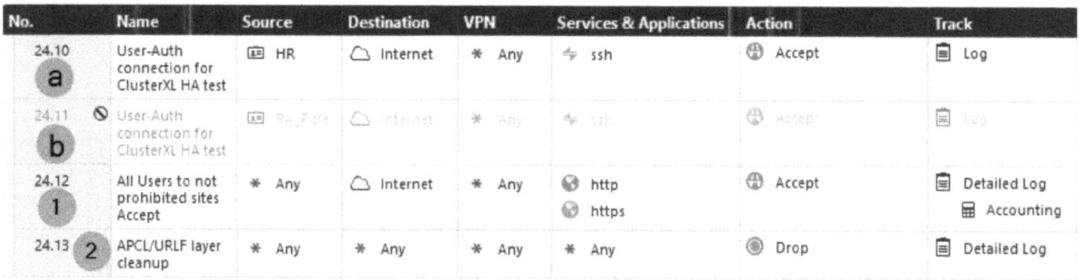

No.	Name	Source	Destination	VPN	Services & Applications	Action	Track
24.10 a	User-Auth connection for ClusterXL HA test	HR	Internet	* Any	ssh	Accept	Log
24.11 b	User-Auth connection for ClusterXL HA test	RA_Role	Internet	* Any	ssh	Accept	Log
24.12 1	All Users to not prohibited sites Accept	* Any	Internet	* Any	http https	Accept	Detailed Log Accounting
24.13 2	APCL/URLF layer cleanup	* Any	* Any	* Any	* Any	Drop	Detailed Log

Figure 14.12 – Creating rules for the HA failover tests

4. Publish the changes and install the policy.

5. SSH into both cluster members from the SmartConsole VM and execute a simple `fw hastat` command. This will let us, at a glance, determine the mode of the HA member for `active` [1], `stand-by` [2], and their status [3]:

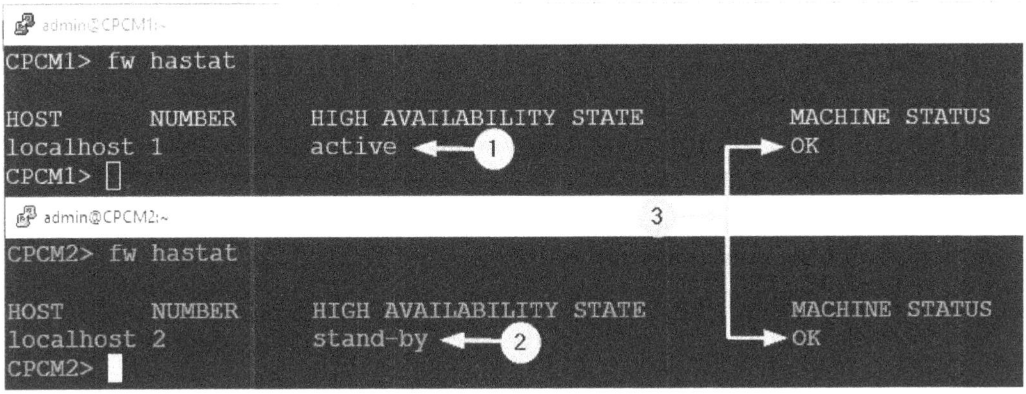

Figure 14.13 – A brief lookup of the HA member status

6. On the same SmartConsole VM, open Chrome and access the
    `https://10.20.20.1/connect/PortalMain` URL. Authenticate as `mycp.`
    `lab\hruser` and `CPL@b8110` to see the **Network Access Granted** message.
    Minimize the browser windows.

7. On the same SmartConsole VM, open a new instance of **Putty** and SSH into our
    VYOS router. Authenticate this as `vyos` and `vyos`. Once logged on, type in `watch`
    `-n 1 date`, and press *Enter* to see the incrementing clock:

Figure 14.14 – The local SSH session for failover process monitoring

8. Click on the rule [1] you have defined for **HR** to **Internet** using **ssh**, and select the
    Logs tab [2] to see the newly established session [3]:

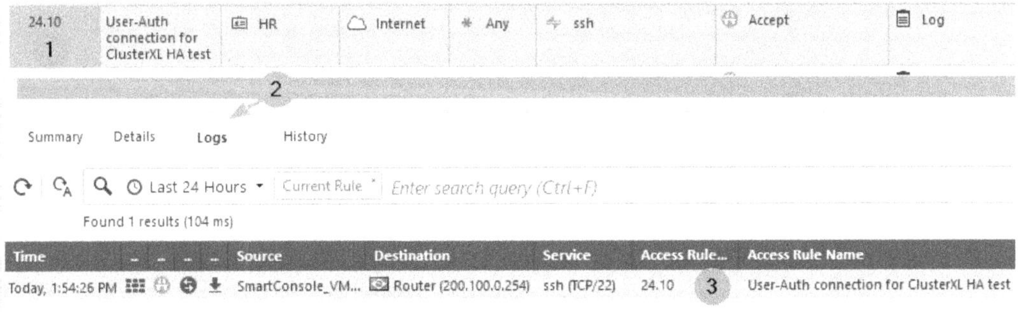

Figure 14.15 – SSH session log to monitored connection

9. On your active cluster member, run `set cluster member admin down`.

10. On both cluster members, run `show cluster state` and verify that the current cluster member in the **ACTIVE** state is the one that was previously in the **STANDBY** state. Verify that the one that was previously in the **ACTIVE** state is now in the **DOWN** state.

11. Check the SSH session to the router with a running clock. Verify that it is still connected and that the time is incrementing.

12. On a cluster member in the **DOWN** state, run `set cluster member admin up`. Then, using `show cluster state`, verify that it is now in the **STANDBY** state.

Catastrophic failure and recovery simulation

The previous test illustrated an orderly manual failover. But what about really nasty catastrophic failure scenarios? Let's create one to find out:

1. From your RightHost VM, connect to the VPN using **EndPoint Security** with username `RA_user1` and password `CPL@b8110`.

2. Once you see the **Connection Succeeded** message in **EndPoint Security**, and the VPN Status is **Connected**, start **Putty** and attempt to SSH into the same router host using IP address `200.100.0.254`. This should fail with a **Connection timed out** message.

3. Now, let's verify that your **ACTIVE** cluster member is **CPCM1**. SSH into **CPCM1** and **CPCM2** and execute `fw hastat` on **CPCM1**. If it is not **ACTIVE**, toggle the **CPCM2** member using the `set cluster member admin down` and `set cluster member admin up` commands to move the **ACTIVE** role.

4. *Pull the plug* on **CPCM1**: On your LabHost PC, in **Oracle VM VirtualBox Manager** [1], right-click on **CPCM1** [2], click on **Close** [3], and then click on **Power Off** [4]:

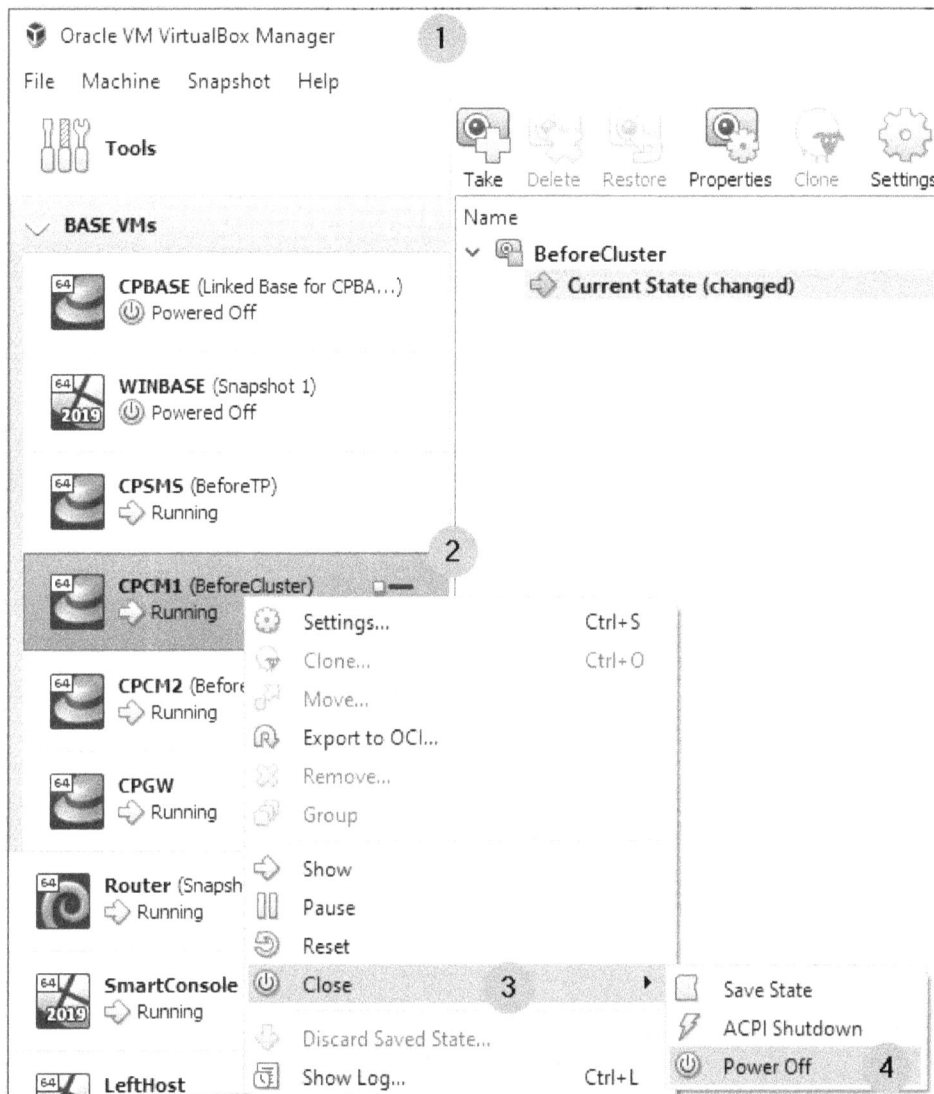

Figure 14.16 – Simulating a power failure for the cluster member

When prompted for confirmation with **This will cause any unsaved data in applications running inside to be lost,** click on the **Power Off** button (not shown).

5. Return to the SmartConsole VM. Enable the second rule for **RA_Role** to **Internet SSH** access that we disabled in the previous section (by right-clicking and unchecking the **Disabled** checkbox). **Publish** the changes and click **Install Policy**.

6. In the **Install Policy** window, we'll see our cluster depicted with an error icon [1] due to one of its members being absent. In the bottom-left corner of this window, under **Install Mode**, uncheck the **For gateway clusters, if installation on a cluster member fails, do not install on that cluster** checkbox [2]. Click on **Install:**

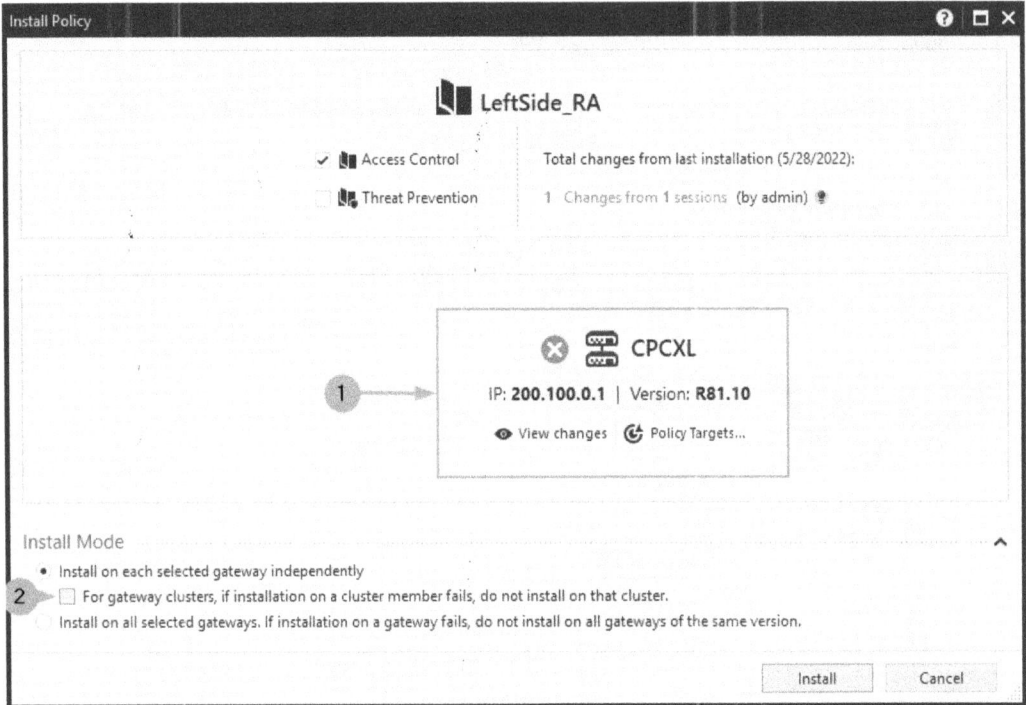

Figure 14.17 – Installing a policy on the remaining cluster members

7. You will be presented with the warning window, but the text explicitly describes a situation when this installation is acceptable. Click **Yes:**

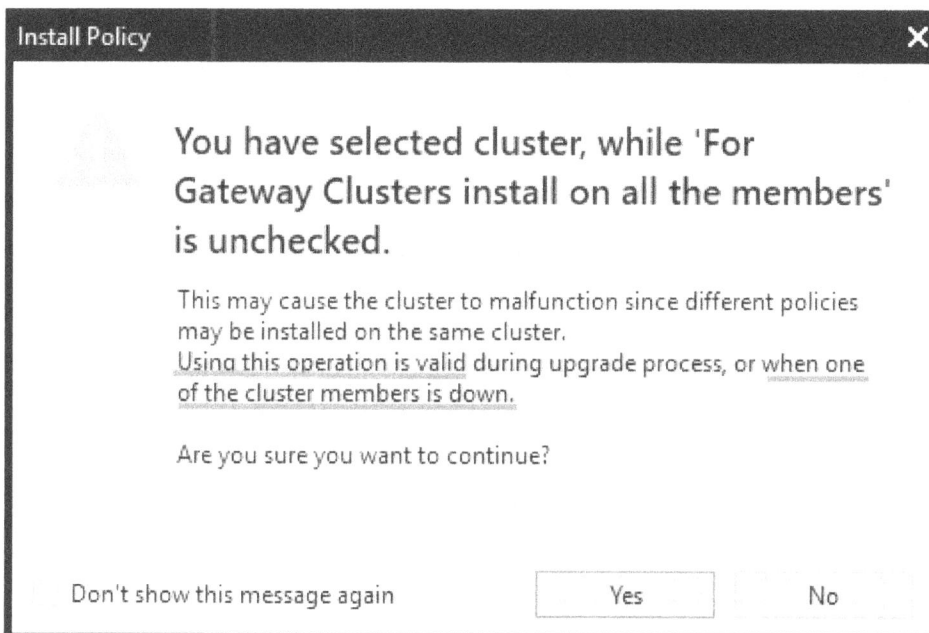

Figure 14.18 – Installation on the cluster with a member down warning

8. The **Install Policy Details** window indicates that the installation partially succeeded on **CPCXL** with the reason for failure to install on **CPCM1** described in the **Status** field (this is visible when the expansion arrow is clicked on):

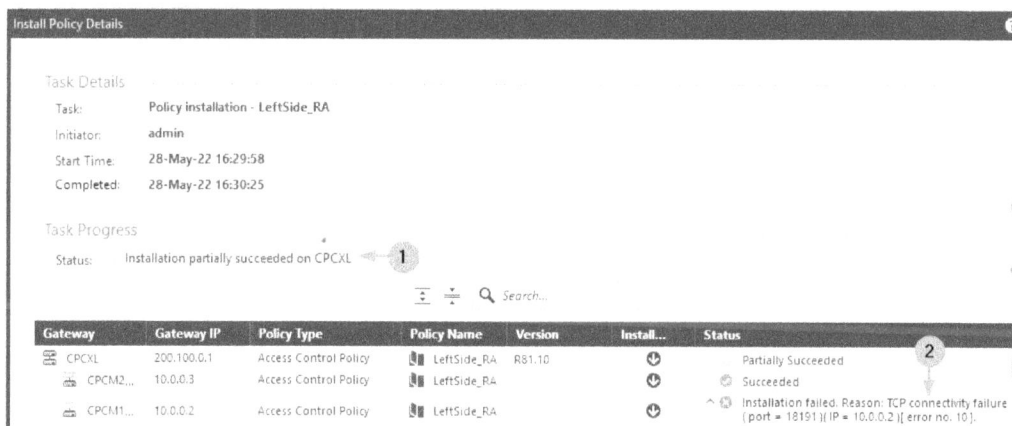

Figure 14.19 – The policy installation details on a partially failed cluster

9. Now, on your RightHost VM, verify that your VPN is still active. Then, start Putty and SSH into our router on 200.100.0.254. Log on using the username vyos and the password vyos. Run the watch -n 1 date command to see the incrementing clock.

10. Recall that, at this point, the policy is only installed on a single remaining cluster member, **CPCM2**. To make things even more interesting, on your LabHost PC, perform the same power failure emulation for our Check Point management server, **CPSMS**.

11. In the SmartConsole VM, we'll get an eventual connectivity error. It might not be immediate, unless we attempt to perform some kind of action, since the session is attempting to recover. If you click on **LOGS & MONITOR**, you can force it to manifest earlier:

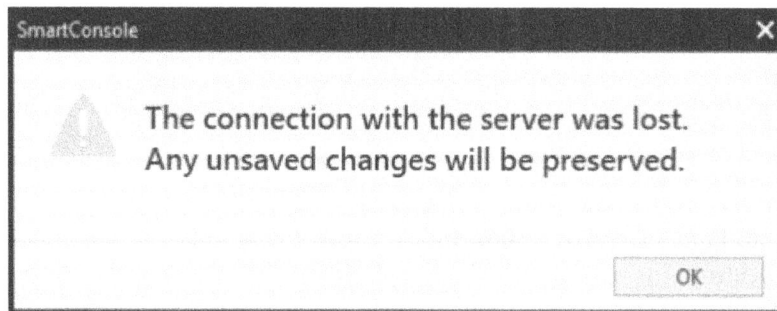

Figure 14.20 – The SmartConsole server connectivity warning

12. Now, with the management server down, let's resurrect our **CPCM1** cluster member. Power it up from **Oracle VM VirtualBox Manager** using *right-click* | **Start | Normal Start**.

13. Wait for **CPCM1** to complete the boot process. Once it is done, SSH into it and verify its status using the show cluster state command:

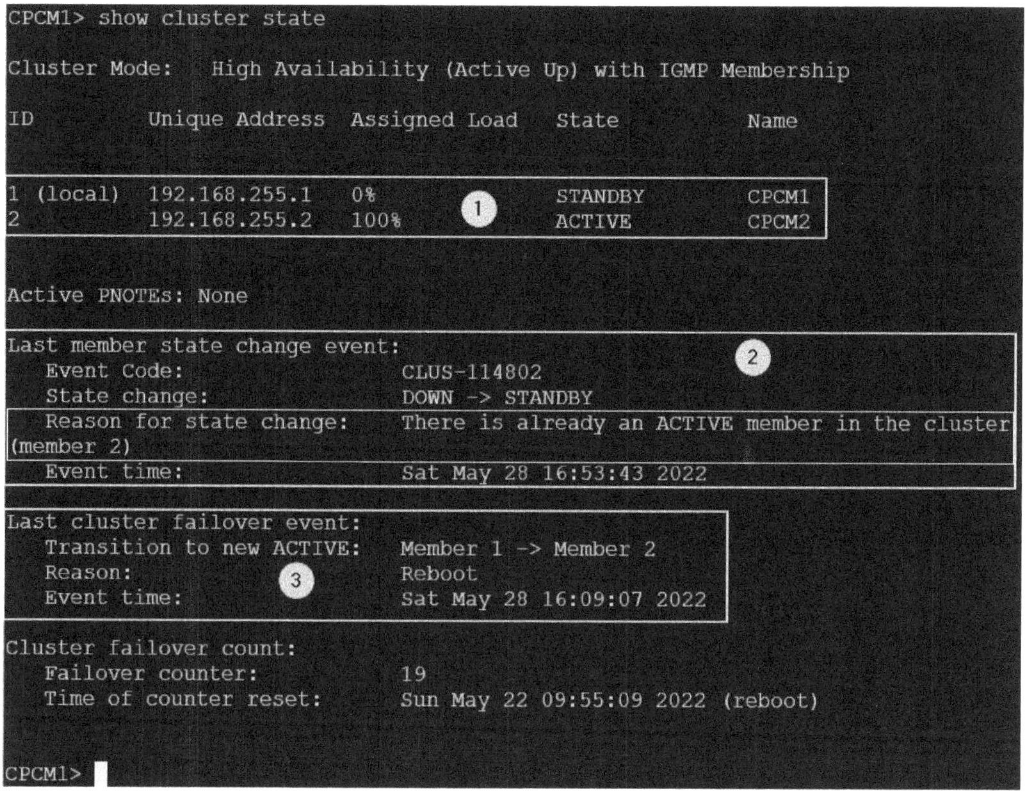

```
CPCM1> show cluster state

Cluster Mode:    High Availability (Active Up) with IGMP Membership

ID          Unique Address  Assigned Load    State          Name


1 (local)   192.168.255.1   0%                  STANDBY        CPCM1
2           192.168.255.2   100%    (1)         ACTIVE         CPCM2


Active PNOTEs: None

Last member state change event:
    Event Code:              CLUS-114802              (2)
    State change:            DOWN -> STANDBY
    Reason for state change: There is already an ACTIVE member in the cluster
(member 2)
    Event time:              Sat May 28 16:53:43 2022

Last cluster failover event:
    Transition to new ACTIVE:  Member 1 -> Member 2
    Reason:          (3)      Reboot
    Event time:              Sat May 28 16:09:07 2022

Cluster failover count:
    Failover counter:        19
    Time of counter reset:   Sun May 22 09:55:09 2022 (reboot)

CPCM1> █
```

Figure 14.21 – Recovered member cluster state report

14. We understand that no policy was installed by us on this cluster member since it was down at the time. Let's check how it'll behave if it has to take over the **ACTIVE** role. To do that, let's simulate a power failure on CPCM2 from our LabHost.

15. With CPCM2 down, check the status of our SSH session on RightHost. It is still up, and the clock is ticking. Start another SSH session to the router, and it will succeed, too. This is an indication that in the absence of the management server, the previously recovered cluster member was able to obtain the current policy from the only remaining member.

16. Perform show cluster state on CPCM1, and you will see output similar to the following. This indicates that the local cluster member is ACTIVE (!) and that communication with the other member is LOST [1]. The last member state change event's reason for change indicates that All other machines are dead [2], and Last cluster failover event describes the transition to a new ACTIVE state [3]:

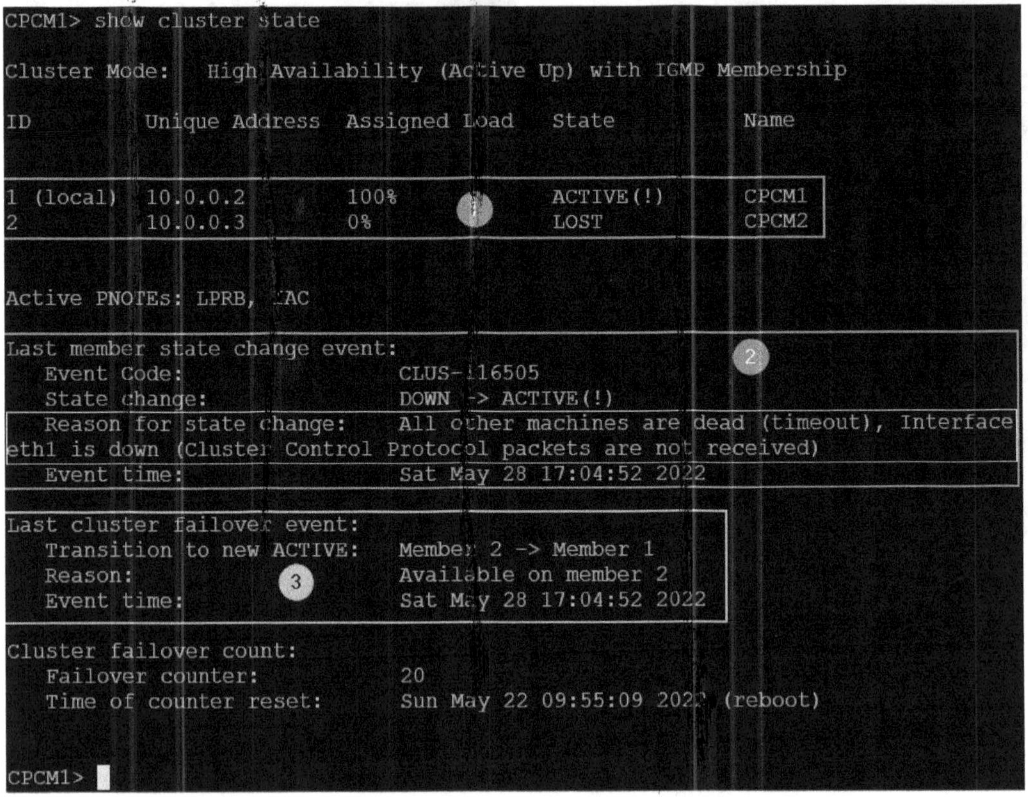

```
CPCM1> show cluster state

Cluster Mode:    High Availability (Active Up) with IGMP Membership

ID           Unique Address   Assigned Load   State           Name

1 (local)    10.0.0.2         100%            ACTIVE(!)       CPCM1
2            10.0.0.3         0%              LOST            CPCM2


Active PNOTEs: LPRB, AC

Last member state change event:
    Event Code:              CLUS-116505
    State change:            DOWN -> ACTIVE(!)
    Reason for state change:  All other machines are dead (timeout), Interface
eth1 is down (Cluster Control Protocol packets are not received)
    Event time:              Sat May 28 17:04:52 2022

Last cluster failover event:
    Transition to new ACTIVE:  Member 2 -> Member 1
    Reason:                  Available on member 2
    Event time:              Sat May 28 17:04:52 2022

Cluster failover count:
    Failover counter:        20
    Time of counter reset:   Sun May 22 09:55:09 2022 (reboot)

CPCM1>
```

Figure 14.22 – Cluster state with the last cluster member remaining

17. Let's power up the components that we so unceremoniously "unplugged" earlier. In your LabHost's Oracle VM Virtual Machine Manager, start CPSMS and CPCM2 and give them a few minutes to recover. If you want to verify that your management server is up, SSH into it, enter expert mode, and run $MDS_FWDIR/scripts/cpm_status.sh. When it returns Check Point Security Management Server is running and ready, start SmartConsole and log in.

18. With SmartConsole connected, open the **GATEWAYS & SERVERS** view and you will see green checks in the **Status** fields of all components of our Check Point infrastructure.

I mentioned that this scenario will be realistic, and that means we are not quite done. Logs generated by a sole surviving cluster member, when our management server was down, are now stored locally on the gateway CPCM1 and will remain there until midnight. Let's get them to our management/log server right away:

1. Navigate back to your SSH session to CPCM1 (or open a new one if the old session has expired). Then, execute the `fw logswitch` command to rotate the active log file:

```
CPCM1> fw logswitch
Log file has been switched to: 2022-05-28_183906.log
CPCM1>
```

Figure 14.23 – Post-management server recovery log rotation on the gateway

2. Now, SSH into our management server, CPSMS, and execute `fw fetchlogs 10.0.0.2` to retrieve the file created in the previous step:

```
CPSMS> fw fetchlogs 10.0.0.2
File fetching in process. It may take some time...
File CPCM1__2022-05-28_183906.log was fetched successfully
CPSMS>
```

Figure 14.24 – Post-management server recovery log extraction

3. Let's run `show cluster state` on both cluster members to verify that they are in a good shape with only the **ACTIVE** and **STANDBY** statuses listed. If you do have a preference for a specific member to be **ACTIVE** and yours is not the one, toggle the currently **ACTIVE** member *down* and *up* to turn it into **STANDBY**.

Only now our recovery is complete.

Conclusion

Now we have experienced a full-cycle disaster recovery for Check Point ClusterXL HA and can confidently state that the established connections have been preserved flawlessly. It is indicated by a still-active SSH session from SmartConsole, an uninterrupted VPN connection from RightHost, with another SSH session through the tunnel running on it. Also, we were able to confirm that Identity Awareness continues to function after failover.

There are other means to achieve HA using Check Point products. In my opinion, some of them, are more suitable than others. Let's take a look at these options next.

Alternative preferred HA options

There is a reason for "preferred" being included in the heading of this section. Check Point offers ClusterXL with **VRRP** and **Load Sharing** using either multicast or unicast as HA modes; however, I cannot recommend them.

The negatives of both **Load Sharing** modes are perfectly described by Timothy Hall in his book, *MAXPOWER: Check Point Firewall Performance Optimization*. My experience with these offerings is aligned with his conclusions, and I will only briefly summarize them here:

- **Load Sharing Multicast** stability is conditional and is based on the compatibility of adjacent routers. In many organizations, some of the routers are provisioned by peers or service providers. Chances are you will run into issues that will affect the stability of this solution sooner or later.

- **Load Sharing Unicast** is inefficient, as one of the cluster members acts as a "pivot" and must forward traffic to other cluster members. This approach creates a poorly balanced cluster with the added downside of certain connections requiring **stickiness** configuration, to force specific inbound and outbound traffic traversing the same cluster members.

- **VRRP** was the original standards-based clustering solution and is perfectly acceptable if your cluster is limited to two nodes and will not participate in VPN communities relying on **Virtual Tunnel Interfaces** (**VTIs**). This last contingency is really what makes me state that VRRP should be avoided, if possible, as most VPN integrations with cloud service providers require the use of **VTIs** with dynamic routing.

Active-Active is an interesting solution, but it is designed for a different purpose. As per Check Point's documentation, *"This mode is designed for a cluster, whose Cluster Members are located in different geographical areas (different sites, different cloud availability zones)."* It is limited to two members, one in each geographical location or availability zone, and supports a limited number of functions.

However, two other Check Point solutions are more than capable of handling both HA and load sharing: **VSX Virtual System Load Sharing** (**VSLS**) and hyperscale solution **Maestro**.

VSLS is an excellent solution allowing near-linear capacity increase with each additional cluster member. Unlike ClusterXL HA, which is limited to five members (with only four being the recommended maximum), VSLS can have as many as 13. There is a combination of added complexity and flexibility in that it requires the creation of virtual systems (virtual gateways). Each virtual system is automatically provisioned with standby and backup instances residing on different cluster members. In the case of a cluster member or virtual system failure, standby members are automatically promoted to active, and backup to standby, ensuring uninterrupted traffic flows. Virtual systems' relative "weights" can be configured to prioritize resource allocation, and active virtual systems can be dynamically moved between cluster members to ensure a continuous rebalancing of the cluster.

Maestro takes HA and load sharing to a whole different level by allowing multiple gateways to be aggregated into security groups with up to 31 per single-site or 28 per dual-site deployment. Once added to a security group, gateways become **Security Gateway Modules** (**SGMs**). As little as two SGMs per security group can be used to provide HA capability.

Additional gateways are auto-provisioned after being plugged into redundant **Maestro Hyperscale Orchestrator** units, assigned to a security group, and become SGMs. This removes the necessity of manually configuring Gaia, defining new objects, and modifying policies and rules, as a security group constitutes a single managed object and becomes a policy installation target.

With each additional system added to a security group, the load is automatically distributed between all members by each connection's state synchronized between only two chosen members. The failure of any member results in an uninterrupted continuous operation with each remaining member taking on an equal load.

SGMs can be moved between different security groups on the fly to dynamically address load issues where they are needed most. Additionally, SGMs of different models but the same generation are supported, that is, they do not all have to be identical.

Both options are deserving of their own book, and I just had to mention them here to make you aware of the possibilities.

Summary

In this chapter, we learned about Check Point's clustering solutions, specifically ClusterXL HA. We had an opportunity to test it using manual and "realistic" failover scenarios and observe different state changes. We learned about a few common commands that are used for cluster operations and diagnostics. Additionally, other clustering options for HA and load sharing were discussed, highlighting their respective shortcomings and advantages.

In the next chapter, we will cover troubleshooting workflows, along with complementary tools and resources. Additionally, we will address the handling of communications with the **Technical Assistance Center** (**TAC**) in order to ensure the timely resolution of problems.

15

Performing Basic Troubleshooting

Inevitably, when using any product of a certain complexity, we run into troubleshooting. Since this book has been written for those new to Check Point, I will refrain from deep dives into troubleshooting procedures, and focus instead on describing a process that will help you resolve the issues you may experience quicker.

In companies with extensively documented infrastructure, change management, dependencies tracking, and well-established cross-teams communication, a lot of possible issues are averted. The opposite is true for companies that do not have those in place.

The success of your troubleshooting attempts depends largely on how well prepared you are at the time of the incident. So, let's figure out how to approach the troubleshooting process.

In this chapter, we are going to cover the following main topics:

- Troubleshooting constraints and your actions
- Typical issues and the tools to solve them
- Service Requests – getting them right every time
- Community resources and engagements
- Postmortems and lessons learned

Troubleshooting constraints and your actions

As a Check Point administrator, who are you? Were you hired by a company with established Check Point infrastructure, to join a team of experts handling it? Or are you a wearer of many hats who is looking to implement and manage Check Point gateways in addition to handling several other security products and, perhaps, the rest of the IT infrastructure? Or is your role somewhere in-between these two scenarios?

If you are working in a highly structured environment, your role and responsibilities will be clearly articulated, and you will likely go through an internal training period before being entrusted with day-to-day administration, let alone troubleshooting responsibilities. If the company you are at is better described by the second scenario, then ready or not, you are a de facto engineer and troubleshooter. If the place you are working at is somewhere in between these two categories, things may get complicated. Read on to better orient yourself.

Determine what you, as a Check Point administrator, are allowed to do according to the company policy (if it exists) and if the policy contains references to specific procedures or guidelines for troubleshooting. If it does, print, read, and keep them close until you know them by heart. Generally, these are high-level documents that do not contain a great deal of technical information, but it is important to either follow the steps outlined in them or (if they are outdated or irrelevant) get your manager to sign off on granting you the necessary permissions in writing (email). It may be tempting to approach troubleshooting from a strictly technical viewpoint, but that may result in a **Resume Generating Event** (**RGE**).

Here are some of the questions that you should have answers to:

- Are you allowed to make configuration changes and install policies during work hours unannounced or is there a process that should be followed?

- What kind of changes should be performed only during maintenance windows and what is the process for scheduling one?

- How intrusive can your troubleshooting activities be? Are you allowed to reboot gateways or restart processes that may interrupt existing sessions during work hours?

- If it becomes necessary, who can authorize intrusive troubleshooting, and what is the process for obtaining such authorization?

- Whose responsibility is it to notify parties that may be affected during troubleshooting in your organization, and what is your role in that process?

- Is there an internal escalation procedure? If so, what is it?

- Are you authorized to open Service Requests with Check Point TAC? If so, necessary rights should be granted to you in User Center.

- What is your support tier with Check Point and its associated response times?

This is by no means a complete list of questions to ask and have answers to, but it should give you an idea of what is involved when determining your course of action in advance.

Typical issues and the tools to solve them

There are typical categories of issues and tools that you can use to resolve them. When troubleshooting, you can do yourself no larger favor than to accurately describe the problem that you are trying to tackle, either on your own or when opening a **Service Request (SR)** with **Technical Assistance Center (TAC)**. The following table shows how you may categorize reported issues when beginning your troubleshooting process:

Issue Categorization		
☐ Anticipated	☐ Unanticipated	
☐ Intermittent	☐ Persistent	
☐ Periodic ☐ Random		
☐ Stability	☐ Performance	☐ Connectivity
Scope of Impact		
Business Criticality		
Severity Level =		

Table 15.1 – Troubleshooting issue categorization

Let's go over what each cell refers to:

- **Anticipated issues** refer to any changes in product configuration, objects, policy, or overall infrastructure that you were aware of, correlated in time with the issue's manifestation.

- **Unanticipated issues** are those that appear seemingly without reason in an otherwise perfectly stable environment, with no one reporting (or admitting) that changes were introduced.

- **Intermittent issues** may be periodic or random (or seemingly random) and are often more difficult to pin down.

- **Persistent issues** are stable and reproducible, and thus, easier to troubleshoot.

- Issues can affect stability, connectivity, or performance. **Stability or performance issues**, while rare, can be caused by:

    A.　Peculiarities specific to your infrastructure

    B.　Abnormal traffic patterns that cause some Check Point services to either underperform or periodically crash

    C.　Hardware problems, corrupt filesystems, or databases

    The first two cases of stability or performance issues are typically resolved by working with TAC to identify specific root causes, make necessary low-level configuration adjustments, or install a custom fix.

    The last one is typically solved by an RMA process and/or reinstalling the product and restoring its configuration from backups.

    > **Important Note:**
    > Before initiating RMA, generate and install an evaluation license for the product slated for replacement using process described in *Appendix-1, Licensing*. When Check Point replacing your appliances, licenses synchronized with User Center may get detached from the unstable unit still in production and attached to that shipped as a replacemnt.

    In no way are you expected to be able to address these issues on your own, but for reference, the *sk97638 Check Point Processes and Daemons* article includes their relevant descriptions, paths, log files and their locations, configuration files, commands to stop/start the processes, and, where applicable, links to debug-specific documentation.

- **Connectivity issues** are generally easier to troubleshoot than those related to performance. While both types can be caused by misconfiguration, more often, performance issues manifest after drastic changes in traffic patterns and volumes. I expect the number of these to decrease with the upcoming release of R81.20, due to the improvement in handling what is known as **Elephant Flows** (a heavy traffic volume per single connection) using HyperFlow functionality for appliances with at least eight CPU cores.

- **Scope of Impact** is derived from the originally reported or identified issue and may evolve as more becomes known throughout the troubleshooting process. It can be limited to specific users, hosts in a specific network, a particular application, or the entire company. Even if the **Scope of Impact** is large (for example, the entire company cannot access YouTube), the **Business Criticality** of this specific issue may be low.

On the other hand, it may be that a single user is unable to access a single application that may be critical for the business (for example, international freight scheduling).

Oftentimes, from a specific user's perspective, the issue may be critical, but from an overall business perspective, it may not warrant the possibility of the larger impact that intrusive troubleshooting may have during normal operating hours.

For you, as an administrator, it is helpful to have a list of business-critical applications, as well as business-critical functions, to assign (or verify) the reported issue's priority.

It may also be that this issue is one of several being reported. In that case, document each reported issue unless it becomes apparent that all of them are symptoms of a larger problem. Still, some useful clues can be derived from these reports, such as common scope(s) of impact or time.

By the time you have checked all the boxes and assigned scope and criticality, your internal description will follow this format:

```
Reported by Jane Doe at 08:15 AM July 12th, 2022, this is
an Unanticipated Persistent Connectivity issue with Scope
of Impact limited to select members of Marketing Group
being unable to access any Internet-based resources. This
issue has a Medium Business Criticality.
```

- **Severity level** is a term used by Check Point for TAC Service Requests and escalations. We'll talk about it in the *Service Requests – getting them right every time* section.

Troubleshooting prerequisites

We must have the necessary tools installed on management workstations or VMs. Most of the common tools have been used throughout this book, but here is a slightly expanded list:

- **SmartConsole**.
- Supported browsers (**Chrome, Firefox**, and **Edge**).
- **PuTTY** or a comparable commercial application.
- **WinSCP** or a comparable commercial application.
- **Adobe Acrobat Reader**.
- A text editor capable of dealing with large files (**Notepad++**).
- An archive management tool (**7-Zip** or an alternative, preferably with a hash verification function).

- A screenshot tool or a combination of screenshot and session recorders and editors that can save and annotate captures (**Greenshot**).

- An Office suite, such as **MS Office** or analog, since you will likely have to work with documents, spreadsheets, and occasional **PowerPoint** presentations.

- The Check Point **IKEView** utility for VPN troubleshooting – *sk30994 What is the IKEView utility?*.

- A network and port scanner, such as **Nmap**.

- **Wireshark**.

- **VISIO** or an alternative diagramming tool, such as the offline version of **Draw.io**.

- Ensure you bookmark `https://codebeautify.org/ip-to-hex-converter` (the site contains a multitude of useful converters and calculators).

- Reference and administration guides for all the versions of the Check Point products you are using. Open *sk170416 Check Point R81.10 | Documentation*, click on the relevant guides to open their online versions and download their PDFs:

  - *Installation and Upgrade Guide*

  - *CLI Reference Guide*

  - *Gaia Administration Guide*

  - *Quantum Security Management Administration Guide*

  - *Logging and Monitoring Administration Guide*

  - *Quantum Security Gateway Guide*

  - *Threat Prevention Administration Guide*

  Feel free to get any other guides that apply to your environment.

- An additional *tool* is the virtual Check Point lab, where you can experiment with a variety of commands, configuration options, and tools before using them in production.

We are used to having access to information online at our fingertips. Unfortunately, if your primary gateway (in extreme cases) is down and you do not have access to the internet, it pays to have a contingency plan. Have at least one laptop that contains all the current software necessary for administration and troubleshooting purposes, as well as a pre-approved process for using alternative internet access (such as a personal cell phone's hotspot) in place. Have a USB-to-RS232 RJ45 console adapter cable and a USB-to-Ethernet adapter with drivers pre-installed on that laptop that are kept in the same bag. It is also a good idea to have a 10 to 25-foot (or 3-8 meter) Ethernet patch cable in the same bag.

Even if you have a robust out-of-band management infrastructure with a dedicated network and **Lights Out Management** (**LOM**) and/or **Console Servers**, this may be handy if you have to pre-stage the appliances before deployment.

Additionally, even if you have an established ticketing and service management system, at the beginning of troubleshooting a newly reported issue, create a new document where you can track the entire progress of the case. This is the place where you may have your running notes, screenshots, comments, questions for follow-up, or subjects and commands to look up after the crisis has been dealt with. These materials are a valuable source of information for postmortems and internal knowledge repositories.

External monitoring tools can aid in identifying possible causes, times, and the Scope of Impact of the issues. Having a NetFlow collector and analyzer in your environment may help you spot traffic pattern anomalies between the last good known state and the time the issue has been reported. Systems monitoring solutions help spot stability issues that otherwise may go unnoticed (this is typical in HA environments with poor monitoring practices).

Equally helpful are automated service monitoring solutions for business-critical applications. If you can confirm that the application that users are experiencing issues with is in a good state, then you can remove it from the list of suspects.

Having Check Point-specific robust test protocols for business-critical applications in place may save you hours of troubleshooting and early gray hair. If test protocols are automated/scripted and either continuously running or executed before and after significant changes in Check Point configuration or other planned infrastructure changes, they may alert you to the problem(s) and let you address those in a timely fashion. Even manual test protocols with checklists, so long as they are regularly updated, are of huge help.

Perhaps close to a third of all troubleshooting sessions I have been involved in have ended up being caused by infrastructure components other than Check Point gateways. That's why, at the beginning of this book, I emphasized the importance of having access to current, accurate diagrams of the environment you are working in, and understanding the dependencies that may cause Check Point gateways to malfunction.

Switching, routing, DNS, timekeeping, and proxy servers have been the culprits in these cases. Accordingly, we should rule those out before diving into Check Point troubleshooting.

Since we hardcode the DNS servers, time servers, and (optionally) proxy servers in a gateway's configuration, check if those are accessible and functioning. I've seen situations where IT has deprecated some instances of these servers or moved them to different addresses in the infrastructure, with mayhem ensuing for firewall administrators.

Additionally, if we rely on variable objects in our policies, the possibility of changes being made outside the Check Point environment should be considered. For example, in the case of Jane Doe described previously, the issue could be caused by the Active Directory administrator moving users to a different AD Security Group. It could – and probably should – be addressed by creating additional rule(s) with a new Access Role in our policies, but we may have to get a request and approval for that first.

Courtesy of Timothy Hall from Shadow Peak Inc., the following table (slightly modified by me), helps break down the functions specific to the Gaia OS and Check Point Product Code running on it:

"Thin Pink Line" of Check Point	
Gaia OS	Check Point Product Code
Interfaces	Firewall Policy
Routing – Static and Dynamic	NAT
DHCP Relay	VPNs
Hostname	APC/URLF
DNS/NTP Servers	Threat Prevention
CPUSE (Update Service Engine)	HTTPS Inspection
	Traffic Logs
Access Methods	
SSH/Console/LOM	SmartConsole GUI
CLISH	Management CLI
Expert Mode	Management API
Gaia Web Interface (WebUI)	GUIdbedit.exe/dbedit
Gaia API	
Central Deployment Tool	

Table 15.2 – Gaia OS and Check Point Product Code functions and access methods

This will help you focus your troubleshooting efforts based on initial data and surface analysis, as well as choose the appropriate tools.

Stability issue troubleshooting example

For a quick and detailed assessment of the possible issues that affect the stability of your gateways or servers, the built-in **HealthCheck Point** (**HCP**) is your first go-to tool.

Extensive information about the tool, the tests it is capable of running, and its revision history are available at *sk171436 HealthCheck Point (HCP) Release Updates.*

In a nutshell, you can easily run a low-impact battery of tests to generate very comprehensive reports and recommendations by simply executing the hcp -r all command [1] in **Expert mode**. Errors [2] and warnings [3] are immediately highlighted and you will be presented with instructions on how to view reports locally in the CLI [4] or how to download a comprehensive interactive version of the report:

```
[Expert@CPCM1:0]# hcp -r all
Test name                                              Status
=================================================================
ARP Cache Limit........................................[PASSED]
Bond Health............................................[SKIPPED]
CPview Diagnostic......................................[PASSED]
Check Point Processes..................................[PASSED]
Cluster................................................[PASSED]
Connectivity to UC.....................................[PASSED]
Core Dumps.............................................[ERROR]
Custom Applications RegEx..............................[PASSED]
Debug flags - FW.......................................[PASSED]
Debug flags - fwaccel..................................[PASSED]
Disk Space.............................................[WARNING]
Dynamic Objects Database...............................[PASSED]
FW Configuration File Sanity...........................[ERROR]
File Descriptors.......................................[PASSED]
Gaia DB................................................[PASSED]
IPv4 forwarding........................................[PASSED]
Identity Awareness - Sharing mechanism error...........[PASSED]
Identity Awareness - tables limit......................[PASSED]
Identity Awareness - tables mismatch...................[PASSED]
Interface Errors.......................................[PASSED]
Kernel crash...........................................[PASSED]
Local Logging..........................................[PASSED]
Memory Usage...........................................[PASSED]
Neighbour table overflow...............................[PASSED]
SIC....................................................[PASSED]
SIM Configuration File Sanity..........................[PASSED]
SecureXL status........................................[PASSED]
Soft lockup............................................[PASSED]
Zombie processes.......................................[PASSED]

Generating Topology....................................[Done]
Generating Story.......................................[Done]
Generating Charts......................................[Done]

To view full report on this machine, run "hcp --show-last-full"

To view report as html file. Copy /var/log/hcp/last/hcp_report_CPCM1_19_06_22_11_2
9.tar.gz to your desktop, extract the tar content and open the index.html via your
web browser
[Expert@CPCM1:0]#
```

Figure 15.1 – HealthCheck Point diagnostics execution and summary

Both options are excellent, but I recommend downloading reports for review, future references, and the ability to share with others on your team without the need to look at them on gateways or servers.

The downloaded report, after being extracted from the .tgz archive, can be opened in a browser [1], presenting you with the opportunity to use a tabbed and searchable interface to easily pinpoint the issues and see zoomable graphs (in the **Charts** tab). Problematic areas are color-coded [2] and by scrolling through each area of concern [3], specific issues can be identified [4]. Recommendations on how to handle them can be found in the **Finding** section [5]:

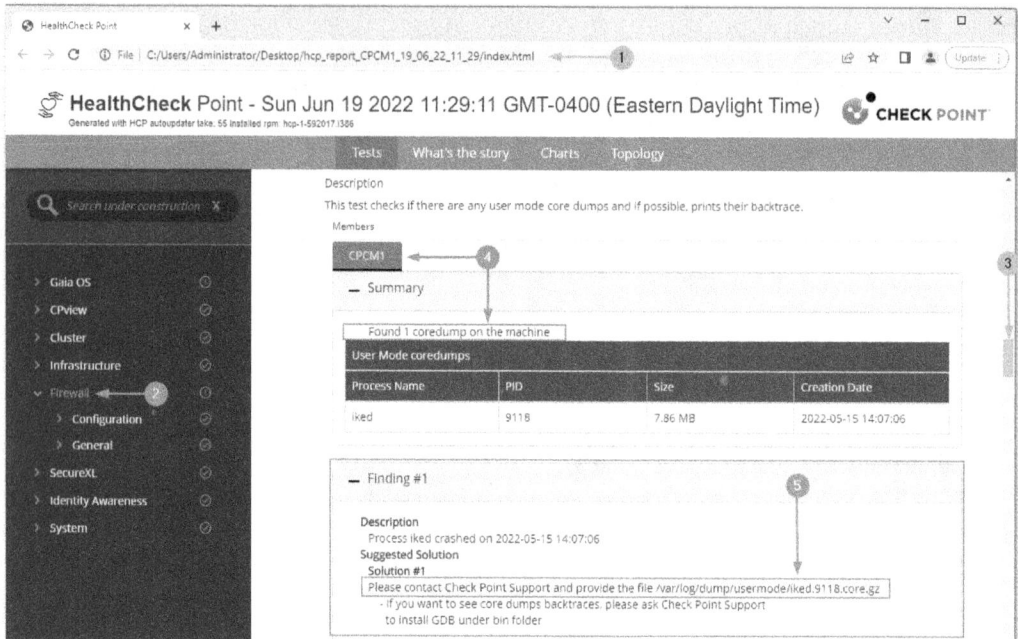

Figure 15.2 – HealthCheck Point HTML report

As you can see, there is a specific recommendation to reach out to Check Point support and explicit instructions on which file should be uploaded to help resolve a particular issue.

Troubleshooting intermittent issues

For troubleshooting intermittent issues, the **HealthCheck point** report's **What's the story** tab allows you to page through the days of the report to see if anything of interest has occurred during the day(s) the issue was observed.

Additionally, the other tool that was mentioned a few times throughout this book, **CPView** (*sk101878 CPView Utility*), provides us with exceptional granularity in terms of both the time of the issue's occurrence and the information available for analysis at that time.

The `cpview` command can be executed in either CLISH or Expert Mode. If executed by itself, you will be presented with the real-time monitoring dashboard for devices, statistics, counters, and alerts. When the same command is executed with the `-t` option, `cpview -t`, you get the option to navigate the state timelines in 1-minute increments. If you're in this mode, which is indicated by the presence of the **HISTORY** banner, and you press *T* on your keyboard, you can enter the target date and time in `[Jan...Dec]  [01...31] [4-digit Year]  [hh:mm:ss]` format – for example, `May 23 2022 00:00:00`. Then, you can use +/- to move around the time of the issue you are investigating while looking at the likely target view.

In the following example, we have used `cpview  -t` to determine that every time, shortly before the reported issues, an application signature update is taking place.

Enter a target date and time [1], navigate to the potential area of interest using the arrow keys [2], and increment the time using +/- to see if any changes have taken place [3]. Note the `Application  Control` blade version change [4 and 5] and the time this happened [6]:

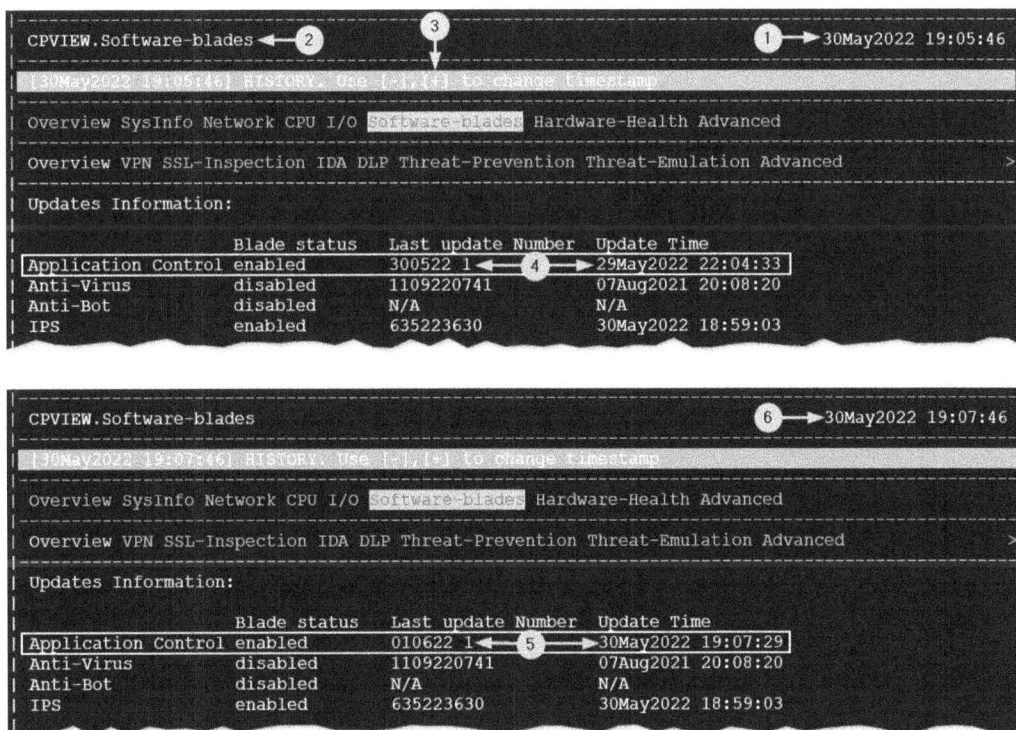

Figure 15.3 – Using cpview -t for intermittent issue troubleshooting

If we see a similar occurrence before the next or a previous time we have experienced an identical issue, this is our culprit.

This was a made-up scenario illustrating the process and is not something you will likely see in production.

Troubleshooting connectivity issues

You can determine the time the issue was noticed and, if it applies to the previously known working connections, the last time it was absent. This will allow you to focus your investigations on a tighter timeframe.

Determine the Scope of Impact. If it is relatively limited, it is unlikely to have been caused by a system-wide issue.

If possible, determine the availability and accessibility of the resource(s) the users are having difficulty accessing. If, for instance, it is a cloud-based service, have `https://downdetector.com/` bookmarked and search for either the service you are experiencing issues with or the cloud platform that you know it is running on. **The results may not be conclusive proof of the root of the problem, but they should be considered.**

Check the logs for the **Accept** or **Drop** actions for relevant traffic. If you do not see anything, check the applicable rule's tracking settings.

Check the logs using the `type:"System Alert"` filter. Here, you can see **Alert**(s) [1] with **Description** set to **Domain resolving error**. Check the DNS configuration on the gateway [2] that started around the time the issue manifested [3]:

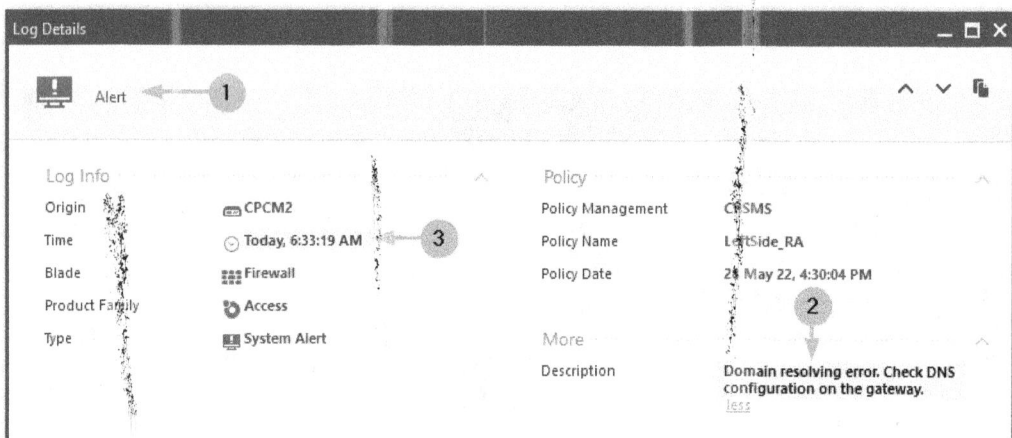

Figure 15.4 – System alert – domain resolving error

There is a high chance that a system-wide issue will affect APCL/URLF, which are variable objects that rely on name resolution as well as Check Point cloud-based services. If, on an affected gateway, you are only using the **Firewall** and **VPN** blades, and very few of your rules contain **Domain** or **Updatable** objects, its impact may be limited and not immediately apparent.

Needless to say, you must fix the DNS on the gateway before moving on.

If you do not see any logs from the client initiating the connection, this may indicate that either you are not logging it or that the traffic does not make it to the gateway. Run a packet capture on the ingress interface to verify this. If no packets are inbound, the issue is between the client and the gateway. If you do see the packets, a more thorough investigation is in order.

Use **Packet Mode** filters in your policy (or all **Ordered Layers** of your policy) to determine which rule is being matched first. If it is not the desired rule, check for the changes to either the policy, variable objects, or **Groups** containing the variable objects present in the matching rule.

If the connection logs indicate the **Accept** action, do the following:

1. Check that the **Policy | Reason** field of **Log Detail** contains **Connection terminated before the Security Gateway was able to make a decision: Insufficient data passed**. To learn more, see *sk113479*.

2. Check that the egress interface in the logs is the appropriate one for accessing the resource(s) in question.

3. Check that the NAT information is correct.

If you see the message mentioned in the first point and that both the latter points are correct, you may conclude that the issue is not on the client side and that no responses are seen from the target resource. You can use `tcptraceroute` with a specific protocol and port from your gateway or active cluster member to attempt to verify its availability and accessibility. For instance, `tcptraceroute -i eth4 -T -n www.google.com -p 80` allows us to check if the host, `www.google.com`, is responding on TCP port `80`:

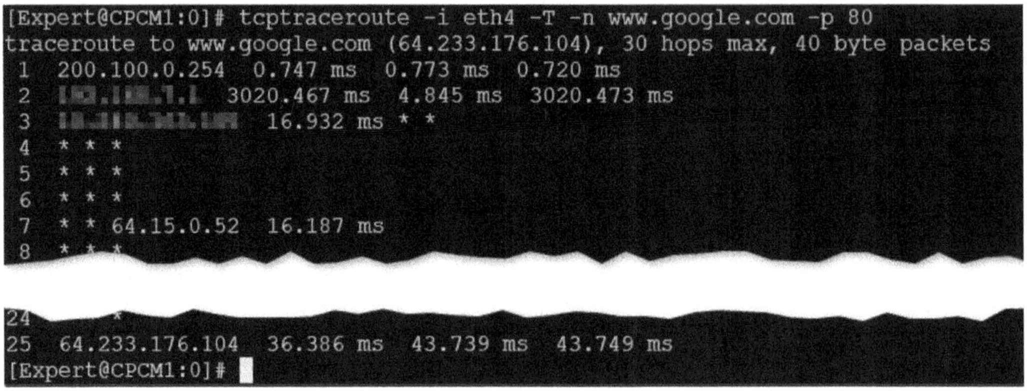

```
[Expert@CPCM1:0]# tcptraceroute -i eth4 -T -n www.google.com -p 80
traceroute to www.google.com (64.233.176.104), 30 hops max, 40 byte packets
 1   200.100.0.254   0.747 ms   0.773 ms   0.720 ms
 2   ███.███.1.█   3020.467 ms   4.845 ms   3020.473 ms
 3   ██████████████   16.932 ms  *  *
 4   *  *  *
 5   *  *  *
 6   *  *  *
 7   *  *  64.15.0.52   16.187 ms
 8   *  *  *
24                    *
25   64.233.176.104   36.386 ms   43.739 ms   43.749 ms
[Expert@CPCM1:0]# ▋
```

Figure 15.5 – Using tcptraceroute to verify the availability and accessibility of resources

Unfortunately, we cannot spoof the source address of the query with that of the original client experiencing issues using this command.

If you must spoof the source address (that is, it is NATed to a specific IP on your side and that IP address is what the remote host is expecting), we can do that too by using the `hping` command coupled with `tcpdump` running in the second session. It is a bit clumsy and was previously possible to accomplish using a tool called `pinj` (Packet Injector), which is no longer supported.

In the following example, in the first session [1], running `fw hastat` [2], we can verify that it is an active cluster member [3] and send a TCP SYN to port `80` of www.google. com using `hping -S -I eth4 -p 80 www.google.com` [4]. In the second session [5], we are running a packet capture on the same interface while looking for responses from www.google.com using `tcpdump -I eth4 -nnv host www.google.com` [6]. We will see our SYN packet [7], as well as a SYN-ACK response [8]:

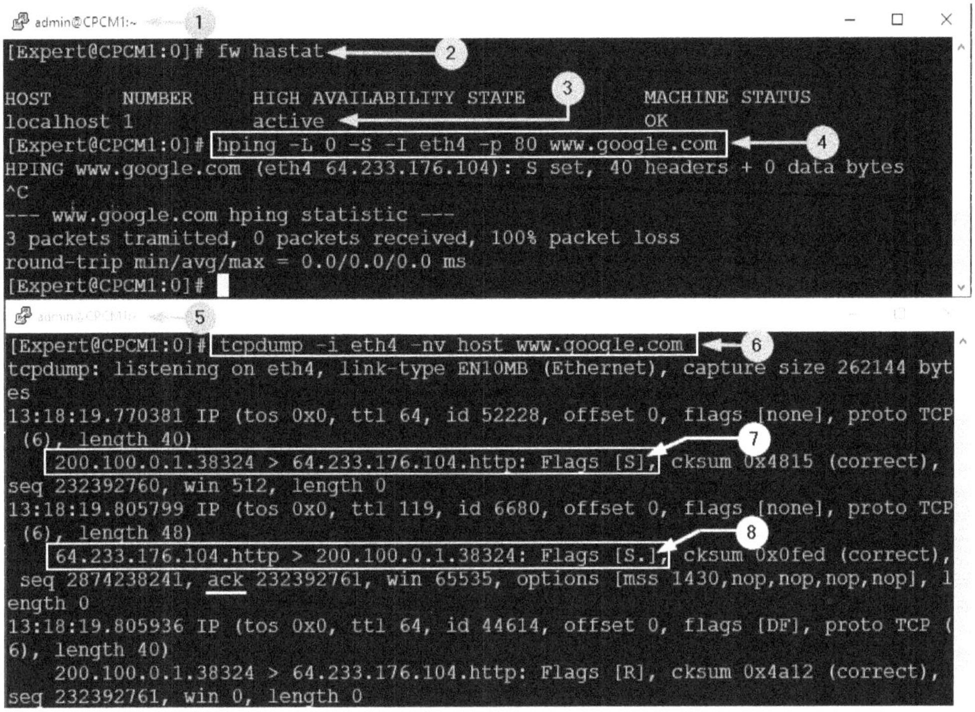

Figure 15.6 – Using hping to access the resource's availability and accessibility

If we need to spoof the source address to simulate traffic from the original client experiencing the connectivity issue, we can use the -a option. Our command will look like this:

```
hping -a 10.0.0.20 -S -I eth4 -p 80 www.google.com
```

If you cannot successfully tcptraceroute to the destination or see SYN-ACKs using hping and tcpdump, the issue is probably on the server side. Check the logs using the original destination as a source in the filter to verify that responses are not being dropped by anti-spoofing. If you do see that happening on the incorrect interface, you have an asymmetric routing issue.

In addition to the standard packet capture tool, tcpdump, Check Point is equipped with the cppcap, fw monitor -e, and fw monitor -F tools. There is an ongoing debate regarding the comparative advantages and shortcomings of each. fw monitor -F is rather special in that it performs an "inline" capture of "accelerated" traffic, tracing the progress of the packets through all inspection chains. If your packets are arriving on one interface and mysteriously disappearing, instead of departing out of the other, this is one of the tools that can be used to understand the reasons for that behavior, as well as the NAT operations at multiple inspection points. My take is that all of these are valid choices under different circumstances.

I suggest to everyone working with Check Point to take a look at Tim Hall's presentation on this subject at `https://community.checkpoint.com/fyrhh23835/` `attachments/fyrhh23835/member-exclusives/484/2/CPX_Preso_` `TimHall_FINAL.pdf`.

> **If You Decide to Skip It, I'll Echo His Warning About One Particular Issue**
>
> In many Check Point documents, older forums, or blogs, you will see the recommendation of executing `fwaccel off` before performing a packet capture or other troubleshooting activities. *DO NOT DO IT!* Executing it in production environments on multicore firewalls may result in significant problems.

I also highly recommend Tim Hall's *Max Capture: Know Your Packets* self-guided video series, which is available for purchase at `http://www.maxpowerfirewalls.com/`.

Additionally, `fw ctl zdebug` drop and its extended version, `fw ctl zdebug + drop`, can be used to determine the reasons for packet drops, even if logging is not configured for the rules or security features responsible for it. Keep in mind that, unlike with packet capture, only the output of debug utilities can be filtered. Given that there is a limited size buffer for debugging, it is possible to miss the relevant data on a busy gateway. Try to time the execution of the tests immediately after debug commands are invoked (kudos to Valeri Loukine for his post reminding me of this).

I suggest bookmarking and studying the following **Secure Knowledge** articles, which are relevant to packet capture and troubleshooting in general:

- *sk141412 cppcap – A Check Point Traffic Capture Tool*
- *sk30583 What is FW Monitor?*
- *sk167457 How to use the fw ctl zdebug command to view drops on the Security Gateway*
- *sk100808 How to use "fw ctl zdebug" command*
- *sk171943 Advanced "fw ctl debug" features*

Another interesting resource is a TCPDUMP101.com [1] online tool that includes, among others, general **TCPDUMP** [2] and Check Point-specific packet capture query builders for **FW MONITOR, CPPCAP,** and **KERNEL DEBUGS** [3]. Read the half-page long landing page description on how to use the tool [4]:

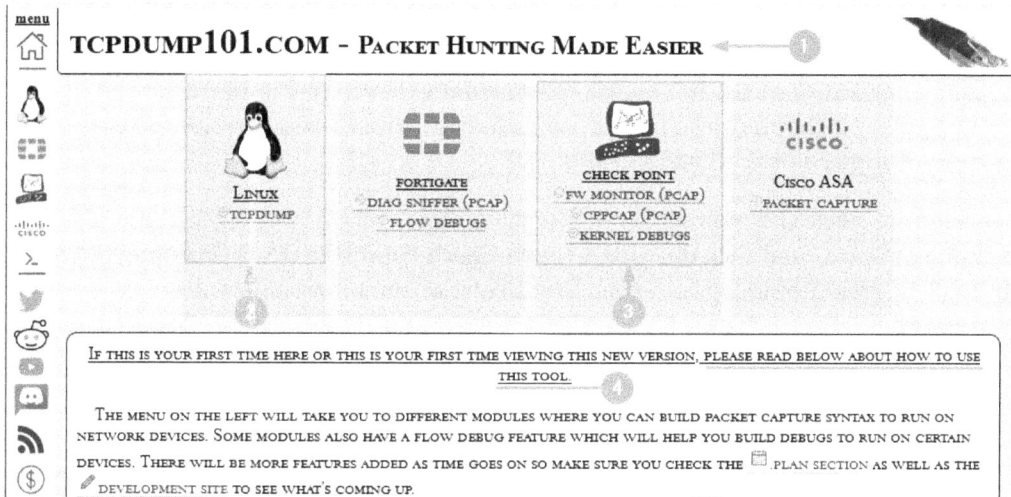

Figure 15.7 – TCPDUMP101, the capture and debug query generator

TCPDUMP101 can be downloaded and run in a browser of your choice.

When you have tried your best to identify the issue and remedy it yourself and are still having no luck, it is time to involve either TAC Ghostbusters by opening a Service Request, which will be described in the next section, or to reach out to your **Check Point Partner**, if they are providing first line support for your organization. It may also be a good time to call on the Check Point community, the CheckMates, for advice and suggestions from your peers.

Service Requests – getting them right every time

Let's go through a few preparatory steps by downloading and bookmarking some resources:

- References to support tiers and response times should be downloaded from `https://www.checkpoint.com/support-services/support-plans/`. Highlight your current support plan.

- The **Check Point Direct Support Program Service Level Agreement** document can be found at `https://www.checkpoint.com/downloads/support-services/support-sla.pdf`. Look for the *"Severity" Definitions for Network Security product(s)* section.

- Align your internal categorization of the issues with the severity levels described in the document. This should help you determine the expected response times and escalation thresholds.

- For escalation, bookmark **Online Escalation Form**, which can be found at `https://www.checkpoint.com/support-services/online-escalation-form/`. In the case of **Immediate/ Severity-1** issues, please go to `https://www.checkpoint.com/support-services/check-point-tac-support-escalation-path/`.

To learn how to create and monitor Service Requests, bookmark `https://help.checkpoint.com/` and click this URL to open **Check Point BEYOND Customer Success Hub**. You'll be prompted to log in to your User Center account if you aren't already:

1. If it is your first time here, take the short, guided video tour. It will only take a few minutes but will save you time later.

2. If this is a new Service Request, click **CREATE A NEW SR**.

3. Click on **Technical Product Issue (On-Premise Product)**:

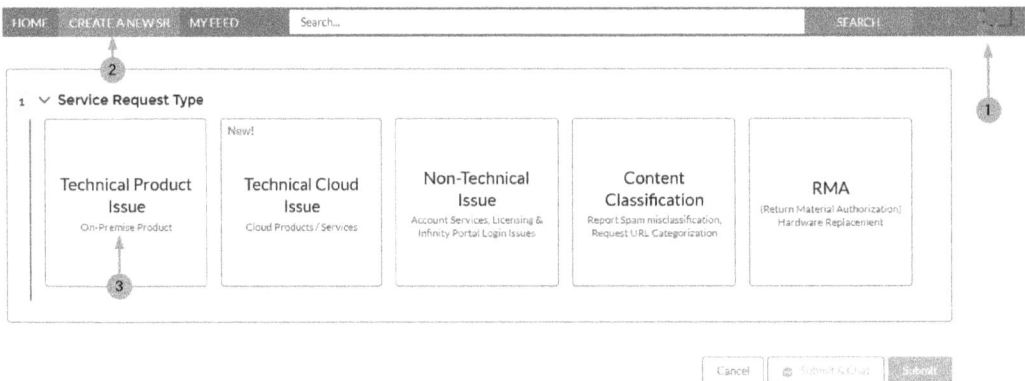

Figure 15.8 – Check Point BEYOND Customer Success Hub

4. Provide a short but concise description of the issue.

5. Select the accurate severity level.

As administrators working in production environments, we have a major imperative to resolve any outstanding troubleshooting issues as soon as possible. If the issue we are troubleshooting does not impact any production traffic, the priority and corresponding severity level of the SR should be low.

If production traffic is impacted and our internal efforts were not enough to resolve the issue, it could be higher. Use Business Criticality as a guideline to claim a specific severity level for SR.

Irrespective of the internal **Business Criticality** level, one thing that should be kept in mind is that the issue should be **isolated to software**. This means that you have performed your due diligence in asserting that it is, indeed, a Check Point problem.

Let's look at Check Point's definitions of **SEVERITY LEVELS**:

- **Severity 1 = Critical**: An error isolated to software that causes the product to fail catastrophically (for example, a major system impact, system down, and so on).

- **Severity 2 = High**: An error isolated to software that substantially degrades the performance of the product (for example, moderate system impact, system hanging, and so on).

- **Severity 3 = Medium**: An error isolated to software that causes only a minor impact on the use of the product.

- **Severity 4 = Low**: An anomaly in the licensed product that does not substantially restrict the use of the licensed product to perform necessary business functions.

Here is my interpretation of *errors isolated to software* and *anomalies in the licensed product*:

- You have configured the product according to the information provided in the user or administration guides that have been written for the currently supported versions of the product in production, and it does not perform in the expected fashion.

- Even if you have been steered by TAC to a specific SK article that contains information that is contrary, or supplemental, to the relevant subjects in these guides, it is still on TAC to provide you with assistance in solving the issue.

- You have configured the product according to the SK article that's been written for the versions of the product running in production and it does not perform in the expected fashion.

- You are having difficulty pinpointing the problem, but the Check Point product is a suspect.

> **Note**
> If you are simply planning to implement new features or are in the process of reconfiguring the existing implementation, this is not a case for TAC but Professional Services. These are provided either directly by Check Point or by your Check Point partner with their assistance. Your local Check Point **Sales Engineer** (**SE**) could also be a good resource to reach out to in these cases.

Continuing with the SR creation process, do the following:

1. Answer questions regarding the product line, JHF take number, version, and operating system.

2. Upload relevant screenshots, command outputs saved as text files, logs, and/or diagnostics files that you may have handy. If you do not have those at the moment, continue creating the SR. You'll be able to upload those later. Keep in mind that, unlike you, TAC engineers are not familiar with your environment. If you are troubleshooting routing or switching-related issues, providing simple diagrams with relevant data will be helpful.

3. In **Add Contact Information**, specify the emails of additional people involved in this issue's troubleshooting, as well as all those you must keep in the loop at a technical level according to internal procedures.

4. Select **Preferred Contact Method**.

5. When opening a new SR, include all individuals that are likely to participate in the troubleshooting process in its notifications list.

6. Unless your SR is of the lowest level (**Question**), the preferred method to initiate it is to click on **Submit & Chat**. This way, you can be sure your request is routed to the correct queue. Do not expect anything to be resolved during this chat session. You will simply be notified that the engineer was assigned to your case.

7. If creating this SR does not warrant a high **Severity** level, but it has a very high **Business Criticality** level, clearly state that in the subject and the description of the case. Additionally, ask the engineer assigned to the case to help you escalate it because of the **Business Criticality** level if it is taking too long to address.

HCP and CPView, tools that we have discussed earlier, produce helpful information that can be uploaded to SR. Generally, TAC prefers to use the files that have been generated by another tool, referenced in the *sk92739 CPInfo utility*. It is executed on the management server and the gateway, active cluster members, or all cluster members that are experiencing the issue.

The output of **CPInfo** provides TAC with as close of a picture of your CheckPoint environment as possible, allowing for a more informed troubleshooting process.

The following is an excerpt from *sk92739*:

> *"Do not run the CPInfo tool if the current CPU utilization of at least one CPU core is greater than 70%. We recommend collecting the CPInfo output during the maintenance window."*

Use either the `top` or `cpview` command to see current utilization data in the **CPU** tab.

You may or may not have the luxury of waiting for the maintenance window. If you are pressed for time and asked to upload CPInfo data but are uncomfortable running this tool because of the aforementioned warning, ask the TAC engineer to do it for you.

If you intend to run `cpinfo` yourself during a maintenance window, ask the TAC engineer for the detailed syntax of the command, along with all the flags necessary to produce the results they require.

SRs have a 25 MB upload limit. CPInfo allows you to directly upload the generated files to SR, but we do not have an idea of how large those files may be. Therefore, I recommend writing the output to the `/var/tmp/` directory. Once you see that they are of acceptable size, you can upload them to SR. Alternatively, if the files are larger than 25 MB, ask TAC to provide you with the SFTP upload target and credentials.

During a typical troubleshooting session, you may be asked to grant the TAC engineer access to console sessions. In this case, it may be prudent to have accounts provisioned for TAC (in Gaia for, relevant components, and in SmartConsole) that will be disabled afterward. Otherwise, you will have difficulties differentiating between your actions and those of TAC engineers.

Ask TAC engineers to briefly narrate all the actions they are performing and take notes on a second computer. This will serve two purposes: you'll learn something, and it will make it easier for you to determine if progress is being made during the troubleshooting session.

If you notice that engineers' actions are taking a circular pattern, politely ask for additional TAC resources to be involved.

If you are asked more than twice to perform debugging sessions on your own (with detailed instructions supplied by TAC) and to upload the results to SR, request a scheduled remote troubleshooting session with explicit instructions for TAC to collect all the necessary debugging data during that session. Ask if there are any specific tests you can execute during that session to provide TAC with all the data necessary for accurate diagnostics.

After each troubleshooting session, if a solution hasn't been provided, ask when you should see an update regarding the status of the SR in writing, for your reference and accurate case progress tracking.

If no timely update is provided or if it is unsatisfactory, politely request an escalation. Your time is as valuable as that of the TAC engineer.

Depending on the complexity of the issue, troubleshooting sessions may take anywhere from a few minutes to many hours. If it looks like the latter case, make internal arrangements to be engaged with TAC for the anticipated duration of the troubleshooting session.

TAC engineers work in shifts. Depending on when you have scheduled your troubleshooting session, your case may be handed over to another engineer less familiar with the progress of the SR. If you want to avoid such situations, pre-arrange the time for a remote session earlier in the shift with the engineer that's been assigned to the SR.

When the issue is conclusively solved, close the SR. No one likes to have open tickets assigned to them longer than necessary.

SRs can be reopened if issues resurface. Try to determine if it is indeed an issue you were experiencing before or a similar one. The difference is that you may not have to go through all the preliminary steps of initial diagnostics unnecessarily.

If you are asked to provide feedback after the SR, be constructive and accurate in your statements.

An additional way to track your SRs on the go is by using *sk91860 User Center Mobile Application*. When signed in, the **Salesforce Beyond** option takes you to the mobile version of **Check Point BEYOND Customer Success Hub**.

TAC and JHFAs

In most cases, especially when you are opening SR with lower severity, Check Point engineers check your currently installed **Jumbo Hotfix Accumulator (JHFA)** take number (version) and will recommend an upgrade to the most recent Recommended General Availability JHFA.

While this is considered best practice, I cannot recommend this course of action without first doing the following:

1. Ask the engineer to produce specific references to the issue you are experiencing being addressed in a specific version of JHFA.

2. Perform a Google search for any problems associated with the Recommended General Availability JHFA. Inquire in CheckMates if anyone has experienced problems with the latest JHFA.

It does not happen often, but there are cases where features that were perfectly stable in one version start experiencing problems in later JHFAs. So, before you undertake the potentially unnecessary task of starting a JHFA upgrade, at least verify that it'll get the current issue resolved.

You can investigate this before opening an SR by going to **Released Hotfixes | Jumbo Hotfix Accumulator for R81.XX** [1], selecting **List of All Resolved Issues and New Features** [2], and searching for the relevant keywords [3]. If Take ## is higher than the one currently installed [4], read the filtered results to see if any of those are relevant to the issue you are troubleshooting [5] and are applicable to the product in question [6]:

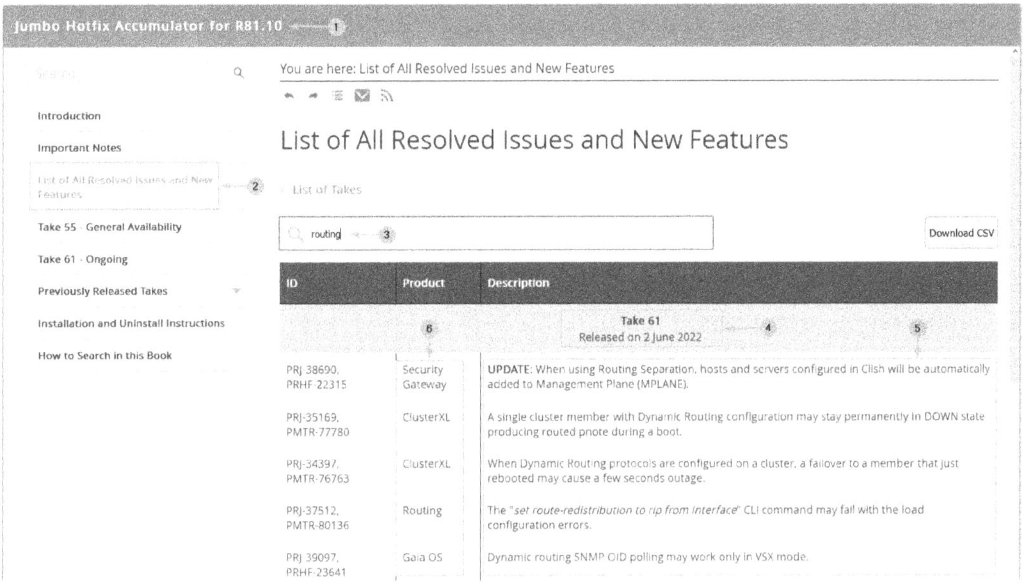

Figure 15.9 – JHFA searching for resolved issues in relevant products

At the end of the troubleshooting process with TAC, politely ask the engineer that's been assigned to the case to provide a one-paragraph summary of how the issue was identified and resolved. They do not have to do this, but it can be helpful to have it as a closing note in SR.

Community resources and engagements

I'll start this section with a brief trip down memory lane. Established in 2017, the Check Point user community known as CheckMates was slowly gaining traction. One of the people responsible for its development, Check Point Chief Evangelist, self-proclaimed and universally acknowledged geek, Moti Sagey, started a thread called *My Top3 Check Point CLI commands*. It tells you something about the community when it became the most visited place with a huge number of responses. Some comments were simple lists, while others expanded on their choices. A smaller number of users were posting scripts by using basic Check Point commands but extended their capabilities by adding formatting and interactivity (referred to as "one-liners"). By page 4, "one-liners" were becoming wrapped into three or four lines. Of course, things got out of hand and some became "one-pagers." To get an idea about the size and complexity of these scripts, I heartily recommend checking out Heiko Ankenbrand's *One-liner collection* at `https://community.checkpoint.com/t5/General-Topics/One-liner-collection/m-p/57994#M11705`.

Then, one of the contributors, Danny Jung, posted, *"I've put all of your commands into this script Common Check Point Commands (ccc)."*

It started as a relatively compact console UI for executing a limited number of commands, but it had such an overwhelmingly positive response from the community with corrections, suggestions, and contributions by other members, as well as Danny's unbelievably fast responses in continuous improvements, that it has evolved into a real powerhouse. To put this into perspective, consider that it contains hundreds of years of the combined wisdom of Check Point administrators, engineers, and master scriptwriters rolled into a simple, user-friendly interface.

The script, at the time of writing, is on version 4.9 and has won *Code Hub Contribution of the Year 2018*, endorsed by Check Point Support, became widely adopted and has become an international hit (I am being quite literal here: CheckMates are now hailing from more than 180 countries).

It can be installed and executed on management servers, gateways, or standalone management and gateway units or VMs. Once installed, ccc is invoked by the same three letters from Expert Mode.

The main screen of the script indicates the host it is running on [1] and it contains a wealth of diagnostics information, including relevant system data [2], management data [3], current VPN statistics, and the most recent signature updates [4], interface types, sync interface(s), backup configurations, and RAID states [5] with areas of concern highlighted in red. And this is just the passive section. Under the MAIN MENU heading, you can find shortcuts to the purpose-specific scripts [6]. For clarification, the Firewall Management & Gateway option [7] refers to a **standalone management and gateway unit**:

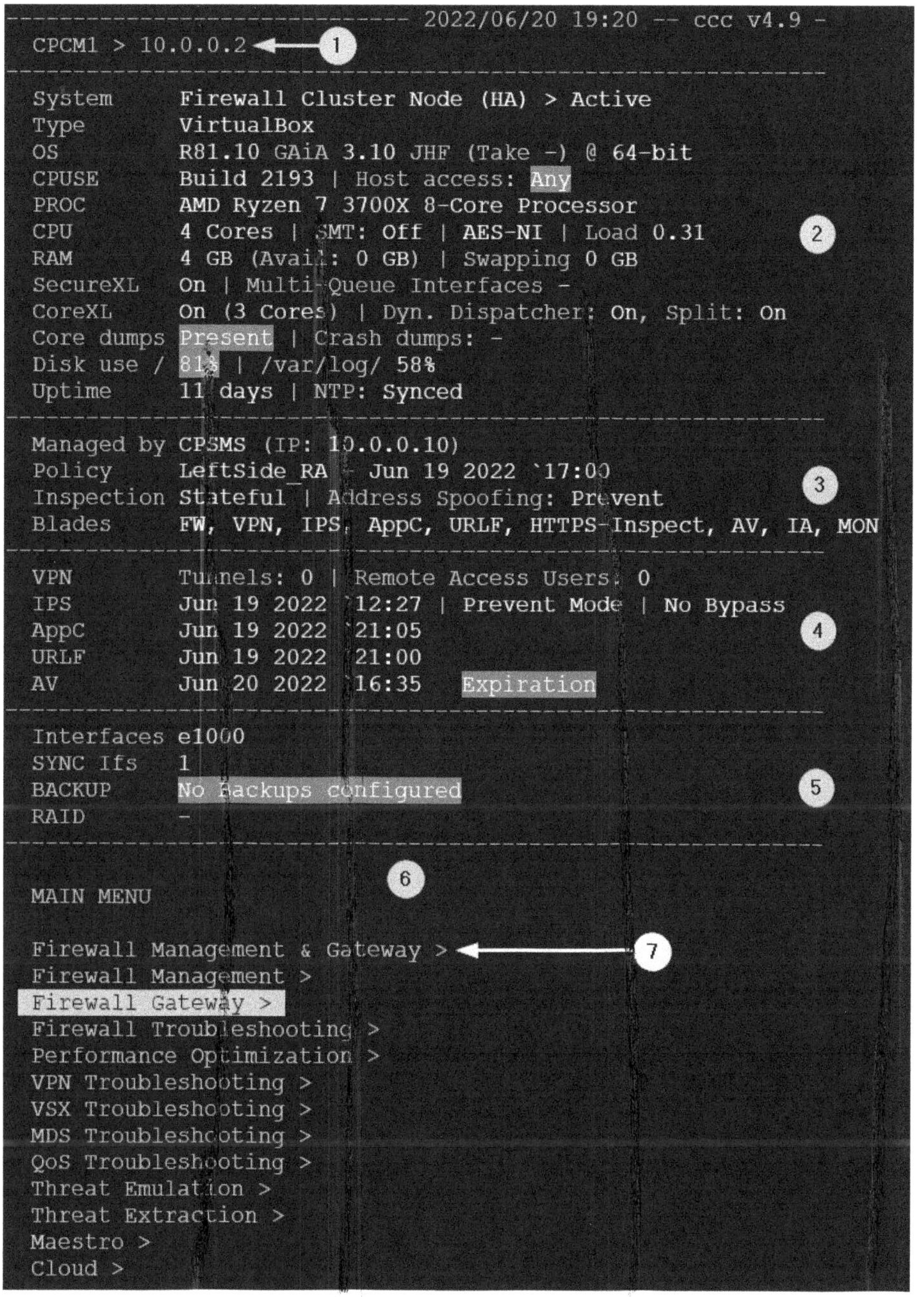

```
---------------------------- 2022/06/20 19:20 -- ccc v4.9 -
CPCM1 > 10.0.0.2 ◄─── 1
---------------------------------------------------------------
System      Firewall Cluster Node (HA) > Active
Type        VirtualBox
OS          R81.10 GAiA 3.10 JHF (Take -) @ 64-bit
CPUSE       Build 2193 | Host access: Any
PROC        AMD Ryzen 7 3700X 8-Core Processor
CPU         4 Cores | SMT: Off | AES-NI | Load 0.31        2
RAM         4 GB (Avail: 0 GB) | Swapping 0 GB
SecureXL    On | Multi-Queue Interfaces -
CoreXL      On (3 Cores) | Dyn. Dispatcher: On, Split: On
Core dumps  Present | Crash dumps: -
Disk use /  81% | /var/log/ 58%
Uptime      11 days | NTP: Synced
---------------------------------------------------------------
Managed by CPSMS (IP: 10.0.0.10)
Policy      LeftSide_RA - Jun 19 2022 `17:00
Inspection Stateful | Address Spoofing: Prevent            3
Blades      FW, VPN, IPS, AppC, URLF, HTTPS-Inspect, AV, IA, MON
---------------------------------------------------------------
VPN         Tunnels: 0 | Remote Access Users: 0
IPS         Jun 19 2022 `12:27 | Prevent Mode | No Bypass
AppC        Jun 19 2022 `21:05                             4
URLF        Jun 19 2022 `21:00
AV          Jun 20 2022 `16:35   Expiration
---------------------------------------------------------------
Interfaces e1000
SYNC Ifs   1
BACKUP     No Backups configured                           5
RAID       -
---------------------------------------------------------------
                        6

MAIN MENU

Firewall Management & Gateway >          ◄─── 7
Firewall Management >
Firewall Gateway >
Firewall Troubleshooting >
Performance Optimization >
VPN Troubleshooting >
VSX Troubleshooting >
MDS Troubleshooting >
QoS Troubleshooting >
Threat Emulation >
Threat Extraction >
Maestro >
Cloud >
```

Figure 15.10 – ccc script – main screen

Let's take a look at the contents of the FIREWALL GATEWAY menu, shown in the previous screenshot, to get an idea of what is there. Here, we can see the path to the current selection of tools [1]. Below are the base commands [2] that are used in each script and an explanation of the functions [3]. Intrusive commands, that may have negative impact on the target, are presented in toggle pairs, with the risky ones colored in purple [4]:

```
------------------------- 2022/06/20 17:56 -- ccc v4.9 -
CPCM1 > 10.0.0.2
---------------------------------------------------------
MAIN < FIREWALL GATEWAY                                                    3

 fw stat; ips stat; fw stat -b AMW; cpstat -f all polsrv; cp_conf sic state   Show FW + IPS/TP + P
olicy Server + SIC status
 fw getifs  Show interfaces, IP addresses + netmask
 fw ctl iflist  List all interface names (for use with connStat - sk85780)
 cpstat blades  Quickly show top rule hits, connections and packets statistics
 cpstat fw -f policy  Show policy information and interface statistics
 netstat -atun  Show established connections
 fw ctl arp -n; arp -n  Show all proxy arp's and active local.arp + normal arp entries + summary
 fw ctl zdebug -T drop  Show dropped connections + reason (with Timestamp)
 fw tab -s -t connections  Show load on FW gateway
 adlog a dc; adlog a s  Identity Awareness > Show Domain Controllers status
 adlog a query all  Identity Awareness > Show this gateway's complete adlog database
 pdp status show  Identity Awareness > Show pdp status information
 pdp monitor all  Identity Awareness > Show information for all connected sessions
 pdp connections pep  Identity Awareness > Show PDP to PEP connection table
 pep show stat  Identity Awareness > Show pep status information
 pep show pdp all  Identity Awareness > Show all connected pdp's
 pep show user all  Identity Awareness > Show all sessions with information summary
 dynamic_objects -l  Show all dynamic objects
 fwaccel stat  Show acceleration status on FW gateway
 fwaccel stats  Show acceleration status on FW gateway
 fwaccel stats -s  Show acceleration status on FW gateway
 cpssh_config istatus  Show status of SSH Inspection
 cpssh_config -q  Show SSH Inspection configuration
 fw tab -t sam_blocked_ips  Show IPs blocked by SAM
 fwaccel off  Disable SecureXL acceleration
 fwaccel on  Enable SecureXL acceleration                            4
 cpssh_config ioff  Disable SSH Inspection
 cpssh_config ion  Enable SSH Inspection
 fw unloadlocal; fw stat  Unload security policy on localhost
 fw fetch localhost; fw stat  Reload security policy from localhost
 fw fetch CPSMS; fw stat  Reload security policy from FW management
 fw ctl set int fw_antispoofing_enabled 0 ; fwaccel off; fwaccel on  Disable Anti-Spoofing
 fw ctl set int fw_antispoofing_enabled 1 ; fwaccel off; fwaccel on  Enable Anti-Spoofing
 ips off; ips stat  Disable IPS
 ips on; ips stat  Enable IPS
 fw amw unload; fw stat -b AMW  Disable Threat Prevention
 fw amw fetch local; fw stat -b AMW  Enable Threat Prevention
 fw ctl set int fw_allow_out_of_state_tcp 1; fw ctl set int fw_allow_out_of_state_icmp 1  Disable
Stateful Inspection
 fw ctl set int fw_allow_out_of_state_tcp 0; fw ctl set int fw_allow_out_of_state_icmp 0  Enable
Stateful Inspection
---------------------------------------------------------
 PANIC MODE  (Disable IPS, Threat Prevention, Anti-Spoofing, SecureXL, Stateful Inspection)
 NORMAL MODE  (Enable IPS, Threat Prevention, Anti-Spoofing, SecureXL, Stateful Inspection)
```

Figure 15.11 – ccc script FIREWALL GATEWAY scripts

When the desired script is highlighted and you've pressed *Enter*, you will be prompted to confirm its execution and will be presented with the entire one-liner derived from the base commands. The following is a condensed split view of the `Show FW + IPS/TP + Policy Server + SIC status` diagnostics combination command. To execute it, simply press *Enter* once more. To exit, use the arrow keys:

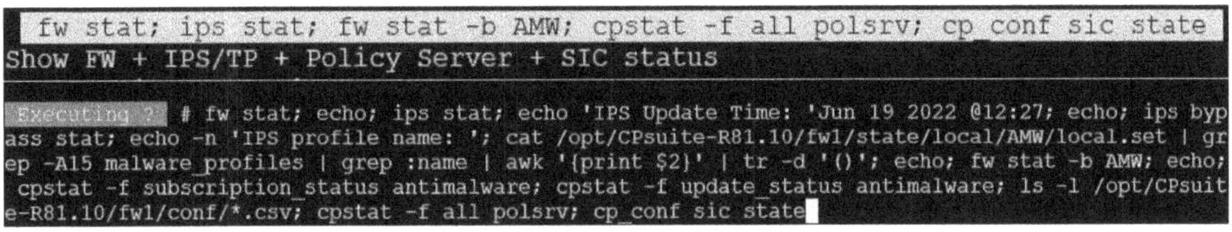

Figure 15.12 – ccc script showing execution confirmation and the full one-liner view

The result is beautifully organized and contains easy-to-read comprehensive diagnostics data. You'll just have to trust me on this one since it's too big for these pages.

Troubleshooting scripts are arranged similarly and presented in terms of non-intrusive and intrusive commands.

You can find ccc script and the necessary installation instructions at `https://community.checkpoint.com/t5/CheckMates-Toolbox/ct-p/CheckMatesToolbox`.

You can find more information at the preceding link since it contains some of the best tools created by community members. These tools are not confined to scripts but include SmartConsole extensions, SmartEvent Views and Reports, cloud deployment and automation templates, and a lot of other features that provide additional functionality not available out of the box. And there are pages and pages of them. Enjoy!

Note the disclaimer on the **CHECKMATES TOOLBOX** site:

> *"Disclaimer: Check Point does not provide maintenance services or technical or customer support for third third-party content provided on this Site, including in CheckMates Toolbox. See also our Third Party Software Disclaimer."*

> **Note**
> Use your virtual lab to test commands, downloaded scripts, and other tools before using them in production!

Be sure to check the other sections of the community displayed in the scrolling tape under the community content search bar. These are as follows:

- **CHECKMATES LAB**
- **TECH TALKS**
- **CP<DEV>**
- **CheckMates Live**
- **CHECKMATESGO**
- **CheckPoint CloudMates**
- **CP4B CHECK POINT FOR BEGINNERS**
- **CHECKFLIX**

Before your first post, spend some time familiarizing yourself with the functionality and rules of the community. When you are logged in on `https://community.checkpoint.com`, on the right-hand side of the screen, find and access **GETTING STARTED ON CHECKMATES**.

Read through the content of this page. It'll help you get oriented and use available resources more effectively.

I'd like to add a few words to those that are already there in the **Posting Messages** section it is a good idea to indicate the version and jumbo hotfix level of the product you are working with since it must be considered by those who are reading your post.

Put yourself in the shoes of those that are looking at the subject of your post. Is it concise enough and informative enough to be of interest and to be taken seriously?

Does it indicate urgency, observation, or a question?

Start the body of your post with a short summary so that it will be the first thing the reader encounters and expand on the details below.

If you want to include lengthy command outputs or text blocks, use the **Insert a Spoiler Tag** option. This will result in a collapsible module and will give viewers or responders going through your post an opportunity to focus on your writing, instead of missing its pieces when they are not segregated from external content.

> **Note**
> Never include your actual public IPs and private keys in the community posts! Use intermediary steps to search for and replace those in the text and use blurring or obfuscation tools for your screenshots.

Unfortunately, the copy/paste functionality is not available for images. If you want to include screenshots, save them beforehand and use the **Insert Photos** option to either drag-and-drop or browse to saved files. Once uploaded, images or screenshots must be approved by a moderator before they become visible in the body of your post to other members. You should mark uploaded images as **Public** for this to work; otherwise, they will only be visible to you:

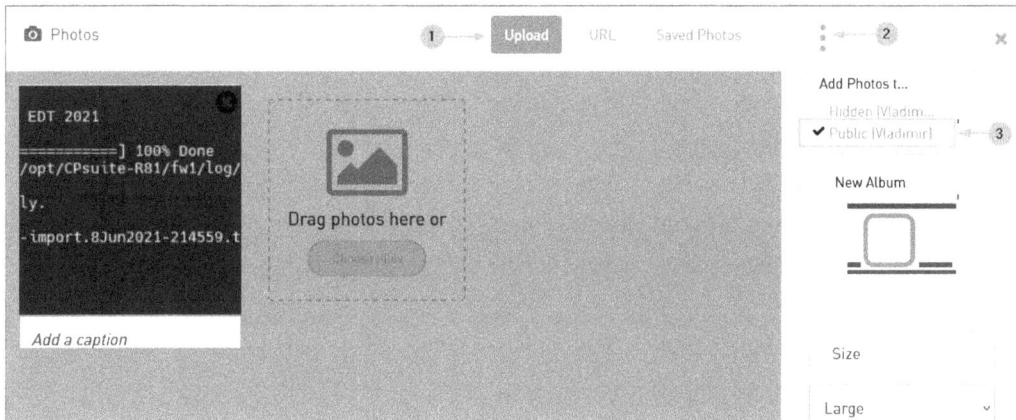

Figure 15.13 – Including images in posts

This limitation is annoying as you will not see any indication in your post if the images are **Hidden**.

> **Note**
>
> I was told that it is kind of possible to copy/paste the images using a very roundabout way, but this is a limitation of the forum hosting platform and is not CheckMates-specific: `https://community.checkpoint.com/t5/General-Topics/CheckMates-Forum-Insert-Screenshot-from-clipboard-is-not-working/m-p/145759/highlight/true#M25693`.

If you have read someone's posts on the relevant subject and think that they may help, consider appending your post with `@User_Name`, asking them if they can offer advice or suggestions. Do not abuse this option randomly.

If your post does not receive any replies, revisit it, and see if you can edit the subject or alter the content to improve it. Doing that will also bump it to the top of the queue and make it more likely that someone looking at the front page may get involved.

Alternatively, you can reply to your post and explicitly ask readers to respond.

Do keep in mind that this is a user group. There are no obligations on anyone to engage with you. How attractive you make that proposition is down to the quality of the subject and the content of your post.

As an example of community interaction when requesting assistance and engagement with members, forum admins, and Check Point employees, take a look at this thread for one of my old posts (`https://community.checkpoint.com/t5/API-CLI-Discussion/Can-someone-put-together-a-script-to-delete-automatically/m-p/9529`):

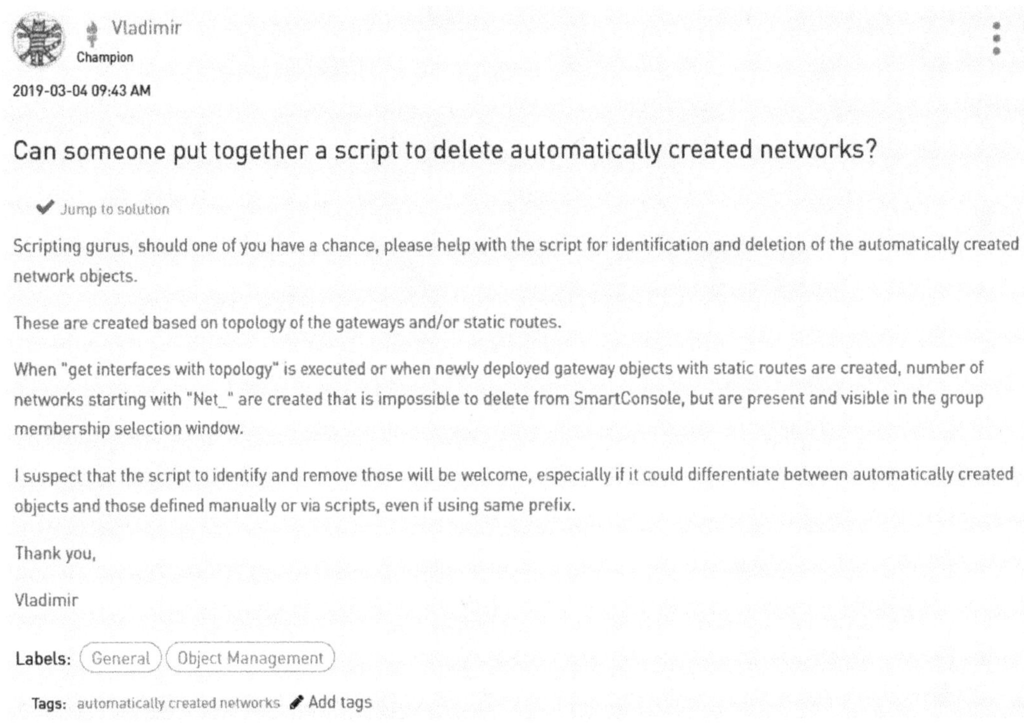

Vladimir
Champion

2019-03-04 09:43 AM

Can someone put together a script to delete automatically created networks?

✔ Jump to solution

Scripting gurus, should one of you have a chance, please help with the script for identification and deletion of the automatically created network objects.

These are created based on topology of the gateways and/or static routes.

When "get interfaces with topology" is executed or when newly deployed gateway objects with static routes are created, number of networks starting with "Net_" are created that is impossible to delete from SmartConsole, but are present and visible in the group membership selection window.

I suspect that the script to identify and remove those will be welcome, especially if it could differentiate between automatically created objects and those defined manually or via scripts, even if using same prefix.

Thank you,

Vladimir

Labels: (General) (Object Management)

Tags: automatically created networks ✎ Add tags

Figure 15.14 – CheckMates post example

Sometimes, you may be referred to a particular SK article that may contain the information you are looking for. Read it and, should you have concerns about applicability, more questions, or need additional clarifications, stay engaged.

If you are working on a long post and had to navigate away from it, it will still be preserved in **Drafts**. To access these, click on your CheckMates avatar (in the top-right corner of the screen), click **My profile**, and scroll to the bottom of the page. You may reopen the draft to continue working on it and post it when you're done.

While you may be new to Check Point, you may be experienced in other areas, such as scripting, automation, cloud technologies, or several other subjects. Spend a bit of time reading posts by other members to see if you may lend a suggestion, explain certain behavior in a specific environment, or offer general troubleshooting tips that may help. The more you interact with the community, the more well-known you become to its members and the more opportunities you have to engage specific members in your later posts.

Welcome to CheckMates!

Postmortems and lessons learned

After concluding the troubleshooting process, a postmortem (short for retrospective analysis) should be performed. Typically, this consists of several questions and short but informative answers to each of them. These questions are as follows:

- What happened?

  We may be able to answer this question better now than when we started the troubleshooting process since, initially, we may have been looking at a subset of effects.

- Why did it happen?

  This is the chain of relevant events that have been backtracked to the root causes of the issue. These may be both technical and procedural. All of them must be addressed. As a technical troubleshooter, you may only be able to cover the technical aspects of this question.

- How did you respond and recover?

  Provide a brief and factual outline of the steps taken to troubleshoot the issue. It may be useful to present this as a timeline.

- What can be done to prevent similar issues from happening again?

  Again, you are answering this question from a strictly technical perspective, but the ultimate solution may be procedural.

Consider a situation where a new feature was enabled or some configuration changes have taken place during the maintenance window. Test protocols are executed to verify that all business-critical functions and applications are in a good state.

The next day, you are informed that a certain business-critical application is inaccessible. This application is not on the list of assets you should be testing and is not included in the test protocol.

In this case, the permanent solution would be to do either of the following:

- Implement the business process, ensuring that such assets are identified promptly and that your department is notified of subsequent augmentation of the test protocols, if such control is not implemented yet.

- Make this process stricter to enforce if it is present but hasn't been followed, and add periodic reviews for critical assets to identify if any are missing from test protocols.

Neither of these is a technical solution. You will encounter the same technical issue, but it won't have an impact on the business if initial troubleshooting was conducted during the maintenance window and the changes have been rolled back.

Looking at this from a technical perspective, whenever issues are encountered and after the fires have been put out, it makes sense to step back, look at the sequence of your actions, and determine if it could be improved.

It could be that a broader look at the infrastructure was in order, or that the number of steps you have taken had no relevance to this kind of problem.

It could also be that you have encountered a different command or several commands used by TAC engineers that should be incorporated into your internal troubleshooting processes.

Whatever the case may be, there is almost always something that could be learned after the fact.

Retaining that information, converting it into structured documents, and making it accessible to others in your organization is as important as solving the issue in the first place.

It is equally important for you to develop your own personal troubleshooting strategies and techniques, preferably maintaining a personal repository of knowledge.

While there are official troubleshooting approaches and sequences that may be vendor or company-specific, my general recommendation is to do the following:

1. Review the steps taken during the last completed troubleshooting process.
2. Identify and cross out those that wouldn't have yielded potentially useful data for this kind of issue.
3. Re-order the remaining steps from most general to most specific, including possible dead-end branches.
4. Visualize it as a flowchart.
5. From the flowchart, generate numbered checklists (including nested checklists) with references to specific tools and notes.

6. Tag or label each checklist so that they can be retrieved by searches when the relevant terms are used.

7. Save them under broad enough names that they won't be overlooked in the future.

Keep doing this every time you run into a new kind of issue.

If you run into the same kind of issue (and are the one responsible for it to begin with) more often than is reasonable, look at a larger picture to determine if there is a way of doing things differently to avoid the repeats.

This is all I can share with new Check Point administrators in a single book.

If you wish to learn more about Check Point troubleshooting, look into the **Troubleshooting Administrator R81.1 (CCTA)** course and its accompanying certification.

Summary

In this chapter, we learned about the variety of environments in which administrators may be called to perform troubleshooting, and some of the questions you should have answered before engaging in this activity. We looked at different types of issues, some of the tools available for troubleshooting them, and sequences of actions that can lead to their successful resolution. We also covered the process of opening a Service Request with TAC and how to engage with support engineers to achieve better outcomes. You were introduced to the resources that are available in the CheckMates community toolbox and given some tips on productive interaction. Finally, we looked at postmortems and how to derive value from lessons learned during the troubleshooting process.

This was the last chapter of this book. I have attached a short appendix on licensing with specifics for the lab environments.

Thank you for reading and welcome to the ranks of Check Point cyber defenders!

Appendix
Licensing

Inevitably, we'll have to license the products we use either in a production or lab environment. Check Point is generous with its evaluation licenses for those learning about its products or using them in lab environments. Generally, you'll be acquiring or renewing production licenses through your Check Point channel partner, who will be very well versed in all the intricacies of the process, but it is helpful to have some understanding of what all of those strings of letters and numbers are referring to. You also have to understand how to apply the licenses to the products you are using.

The following topics will be covered in this appendix:

- Licensing
- Licensing for gateways
- Licensing for management servers
- Central and local licenses
- License activation
- Evaluation licenses for the lab
- SmartUpdate and additional information

Licensing

The terminology Check Point uses for licensing is a bit weird, but that's because it was using the term "containers" way before its virtualization namesake existed. The same goes for the term "blade," which originally referred to pluggable modular servers dedicated to running specific applications.

Both terms seem a bit outdated since, with the resurgence of **hyper-converged infrastructure** (**HCI**), there are now chassis of blades. These are provisioned as universal resource nodes and are often used to run container hosts which, in turn, running containers rather than being function-specific. This is the exact opposite of the terminology that's used by Check Point and is a source of increasing consternation among younger technologists not familiar with the terms' historical connotations.

Containers and blades

Containers are defined by their size and the types of blades they can contain.

There are three types of software containers, each associated with specific **Stock Keeping Unit** (**SKU**) types:

Container	Hardware SKU	Software SKU
Security Management	CPAP-NGSM###	CPSM-NGSM###
Security Gateway	CPAP-SG####-NGXX	CPSG-P##
Endpoint Security	n/a	CPEP-XX-#####

Table A.1 – Software containers

Here, the ## instances are used to denote the model of the appliance, the capacity of the software-only gateway products for CPUs, the number of gateways that can be managed, or the number of endpoints. The XX instances are used to denote a specific package or feature.

The abbreviations that are used in SKUs are easy to decipher:

- CP: Check Point
- AP: Appliance
- SM: Security Management
- SG: Security Gateway
- NG: Next Generation
- EP: Endpoint

There may be more than one container attached to a product.

SKUs for security blades (functions) for all containers begin with **Check Point Security Blade** (**CPSB**).

When you purchase Check Point hardware or software solutions, containers are attached in User Center in your account for each product [1], with several **Built-in Blades**. Containers and built-in blades [2] licenses are **Perpetual** (except for cloud-based **pay-as-you-go** subscriptions). Built-in blades cannot be detached from their container, but additional individual blades, referred to as **a la carte**, can be purchased separately and can be attached, detached, and re-attached to comparable containers.

Additionally, several subscription-based blades [3] are attached as a part of the **Enterprise Based Protection** packages [4], with SKUs of **CPEBP-NGXX**. These blades remain active so long as the associated service contracts are maintained:

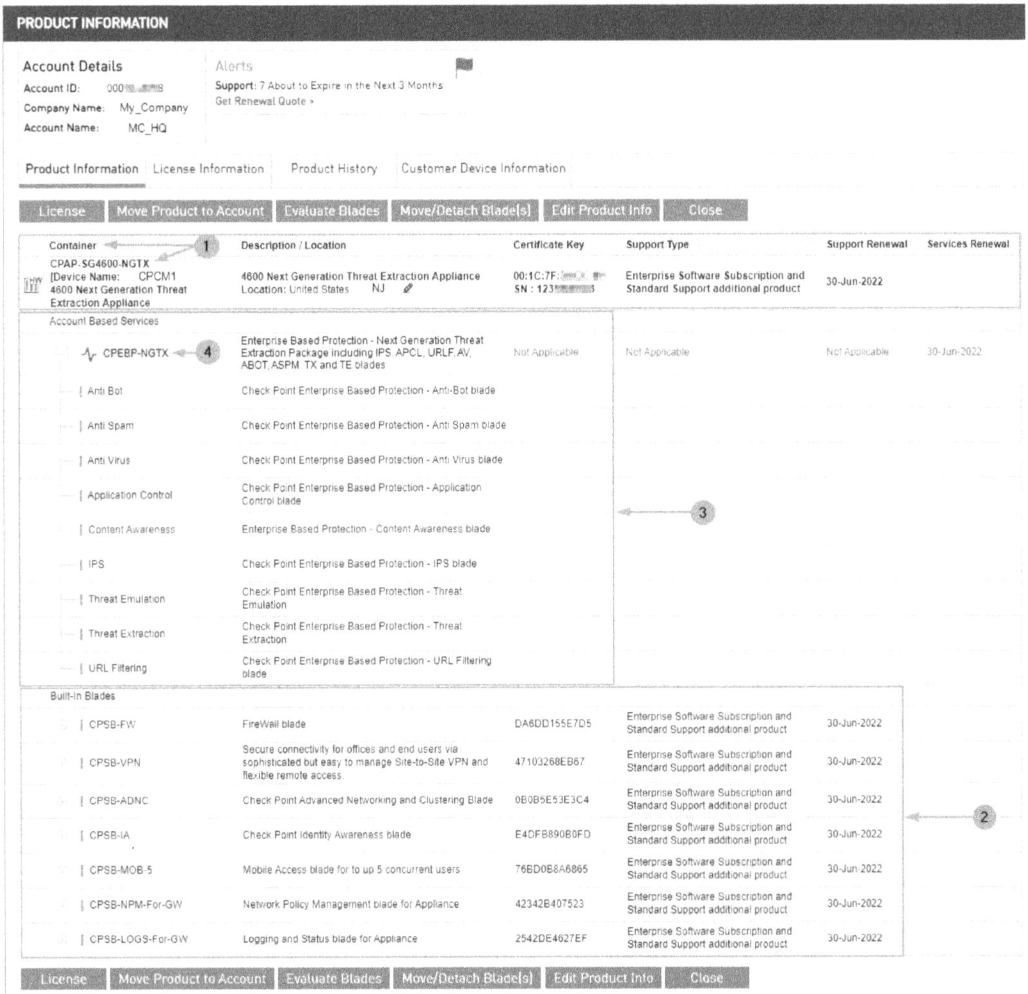

Figure A.1 – The containers, built-in, and subscription-based blades

Another way to look at the containers, perpetual blades, and subscription blades associated with your products is to hover your mouse cursor over any product in the **ASSSET /INFO | MY ACCOUNTS | Product Center** section of User Center:

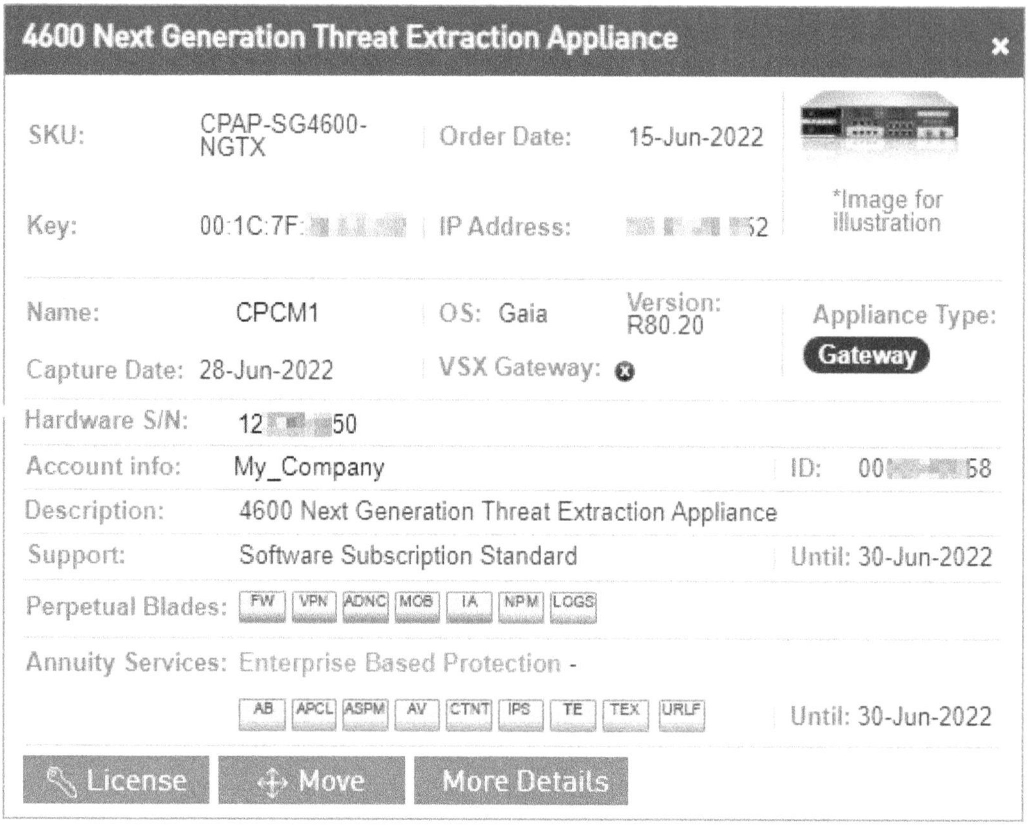

Figure A.2 – Product card

Yet another type of blade you may encounter when working with hardware appliances is an **Accessory** blade. These are issued for physical items, such as additional memory modules, network interface cards, PSUs, and more. Their SKUs begin with **CPAC**.

Definition of Perpetual Licenses

The term **perpetual** refers to your right to continue operating an appliance (physical or virtual) with up-to-date supported versions of the software on the appliance, so long as you maintain at least a minimum required support contract.

Should your support contract lapse, you are entitled to continue running versions of the software and JHFAs that are released during your active support contract, but you will be limited to blades that are defined as **perpetual**.

Licensing for gateways

Check Point hardware gateway appliances are sold with licenses that reflect their capacity, desired functionality, and additional components. Here, functionality refers to NGTP or NGTX packages, high-performance configuration, and clustering. Additional components include redundant power supply units, extra (or different) network interface cards, memory, and dedicated acceleration modules.

Open Server (and virtual machines defined as Open Server) gateways are licensed using core-pack bundles. These licenses are available in 2, 4, 8, 16, 24, and 32-core denominations and are sold as either NGTP or NGTX packages. The only exception is a 2-core license, which can be sold with the NGFW package (containing the Firewall and IPSec VPN blades). These licenses are not additive; for example, you may use a 4-core license and replace it with an 8-core license, but these cannot be combined to produce a single 12-core license. The list of supported hardware for Open Servers, network interface cards, and virtual machines can be found here: `https://www.checkpoint.com/support-services/hcl/`.

NGTP and NGTX licenses for Check Point hardware and Open Servers are comprised of perpetual and service subscription license components.

FW and VPN blades are perpetual. The same goes for other features that may be included in the license at the time of purchase, such as the **Identity Awareness** and **Advanced Networking & Clustering** blades (**Dynamic Routing**, **SecureXL**, **QoS**, **ClusterXL**, and **Monitoring**).

The **Application Control & URL Filtering**, **Content Awareness**, and **Threat Prevention** blades (**Anti-Virus**, **Anti-Bot**, and **IPS**), as well as the **Threat Extraction** blades (**Threat Emulation** and **Sandblast**), are subscription-based. With each newly purchased gateway, you are automatically granted a free first year of all **NGTX** blades.

Threat Emulation and **Threat Extraction** can be implemented locally on gateways, on dedicated appliances, or using Check Point cloud services (this is the default option, which I recommend). Note that with Check Point cloud services, there is a limit on the number of files per month these blades are allowed to process.

This limit may vary, depending on the model of the appliance you have purchased, but should you run close to or into it, additional licenses are available, starting from 50,000 files and up to 2,000,000 files per month. These licenses are additive.

For true virtual implementations, where VMs are recognized as such (that is, all major Level-1 hypervisors and cloud platforms), Check Point gateways are licensed under the name **CloudGuard Network Security** on a per-vCPU basis and are available in the **NGTP** and **NGTX** packages. These are typically purchased as pools of single-core licenses and are assigned to VMs using **CloudGuard Controller**, an integral part of Security Management servers. These types of licenses are suitable for static cloud assets. For dynamic cloud assets, such as gateways used as members of scalable groups, **pay-as-you-go** (**PAYG**) Check Point licenses are available in each cloud vendor's catalog.

In addition to the **NGTP** and **NGTX** packages, a la carte licenses are available that can be attached to or removed from a **Gateway Container**, such as **DLP**.

Licensing for management servers

Hardware appliance models are sold in 5,10, 25, 50, 75, 150, 200, and 400 gateways under management capacities with corresponding amounts of compute, memory, and storage resources.

Open Server management server licenses do not have per-CPU license limitations and are only restricted by the size of the container, which reflects the maximum number of gateways they are capable of managing. At the time of writing, licenses for management servers on Open Server Management software are sold in the same gateways under management capacity increments, with a maximum base capacity of 150 gateways. This can be further expanded with additional 50-gateway licenses.

Security Management servers can be licensed as any of the following:

- Management (including Log) servers
- Dedicated Log servers
- Dedicated SmartEvent servers

Each of these licenses is perpetual.

Management servers' capabilities can be extended with the following:

- À la carte blades for perpetual SmartEvents, additional correlation units, and unlimited **Large-Scale Management** (**LSM**), which is used for managing hundreds of thousands of gateways
- Annual SmartEvent and Compliance, as well as subscriptions (service extensions)

SmartEvent and compliance first-year licenses are included by default for Management Server and will become inactive when they expire, if not renewed.

Central and local licenses

While licenses are pre-provisioned at the time of purchase, they are not pre-activated. To activate the licenses, they must be assigned to either of the following:

- The individual appliances' interface IP address and primary Security Management server's IP address (for a local license)

- The primary Security Management server's IP address (for a Central license)

Local licenses are tied to a specific IP address and require you to make changes in User Center if this address changes. You can do this a limited number of times, but Check Point support can extend it if exceeded. The upside of local licenses is that they can be attached automatically to managed gateways or cluster members based on predefined IP addresses.

Central licenses are not tied to a gateway's IP addresses; instead, they are tied to the IP address of your primary management server and require an administrator to attach them to a specific gateway. You have more flexibility in attaching these licenses to different IPs, and this may be your preferred method if your environment is that fluid. The downside here is that they must be attached manually.

License activation

All Check Point appliances, physical or virtual, after completing the First Time Configuration Wizard, have an automatically-activated 15-day trial license for every blade in the container(s) attached to the product. This license is sufficient for the following purposes:

- Quickly testing a feature, configuration, or behavior

- Performing the initial configuration of the Check Point infrastructure and Access Control policies

- Assigning and activating either production or evaluation licenses

Your first action, in terms of licensing, is to generate a license associated with the IP address of your primary **Security Management Server** appliance. In **ASSET /INFO | MY ACCOUNTS | Product Center**, open the **Product Center** page. On that page, click on **Smart-1 ### NGSM** (for physical appliances) or **Next Generation Security Management** (for virtual appliances). This will be used as a primary management server.

> **Note**
>
> In **Product Center**, the hardware appliance you are interested in generating the license for is identified by the value of its **Key** filed, which is also a MAC address of its **Mgmt** interface. Each hardware appliance is equipped with a pull tab containing its serial number, MAC address, and model. The **Key** value for virtual appliances (referred to as *software* in Check Point licensing lingo) is only important from the point of view of capacity and the features assigned to it.

Clicking on the chosen product opens the **PRODUCT INFORMATION** page, similar to the one shown in *Figure A.1*. At the top and bottom of that page, you will see two identical rows of action buttons. Click on **LICENSE**. Enter the relevant information in the mandatory fields (this will include the IP address) and click **LICENSE** again. You will be notified that an email containing the license information (that is, the license file) was sent to your email and will be presented with three buttons (**Get License File**, **Show License Info**, and **Back To Products**) below the licensed product.

Clicking on **Show License Info** displays instructions for installing the license using SmartConsole and the `cplic put` command.

If your Security Management server is a Check Point hardware appliance and has either direct or via proxy internet access, it will retrieve the licenses and associated contracts' data from User Center automatically. If it is a virtual appliance or does not have internet access, continue with offline activation, as described here.

If you are still within the first 15 days of a **Trial License**, connect to the management server using SmartConsole. Then, in **GATEWAYS & SERVERS**, select your primary management server. After that, click the **Licenses** tab in the bottom portion of the screen, click **Add**, and choose between a **License File** or **License String**. If you have chosen a **License File**, browse to its location, select it, and click **Open**. If you have chosen a **License String**, copy and paste it from the **Show License Info** page into the **Enter your license here:** field and click **OK**.

If you are attempting to apply the license after your initial **Trial license** has expired, you will not be able to connect to the management server using SmartConsole. In this case, SSH to your Security Management server and use the `cplic put` command option to activate the license.

Offline activation

Offline activation works for all Check Point appliances (physical or virtual) and can be accomplished in several different ways. For instance, you can use the following:

- The Gaia WebUI

- The `cplic put` command from the CLISH or Expert Mode shells

- The SmartConsole's **GATEWAYS & SERVERS | License** tab for the selected appliance

- The SmartUpdate tool, which is accessible from the main menu (icon in the top-left corner of SmartConsole) and clicking **Manage licenses and packages**

Licensing options for hardware appliances

There are some license activation and renewal options that are specific to hardware appliances. Let's take a look.

Online activation

Licenses for hardware appliances with internet access can be activated online once a primary Security Management server has been licensed (if you have consented to sending data to Check Point during FTW).

Automatic licensing

If you have checked the main menu (the icon in the top-left corner of SmartConsole) and ticked the **Global Properties | Security Management | Internet Access** checkbox for **Automatically download and install Blade Contracts, new software, and other important data (highly recommended)**, licensing and service contract data will be automatically synchronized between User Center and the rest of your Check Point hardware appliances (through your Security Management server).

Evaluation licenses for the lab

As a User Center account holder, you may request Product Evaluation licenses. Unlike Trial licenses, these are issued for 30 days and are available for a wider range of products. Evaluation licenses are accessible from **User Center | TRY OUR PRODUCTS | Product Evaluation** [1].

For our lab, we are going to request **ALL-IN-ONE EVALUATION** instances [2] and click **Next** [3]. Using the drop-down menu, choose your **User Center Account** [4], enter the IP address of our **Security Management Server** [5], and, under **Purpose of Evaluation**, choose **Security Gateway in lab environments** [6]. Check the confirmation and acknowledgment of EULA checkbox [7] and click **Get Evaluation** [8]:

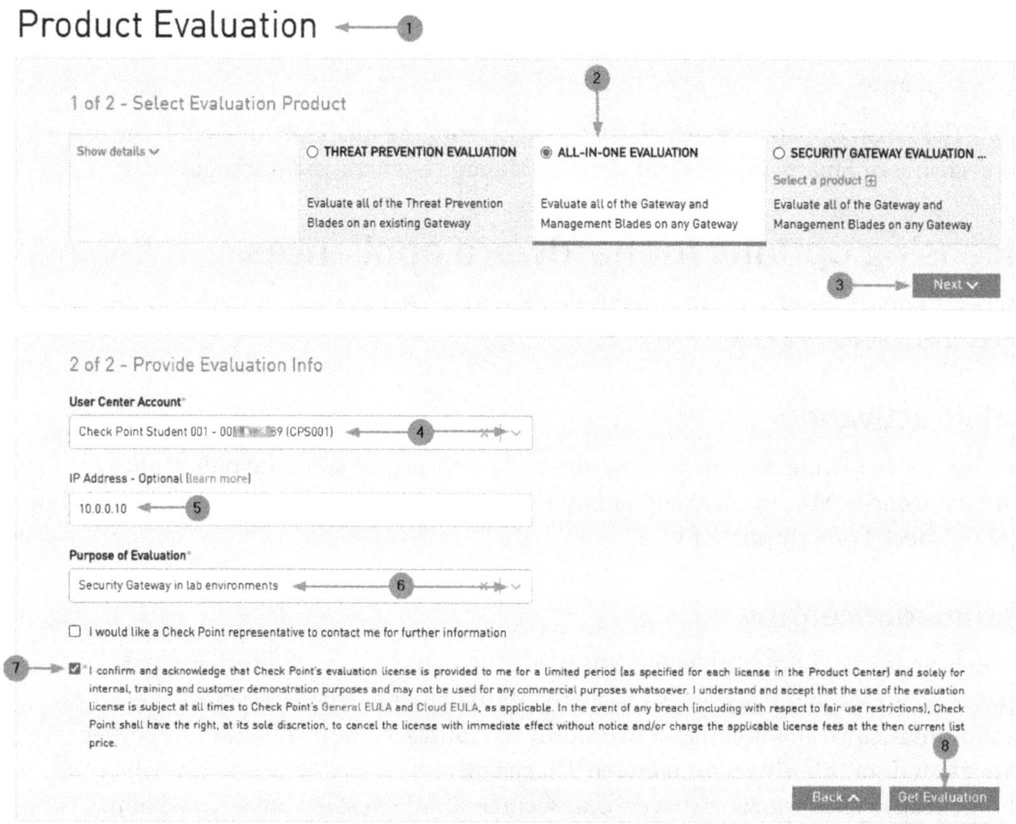

Figure A.3 – Request for product evaluation

When presented with **Congratulations! Your evaluation product has been successfully added to specified UC account**, click on the **Back to Product Evaluation** button (not shown) and repeat the process two more times.

Return to **Product Center** [1], click on the **Evaluations** tab [2], and then click **Valid #** [3] to see **Details** regarding the active evaluations. Check one box at a time [4] and click on **License Instructions** [5] to see the detailed description and license strings for individual All-in One licenses:

Product Center ① ②

| Selected Accounts | Products | Blades | Services | Accessories | Evaluations | Support | Training |

▲ Summary

Issue Date ▲	Total	Not Licensed Yet	Valid	Expired
Last 1-6 Months	41	24	3 ③	13
Last 7-12 Months	8	4	0	4
Previous Periods	2	1	0	1
Total	51	29	4	18

⑤

▲ Details

[License] [Move] [Edit Info] [Export] [License Instructions] [Get Contracts]

All-In-One   × 🔍 3 Evaluations   Last 1-6 Months ⊗   Valid ⊗   Showing 1 to 3 of 3 evaluations (filtered from 51 evaluations)

	Product Evaluation Name	SKU	Account ID	Key	IP	Issue Date ▼	Expiration Date	Comment
Issue Date: 27-Jun-2022 ④								
☑	All-in-One Security Bundle Eval	CPSG-CPSM-EVAL	8364389	C0AD661F3E8A	10.0.0.10	27-Jun-2022	01-Aug-2022	
☐	All-in-One Security Bundle Eval	CPSG-CPSM-EVAL	8364389	C0AF8F346664	10.0.0.10	27-Jun-2022	01-Aug-2022	
☐	All-in-One Security Bundle Eval	CPSG-CPSM-EVAL	8364389	C19FBA028C4F	10.0.0.10	27-Jun-2022	01-Aug-2022	

Previous 1 - 3 of 3 evaluations Next

Figure A.4 – Lab product evaluation licenses

All-in-One licenses are comprised of two sections – one for applying the license to the management server and another, for the gateway.

From each of the license instructions, apply management server licenses to our CPSMS server and single gateway licenses to CPCM1, CPCM2, and (when you get to *Chapter 12, Configuring Site-to-Site and Remote Access VPNs*), the last gateway license, to CPGW.

Should you elect to extend the life of your lab further, request three more All-in-One evaluation licenses and repeat the license application process.

SmartUpdate and additional information

For flexible bulk license and contract management, a legacy application called SmartUpdate can be launched either from SmartConsole's main menu (via the icon in the top-left corner of SmartConsole) and clicking **Manage Licenses and Packages** or (if you cannot log in to SmartConsole after the trial license has expired) by opening the `C:\ Program Files (x86)\CheckPoint\SmartConsole\R81.10\PROGRAM\ SmartDistributor.exe` executable.

There is a **SmartUpdate** executable in the same directory, but it seems like it was supplemented by **SmartDistributor** in version R81.10.

sk11054 Check Point License Guide contains videos describing various license generation, activation, and management processes.

Index

W

X

Z

Other Books You May Enjoy

If you enjoyed this book, you may be interested in these other books by Packt:

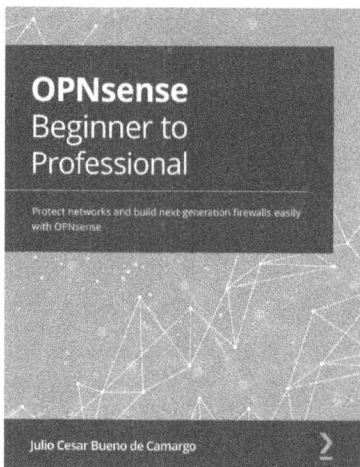

OPNsense Beginner to Professional

Julio Cesar Bueno de Camargo

ISBN: 978-1-80181-687-8

- Understand the evolution of OPNsense
- Get up and running with installing and setting up OPNsense
- Become well-versed with firewalling concepts and learn their implementation and practices
- Discover how to apply web browsing controls and website protection
- Leverage Sensei to implement next-generation firewall features
- Explore the command-line interface (CLI) and learn the most relevant FreeBSD commands

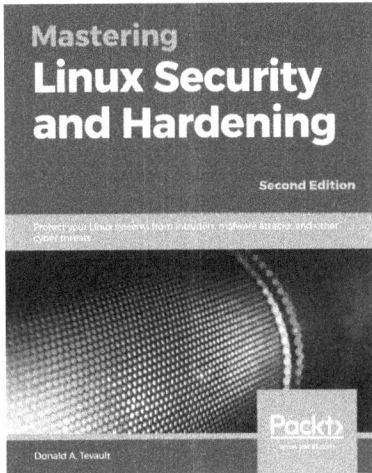

Mastering Linux Security and Hardening - Second Edition

Donald A. Tevault

ISBN: 978-1-83898-177-8

- Create locked-down user accounts with strong passwords
- Configure firewalls with iptables, UFW, nftables, and firewalld
- Protect your data with different encryption technologies
- Harden the secure shell service to prevent security break-ins
- Use mandatory access control to protect against system exploits
- Harden kernel parameters and set up a kernel-level auditing system
- Apply OpenSCAP security profiles and set up intrusion detection
- Configure securely the GRUB 2 bootloader and BIOS/UEFI

Packt is searching for authors like you

If you're interested in becoming an author for Packt, please visit `authors.packtpub.com` and apply today. We have worked with thousands of developers and tech professionals, just like you, to help them share their insight with the global tech community. You can make a general application, apply for a specific hot topic that we are recruiting an author for, or submit your own idea.

Share your thoughts

Now you've finished *Check Point Firewall Administration R81.10+*, we'd love to hear your thoughts! Scan the QR code below to go straight to the Amazon review page for this book and share your feedback or leave a review on the site that you purchased it from.

`https://packt.link/r/180107271X`

Your review is important to us and the tech community and will help us make sure we're delivering excellent quality content.

www.ingramcontent.com/pod-product-compliance
Lightning Source LLC
Chambersburg PA
CBHW080348220326
41598CB00030B/4634